Natural products: their chemistry and biological significance

J. Mann,
R. S. Davidson,
J. B. Hobbs,
D. V. Banthorpe
and J. B. Harborne

Prentice Hall

An imprint of **Pearson Education**

Harlow, England · London · New York · Reading, Massachusetts · San Francisco · Toronto · Don Mills, Ontario · Sydney
Tokyo · Singapore · Hong Kong · Seoul · Taipei · Cape Town · Madrid · Mexico City · Amsterdam · Munich · Paris · Milan

Pearson Education Limited
Edinburgh Gate
Harlow
Essex CM20 2JE
England

and Associated Companies throughout the world

Visit us on the World Wide Web at :
http://www.pearsoneduc.com

First edition 1994

British Library Cataloguing in Publication Data

A catalogue record of this book is available from the British Library

ISBN 0-582-06009-5

Library of Congress Cataloging-in-Publication Data

Natural products : their chemistry and biological significance / J.
  Mann . . . [et al.]. -.- 1st ed.
      p.   cm.
   Includes bibliographical references and index.
   ISBN 0-470-20002-2
   1. Natural products.   I. Mann, J.
QD415.N365   1993
547.7--dc20   93-28148
                                                               CIP

10 9 8 7 6
05 04 03 02 01

Set in 10/12 Times roman by 6

Printed in Malaysia, PP

Natural products:
their chemistry and biological significance

Contents

List of contributors   x

Foreword   xi

INTRODUCTION   *1*

1   CARBOHYDRATES by J. Mann   *7*

1.1   Introduction   *7*
1.2   Structural types   *8*
1.3   Sources and functions   *12*
1.4   Chemistry of monosaccharides   *26*
    1.4.1   Reactions of the hydroxyl group   *26*
    1.4.2   Reactions at the anomeric centre   *35*
1.5   Structure elucidation   *41*
1.6   The total synthesis of natural products and related compounds using carbohydrates   *43*
    1.6.1   Synthesis of thromboxane B_2 *45*
    1.6.2   Synthesis of (−)-shikimic acid   *49*
    1.6.3   Synthesis of (+)-showdomycin   *51*
    1.6.4   Synthesis of (+)-*exo*-brevicomin   *53*
    1.6.5   Synthesis of (+)-muscarine   *55*
1.7   Synthesis of carbohydrates   *57*
    1.7.1   Synthesis of L-(−)-daunosamine   *57*
    1.7.2   Synthesis of methyl-L-ribofuranoside   *59*
    1.7.3   Synthesis of lincosamine   *60*
    1.7.4   Synthesis via asymmetric epoxidation of allylic alcohols: the Sharpless epoxidation   *62*
    1.7.5   Synthesis of glycopeptides   *62*
    1.7.6   Synthesis of an artificial antigen   *64*
Further reading   *64*

2   NUCLEOSIDES, NUCLEOTIDES, AND POLYNUCLEOTIDES by J. B. Hobbs   69

2.1   Introduction   *69*
2.2   Nucleosides   *70*
    2.2.1   Nucleoside conformation   *72*
    2.2.2   Nucleoside synthesis   *74*
2.3   Nucleotides   *89*
    2.3.1   Nucleotide biosynthesis   *91*
    2.3.2   Nucleotide synthesis   *93*
    2.3.3   P-chiral nucleotides   *97*
    2.3.4   Some applications of ^{31}P NMR in nucleotide research   *100*
2.4   Oligo- and polynucleotides   *101*
    2.4.1   Biosynthesis   *104*
    2.4.2   Oligonucleotide synthesis   *108*
    2.4.3   Assembly of longer oligonucleotides and genes   *119*
    2.4.4   Nucleic acid sequencing   *119*
    2.4.5   Recombinant DNA   *123*
    2.4.6   Copying DNA: the polymerase chain reaction   *126*
Further reading   *128*

3   AMINO ACIDS AND PEPTIDES by R. S. Davidson and J. B. Hobbs, with D. O. Smith   *131*

3.1   Introduction   *131*
3.2   Synthesis of α-amino acids   *131*
    3.2.2   Newer synthetic routes   *135*
    3.2.3   Syntheses based on α-amino acids as chiral building blocks   *139*
    3.2.4   Resolution of racemic mixtures of α-amino acids   *143*
3.3   Biodegradation of the amino acids   *145*
    3.3.1   Transamination   *146*
    3.3.2   The metabolic fate of the α-ketoacids   *147*
    3.3.3   Biosynthesis of the amino acids   *150*
3.4   Chemical synthesis of peptides   *153*
    3.4.1   N-protecting groups   *155*
    3.4.2   Selective protection of α,ω-diaminocarboxylic acids   *159*
    3.4.3   C-protecting groups   *160*
    3.4.4   Selective protection of carboxyl groups of the mono amino dicarboxylic acids   *163*
    3.4.5   Protection of other functional groups   *164*
    3.4.6   Activation and coupling   *165*
    3.4.7   Summary of strategies and methods available for synthesizing peptides in solution   *170*
    3.4.8   Example of peptide synthesis using the solution method   *171*
    3.4.9   Use of solid supports in peptide synthesis: the Merrifield approach   *172*
    3.4.10   Use of solid supports in peptide synthesis: the use of polyacrylamide resins   *179*

      3.4.11 Application of protease-catalysed peptide bond formation in peptide synthesis *182*

      3.4.12 Polypeptide synthesis using recombinant DNA *184*

3.5   Some specific peptides of interest *185*

      3.5.1 Linear peptides *185*

      3.5.2 Cyclic peptides *187*

3.6   The biosynthesis of proteins *191*

3.7   Structure determination of proteins and polypeptides *195*

      3.7.1 Purification of peptides *196*

      3.7.2 Evaluation of purity *198*

      3.7.3 Strategy for polypeptide sequencing *198*

      3.7.4 Methods involved in sequencing *200*

3.8   Nuclear magnetic resonance spectroscopy of peptides *223*

3.9   Polypeptide conformation *229*

      3.9.1 Secondary structure in polypeptides *230*

      3.9.2 The prediction of protein folding *235*

      3.9.3 X-Ray diffraction methods of structural analysis *231*

Further reading *237*

4 FATTY ACIDS AND THEIR DERIVATIVES by J. B. Hobbs *239*

4.1   Introduction *239*

4.2   Fatty acid structures *240*

      4.2.1 Straight chain saturated fatty acids *240*

      4.2.2 Unsaturated fatty acids *241*

      4.2.3 Fatty acids of unusual structure *242*

4.3   Fatty acid biosynthesis *244*

      4.3.1 Mono-unsaturated fatty acids *246*

      4.3.2 Polyunsaturated fatty acids *247*

      4.3.3 'Odd' fatty acids *248*

4.4   Fatty acid catabolism *250*

4.5   Fatty acid synthesis *252*

4.6   Prostaglandins *259*

      4.6.1 Prostaglandin structures *260*

      4.6.2 Occurrence and significance of the prostaglandins *260*

      4.6.3 Prostaglandin biosynthesis *262*

      4.6.4 Prostaglandin catabolism *264*

      4.6.5 Prostaglandin synthesis *264*

4.7   Thromboxanes *274*

      4.7.1 Thromboxane synthesis *275*

4.8   Leukotrienes *277*

      4.8.1 Biological effects of the leukotrienes *280*

      4.8.2 Leukotriene synthesis *281*

Further reading *287*

5   TERPENOIDS by D. V. Banthorpe   *289*

5.1   Introduction   *289*
5.2   General routes of biogenesis   *291*
5.3   Structure determination   *296*
5.4   Hemiterpenoids – C_5 compounds   *304*
5.5   Monoterpenoids – C_{10} compounds   *306*
5.6   Sesquiterpenoids – C_{15} compounds   *316*
5.7   Diterpenoids – C_{20} compounds   *327*
5.8   Sesterpenoids – C_{25} compounds   *331*
5.9   Steroids – C_{18} to C_{29} compounds   *331*
       5.9.1   Bile acids   *336*
       5.9.2   Sex hormones   *336*
       5.9.3   Adrenocortical hormones   *338*
       5.9.4   Saponins   *339*
       5.9.5   Cardiac glycosides   *340*
       5.9.6   Vitamin D   *341*
       5.9.7   Phytosterols   *342*
       5.9.8   Miscellaneous   *343*
5.10   Non-steroidal triterpenoids – mainly C_{30} compounds   *344*
5.11   Carotenoids – C_{40} compounds   *345*
5.12   Polyisoprenoids   *349*
5.13   Synthesis   *350*
Further reading   *358*

6   PHENOLICS by J. B. Harborne   *361*

6.1   Introduction   *361*
6.2   Structural types   *362*
6.3   Natural occurrence   *365*
6.4   Isolation and structure elucidation   *367*
       6.4.1   Isolation and purification   *367*
       6.4.2   Chromatography and absorption spectroscopy   *367*
       6.4.3   Structural elucidation   *369*
6.5   Biosynthesis   *372*
       6.5.1   General principles   *372*
       6.5.2   Biosynthesis of usnic acid   *375*
       6.5.3   Biosynthesis of lignin   *377*
       6.5.4   Biosynthesis of cyanidin 3-glucoside   *379*
       6.5.5   Biosynthesis of rotenone   *380*
       6.5.6   Biosynthesis of plant naphthoquinones   *381*
6.6   Laboratory synthesis   *383*
Further reading   *388*

7   ALKALOIDS by J. Mann   *389*

7.1   Introduction   *389*
7.2   Structural types   *390*

7.3   Occurrence *391*
7.4   Isolation and structure elucidation *395*
    7.4.1   Isolation and purification *395*
    7.4.2   Structure elucidation *396*
7.5   Biosynthesis *403*
    7.5.1   Alkaloids derived from ornithine and lysine *403*
    7.5.2   Alkaloids derived from phenylalanine and tyrosine *413*
    7.5.3   Alkaloids derived from tryptophan *420*
7.6   Synthesis *427*
    7.6.1   Classical era *427*
    7.6.2   Modern era *428*
7.7   Asymmetric synthesis *441*
    7.7.1   Synthesis of deoxynojirimycin *441*
    7.7.2   Synthesis of indolactam V *443*
Further reading *445*

INDEX *448*

List of contributors

Dr Derek V. Banthorpe, Department of Chemistry, University College, London

Professor R. Stephen Davidson, Chemical Laboratory, The University of Kent at Canterbury

Professor Jeffrey Harborne, School of Plant Sciences, University of Reading

Dr John Hobbs, Biotechnology Laboratory, University of British Columbia

Professor John Mann, Department of Chemistry, University of Reading

Foreword

Naturally occurring organic chemical compounds (natural products) have always fascinated chemists. Interesting and intriguing chemistry is involved in their *in vivo* production and in their laboratory utilization, and their importance as structural materials and biologically active molecules (substrates for life processes, toxins, hormones, drugs, etc.) is of unparalleled importance. Yet they are largely ignored by the major textbooks of organic chemistry. This book seeks to remedy this omission.

Each chapter is devoted to a particular class of compounds. Some, like carbohydrates, nucleosides and nucleotides, and amino acids and proteins, are products of primary metabolism and are vital for the maintenance of life processes. Others, like terpenoids, phenolics, fatty acid metabolites and alkaloids, are products of secondary metabolism and have toxicological, pharmacological and ecological importance.

The authors of the various chapters were given a free rein in terms of the contents of their chapters, though each chapter contains information about structural types, structure elucidation, synthesis, biosynthesis, and biological significance. This is not a research text and no attempt has been made to include the very latest (and possibly rather esoteric) advances. It is, however, intended to provide a comprehensive and readable introduction to most classes of natural products, and thus complements the standard texts on basic organic chemistry.

Introduction

It is just over a hundred years since Emil Fischer announced (in 1891) his elucidation of the structure of glucose. The complex alkaloids vinblastine and vincristine, which are highly potent anti-tumour agents, only succumbed to structure elucidation in 1964. The three compounds are clearly natural products, yet glucose is ubiquitous and essential for life, whilst the two alkaloids are only produced by a few species of plants, notably the rosy periwinkle *Catharanthus roseus*, and have no apparent functions in the plants.

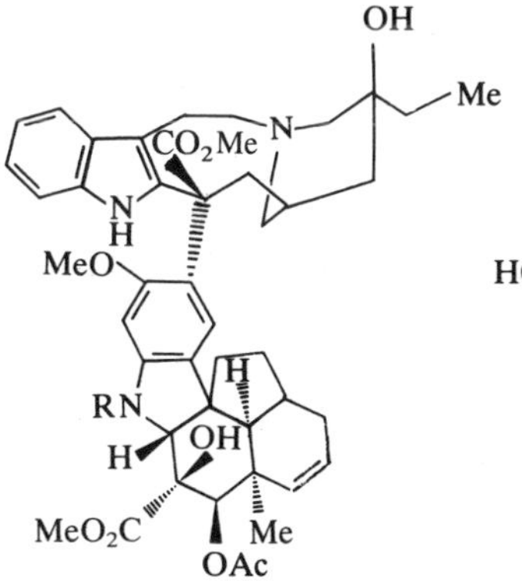

β-D-glucose

R = −Me:vinblastine
R = −CHO:vincristine

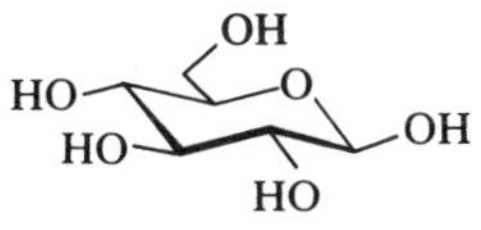

The two types of compounds are, however, connected via complex metabolic pathways as shown in Fig. 1. Admittedly the link is somewhat tenuous, but carbohydrate metabolism does give rise to the aromatic amino acid tryptophan and to the terpenoid precursor mevalonic acid. These two building blocks provide most of the skeleton of vinblastine and vincristine.

A more detailed examination of Fig. 1 reveals the numerous metabolic interconnections between the various classes of natural products that are included in this book. Usually the compounds are identified as products of primary metabolism, i.e. carbohydrates, nucleosides, amino acids and the polymers derived from them, or as products of secondary metabolism, i.e.

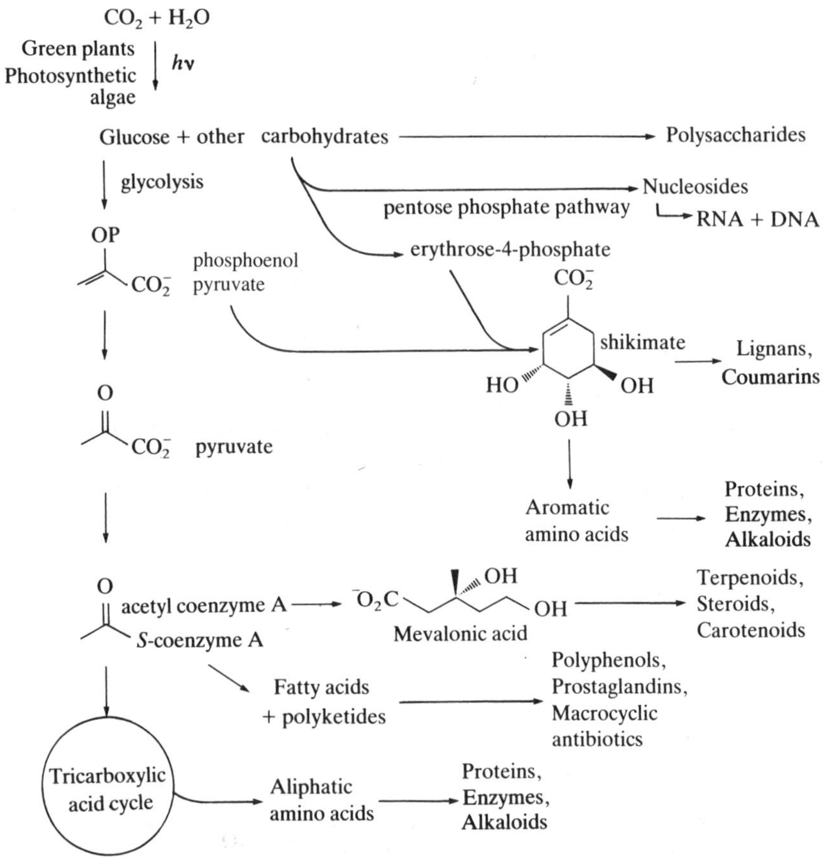

CO₂ + H₂O

Green plants
Photosynthetic | *h*ν
algae

Glucose + other carbohydrates ⟶ Polysaccharides

| glycolysis

Nucleosides

pentose phosphate pathway ⤷ RNA + DNA

OP

phosphoenol
pyruvate

erythrose-4-phosphate

shikimate   Lignans, Coumarins

pyruvate

Aromatic
amino acids ⟶ Proteins, Enzymes, Alkaloids

acetyl coenzyme A ⟶

Mevalonic acid ⟶ Terpenoids, Steroids, Carotenoids

S-coenzyme A

Fatty acids
+ polyketides ⟶ Polyphenols, Prostaglandins, Macrocyclic antibiotics

Tricarboxylic
acid cycle

Aliphatic
amino acids ⟶ Proteins, Enzymes, Alkaloids

Fig. 1.

phenolics, terpenoids and steroids, and alkaloids. Primary metabolites are essentially ubiquitous and certainly essential for life, whilst the secondary metabolites are of restricted occurrence and of no apparent utility. This division is useful but somewhat arbitrary. Testosterone, for example, is of limited occurrence, but has vital hormonal activity, and nicotine from the tobacco plant (and a few other species) has a definite role as an insect feeding deterrent. It is thus much more important to appreciate the chemical and biochemical interconnections between the various classes of natural products, than to worry about whether they are primary or secondary metabolites.

Although the book is primarily concerned with the chemistry of natural products, most of the chapters do include some discussion of the biogenesis of the various natural products. The metabolic pathways by which they are produced and subsequently transformed are almost invariably enzyme-controlled, so a brief discussion on enzymes and their cofactors is essential.

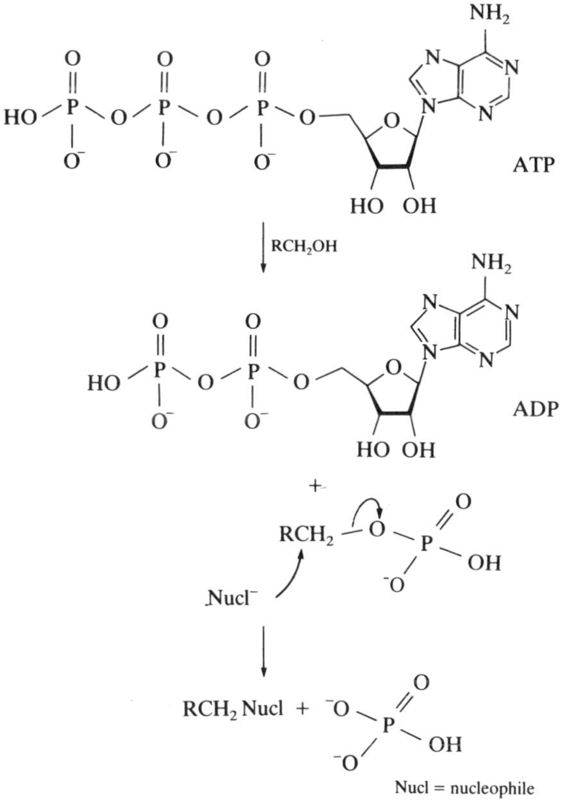

Fig. 2.

Enzymes and cofactors

There are two major classes of proteins: structural proteins such as keratin of skin and collagen of tendon; and enzymes such as chymotrypsin or trypsin, which catalyse the breakdown of ingested proteins in the gut. The former are usually fibrous, whilst the latter are often globular in shape.

One feature that all enzymes possess is an active site, which is the 'cavity' in which the chemical interconversions that are catalysed by the enzyme actually take place. In general, the chemistry that occurs is mechanistically similar to that which could be accomplished in the laboratory, but the rates of reaction are up to 10^{12} times faster. Most of the reactions are also stereospecific, or at least highly stereoselective, and the rate acceleration and this stereospecificity are a direct result of the way in which the substrate(s) and the enzyme interact. The substrate(s), specific participating amino acids, and necessary cofactors (small organic molecules or metal ions), are held in a pseudo transition state configuration, so that the activation energy for the reaction is reduced and the stereospecificity ensured.

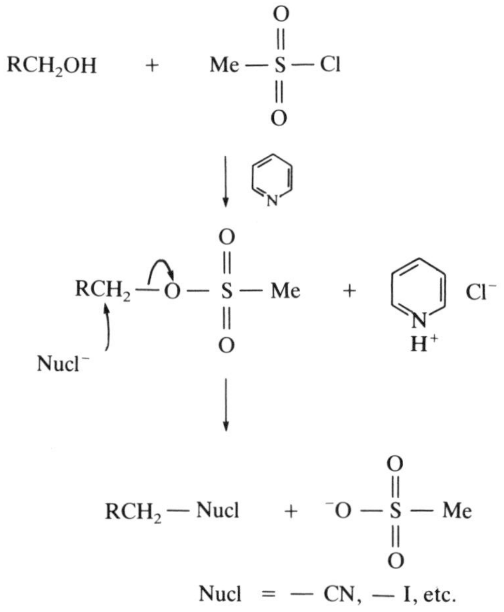

Fig. 3.

$$RCH_2OH \quad + \quad Me - \overset{\overset{O}{\|}}{\underset{\underset{O}{\|}}{S}} - Cl$$

$$RCH_2 - O - \overset{\overset{O}{\|}}{\underset{\underset{O}{\|}}{S}} - Me \quad + \quad \text{(pyridinium)} \; Cl^-$$

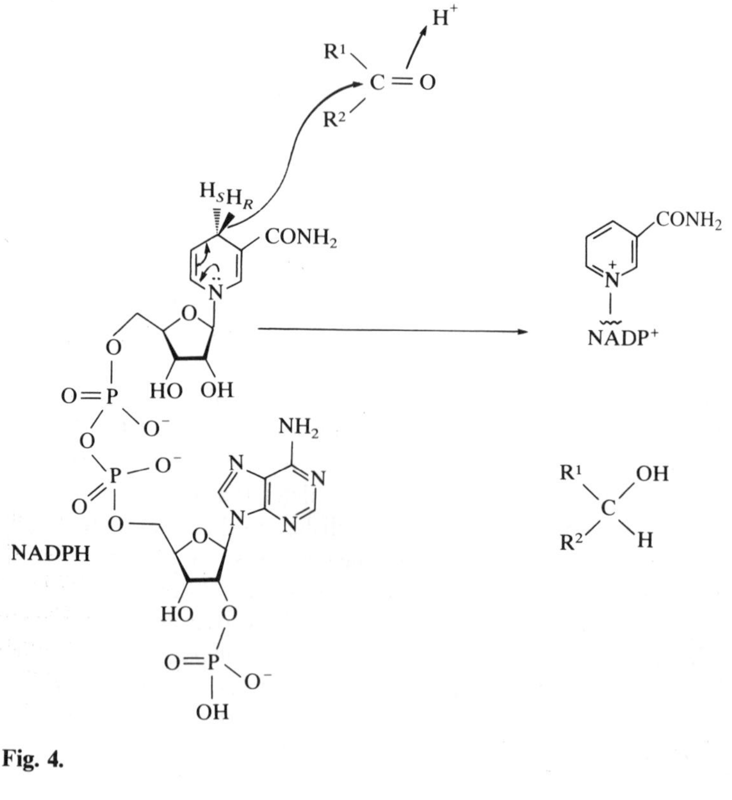

$$Nucl^-$$

$$RCH_2 - Nucl \quad + \quad {}^-O - \overset{\overset{O}{\|}}{\underset{\underset{O}{\|}}{S}} - Me$$

$$Nucl \; = \; - CN, \; - I, \text{etc.}$$

Fig. 4.

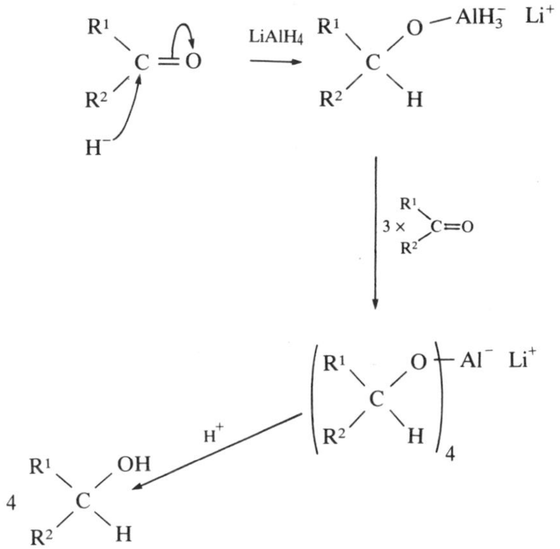

Fig. 5.

Numerous cofactors have been identified, but two are of particular importance. These are adenosine triphosphate (ATP) and nicotinamide adenine dinucleotide phosphate (NADPH). The general mode of reaction of ATP is shown in Fig. 2, and involves activation of a substrate molecule for nucleophilic displacement. Usually a hydroxyl function is converted into the phosphate or pyrophosphate, and this then suffers nucleophilic displacement. This type of chemistry is familiar to anyone who has ever made a methanesulphonate or toluenesulphonate in order to convert an alcohol into an iodide, nitrile, etc. (Fig. 3).

The cofactor NADPH is a mediator of numerous reductions that occur during biosynthesis, and the mechanism of action is shown in Fig. 4. Once again this chemistry is analogous to laboratory chemistry, since reductions effected with lithium aluminium hydride, sodium borohydride, and other hydride transfer reagents also result in addition of hydride followed by addition of a proton (upon work-up with water) (Fig. 5). One major difference is that the two hydrogens of NADPH are pro-chiral (i.e. if one is replaced with deuterium or tritium, the carbon centre becomes chiral), and the reductions mediated by the cofactor are usually stereospecific. This means that either the H_R or H_S hydrogen is used by a particular enzyme for a specific reduction. Other cofactors will be encountered in the main body of the text, and their chemistry will be discussed at that point.

An examination of Fig. 1 will reveal that photosynthesis and the production of carbohydrates are central to all primary and secondary metabolic pathways, and the book thus commences with a discussion of the chemistry and biological significance of carbohydrates.

1 Carbohydrates

J. Mann

1.1 Introduction

Photosynthesis is probably the most fundamental of all life processes, and provides a means of converting 'inorganic carbon', in the form of carbon dioxide, into carbohydrates and thence into other organic compounds. The early photosynthetic bacteria first appeared about 3000 million years ago, and were joined much later by blue-green algae (about 2000 million years ago) and the first vascular land plants (about 400 million years ago). Together these organisms now produce approximately 14×10^{10} tonnes of organic matter every year according to the process shown in equation 1.1:

$$CO_2 + H_2O \xrightarrow[\text{photosynthetic organisms}]{hv} \text{Carbohydrates} + O_2 \qquad \text{[eqn 1.1]}$$

The complexities of 'CO_2 fixation' will be dealt with in a later section, but suffice to state at present that a diverse array of structural types are produced. Many of these, for example D-glucose (1.1) and D-ribose (1.2), are of prime importance for a whole range of essential biochemical processes, and as structural materials or energy stores. They are true 'primary metabolites' in the sense that they are ubiquitous and essential for all lifeforms. Other carbohydrates are components of complex antibiotics, and of other so-called 'secondary metabolites' which are not widespread and do not necessarily have a proven function.

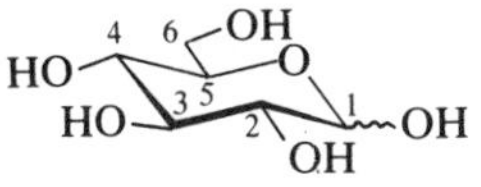

(1.1) 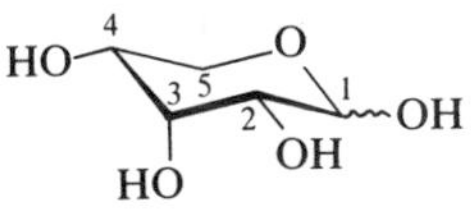 (1.2)

To the chemist carbohydrates were initially of interest because of their complex structural and stereochemical relationships, and more recently their utility as stereochemically defined starting materials for organic synthesis has been widely explored. This aspect will be covered in depth later in the chapter.

1.2 Structural types

The term 'carbohydrate' was originally introduced in the 19th century because the molecular formulae of the simpler carbohydrates were shown to be of the form $C_nH_{2n}O_n$ or $C_n(H_2O)_n$ – hence 'hydrate of carbon'. Many of these compounds were sweet, and the term 'sugar' was also adopted. These simple sugars or 'monosaccharides' of known molecular weight were shown to be polyhydroxy-aldehydes or polyhydroxy-ketones, and the interconversions shown in equation 1.2 are typical of the experiments carried out which provided evidence for these functionalities:

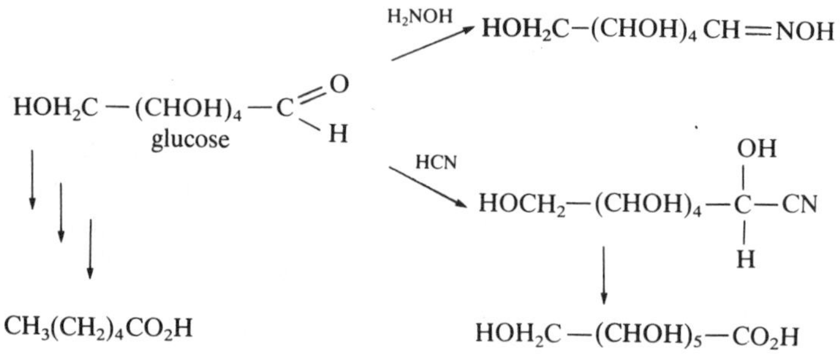

$$H_2NOH \longrightarrow HOH_2C-(CHOH)_4\,CH{=}NOH$$

$$HOH_2C-(CHOH)_4-C{\overset{O}{\underset{H}{\diagdown}}}$$
glucose

$$HCN \longrightarrow HOCH_2-(CHOH)_4-\underset{\underset{H}{|}}{\overset{\overset{OH}{|}}{C}}-CN$$

$$CH_3(CH_2)_4CO_2H \qquad\qquad HOH_2C-(CHOH)_5-CO_2H$$

[eqn 1.2]

Glucose is by far the most common carbohydrate, and although it occurs free in a variety of fruit juices, honey, etc., it is more commonly encountered in polymers termed 'polysaccharides', such as cellulose and starch (see Section 1.3). Much of the classical structure elucidation was carried out by Emil Fischer using glucose, and he correctly assigned the relative configurations at all four asymmetric carbon atoms. He had no way of knowing or being able to determine its absolute stereochemical configuration, but made an inspired guess and chose the D-form (**1.3**), and this was subsequently confirmed by an X-ray diffraction study in 1951. Emil Fischer was awarded the Nobel prize for chemistry in 1902 in recognition of his immense contributions to the structural elucidation of carbohydrates.

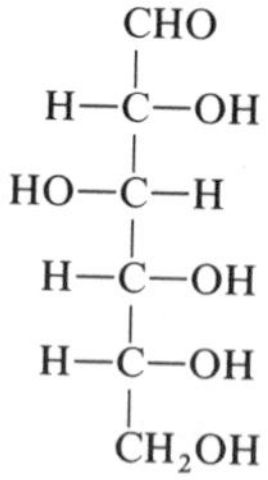

$$
\begin{array}{c}
CHO \\
| \\
H-C-OH \\
| \\
HO-C-H \\
| \qquad\qquad (1.3)\\
H-C-OH \\
| \\
H-C-OH \\
| \\
CH_2OH
\end{array}
$$

The relationship of D-glucose to all other simple aldehydo-sugars (aldoses) is shown in Fig. 1.1. A similar series can be drawn for the keto-sugars (ketoses) of which D-fructose (**1.4**) – an isomer of D-glucose – is the most

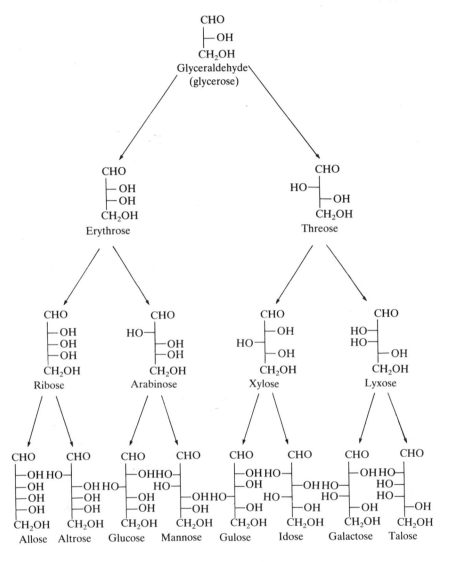

Fig. 1.1 Acyclic forms of the D-series of aldoses.

important member, and this series is shown in Fig. 1.2. These acyclic structures are useful as a means of representing the relationships between the monosaccharides, but bear little relation to their true structures under most conditions. The evidence for this assertion came initially from chemical studies, and more recently from spectral and X-ray crystallographic investigations. For example, although glucose forms an oxime, it does not react with Schiff's reagent (a complex formed between SO_2 and the reduced form of the dye magenta) to abstract SO_2 and liberate the pink dye, as an aldehyde should; its polyacetylated and polybenzoylated derivatives do not

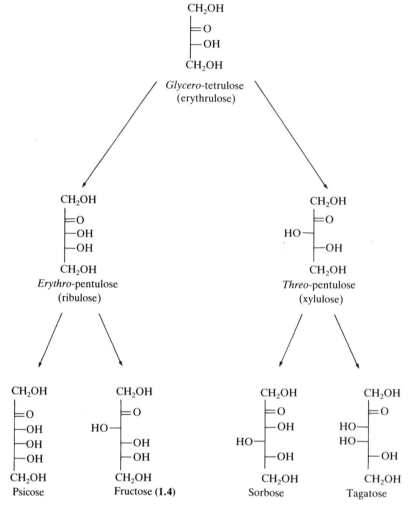

Fig. 1.2.

behave as aldehydes in any of their reactions; and glucose does not form a dimethyl acetal when treated with methanol and an acid catalyst, but forms instead two isomeric mono-methylated derivatives. All of these findings, and the absence of a carbonyl absorption in the IR spectrum of glucose, can be explained if glucose exists primarily in the form of a cyclic hemiacetal, or rather a mixture of several possible cyclic hemiacetals (Fig. 1.3).

The cyclizations create 'furanose' and 'pyranose' forms, and a new chiral centre at what was the aldehyde (or ketone) carbon, and this is termed the 'anomeric centre'. When the oxygen substituent at this centre is *trans* to the group emanating from the highest numbered chiral carbon atom, the isomer is called the 'α-anomer', and when they have *cis* arrangement the isomer is called the 'β-anomer'. Structural formulae depicting the various cyclic forms

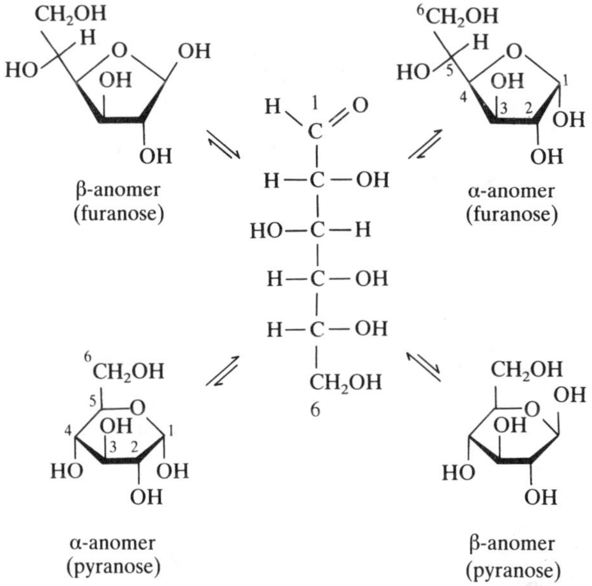

Fig. 1.3.

of the aldoses and ketoses are shown in Fig. 1.4. These representations were first devised by Haworth in the 1920s, and although very useful, do not provide any indication of the possible conformation of the various furanose and pyranose forms. Thus the pyranose form of β-D-glucose can exist in two chair conformations and numerous other interconvertible skew and boat forms (Fig. 1.5). Here the symbols $^4C_1$ (D) and $^1C_4$ (D) indicate that the chair conformations have carbon atoms C-4 and C-1 above, and atoms C-1 and C-4 below, the planes of the rings.

As would be anticipated from conformational analysis, the chair form with all substituents equatorial is favoured over all of the other forms, primarily because of reduced steric interactions, and the need to avoid 1,3-*cis*-diaxial interactions. Unfavourable interactions between polar groups are also important.

Rather less is known about the conformations of the furanoses, but the available evidence suggests that an envelope form (**1.5**) and a twist form (**1.6**) are normally favoured. It is well established that 1,2-*cis*-interactions are disfavoured, and as far as possible furanose sugars appear to adopt the conformation that has the fewest (or none) of these interactions.

(1.5)

(1.6)

The predominant species present under a given set of conditions is often hard to predict, since complex equilibria exist in aqueous solutions of sugars,

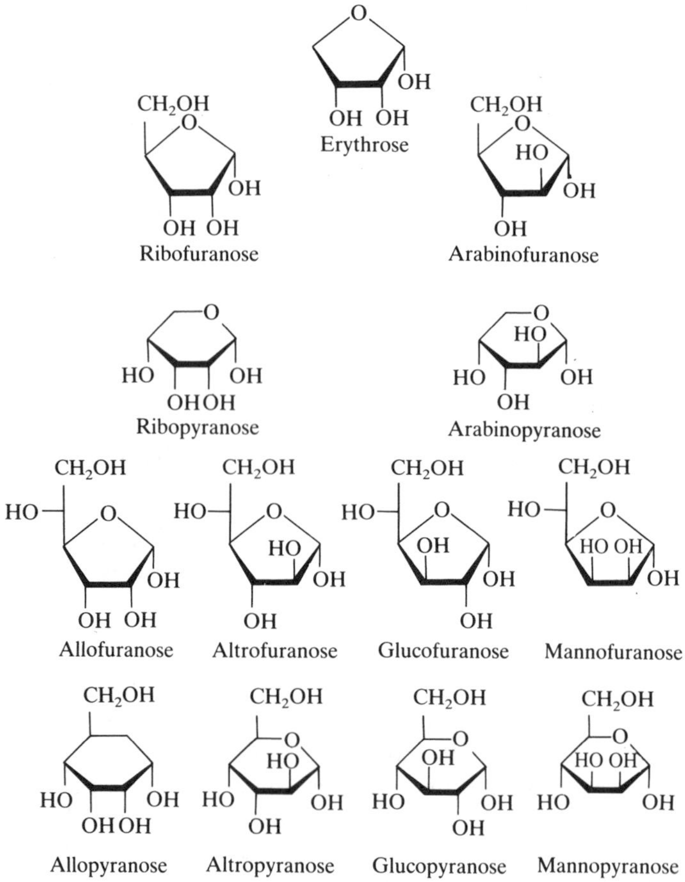

Fig. 1.4 Cyclic forms of the α-D-aldoses.

and acyclic, furanose, and pyranose forms may all be present. If chemistry is carried out on the sugars to form derivatives, or if acid or base is added to the solutions, or if the solvent or temperature is changed, the predominant species may change. This feature of carbohydrate chemistry will be discussed in later sections.

1.3   Sources and functions

The process of photosynthesis, which was described in very simple terms in equation 1.1, is in reality extremely complex, but can be conveniently divided into two main parts: the use of light energy to produce the reactive phosphorylating species adenosine triphosphate (ATP, **1.7**) – the so-called 'light reactions'; and carbon dioxide assimilation *en route* to sugars – the so-called 'dark reactions'.

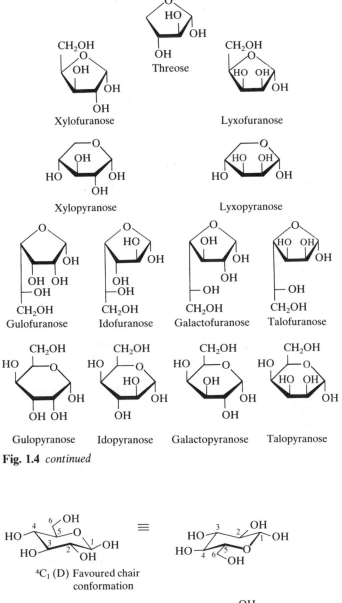

Fig. 1.4 *continued*

$^4C_1$ (D) Favoured chair conformation

One of several boat conformations

$^1C_4$ (D) Disfavoured chair conformation

Fig. 1.5.

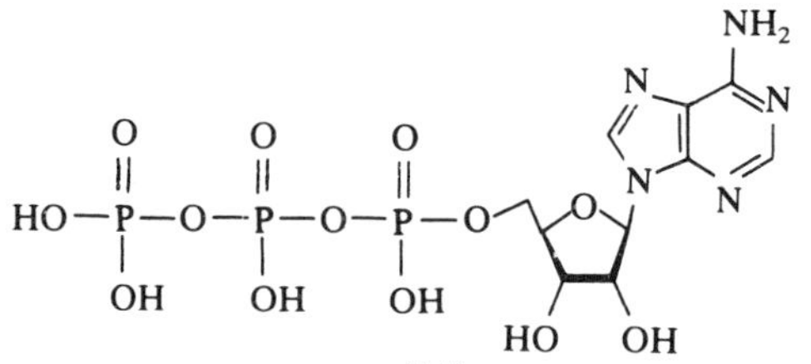

(1.7)

In the light reactions photons are captured by chlorophyll molecules to produce excited state molecules, and these then donate electrons which pass along an electron transport chain via a series of discrete molecular species (mainly quinones and cytochromes) in a sequence of one-electron oxidation/reduction reactions (e.g. equation 1.3):

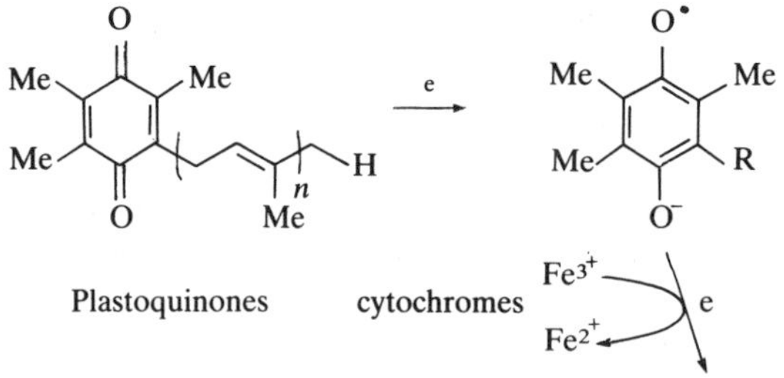

Plastoquinones   cytochromes   [eqn 1.3]

The ultimate acceptor is the enzyme cofactor nicotinamide adenine dinucleotide phosphate ($NADP^+$), and the chemistry involved is shown in equation 1.4:

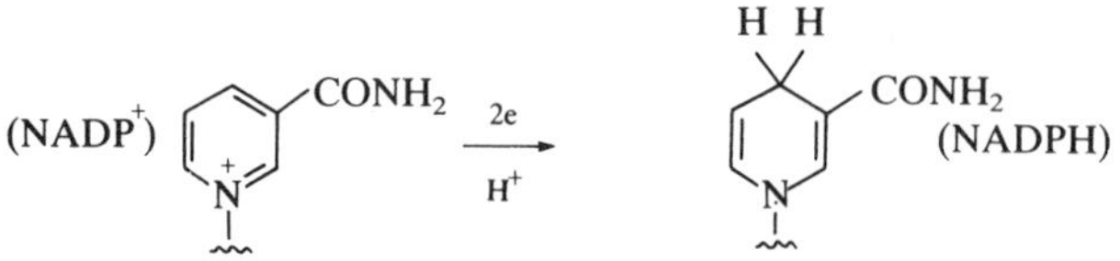

[eqn 1.4]

This process can be presented in an alternative way which emphasizes the attendant phosphorylation of adenosine diphosphate and production of oxygen (equation 1.5):

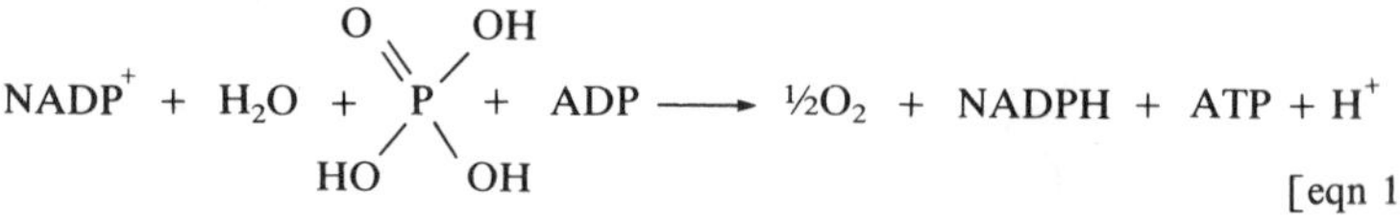

$$NADP^+ + H_2O + \underset{HO}{\overset{O}{\underset{}{P}}}\overset{OH}{\underset{OH}{}} + ADP \longrightarrow \tfrac{1}{2}O_2 + NADPH + ATP + H^+$$

[eqn 1.5]

These two species, NADPH and ATP, play vital roles in the associated dark reactions.

Much of our knowledge of the mode of CO_2 assimilation arises out of the work of Melvin Calvin and his coworkers, who followed the incorporation of ^{14}C-labelled CO_2 as it was assimilated by algae. They were able to isolate and identify the first-formed intermediates, and to delineate the enzyme-mediated pathway shown in Fig. 1.6. In step 1, ribulose-1,5-diphosphate is carboxylated to produce an unstable intermediate which fragments to yield two molecules of 3-phosphoglycerate. This is formally a reverse Claisen condensation. ATP is then used to phosphorylate this with obtention of 3-phosphoglyceryl phosphate, and reduction of this through the intervention of NADPH yields 3-phosphoglyceraldehyde (step 2). Rearrangement via the common enol of 3-phosphoglyceraldehyde and of dihydroxyacetone phosphate provides a mixture of both of these species, and an aldol condensation between them ensues (steps 3 and 4). The sequence is then completed by phosphate hydrolysis, another interconversion via a common enol form, and a second phosphate hydrolysis to produce glucose (steps 5, 6, and 7). Overall the constructive process (or 'anabolic' process) from step 2 through to step 7 is formally the reverse of glycolysis – the process by which glucose is broken down (a 'catabolic' process) to provide 3-phosphoglycerate and ultimately pyruvate (2-ketopropionate) for the citric acid cycle. The direction of the reactions is determined by a variety of factors, not least of which is the availability of particular enzymes, and also by the types of cofactors involved. Thus the success of CO_2 assimilation depends upon the use of the oxidation/reduction couple NADPH/NADP$^+$, which is kept mainly in the reduced form (i.e. NADPH), and upon the free energy of reaction released in steps 5 and 7.

In contrast, the catabolic pathway (glycolysis) employs the couple NAD$^+$/NADH maintained primarily in the oxidized form (i.e. NAD$^+$), and must also involve two phosphorylations which demand energy input into the system. Obviously the balance between these two pathways – catabolism and anabolism – is of vital importance to all photosynthesizing organisms.

Ribulose-1,5-diphosphate, the prime acceptor for CO_2, is produced in a separate pathway (Fig. 1.7) from fructose-6-phosphate. In the first step fructose-6-phosphate reacts with the carbanionic form of the cofactor thiamine pyrophosphate (**1.8**) (a nitrogen ylid – denoted as R$^-$ in Fig. 1.7). Fragmentation (step 2) then yields erythrose-4-phosphate and another carbanion which further reacts with glyceraldehyde-3-phosphate (step 3) to yield, after fragmentation and loss of thiamine pyrophosphate (step 4), xylulose-5-phosphate. This then yields ribulose-5-phosphate via epimerization, and ribulose-1,5-diphosphate is formed after phosphorylation.

The role of thiamine pyrophosphate is amplified in Fig. 1.8, and this process is usually known as 'transketolization'. Other processes which combine the use of aldol condensations and transketolizations provide additional molecules of ribulose-1,5-diphosphate, and other sugar precursors having three, four, five, six, and seven carbon atoms. In these reactions, a two-carbon unit is transferred from one carbohydrate to another.

The other hexoses are formed from glucose (and pentoses from ribose) after activation, usually by conversion into uridine diphosphoglucose (Fig. 1.9). Chemical interconversions and epimerizations then afford the

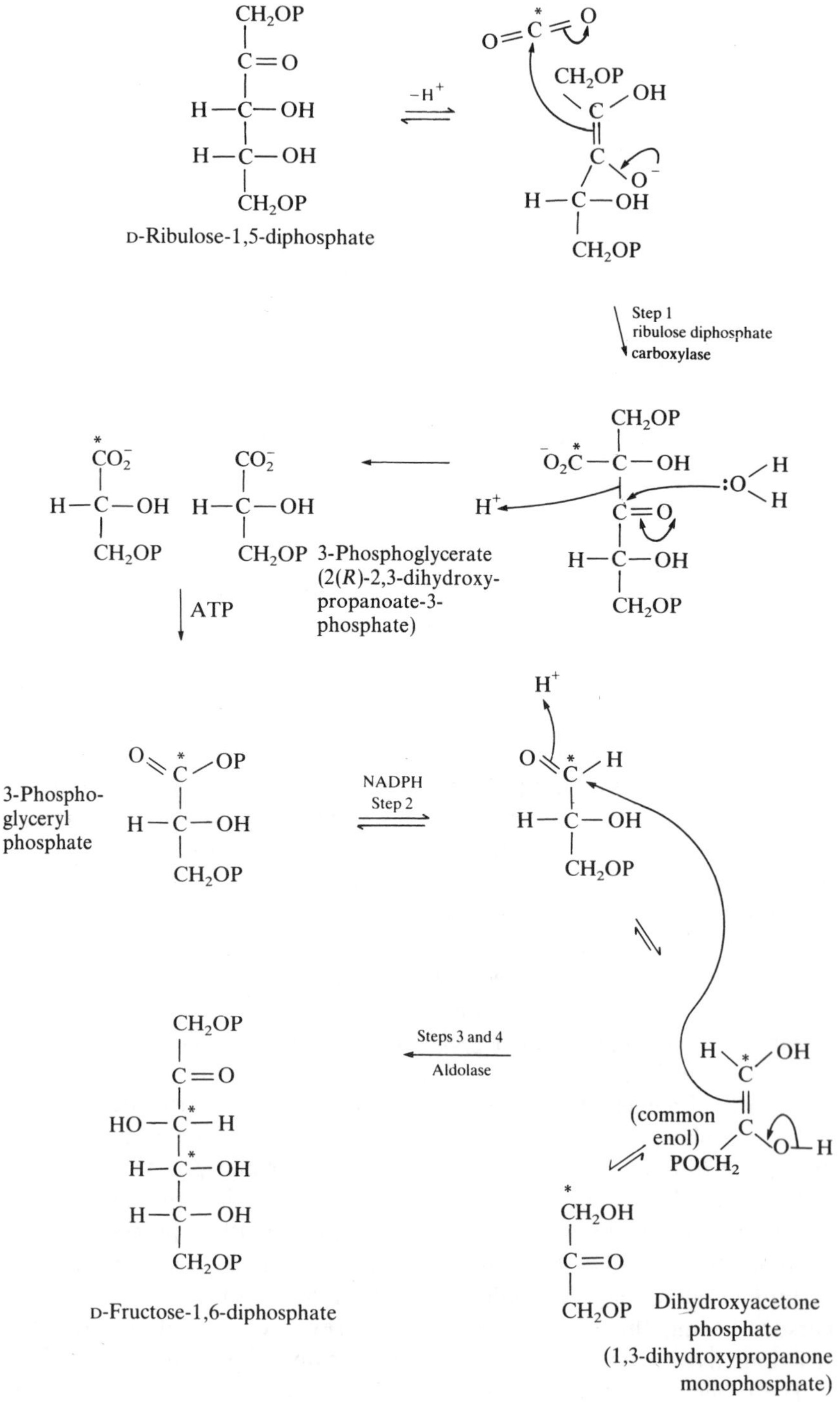

CH$_2$OP
|
C=O
|
H—C—OH
|
H—C—OH
|
CH$_2$OP

D-Ribulose-1,5-diphosphate

$-H^+$ ⇌

O=C=O

CH$_2$OP   OH
\ /
C
|
C
/ \
O$^-$
|
H—C—OH
|
CH$_2$OP

Step 1
ribulose diphosphate
carboxylase

CH$_2$OP
|
$^-O_2$C—C—OH
| H
C=O   :O
| H
H—C—OH
|
CH$_2$OP

H^+ ←

CO$_2^-$ CO$_2^-$
| |
H—C—OH   H—C—OH
| |
CH$_2$OP   CH$_2$OP

3-Phosphoglycerate
(2(R)-2,3-dihydroxy-
propanoate-3-
phosphate)

| ATP

3-Phospho-
glyceryl
phosphate

O=C—OP
|
H—C—OH
|
CH$_2$OP

NADPH
Step 2
⇌

H^+

O=C—H
|
H—C—OH
|
CH$_2$OP

CH$_2$OP
|
C=O
|
HO—C—H
|
H—C—OH
|
H—C—OH
|
CH$_2$OP

D-Fructose-1,6-diphosphate

Steps 3 and 4
Aldolase

H   OH
\ /
C
||
C
/ \
POCH$_2$ O—H

(common
enol)

CH$_2$OH
|
C=O
|
CH$_2$OP

Dihydroxyacetone
phosphate
(1,3-dihydroxypropanone
monophosphate)

Fig. 1.6.

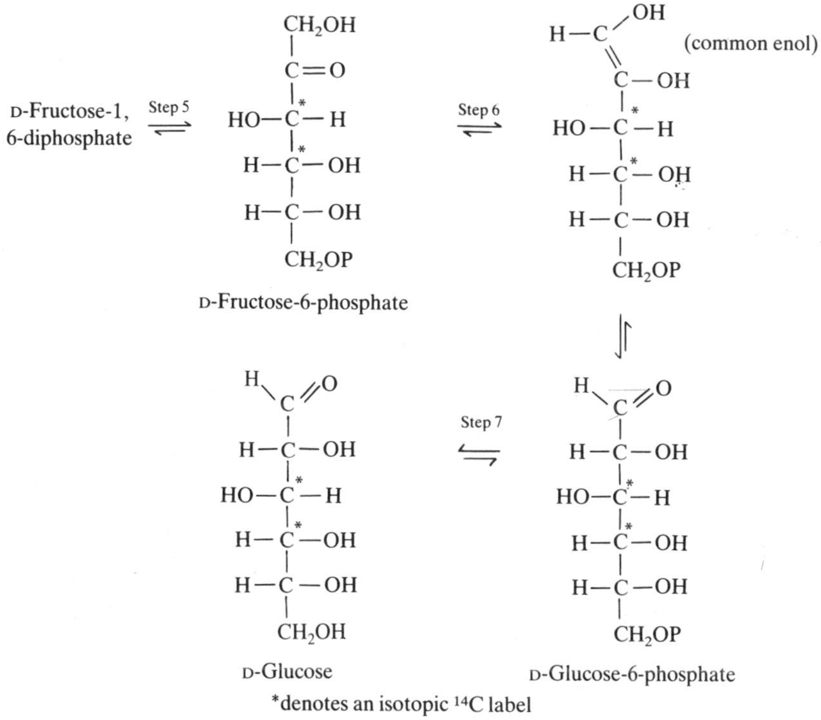

D-Fructose-1,6-diphosphate $\underset{\Longleftarrow}{\overset{\text{Step 5}}{\Longrightarrow}}$ D-Fructose-6-phosphate $\underset{\Longleftarrow}{\overset{\text{Step 6}}{\Longrightarrow}}$ (common enol)

D-Glucose $\underset{\Longrightarrow}{\overset{\text{Step 7}}{\longleftarrow}}$ D-Glucose-6-phosphate

*denotes an isotopic ^{14}C label

Fig. 1.6 *continued*

various hexoses, and the route to galactose, shown in Fig. 1.9, is exemplary. Of the simple sugars, glucose and ribose (together with 2-deoxyribose – see **1.2**) are probably the most important. The former occurs as the monomeric unit in numerous polysaccharides (e.g. cellulose, starch, and glycogen – see below), but is also a common constituent of secondary metabolites as in salicin (**1.9**), a natural anti-inflammatory agent from willow bark, and in hesperidin (**1.10**), the dominant flavonoid in lemons and sweet oranges where it co-occurs with the more unusual sugar L-rhamnose (6-deoxy-L-mannose).

D-Ribose and 2-deoxy-D-ribose are key constituents of the essential ribonucleic acids (RNAs) and deoxyribonucleic acids (DNAs) respectively, but the ribosyl unit is also found in those secondary metabolites which resemble the monomeric units of DNA and RNA, so-called C-nucleosides like showdomycin and pyrazofurin (**1.11** and **1.12**). These are both mould metabolites and the former has anti-bacterial and anti-tumour activities, while the latter has anti-viral properties. The chemistry and biological properties of these and other C-nucleosides will be dealt with in Chapter 2.

Apart from the common monosaccharides (see Figs 1.1 and 1.2), there are many rare sugars and these are usually components of antibiotics. Most are amino-sugars, and three examples are to be found in erythromycin (**1.13**)

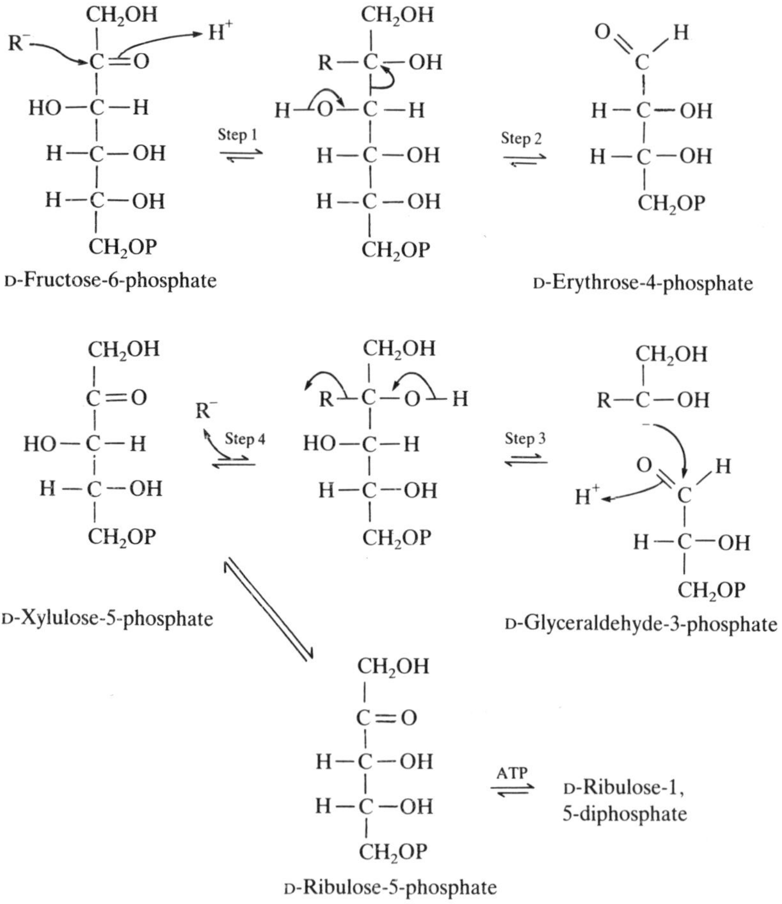

D-Fructose-6-phosphate
Step 1
Step 2
D-Erythrose-4-phosphate
Step 4
Step 3
D-Xylulose-5-phosphate
D-Glyceraldehyde-3-phosphate
ATP
D-Ribulose-1, 5-diphosphate
D-Ribulose-5-phosphate

Fig. 1.7.

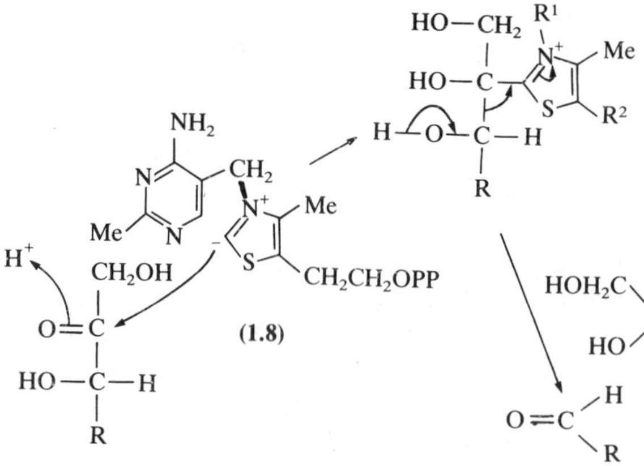

(1.8)

Fig. 1.8.

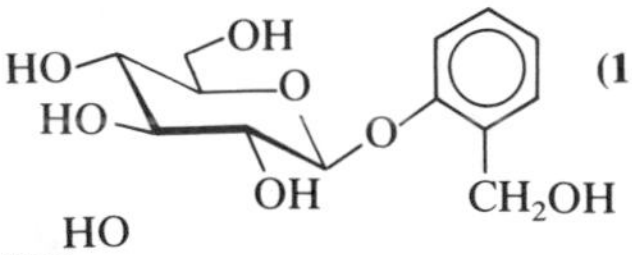

(1.9)

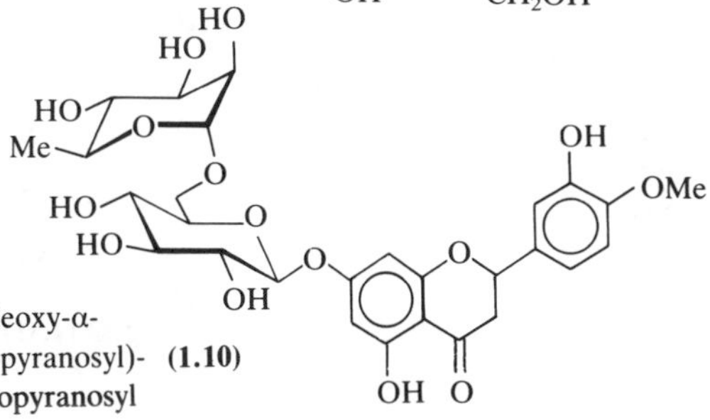

6-O-(6-deoxy-α-
L-mannopyranosyl)- (1.10)
β-D-glucopyranosyl

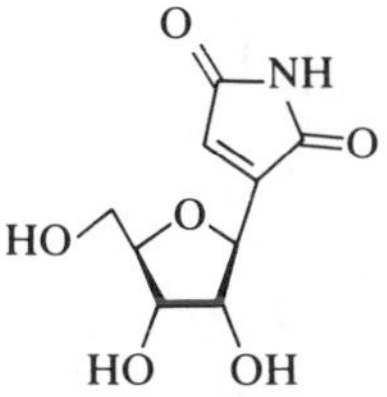

(1.11)

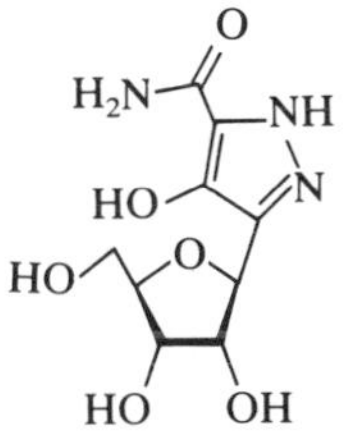

(1.12)

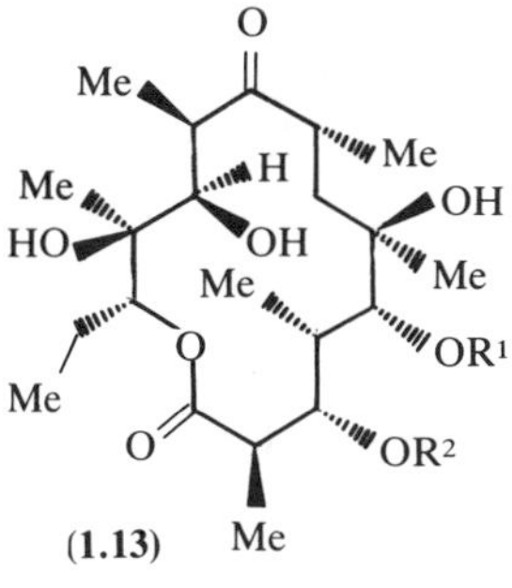

(1.13)

$R^1 =$

D-desosamine

$R^2 =$

L-cladinose

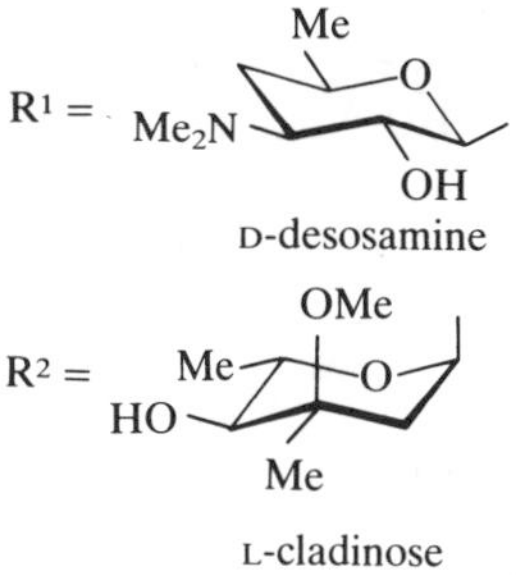

from *Streptomyces erythreus*, one of the macrolide antibiotics, and particularly useful against penicillin-resistant organisms and against myco-plasms; adriamycin (1.14) from *Streptomyces achromogenes*, a broad-spectrum anti-tumour agent; and streptozotocin (1.15), a naturally occurring nitrosourea which again has some utility as an anti-tumour agent. The synthesis and biosynthesis of these sugars are currently areas of active investigation.

These compounds all have the non-sugar (aglycone) attached to the anomeric centre of the sugar, and are termed 'glycosides'. The methods and mechanism for their formation will be discussed in Section 1.4.2.1. When

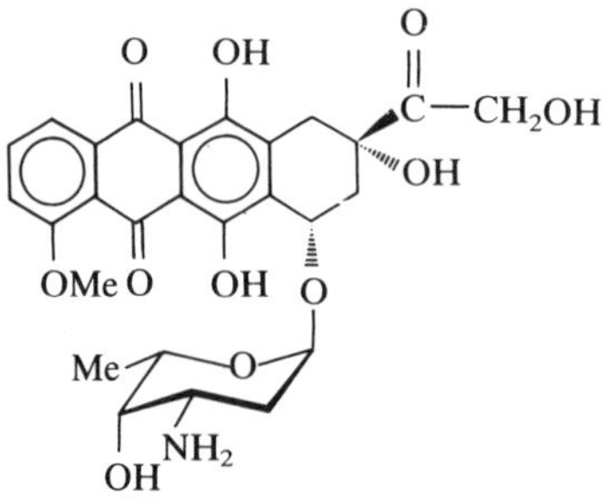

daunosamine
(1.14)

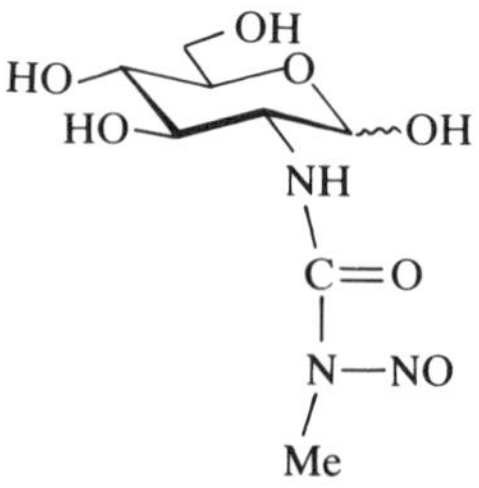

(1.15)

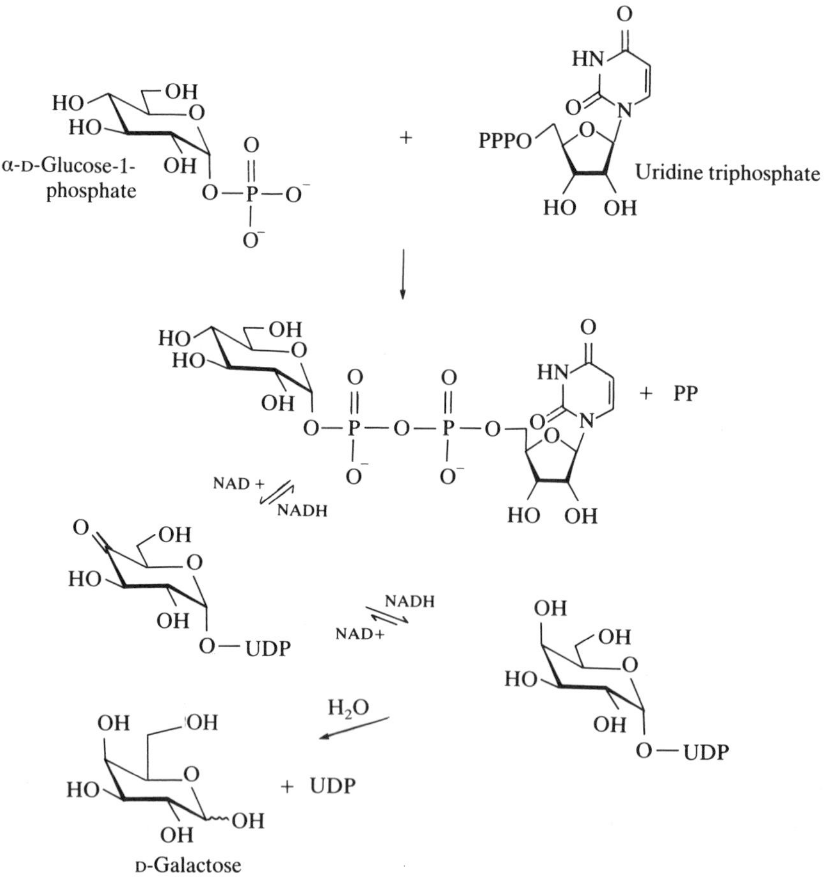

α-D-Glucose-1-phosphate

Uridine triphosphate

+ PP

NAD +
NADH

NADH
NAD+

H_2O

+ UDP

D-Galactose

Fig. 1.9.

two sugars are joined in this way, 'disaccharides' are produced, and of these, four are of particular importance: sucrose (**1.16**), trehalose (**1.17**), lactose (**1.18**), and maltose (**1.19**). The biosynthesis of lactose shown in equation 1.6 is typical of the pathways leading to the disaccharides.

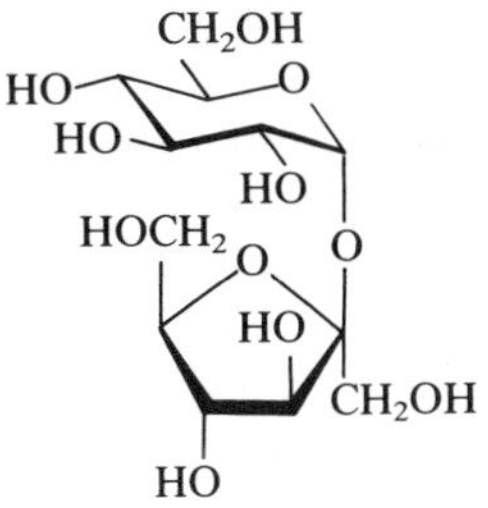

(1.16) 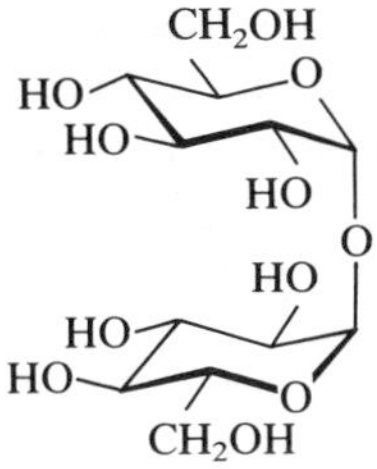   **(1.17)**

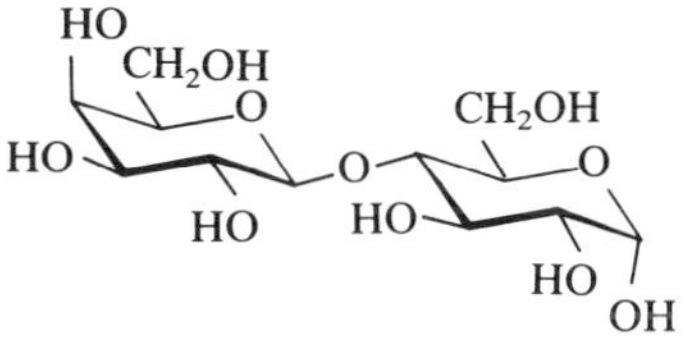

(1.18) 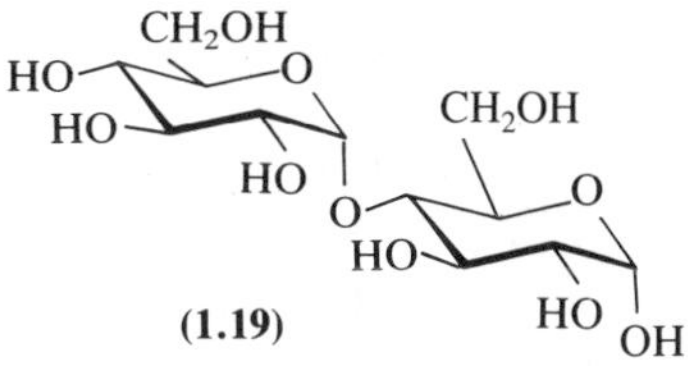   **(1.19)**

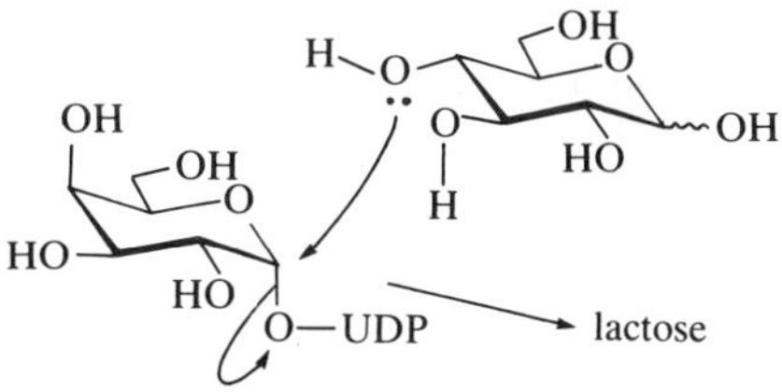

[eqn 1.6]

Sucrose has obvious importance as a sweetener, and around 5×10^7 tonnes of the pure compound are produced annually – about two-thirds from sugar cane and the rest from sugar beet. Although there are a number of theories concerning the perceptions of sweetness, none is entirely satisfactory, but the HA—B system (see **1.20**) first described by Shallenberger is now widely accepted. This envisages a hydrogen bond donor (HA—) in close proximity to a proton acceptor (—B), where the three-dimensional relationship between these two moieties and an apparently essential hydrophobic entity are as shown in (**1.20**), which could represent part of the glucose molecule.

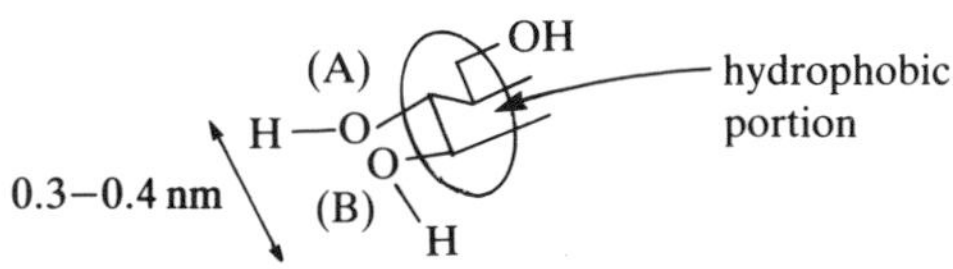

(1.20)

At present much effort is being expended with the aim of producing non-cariogenic (i.e. dentally safe) and non-calorific (i.e. non-fattening) sweeteners. One promising poly-halosugar is the galactose derivative (**1.21**), which is claimed to be 650 times sweeter than sucrose, and is presently undergoing exhaustive toxicological testing. At present a large proportion of the market for soft-drink sweeteners is held by high-fructose corn syrup. This is produced by tandem use of immobilized enzymes, firstly to depolymerize the corn starch producing D-glucose, and then to isomerize the D-glucose into a mixture enriched in D-fructose. This latter sugar is almost twice as sweet as glucose.

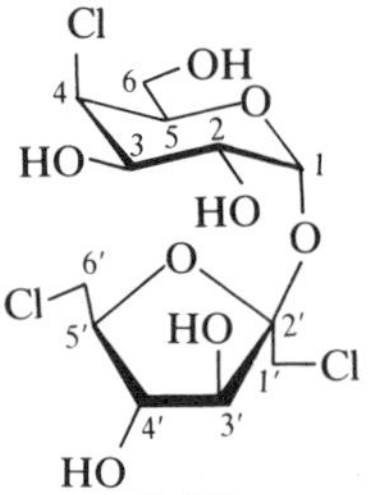

1',6'-dichloro-1',6'-dideoxy-β-D-fructo-
furanosyl-4-chloro-4-deoxy-α-D-
galactopyranoside

(**1.21**)

Trehalose (**1.17**) occurs in certain plants, fungi (e.g. fly agaric, *Amanita muscaria*), algae, and yeasts, and to the extent of up to 30 per cent in the cocoons of a parasitic beetle of the *Larinus* species. In most of these species it is used as the major carbohydrate source. Lactose (**1.18**) is the major sugar of mammalian milk and is obtained as a by-product in cheese-making (in whey). Maltose (**1.19**) is produced by enzymatic hydrolysis (by the enzyme diastase) of starch (see below).

The polymeric sugars – 'polysaccharides' – can be subdivided into two main classes: structural materials and energy stores. In the first category is the ubiquitous cellulose (**1.22**) – the main constituent of plant cell walls and in consequence of such plant products as cotton (essentially pure cellulose), flax (80 per cent), and straw from cereals (*c.* 45 per cent); and the other structural polysaccharide of widespread occurrence is chitin (**1.23**), a major component of arthropod shells (crabs, lobsters, etc.) and of insect wings.

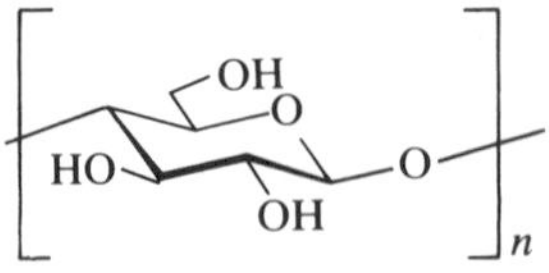

(**1.22**)

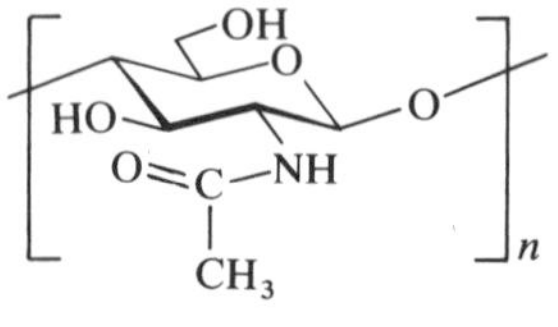

(**1.23**)

Indeed, chitin has been isolated from very ancient shells, for example from the shells of trilobites, which lived around 550 million years ago and were among the first arthropods on the planet.

Bacterial cell walls usually comprise a polysaccharide of some kind, normally bound to polypeptide chains, with the whole constituting a complex three-dimensional structure. The repeating unit of the cell wall of the bacterium *Staphylococcus aureus* is shown in (**1.24**), and the mode of action of the β-lactam antibiotics (penicillins and cephalosporins) is to disrupt the formation of cross-links between the peptide chains, and thus prevent the assembly of the three-dimensional matrix. These bacterial polysaccharides, and the much more complex ones found in the cell walls of blood cells, are also responsible for the immunological characteristics of the cells, that is they account for the virulence of the bacteria and provide the so-called blood-group antigens for the red blood cells.

(**1.24**)

The molecular basis of the ABO blood-group specificity is now well understood, and depends upon the presence of different carbohydrates in certain glycoproteins and glycolipids. The so-called ABO genetic locus carries information for the production of proteins with glycosyl transferase activity. These proteins catalyse the transfer of *N*-acetyl-D-galactosamine from UDP-*N*-acetylgalactosamine to a basic core structure (A determinant); or the transfer of D-galactose from UDP-galactose onto the same site on the core structure (B determinant) (Fig. 1.10). These blood-group substances are present not only on red blood cells, but also in various secretions.

Numerous other oligosaccharides (i.e. short polysaccharide sequences) are important components of membrane glycoproteins, glycolipids, and glyco-phospholipids. During the last 20 years it has become apparent that the oligosaccharide portion is essential for 'communication' between cells, i.e. for intercellular recognition and interaction. They act as cell-surface antigens (as in the case of the blood-group determinants), but also as receptors for viruses, hormones, toxins, and bacteria. The key to all of these roles is the

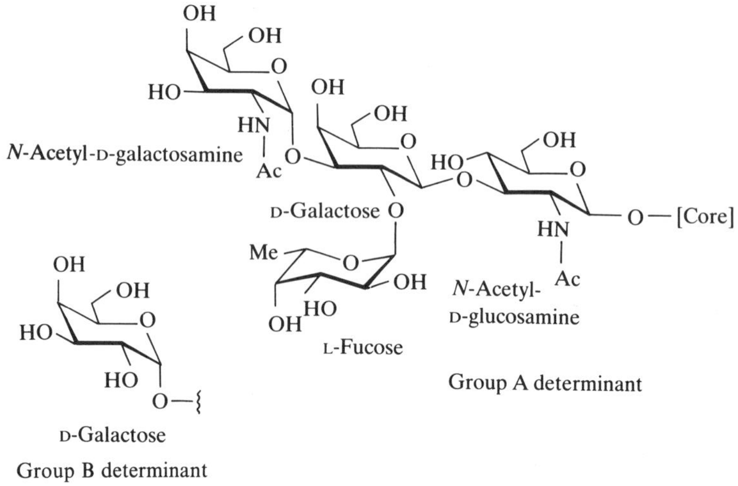

N-Acetyl-D-galactosamine

D-Galactose

N-Acetyl-D-glucosamine

L-Fucose

Group A determinant

D-Galactose

Group B determinant

Fig. 1.10.

'information potential' of oligosaccharides. Unlike the polypeptides and polynucleotides, their structural complexity depends upon points of linkage as well as monomer content. So with a dimer of D-glucose there are 11 possible structural isomers: glucose-β(1 → 2, 3, 4 or 6) glucose; glucose-α(1 → 2, 3, 4 or 6) glucose; and three possible isomers linked via C-1: α,α; α,β; β,β. An amino acid dimer, in contrast, has only one possible structure. A pentasaccharide with five different carbohydrate moieties can provide a staggering 2 144 640 possible structures – the corresponding figure for a pentapeptide is only 120. The structure elucidation and synthesis of oligosaccharides is thus of increasing importance, and mention will be made of these two areas in Section 1.4.

Before leaving this section it is worth emphasizing the importance of the glycopeptides and glycoproteins. They are intimately involved in biological recognition, especially where this involves cell-to-cell, cell-to-molecule, or molecule-to-molecule contact. For example, the structures of glycoproteins on the surface of cancer cells are very different from those on the surface of untransformed normal cells. This presumably allows the cancer cells to escape recognition by the immune system, and also to grow without the constraining influence of inhibitors that associate with the normal glycoproteins.

Clearly a knowledge of the molecular structures and interactions of these glycoproteins should assist in the design of specific chemotherapeutic agents for treating cancer and other diseases. Human immunodeficiency virus (HIV) is the organism that causes AIDS (acquired immune deficiency syndrome), and it is now known that it binds preferentially to T4 lymphocytes, which are key regulatory cells of the immune system. This binding involves interaction between the major surface glycoprotein of the virus, so-called gp120, and the cell surface glycoprotein of the T4 cells, CD4. A subsequent structural change at the cell surface allows uptake of the virus particle, and the train of events leading ultimately to viral reproduction and cell death is initiated. One obvious strategy for chemotherapy would thus involve agents that changed the structure of either CD4 or gp120, and some recently described glycosidase inhibitors (i.e. compounds that inhibit the enzymes that cleave glycoside linkages) show promise in this area (see Section 7.7.1).

Other plant polysaccharides include the pectic acids and the alginic acids, polymers containing sugar acids which are valued for their gelling properties and are much employed in the food industry (jams, preserves, and ice-creams). Another polysaccharide containing carboxylic acid groups is hyaluronic acid (**1.25**), which seems to function as a lubricant in animal joints.

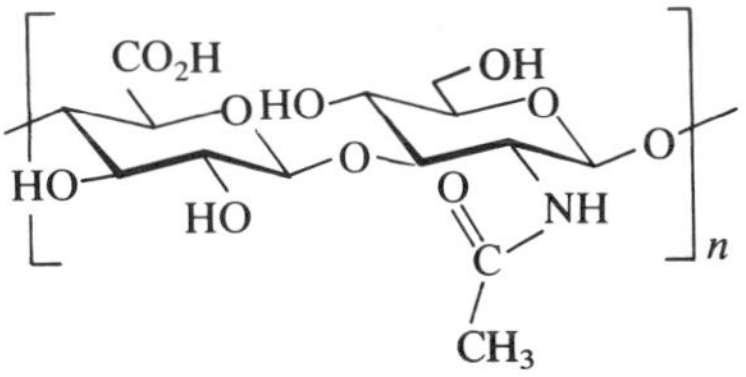

(**1.25**)

The common energy stores are starch (a mixture of amylose and amylopectin; **1.26**), the carbohydrate reserve of plants, especially for tuberous plants such as the potato and in cereals; and glycogen which is the carbohydrate reserve of mammals and is stored in the liver. During muscular exercise it is converted into D-glucose and thence into pyruvic acid and on towards the citric acid cycle. The structure is similar to that of starch, but the degree of branching is more extensive.

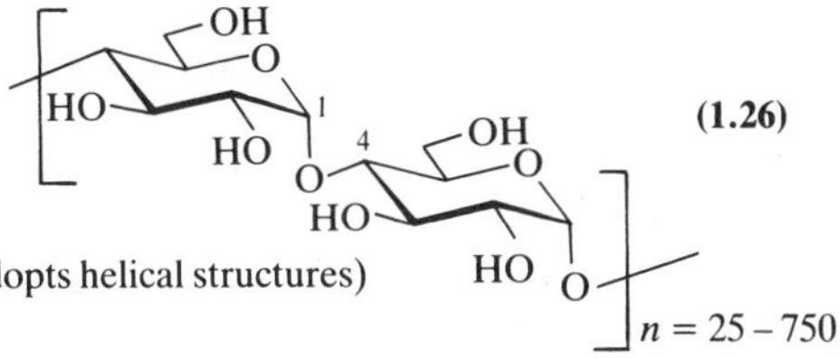

(**1.26**)

Amylose (adopts helical structures)

$n = 25 - 750$

Amylopectin: (random structure)   α-1-4 and α-1-6 linkages

Two less common energy stores are inulin (**1.27**), a D-fructose polymer from certain plants, and the dextrans which are produced by a number of

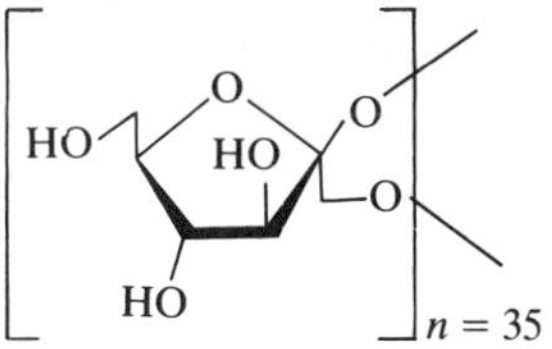

$n = 35$

(**1.27**)

microorganisms. They contain 1,6-linked α-D-glucopyranose units and also other more complex linkages of glucose units, and are of interest because they can serve as blood plasma substitutes.

1.4   Chemistry of monosaccharides

Now that the main structural types and their modes of formation and functions have been described, the basic chemistry of carbohydrates can be considered. The monosaccharides are usually polyhydroxyaldehydes or polyhydroxyketones, and in consequence they display most of the chemical reactions of these groups. This chemistry can be conveniently divided into two classes: reactions of the hydroxyl group, and reactions at the anomeric centre (i.e. pseudo-carbonyl chemistry).

1.4.1   Reactions of the hydroxyl group

All of the characteristic reactions of the hydroxyl group are displayed, including ether, ester and acetal formation, and a variety of oxidative changes can also be accomplished. None of the processes is unique to the carbohydrates, and much of the chemistry to be described is applicable to any alcohol. The presence of two, three, and sometimes four hydroxyl groups within the same molecule does, however, enhance the likelihood of intramolecular reactions, e.g. the formation of cyclic acetals and anhydro ethers.

In recent years, with the emphasis turning increasingly towards the synthetic utility of carbohydrates, selective reactions of one or other hydroxyl groups have become important, and in particular numerous methods are now available for the selective protection, or substitution, of particular hydroxyls.

1.4.1.1   Ethers

Historically, poly-methyl ethers of carbohydrates were much used in structure elucidation of oligosaccharides by degradation. They can be prepared in a variety of ways ranging from the classical methods which utilized methyl iodide and silver oxide or dimethyl sulphate and aqueous sodium hydroxide, through to more modern methods such as methyl iodide in *N,N*-dimethylformamide (DMF) with sodium hydride as base, diazomethane and boron trifluoride etherate (for base-sensitive sugars). With β-D-glucose and excess methyl iodide, the penta-methyl derivative (**1.28**, R = Me) is usually obtained:

RO—⟨ring structure⟩—OR, RO, OR, OR

(1.28)

The overall result of this chemistry is to turn a water-soluble, involatile carbohydrate into a water-insoluble (and incidentally organic-soluble), volatile derivative, and this has obvious value for gas–liquid chromatography (GLC)/mass spectroscopic (MS) analysis, and for purification by distillation, or by the use of silica and alumina chromatography. Methyl ethers are stable to acid and base under reasonable conditions, and can be converted into the free hydroxyl using boron trichloride.

Benzyl ethers (**1.28**, $R = CH_2Ph$) are prepared by methods closely similar to those employed to make methyl ethers, and are valued because the ethers can be cleaved by hydrogenolysis (i.e. H_2 and palladium on charcoal).

Silyl ethers are even more volatile than the methyl ethers, and are often prepared for GLC/mass spectroscopic analysis. They are also easy to purify by chromatography or distillation. A number of silylating reagents are in common use, and these include chlorotrimethylsilane or hexamethyl disilazane, both used in pyridine, for the formation of per-trimethylsilyl ethers (**1.28**, $R = SiMe_3$); and the 'bulky' reagents chloro-*t*-butyldimethylsilane, or chloro-*t*-butyldiphenylsilane, both used with DMF as solvent and imidazole as base, for the selective silylation of primary and sterically unhindered secondary hydroxyls.

The trimethylsilyl ethers are usually cleaved using aqueous alcohol, while the other silyl ethers are cleaved with aqueous acids (including aqueous HF) or with tetrabutylammonium fluoride in tetrahydrofuran (THF).

Although the primary hydroxyls usually react faster than the secondary hydroxyls, if complete selectivity for the protection of a primary hydroxyl is required, chlorotriphenylmethane in pyridine is usually employed. The triphenylmethyl (trityl) ethers thus formed can be cleaved at the appropriate time by hydrogenolysis or by using acids (e.g. HBr in glacial acetic (ethanoic) acid).

1.4.1.2 Anhydro-sugars

Intramolecular ether formation can be accomplished either by loss of water from a pair of hydroxyls, as in the formation of 1,4-anhydro-D-glucitol shown in equation 1.7, or through displacement of a leaving group (often tosylate

[eqn. 1.7]

or mesylate) by a proximate hydroxyl, as in the formation of the epoxide 5,6-anhydro-1,2-*O*-isopropylidine-α-D-glucofuranose (equation 1.8). (The utility of the isopropylidene grouping will be discussed shortly.)

As will become apparent in later sections, many of these anhydro-sugars (especially the epoxides) are key synthetic intermediates *en route* to complex natural products.

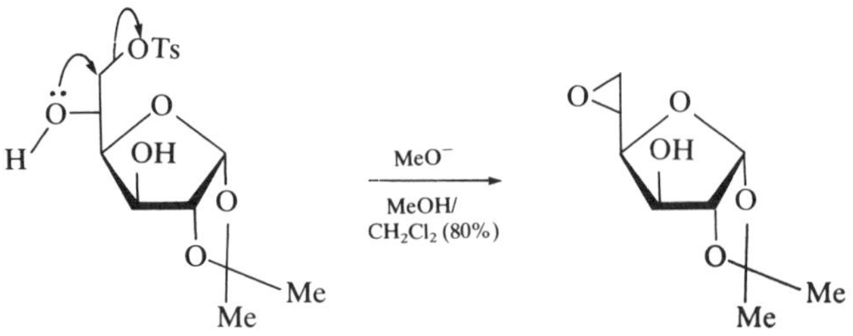

[eqn 1.8]

1.4.1.3   Esters

Carbohydrates will form esters with both organic and inorganic acid derivatives. Important examples include acetates (usually from acetic anhydride), benzoates (benzoyl chloride), sulphonates (sulphonyl chlorides), sulphates (chlorosulphonic acid), phosphates (substituted phosphoro-chloridates, e.g. $(PhO)_2PO \cdot Cl$), carbonate (phosgene), and borates (boric acid). Obviously with excess reagent complete esterification is possible, especially with simple reagents such as acetic anhydride, but under mild conditions or with the more bulky reagents it is sometimes possible to esterify the primary hydroxyls selectively. As with the ethers, sugar esters are often formed to facilitate purification and for characterization purposes, and since they can be hydrolysed by base to yield the parent sugars, they are complementary to the acid-labile ethers as protecting groups.

One complication that occurs with esterification is that a change of stereochemistry may occur at the anomeric centre. This is illustrated for glucose in Fig. 1.11. The isomer obtained can be controlled to some extent

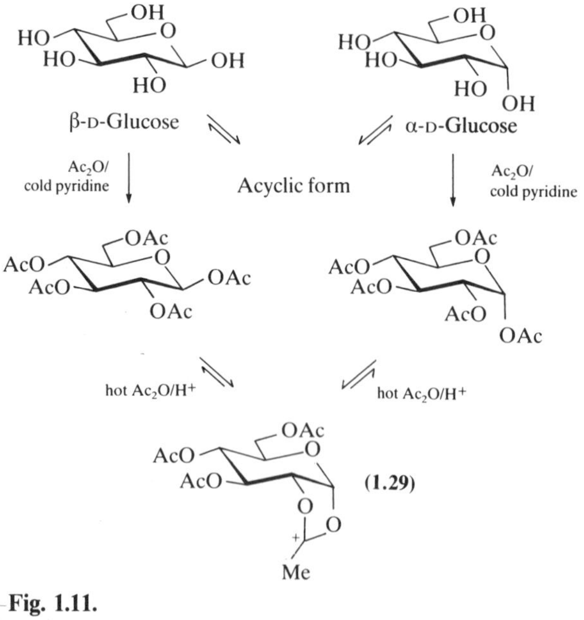

—**Fig. 1.11.**

by selection of specific reagents and conditions. Thus under mild conditions and without acid catalysis, acetylation proceeds faster than interconversion via the acyclic form (so called 'mutarotation'), and the penta-acetate retains the stereochemistry of the starting sugar. If, on the other hand, hot acetic anhydride is employed in the presence of the weak base sodium acetate, equilibration via the acyclic form is faster than acetylation and the β-form predominates. Finally, acid catalysts effect equilibration of the acetates, probably via the intermediacy of the acetoxonium species (**1.29**). For glucose this then provides an 84:16 mixture of the α- and β-forms. This ratio should be compared with the approximately 36:64 ratio of α- and β-forms found when the pure anomers of glucose are allowed to equilibrate in water. An explanation for this apparent anomaly is provided by the so-called 'anomeric effect'.

It is found that tetrahydropyrans, in general, exhibit a preference for the axial orientation of a 2-substituent if it is an electronegative group, e.g. (**1.30**). This is probably due to the unfavourable electronic interactions between the lone pairs on the ring oxygen and on the group X in the alternative equatorial orientation (**1.31**).

X = OH, OAc, halide, etc.

(**1.30**) X

(**1.31**)

Since this effect is polar in nature, it decreases as the polarity of the solvent increases, and for glucose in water, extensive solvation of the hydroxyls means that steric 1,3-interactions between the axial hydroxyl at C-1 and axial hydrogens (for the α-anomer **1.32**) are more serious than the electronic interactions discussed above as applied to the β-anomer. This latter form thus predominates.

(**1.32**)

Although this discussion has been confined to glucose, similar consider-ations apply to the other sugars and their peresters.

Before leaving the esters it is worth noting their natural abundance and industrial importance. Sugar phosphates are of pivotal importance in biochemistry, for example adenosine triphosphate (**1.7**) and the glucose and fructose phosphates already mentioned in connection with photosynthesis and glycolysis. Several glycoproteins contain oligosaccharides that contain sulphate groups, for example (**1.33**) and (**1.34**), the dimeric subunits of keratan sulphate (cornea) and heparin (anti-coagulant secreted by mast cells in the lungs, liver, and other tissues).

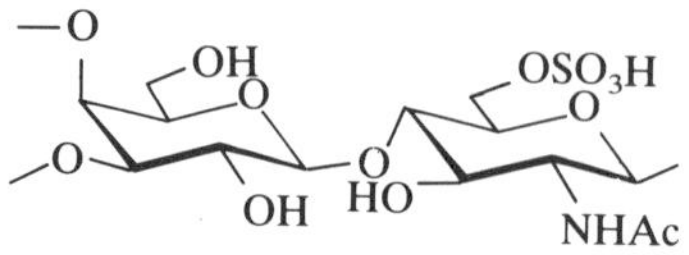

[→ 3)-*O*-β-D-galactopyranosyl-(1→ 4)
O-(2-acetamido-2-deoxy-β-D-glycopyranosyl-6-sulphate-(1 →]

(**1.33**)

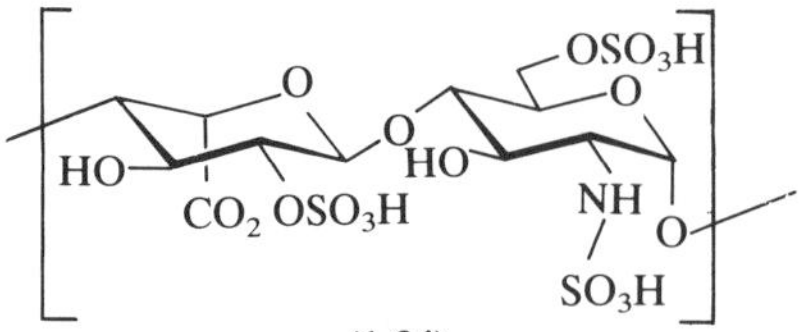

(**1.34**)

Heparin is of particular contemporary interest because it prevents the interaction of the glycoprotein gp120 of the human immunodeficiency virus (HIV) with the CD4 glycoprotein molecules on the surface of T4 lymphocytes. Clinical trials with other sulphated polysaccharides are underway.

Industrially, cellophane and rayon are produced from hydrolysis of xanthates of cellulose as shown in equation 1.9:

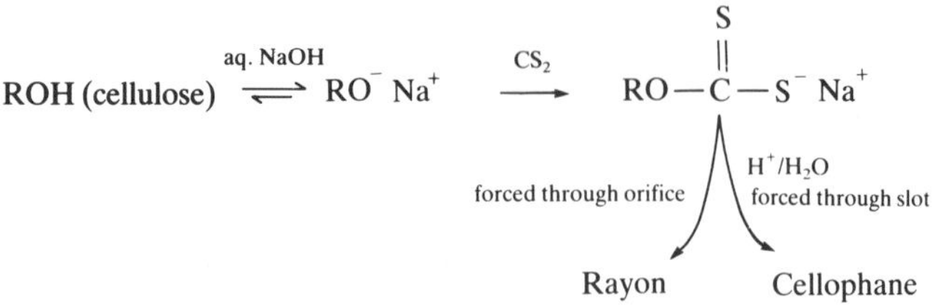

1.4.1.4   Acetals

Acetals may be formed by the reaction between the carbonyl group of the acyclic forms and alcohols, but the products of these reactions are usually glycosides (see Section 1.4.2.1 and equation 1.10), e.g. methyl glucoside (**1.35**) from glucose and methanol in the presence of an acid catalyst.

Sugar acetals are also formed when two hydroxyls of the sugar react with an aldehyde or ketone. Once again the product or products obtained depend upon the reaction conditions, and upon the stabilities of the various possible conformations of these products. The reactions normally proceed under thermodynamic control, and several generalizations may be made:

(i)   Ketones tend to favour formation of five-membered-1,3-dioxolane

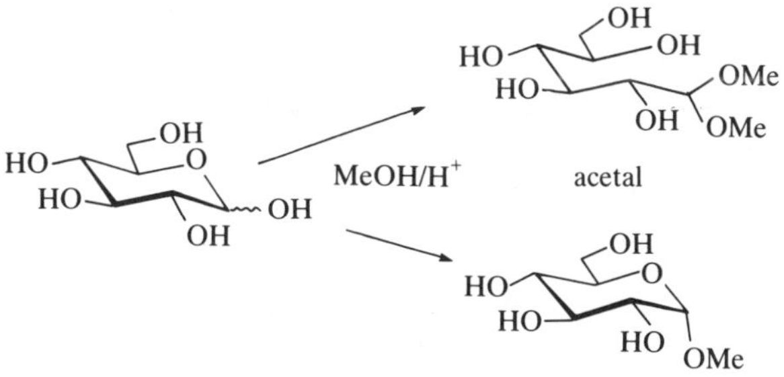

[eqn 1.10]

(1.35)

rings, as in the reaction of glucose and acetone (with sulphuric acid as catalyst) to produce 'diacetone glucose' – 1,2:5,6-di-*O*-isopropyl-idene-α-D-glucofuranose (1.36), which is the form that possesses the most 1,2-*cis*-diol relationships. In addition, selective hydrolysis of the 5,6-isopropylidene acetal is possible since the residual *cis*-fused structure constitutes a relatively stable bicyclic system (equation 1.11). Diacetone glucose is thus a useful (and cheap) starting material for the construction of more complex chiral molecules, especially those containing polysubstituted tetrahydrofuran rings.

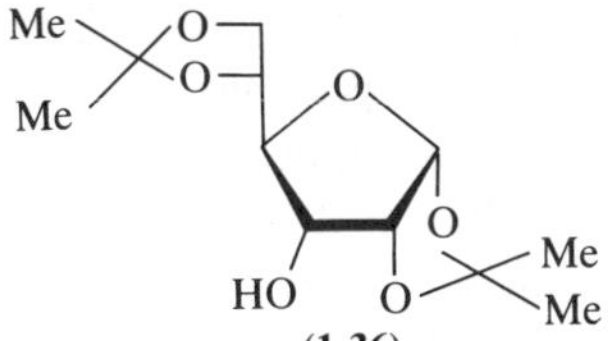

(1.36)

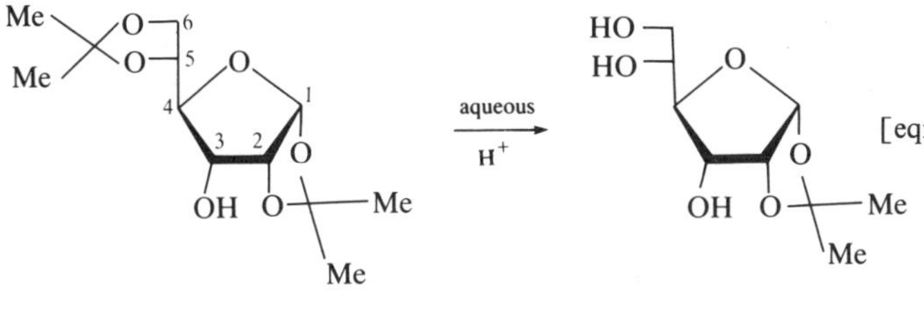

[eqn 1.11]

(ii) If more than one *cis*-1,2-diol grouping is possible, the acetal formed will be that with the minimum number of *endo* groups in order to minimize steric interactions. Thus, with ribose, three *cis*-fused acetals are possible upon reaction with acetone (structures 1.37–1.39), but (1.39) is favoured since it only bears one *endo* group.

(iii) Aldehydes tend to favour formation of 1,3-dioxane rings as the reaction of methyl-α-D-glucopyranoside (1.35) with benzaldehyde to produce methyl-4,6-*O*-benzylidene-α-D-glucopyranoside (1.40) demonstrates. The alkyl or aryl group is usually found to be equatorial for steric reasons. More complex products and mixtures of epimers at the acetal carbon are, however, often produced, and careful analysis of nuclear

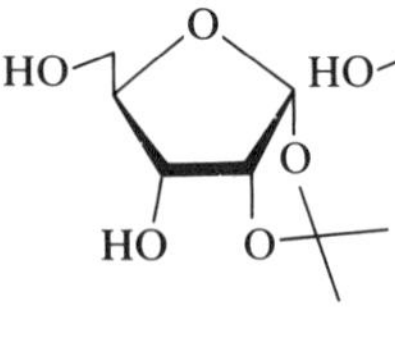

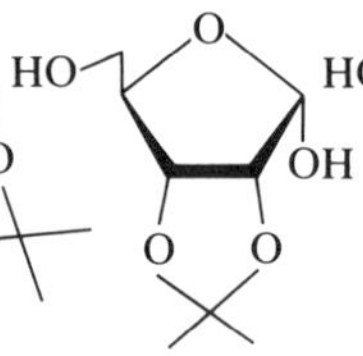

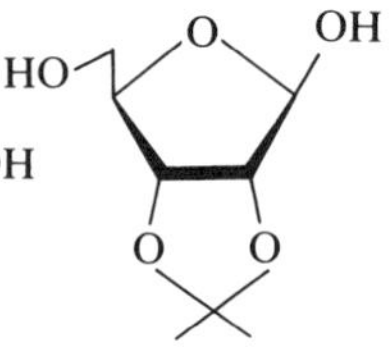

(1.37)   (1.38)   (1.39)

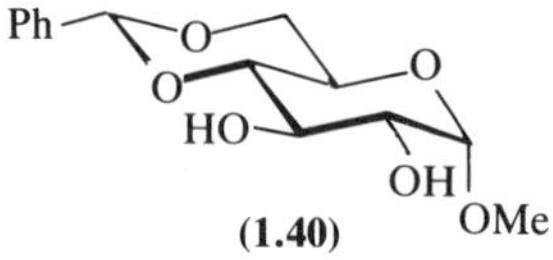

(1.40)   OMe

magnetic resonance (NMR) spectra and X-ray structural data are required for structure assignment.

In contrast to the processes just described, it is also possible to carry out acetalization under kinetic control. For example, standard acetonide formation with D-glucose (excess acetone and acid catalysis) provides the di-isopropylidene derivative (1.36). The various possible acetonides, both cyclic and acyclic, are formed and suffer cleavage under the relatively harsh reaction conditions, and the ultimate product is the thermodynamically favoured one. Through the use of 2-alkoxypropenes and a very limited quantity of an acid catalyst, the reaction can proceed under kinetic control, and in the case of D-glucose, the 4,6-*O*-isopropylidene-α-D-glucopyranose (1.41) is formed (Fig. 1.12). This methodology allows access to partially protected carbohydrates that would otherwise be difficult to prepare.

A useful summary (Table 1.1) of the various methods of protection has been provided by Garegg (reproduced with permission from *Chemistry in Britain*).

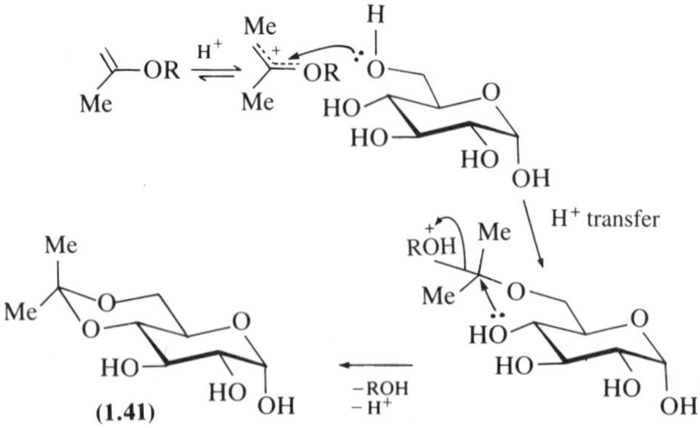

Fig. 1.12.

Table 1.1 Some commonly used O-protecting groups.

O-Protecting group	Usual method of attachment	Comment	Usual removal
OAc	Ac_2O or AcCl/pyridine	Good reactivity for neighbouring OH	$NaOCH_3/CH_3OH$
OBz	BzCl/pyridine	Decreases reactivity for neighbouring OH	$NaOCH_3/CH_3OH$
OBn	BnBr/DMF/NaH	High reactivity for neighbouring OH 'persistent' blocking group	H_2/Pd
$OCH_2CR^1{=}CR^2R^3$ (R = CH_3 or H)	$ClCH_2CR^1{=}CR^2R^3$/ DMF/NaH	Wide variation in ease of removal allows flexible strategies	$KOBu^t$–DMSO or $(Ph_3P)_3RhCl$, then mild acid
$OCPh_3$	$ClPh_3$/pyridine	High O-6 selectivity 'transient' blocking group	Mild acid
$OSi(CH_3)_2Bu^t$	$ClSi(CH_3)_2Bu^t$/pyridine	Good O-6 selectivity	H_2O-pyridine or $Bu_4N^+F^-$
$OSiPh_2Bu^t$	$ClSiPh_2Bu^t$	Extreme O-6 selectivity	$Bu_4N^+F^-$
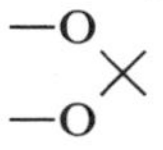 $-O{\diagdown}{\diagup}{\diagup}{\diagdown}{-}O$	$(CH_3)_2C(OCH_3)_2$, TsOH	Cyclic acetal	Mild acid
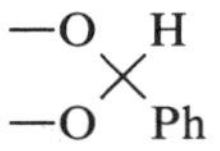 $-O{\diagdown}^H{\diagup}_{Ph}{-}O$	PhCHO, TsOH	Cyclic acetal	Mild acid or H_2/Pd

Source: *Chemistry in Britain*, July 1990.

1.4.1.5   Oxidative reactions

Aldoses on mild oxidation (e.g. with bromine water) suffer oxidation of the aldehyde function to yield aldonic acids, and these are usually isolated as the γ-lactones. For example, D-gluconic acid (**1.42**) is formed from D-glucose (equation 1.12):

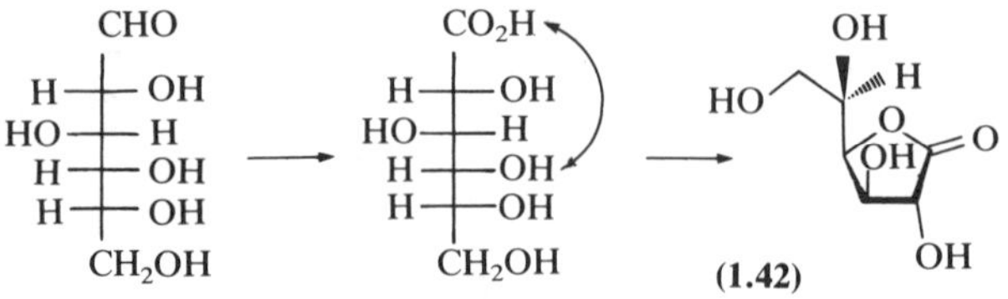

[eqn 1.12]

More vigorous oxidation (e.g. with nitric acid) yields dicarboxylic acids known as aldaric acids, and these are often obtained as dilactones, for example (**1.43**) from D-glucose (equation 1.13):

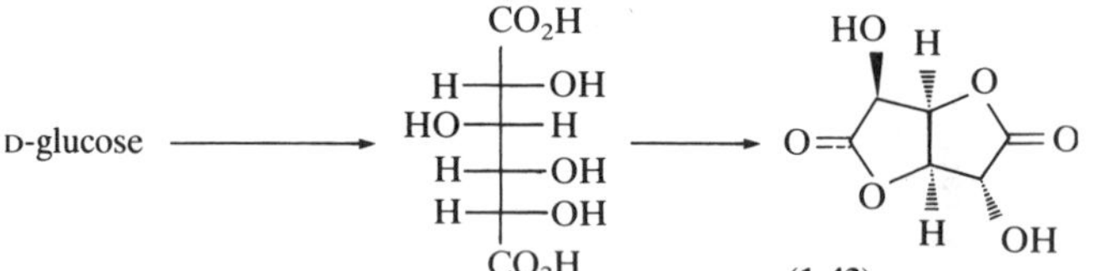

[eqn 1.13]

When partially protected sugar derivatives are employed, selective oxidations may be accomplished with a variety of reagents. These include the use of Moffatt's oxidant (dimethylsulphoxide (DMSO) and dicyclohexyl carbodiimide with phosphoric acid or pyridinium trifluoroacetate as catalysts), or various modified forms of this reaction (e.g. the Swern oxidation which employs DMSO and oxalyl chloride); the use of RuO_4 and RuO_2^- and HIO_4 (periodic acid); and the use of various Cr^{VI} reagents such as pyridinium chlorochromate and pyridinium dichromate. The utility of these will become apparent in the sections on synthesis.

Treatment of sugars containing 1,2-diol groupings with periodate (as the free acid or as the sodium salt and potassium salt), or with lead tetra-acetate (in benzene) leads to cleavage as shown in equation 1.14:

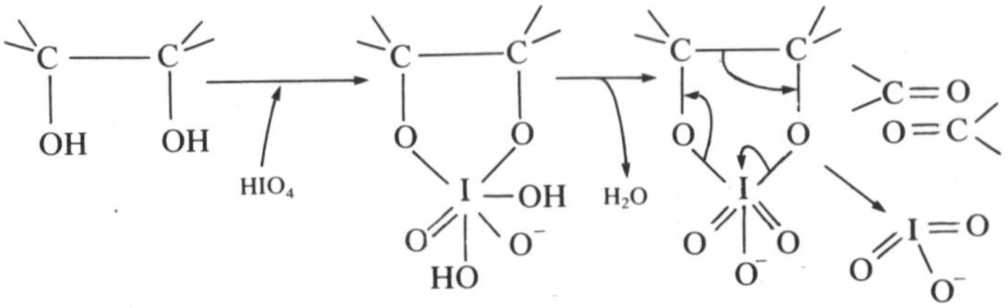

[eqn 1.14]

This oxidative cleavage has obvious synthetic utility (it is equivalent to the ozonolysis of a double bond), but it has also been used to determine the ring sizes of sugars. Thus, a methyl-furanoside consumes two moles of periodate and liberates methanal (equation 1.15):

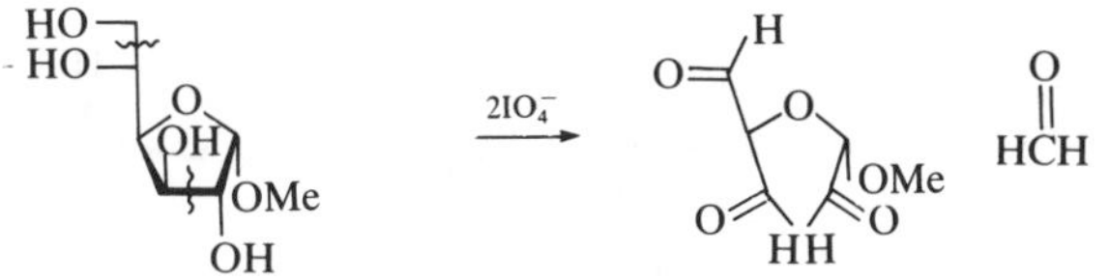

[eqn 1.15]

whereas a methyl-pyranoside, under similar conditions, gives a hydroxy-aldehyde and methanoic acid (equation 1.16):

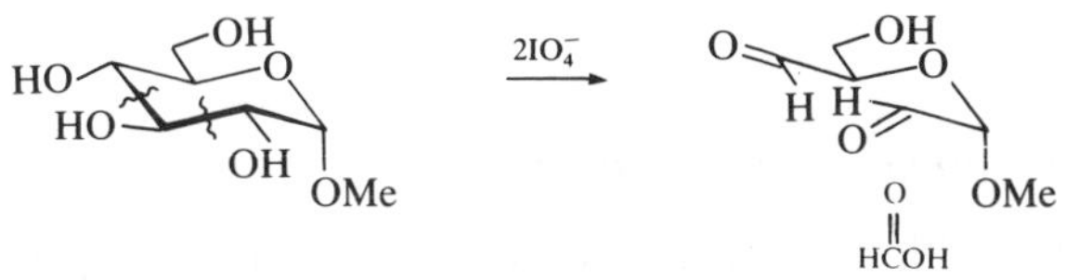

[eqn 1.16]

The mechanism shown in equation 1.14 involving a cyclic intermediate implies that steric factors control the rates of cleavage, and the two hydroxyls must be either axial–equatorial or equatorial–equatorial in their relationships. The reaction fails when the hydroxyls are rigidly *trans*-diaxial as in 1,6-anhydro-β-D-glucofuranose (**1.44**).

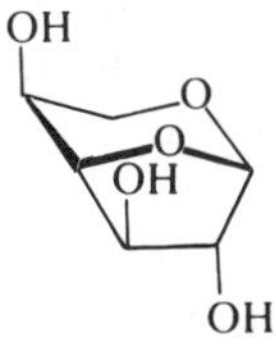

(**1.44**)

1.4.2   Reactions at the anomeric centre

Since the aldoses and ketoses are hemiacetals, most of the chemistry involving the anomeric centre is closely related to aldehyde and ketone chemistry. Indeed, much classical sugar chemistry involved reactions of nitrogen nucleophiles at the anomeric carbon to produce oximes (with hydroxyl-amine), arylhydrazones (with arylhydrazines), and imines (with amines) (Fig. 1.13).

Of greater relevance to modern synthetic organic chemistry are the reactions of alcohols, thiols, and carbon nucleophiles at the anomeric centre, and these will be considered in subsequent sections.

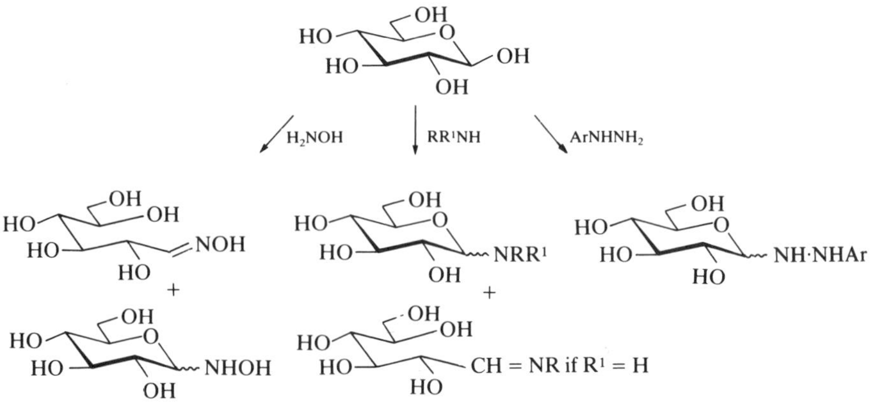

Fig. 1.13.

1.4.2.1   *Alcohols: formation of glycosides*

Reaction of sugars with alcohols in the presence of catalytic quantities of anhydrous acids (often HCl gas) usually provides complex mixtures of glycopyranosides (six-membered rings) and glycofuranosides (five-membered rings), although one major product may be obtained through preferential crystallization from the reaction medium. An example of the complexity of these processes and their variability is provided by the reaction of D-galactose in methanolic HCl (Fig. 1.14).

(i)   After attainment of equilibrium (12 hours at reflux): about 40 per cent yield of the α-pyranoside was isolated following crystallization from water, and various amounts of the β-pyranoside and α- and β-furanosides were present in the mother liquors.

(ii) After 6 hours at reflux (equilibrium not attained): the β-furanoside was isolated in about 55 per cent yield by crystallization, and the mother liquors contained various amounts of the other possible isomers, and some acyclic dimethyl acetal.

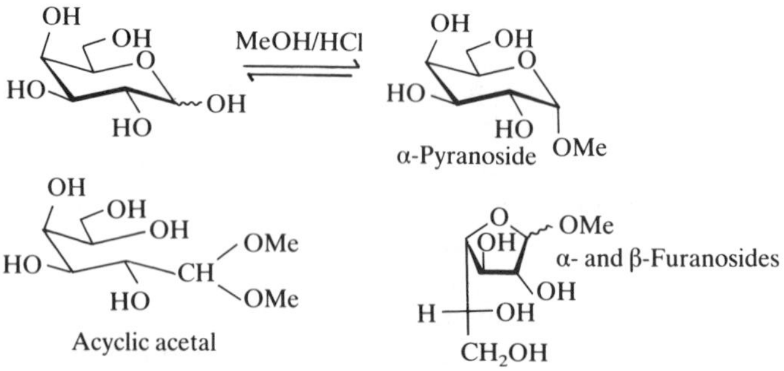

Fig. 1.14.

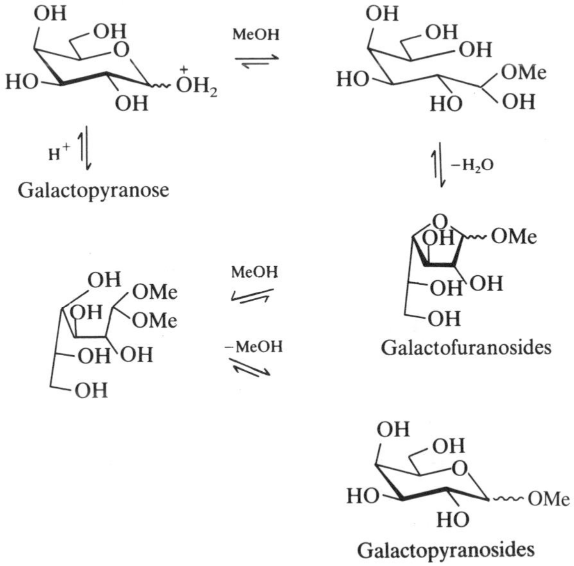

Fig. 1.15.

The furanosides thus appear to be 'kinetic products' and the pyranosides 'thermodynamic products', and these results suggest that equilibria of the kind shown in Fig. 1.15 may be operating. Glycosidation may proceed via the acyclic hemiacetal, galactofuranoside, acyclic dimethyl acetal, and thence to the galactopyranoside. However, even this scheme is probably too simple, since the pyranosides can anomerize via cyclic carbocations such as (**1.45**), and the furanosides may anomerize via acyclic cations such as (**1.46**).

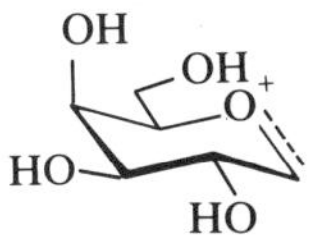

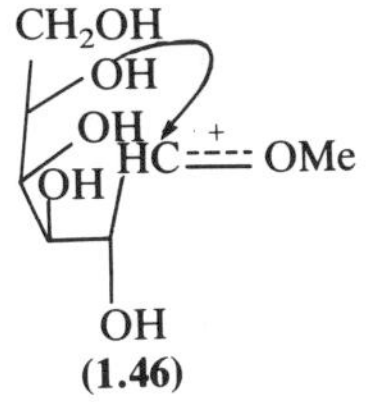

Suffice to say, glycoside formation provides a useful method for selective protection of the anomeric hydroxyl (and one other hydroxyl), and since hydrolysis can be effected with aqueous acid (formally a reverse of the glycosidation process), selective deprotection can also be effected if the other hydroxyls are protected as esters or other acid-stable derivatives.

Many naturally occurring compounds are glycosides, and several have already been mentioned, for example salicin (**1.9**), hesperidin (**1.10**), and adriamycin (**1.14**). Where the aglycone (i.e. non-sugar moiety) is complex or expensive to produce, the glycosides are usually prepared via the so-called Koenigs–Knorr method which employs the glycosyl halide (most often the bromide) in conjunction with heavy metal salts, as shown in equation 1.17.

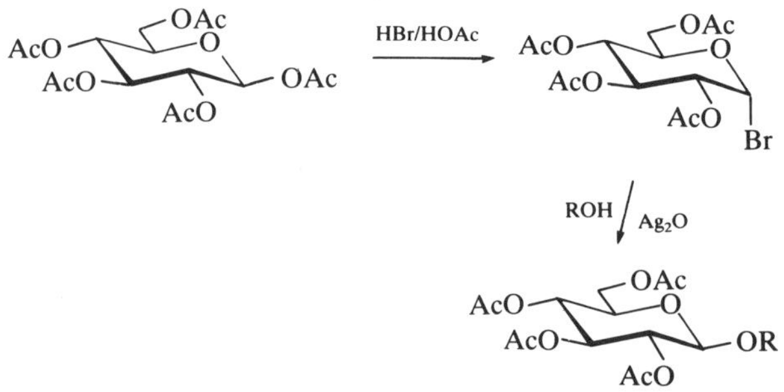

[eqn 1.17]

A facile alternative synthesis of bromides is shown in equation 1.18. The classical Koenigs–Knorr method was first introduced in 1901, and has been widely used during the ensuing 90 years. The major difficulties inherent in the reaction are the instability of the glycosyl halides and the expense of the various heavy metal salts (primarily silver and mercury salts) that have been employed. However, the wealth of experience gained over the years often allows the synthetic chemist to adjust the reagents and reaction conditions to favour the formation of one or other anomeric product.

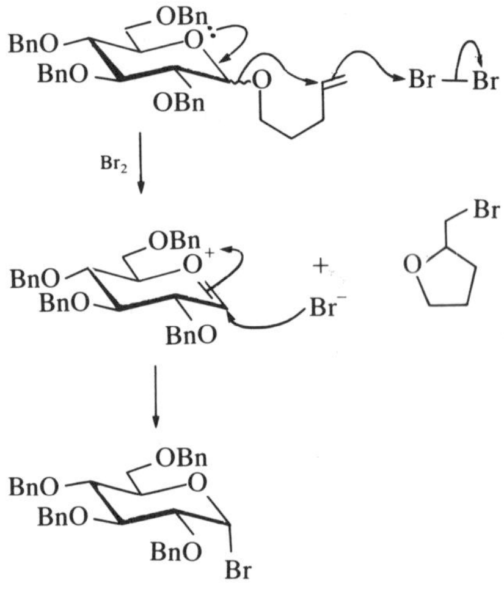

[eqn 1.18]

With the increasing interest in oligosaccharide synthesis, due to the important biological roles of these species, alternative methods have been developed, and various examples of more recent methodology are shown in equation 1.19:

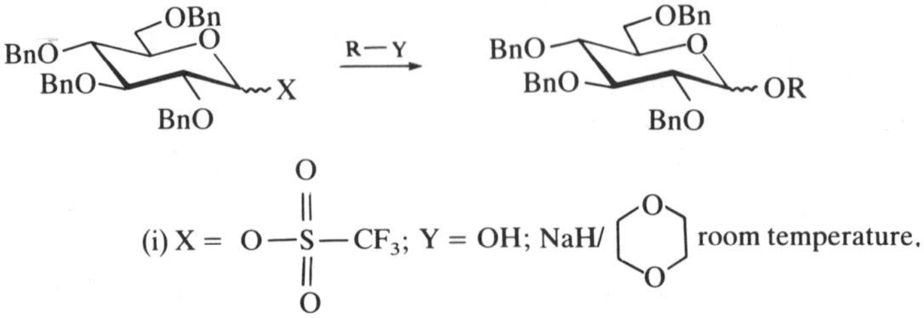

[eqn 1.19]

(i) $X = O-\overset{\overset{O}{\|}}{\underset{\underset{O}{\|}}{S}}-CF_3$; $Y = OH$; NaH/ room temperature.

$$\text{(ii) } X = O - \overset{\overset{\displaystyle NH}{\displaystyle \cdot \|}}{C} - CCl_3; Y = OH; \text{ acid catalyst.}$$

(iii) $X = F$; $Y = OH$; $SnCl_2/AgClO_4$.

It is beyond the scope of this chapter to discuss the subtleties of these processes further; suffice to state that it is usually possible to select reagent and reaction conditions that provide a predominance of the β-glycoside, or, with more difficulty, the α-glycoside.

1.4.2.2 Thiols

The reaction of sugars with alkyl or aryl thiols under acid catalysis usually produces the dialkyl or diaryl thioacetals as the major products (equation 1.20), though these may react further to yield thioglycosides, e.g. (1.47). The chemistry here should be contrasted with that observed in glycoside formation, and although a variety of factors are involved, the dithioacetal will be favoured because thiols are better nucleophiles than alcohols. This can be appreciated if one considers the different bond energies for —S—H (343 kJ mol^{-1}) and —O—H (460 kJ mol^{-1}).

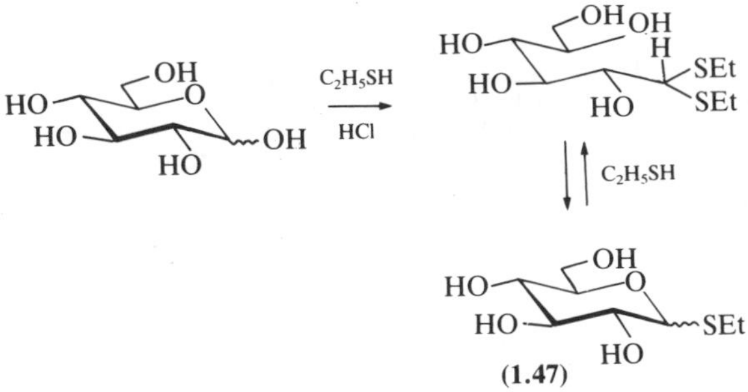

[eqn 1.20]

(1.47)

The thioacetals are most frequently employed when the acyclic form of the sugar is required, since the thioacetal moiety can be readily removed to produce an aldehyde, following protection or other reactions at the other hydroxyls. In addition, Raney nickel desulphurization can be used on the thioacetals and thioglycosides in order to prepare deoxysugars.

Naturally occurring thioglycosides are prevalent in mustard oil and in horseradish. Sinigrin (1.48) is responsible for much of the flavour of the latter.

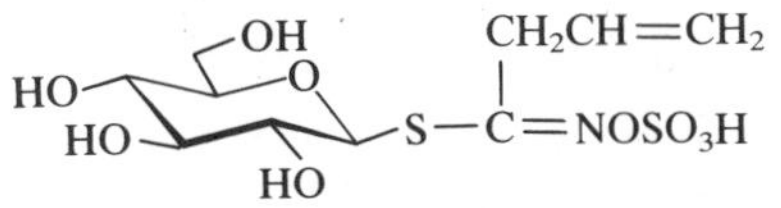

(1.48)

1.4.2.3   Carbon nucleophiles

Grignard and Wittig reagents are the most commonly employed carbon nucleophiles for attack at the anomeric centre, and typical reaction products are shown in equation 1.21:

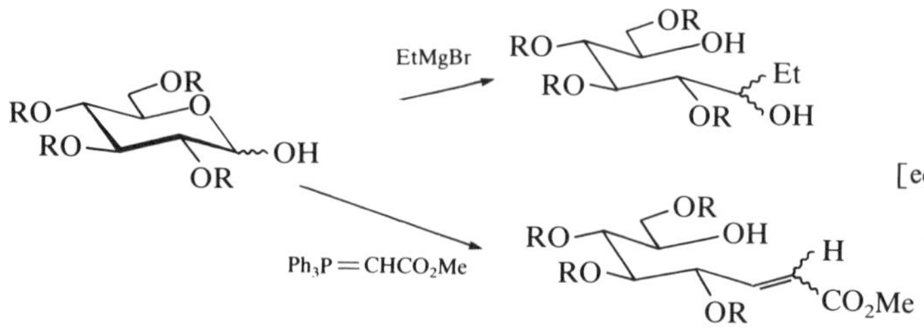

[eqn 1.21]

Both types of product can be cyclized to produce a mixture of stereoisomers at what was the anomeric centre, although in certain cases one stereoisomer predominates. Thus in the classical work of Moffatt and coworkers, 2,3-*O*-isopropylidene-5-*O*-protected-ribose derivatives were reacted with stabilized phosphorus ylids in refluxing acetonitrile with obtention of 1-deoxy-1-alkylated ribose derivatives directly (equation 1.22):

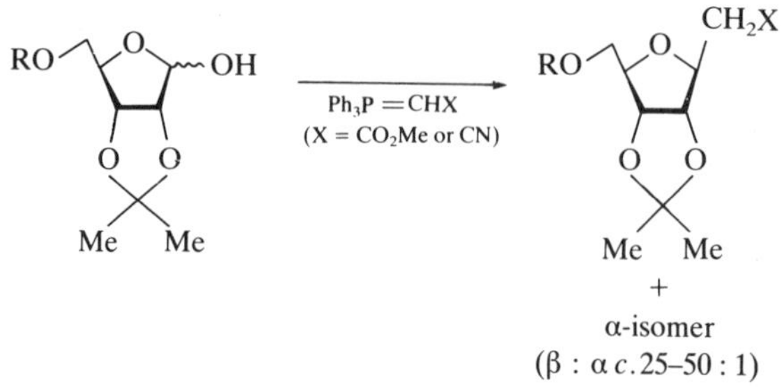

[eqn 1.22]

In this type of chemistry, it appears that the steric bulk of the isopropylidene group has a profound effect upon the course of the ring closure.

Both ^{1}H and ^{13}C NMR are of particular utility for establishing the stereochemistry of ribose derivatives. For example, in the proton spectra the difference between the δ values for the isopropylidene methods is greater for the β-isomer than for the α-isomer (>15 Hz and usually considerably <15 Hz respectively); while in the ^{13}C-spectra these same methyls show carbon signals separated by 2 p.p.m. versus just over 1 p.p.m. for the β- and α-forms respectively.

A good example of the use of a C-nucleophile, and of related chemistry, is provided by the synthesis of the C-nucleoside antibiotic formycin by Buchanan's group (Fig. 1.16). Obviously, the NMR criteria just described are of great value in syntheses of this kind.

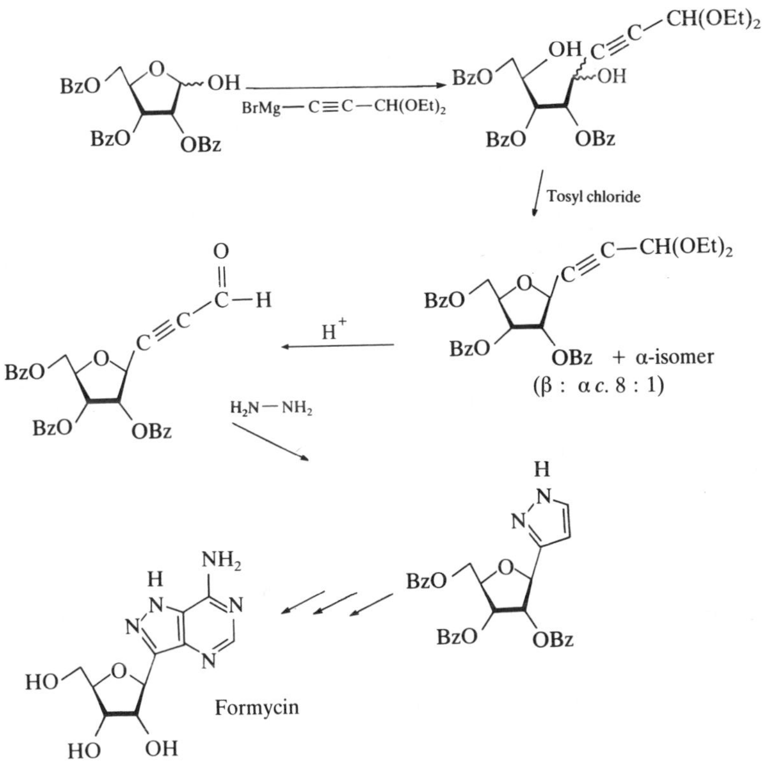

Fig. 1.16.

1.5 Structure elucidation

Brief mention has been made of the use of modern spectroscopic techniques for structural studies on carbohydrates. In this section one recent example of a structure elucidation using a range of techniques is provided.

Extracts of the dried root of the Korean plant *Cynachum wifordi* have been used as a tonic for centuries. Recently, a major constituent, wilforibiose (Fig. 1.17), has been isolated and its structure determined by a combination of chemical and spectroscopic methods. A summary is given in the figure. Elemental analysis and mass spectral measurements on wilforibiose (R = H) and on its α- and β-methylglycosides (R = Me) gave $C_{12}H_{20}O_8$ (molecular weight 292) and $C_{13}H_{22}O_8$ (306) respectively. The glycosides consumed one mole of periodate (see Section 1.4.1.5), establishing that only one 1,2-diol was present. Finally, ^1H and ^{13}C NMR spectra (Tables 1.2 and 1.3) were of particular value, and it was quickly established that the shift values and coupling patterns were similar to those of D-glucose and D-olivose. The actual data for the α-tetra-acetate of wilforibiose are shown in the figure, and severe deshielding of five protons (relative to those in the parent sugars) was noted, indicating that acetylation had occurred at C-1 of olivose and

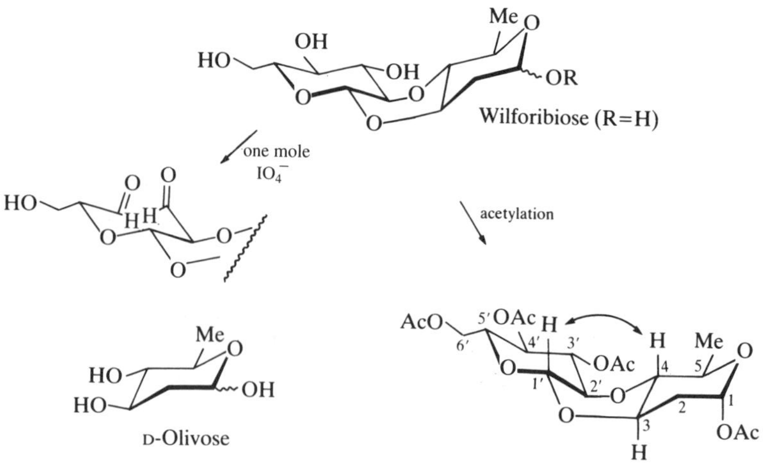

Fig. 1.17.

Table 1.2 ^{1}H NMR chemical shifts (500 MHz) for wilforibiose.

⟶	1-CH 6.36 (1H, dd, $J = 3.7$, 1 Hz)
	2-CH$_{ax}$ 2.15 (1H, ddd, $J = 11.9$, 4.6, 1 Hz)
	2-CH$_{eq}$ 1.91 (1H, dt, $J = 11.9$, 3.7 Hz)
	3-CH 3.95 (1H, ddd, $J = 11.9$, 9.5, 4.6 Hz)
	4-CH 3.34 (1H, t, $J = 9.5$ Hz)
	5-CH 4.08 (1H, dq, $J = 9.5$, 6.4 Hz)
	6-CH$_3$ 1.30 (3H, d, $J = 6.4$ Hz)
	1'-CH 4.86 (1H, d, $J = 7.9$ Hz)
	2'-CH 3.74 (1H, dd, $J = 9.8$, 7.9 Hz)
⟶	3'-CH 5.75 (1H, t, $J = 9.8$ Hz)
⟶	4'-CH 5.50 (1H, t, $J = 9.8$ Hz)
	5'-CH 4.23 (1H, ddd, $J = 9.3$, 5.2, 2.1 Hz)
⟶	6'-CH$_2$ 4.44 (1H, dd, $J = 12.5$, 2.1 Hz)
	4.53 (1H, dd, $J = 12.5$, 5.2 Hz)
	—OCOCH$_3$ 2.08, 2.05 × 2, 2.00 (each 3H, s)

C-3', C-4', and C-6' of glucose. Since no free hydroxyl can have been present at C-3 and C-4 of olivose, or at C-1' and C-2' of glucose, linkages must be present at these positions. Of the structures then possible, the one shown in the figure was chosen on the basis of a positive nuclear Overhauser effect (n.O.e.) between protons at C-4 and C-1'. This technique involves the irradiation of one particular proton resonance, while observing the variation of the intensities of other signals using an integrator. In general, only those protons which are close in space to the one being irradiated will exhibit an enhanced signal intensity. The structure of wilforibiose was thus shown to be β-D-glucopyranose-D-olivopyranose-1',3:2',4-dianhydride.

Table 1.3 ^{13}C NMR chemical
shifts for wilforibiose.

C-1	91.8
2	34.1
3	73.2
4	80.9
5	68.7
6	17.3
C-1'	99.0
2'	78.0
3'	72.4
4'	69.7
5'	74.0
6'	62.6
—OCOCH$_3$	170.4
	170.3
	169.9
	169.1
—OCOCH$_3$	20.8
	20.7
	20.5 × 2

1.6 The total synthesis of natural products and related compounds using carbohydrates

It will by now be apparent that carbohydrates have a number of attributes that make them ideal starting materials for the synthesis of natural products. They have a defined stereochemistry, they may adopt a number of ring sizes and conformations, their hydroxyls are easily protected or converted into other functionalities often with great selectivity, and last, but by no means least, they are usually cheap and readily available.

The key to the use of carbohydrates for synthesis is the recognition of their stereochemical relationship to particular portions of the target molecule. A good example of this is provided by an analysis of the molecule thromboxane B$_2$ (see Fig. 1.18). This is one of the products of the so-called 'arachidonic acid cascade', whereby the polyunsaturated fatty acid arachidonic acid (**1.49**) is converted into the prostaglandins and other metabolites, many of which have potent biological activities. Thromboxane B$_2$ itself is believed to be a chemotactic factor for a variety of cell types.

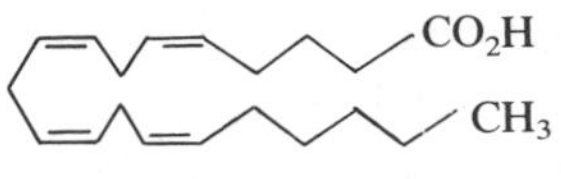

(**1.49**)

The first step in the analysis is the removal of the two side-chains – formally retro-Wittig reactions in the sense that we should envisage carrying

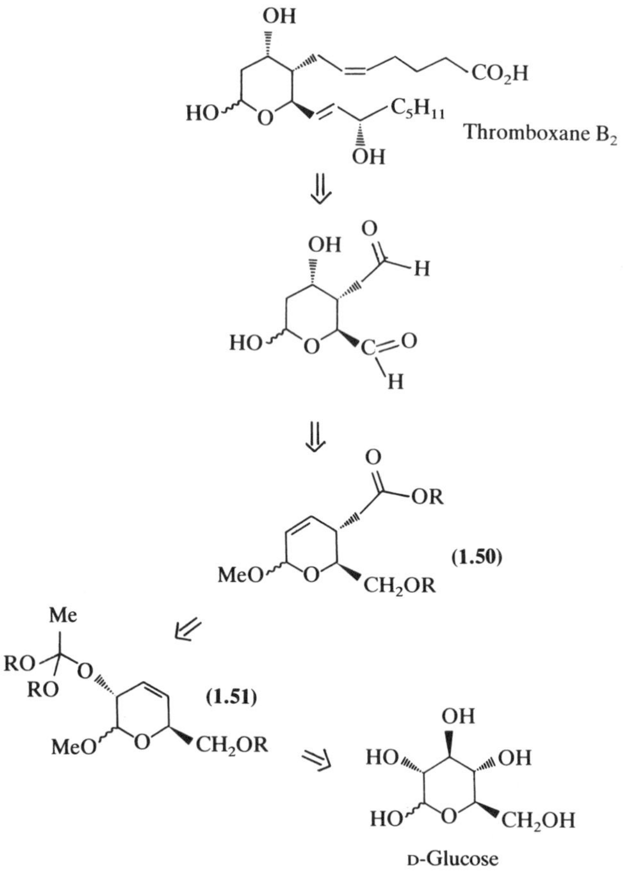

Fig. 1.18.

out two Wittig reactions in the actual synthesis (the terminology and the general ideas of retro-synthetic analysis were first introduced by E. J. Corey in 1969). A few functional group changes then provide the potential intermediate (**1.50**). Transfer of stereochemistry, formally via a retro-Claisen ortho ester rearrangement, produces (**1.51**) as the new target, and this has stereochemistry that can be related to D-glucose.

Recently Hanessian has introduced the term 'chiron' to describe potential synthetic intermediates, like (**1.50**) and (**1.51**), generated during retro-synthetic analysis. These may be derived from a 'chiral template' or stereochemically defined starting materials – usually a carbohydrate but also perhaps a terpene or amino acid. The value of this kind of stereochemical recognition will become apparent during the remainder of the chapter.

The syntheses that follow all employ cheap and readily available carbohydrates, and most of the reagents and reactions are not exotic or unique to carbohydrate chemistry. All provide products of the correct stereochemistry without recourse to resolution or extensive chromatography, and most of them are fairly short.

1.6.1 Synthesis of thromboxane B_2

Thromboxane B_2 is a major product of the metabolism of the biologically more interesting thromboxane A_2 (**1.52**, equation 1.23). This potent contractile agent is produced in arterial walls and causes blood platelets to aggregate and to adhere to blood vessel walls. However, its biological half-life under physiological conditions (pH 7, water, 37°C) is only about 30 seconds, and this precludes isolation and structure elucidation. Until recently the structure proposed (**1.52**) rested solely upon an understanding of its chemistry – hence the importance of metabolites such as thromboxane B_2 and of other compounds produced in trapping experiments (e.g. **1.53** in equation 1.23).

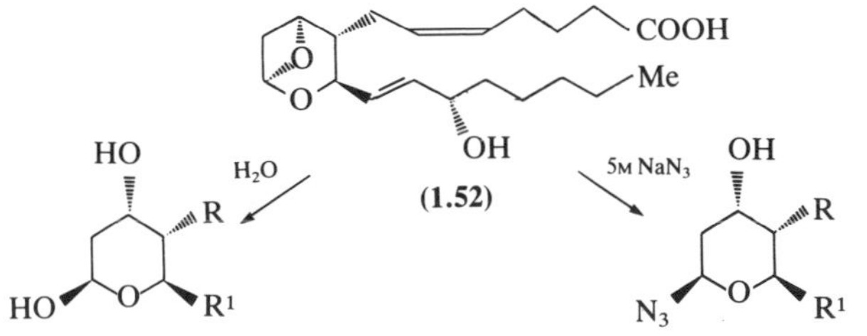

Thromboxane B_2 (TXB$_2$)   (**1.53**)

Several syntheses of TXB$_2$ have been accomplished, but the two illustrated (Figs 1.19 and 1.20) are particularly efficient in the way that chirality is transferred from the starting materials to the products.

The route used by Corey and coworkers (Fig. 1.19) arises from the analysis shown in Fig. 1.18 and utilized the key unsaturated sugar derivative (**1.54**). This had been previously prepared by Fraser-Reid from α-methylglyco-pyranoside via the sequence:

(i)   careful benzoylation at C-2 and C-6;
(ii)   mesylation at C-3 and C-4;
(iii) elimination, probably via the diiodide, and reaction of this with zinc; and
(iv) removal of the benzoates using base.

The most elegant step of the whole sequence then ensued, and this employed a modified Claisen ortho ester rearrangement to transfer chirality from C-3 to C-5 with introduction of a dimethylacetamido moiety. An iodolactoniz-ation (equation 1.24) and subsequent removal of the iodine atom completed the construction of the required bicyclic system (**1.55**).

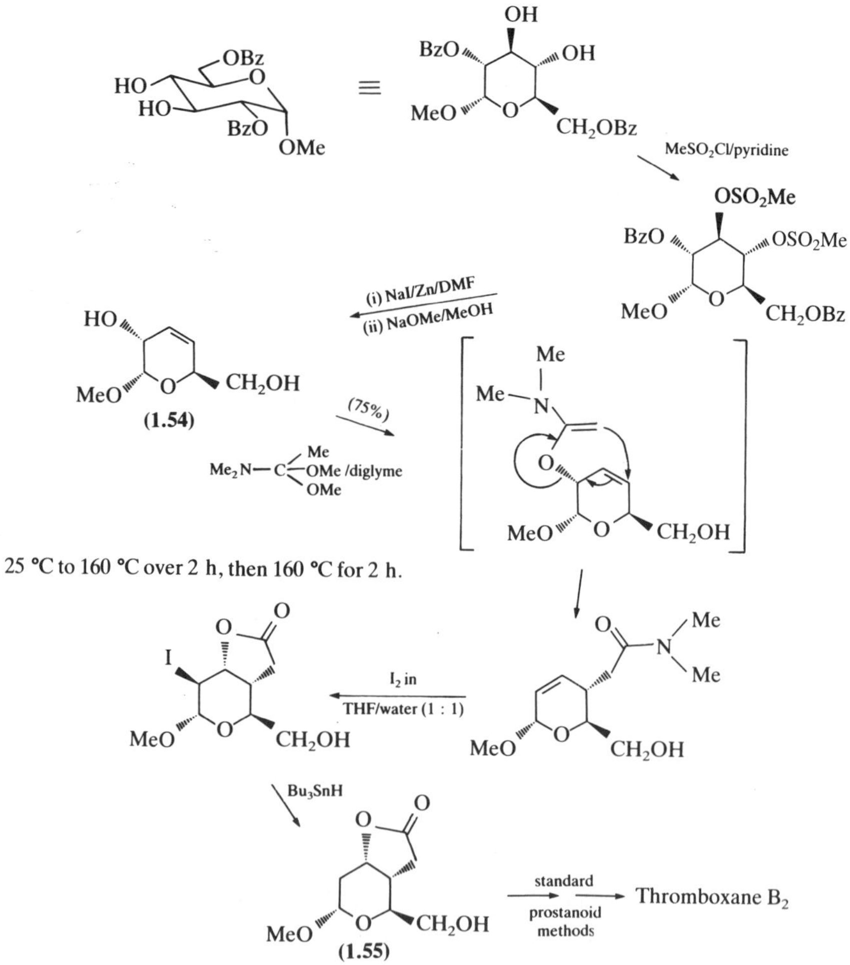

Fig. 1.19.

[eqn 1.24]

To complete the synthesis, the appropriate side-chains were introduced using standard prostaglandin methodology, and hydrolysis of the methyl acetal (formerly methyl glucoside) produced TXB$_2$.

Roberts and Kelly also prepared the intermediate (**1.55**), and the sequence they employed is shown in Fig. 1.20 (Ts, tosyl group, toluene-4-sulphonyl). They commenced with laevoglucosan (1,6-anhydroglucose) which can be prepared by the controlled pyrolysis (dehydration) of starch. Selective tosylation of the two hydroxyls which are on the least hindered face of the

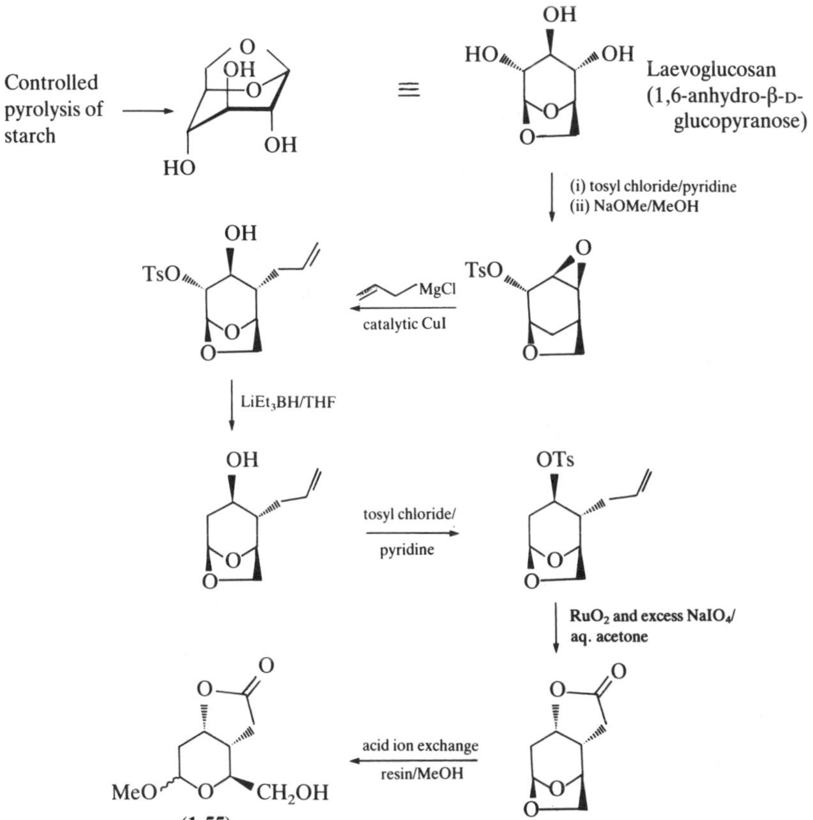

Fig. 1.20.

molecule (*anti* to the ether bridge) was followed by epoxide formation via intermediate (**1.56**). This epoxide could then be opened regio- and stereo-selectively using an allylcopper reagent. Reduction with the bulky reducing agent lithium triethylborohydride yielded only one alcohol, though the mechanism of this reduction was probably more complex than a simple displacement of tosylate by hydride, and may have involved the sequence shown in equation 1.25:

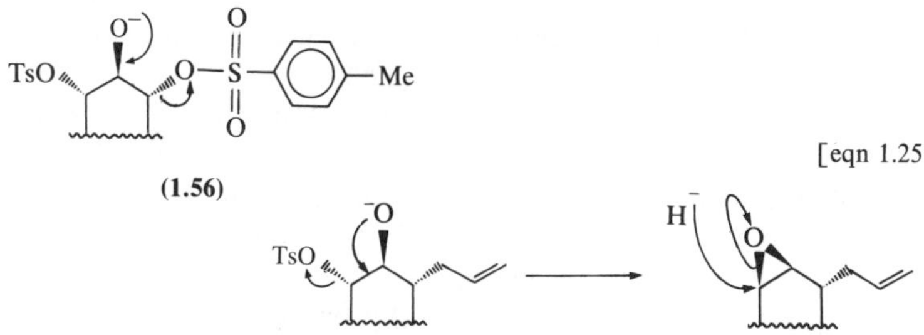

[eqn 1.25]

(**1.56**)

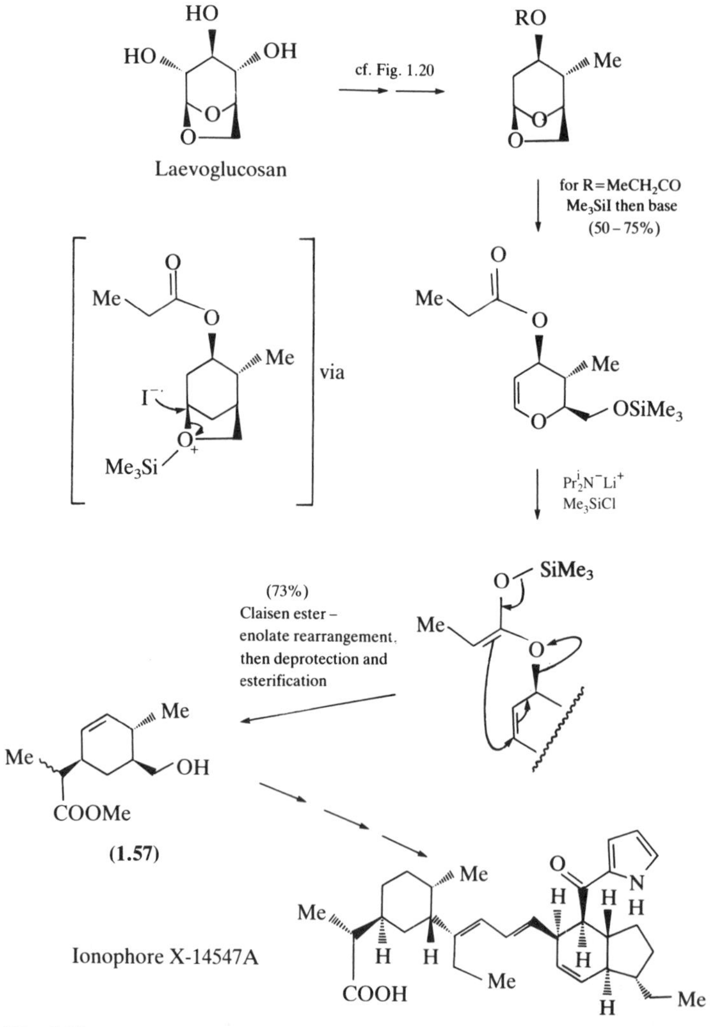

Fig. 1.21.

A further tosylation and subsequent oxidative side-chain cleavage (RuO_2/RuO_4 with excess periodate) provided an ideal positioning of carboxylate and leaving group for production of the required lactone. Finally, cleavage of the anhydro-bridge was achieved using an acid ion exchange resin to produce a mixture of acetals (**1.55**).

A related sequence of reactions was carried out by Ley and coworkers to produce compound (**1.57**), an intermediate in their synthesis of the ionophore X-14547A indanomycin (Fig. 1.21). This structurally unique compound can transport both mono- and divalent cations, and possesses a range of biological activities including anti-bacterial, anti-tumour, and anti-hypertensive (blood pressure lowering) properties.

1.6.2 Synthesis of (−)-shikimic acid

This synthesis (Fig. 1.22) by Fleet and Shing is notable for its brevity and minimal use of protecting groups. Shikimic acid is a key biosynthetic intermediate for the formation of the phenylpropanoid secondary metabolites, and is converted into the essential amino acids phenylalanine and tyrosine, and thence by deamination into 3-arylpropenoic acids (cinnamic acids). All

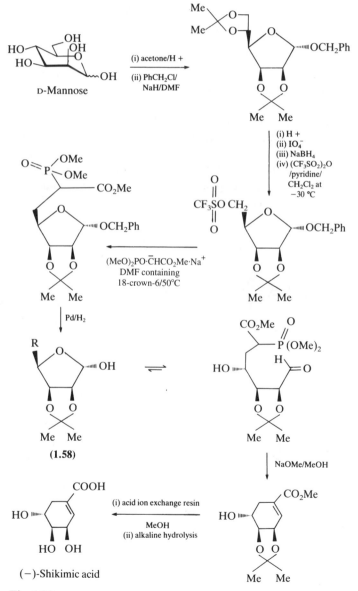

Fig. 1.22.

lignins, lignans, and flavonoids, as well as a host of less complex secondary metabolites, incorporate within their structures arylC$_3$ moieties derived from cinnamic acids. The importance of phenylalanine and tyrosine as precursors of various classes of alkaloids is discussed in Chapter 7.

A retrosynthetic analysis is given in Fig. 1.23 and shows how D-mannose was identified as a potential starting material. In the synthesis, this sugar was firstly converted into the crystalline diisopropylidene derivative, and thence into benzyl-2,3:5,6-di-*O*-isopropylidene-α-D-mannofuranoside. Selective acetal cleavage (compare with Section 1.4.1.4) was followed by cleavage of the 5,6-diol grouping with metaperiodate (compare with Section 1.4.1.5) to provide an aldehyde. This was reduced to the alcohol and converted into the trifluoromethanesulphonate (triflate). It proved necessary to prepare this highly reactive sulphonate, because the desired displacement with the anion of trimethylphosphonoacetate could not be accomplished with either tosylate or mesylate. Reductive removal of the benzyl ether protecting group provided the key hemiacetal (**1.58**), and this underwent an intramolecular Wittig reaction via its acyclic form. Finally, removal of the isopropylidene group and hydrolysis of the ester yielded (−)-shikimic acid. If the *t*-butyl ester was used in place of the methyl ester, both acetal protecting group and the *t*-butyl group could be removed simultaneously using acid.

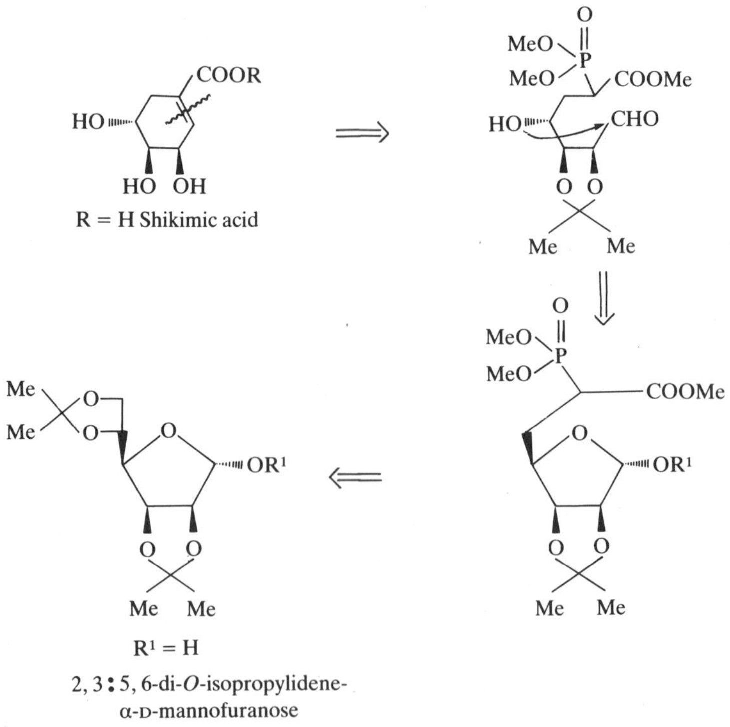

R = H Shikimic acid

R^1 = H

2, 3 : 5, 6-di-*O*-isopropylidene-
α-D-mannofuranose

Fig. 1.23.

1.6.3 Synthesis of (+)-showdomycin

Wittig methodology is also pre-eminent in the synthesis of the showdomycin derivative (**1.59**) by Mann and Kane shown in Fig. 1.24. In 1974 Moffatt and coworkers demonstrated that 2,3-isopropylidene derivatives of D-ribose reacted with stabilized phosphorus ylids in hot acetonitrile to yield 1-β-alkyl-1-deoxyribose derivatives (cf. equation 1.22). This stereochemical preference was ascribed to preferential formation of the *cis*-olefines in the Wittig reactions, and subsequent ring-closure to form the least hindered

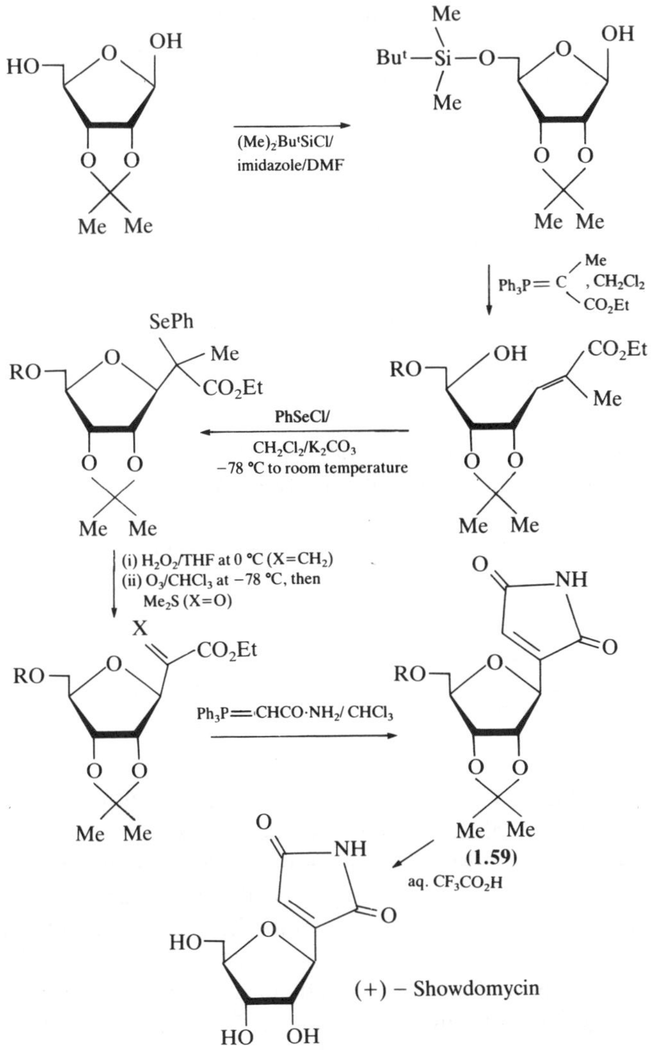

Fig. 1.24.

bicyclic system (equation 1.26).

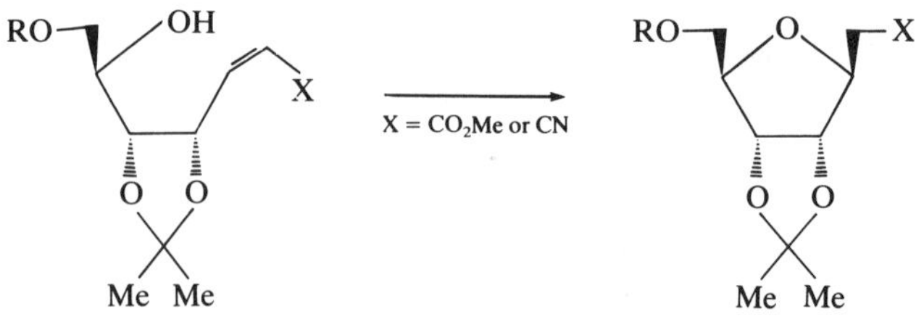

[eqn 1.26]

In contrast, when the reaction was carried out in dichloromethane at ambient temperature, only the Wittig product was obtained (mainly the *trans* isomer), and in the synthesis under discussion, cyclization was effected via phenylselenoetherification (equation 1.27):

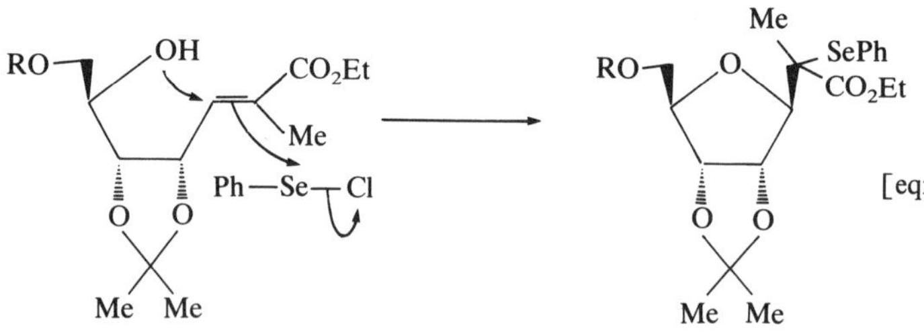

[eqn 1.27]

This ring-closure was also stereoselective, and the required β-substituted product was the only isolated product (together with some recovered starting material). Alpha-methylene ester was then produced via the selenoxide (equation 1.28), and ozonolysis provided a keto ester:

[eqn 1.28]

Finally, another Wittig reaction – this time with carboxyamido-methylidene triphenylphosphorane – yielded the showdomycin derivative (**1.59**). This can be converted into (+)-showdomycin in one step using aqueous trifluoroacetic acid (hydrolysis of isopropylidene and silyl ether groupings). This structurally unique metabolite is produced by the mould *Streptomyces showdoensis*, and has stimulated much interest due to its interesting antibiotic and anti-cancer activities. A number of other syntheses have been accomplished, but the present one has the advantages of brevity and minimal use of protection and deprotection steps. A retro-synthetic analysis is rather simple, since ribose is such an obvious starting material.

1.6.4 Synthesis of (+)-exo-*brevicomin*

The target of the two syntheses described in this section, (+)-*exo*-brevicomin, is a good example of a secondary metabolite with a proven biological role. It is produced by females of the species *Dendroctonus brevicomins*, a major pest of pine forests in the western states of North America. The beetle is attracted to the trees by the resin exudate, mainly β-myrcene (**1.60**) and β-pinene (**1.61**). She then releases a complex aggregation pheromone which contains (+)-*exo*-brevicomin as a major component, and other beetles, males and females, congregate. The primary cause of tree death is, however, due to the inoculation of a pathogenic fungus as the beetles bore into the tree bark.

(**1.60**)   (**1.61**)

A retro-synthetic analysis is shown in Fig. 1.25, and although perhaps not immediately apparent, the final structure (**1.62**) possesses much of the structure of 2-deoxyglucose, and both of the syntheses described here produced intermediates of this type from D-glucose. Ferrier and Prasit (Fig. 1.26) employed the highly functionalized glucose derivative (**1.63**), which they prepared from glucose via the sequence:

(i)   formation of the methyl-α-D-glucopyranoside;
(ii)   reaction with tosyl chloride in pyridine overnight, followed by addition of benzoyl chloride to produce methyl-2,6-di-O-tosyl-3,4-di-O-benzoyl-α-D-glucopyranoside; and finally
(iii)   displacement of the primary tosylate with iodide using sodium iodide in acetic anhydride.

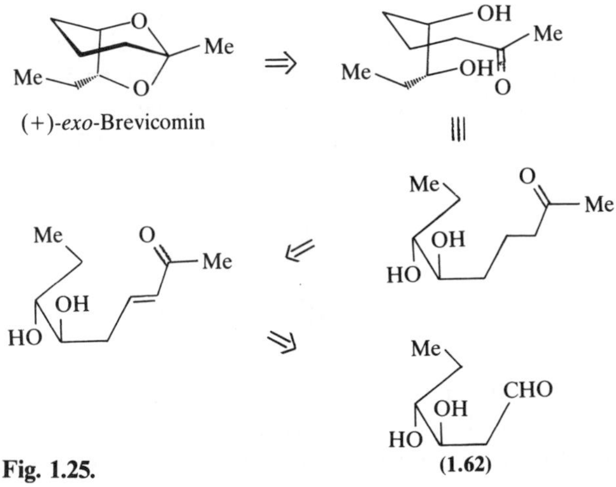

(+)-*exo*-Brevicomin

Fig. 1.25.   (**1.62**)

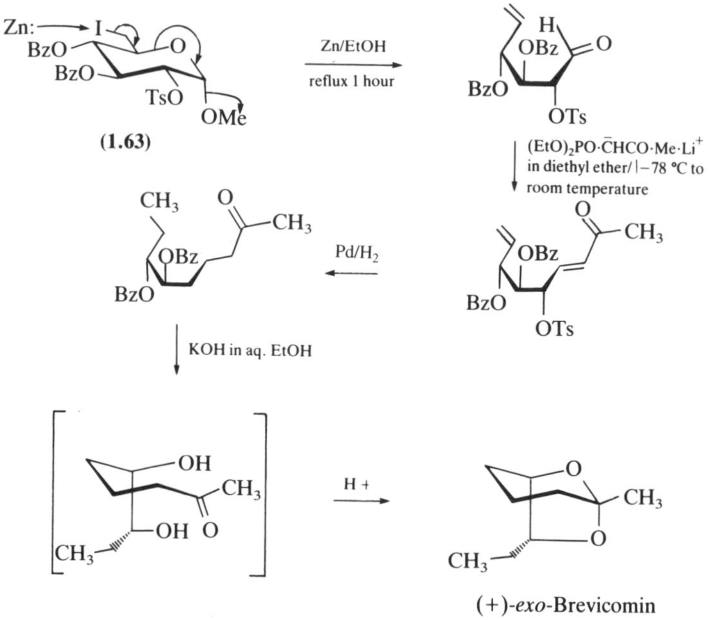

Fig. 1.26.

Treatment with zinc resulted in fragmentation, and obtention of an intermediate related in structure to (1.62). A Wittig reaction with the anion of diethyl 2-oxopropylphosphonate provided three extra carbon atoms, and catalytic hydrogenation resulted in saturation of the double bonds but also resulted in removal of the tosylate group. This chance discovery obviated the need to deoxygenate at what was formerly the C-2 of glucose. Hydrolysis of the benzoates with base yielded a diol which was able to participate in an intramolecular ketalization to produce (+)-brevicomin.

Fraser-Reid and Sherk employed similar intermediates (Fig. 1.27), but were forced to effect deoxygenation in a separate sequence of steps. The 3-benzyl derivative of 1,2:5,6-di-*O*-isopropylidene-D-glucofuranose was converted (selective acetal cleavage and mesylation) into the 5,6-dimesylate, which upon reaction solely with sodium iodide produced an alkene, probably via initial formation of an iodo, mesylate (mes; equation 1.29):

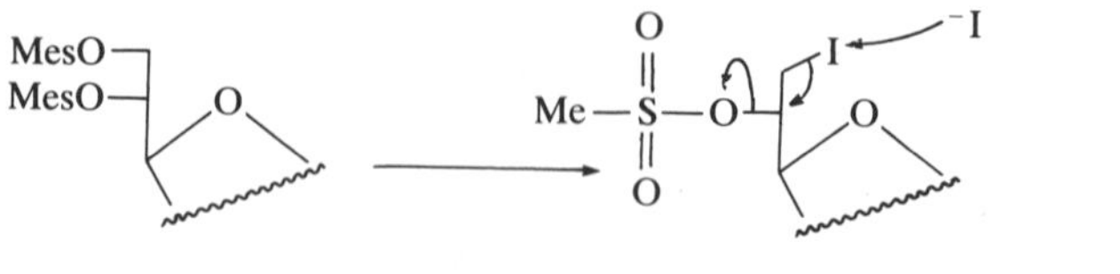

[eqn 1.29]

Isopropylidene cleavage and concomitant methyl glycoside formation was effected with methanolic HCl, and the 2-hydroxyl was then removed via the xanthate and reaction with tributylstannane. Hydrolysis then produced the

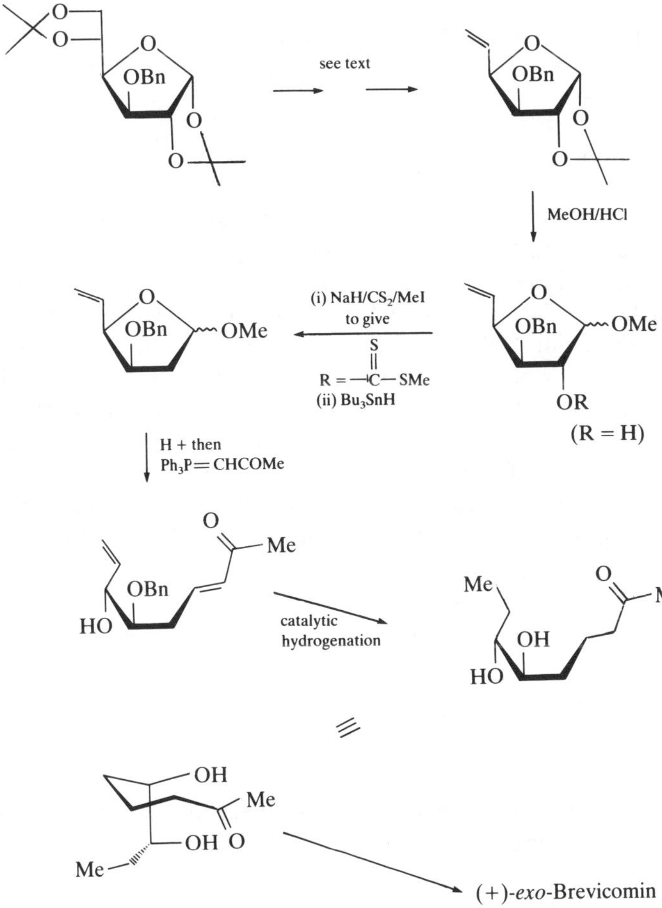

Fig. 1.27.

expected hemiacetal, and the remaining steps were similar to those in the Ferrier synthesis. It is pertinent to note that the success of the first route depended upon the fortuitous discovery that the tosylate was lost during catalytic hydrogenation, since more conventional chemistry had been investigated without success.

1.6.5 Synthesis of (+)-muscarine

The mushroom *Amanita muscaria* (fly agaric) has a long association with folklore, and many believe that it was the 'soma' of ancient civilizations. It produces two hallucinogens, ibotenic acid (**1.64**) and muscimol (**1.65**), and one major toxic compound, muscarine (**1.66**). This stimulates the para-sympathetic nervous system, with resultant contraction of the pupils,

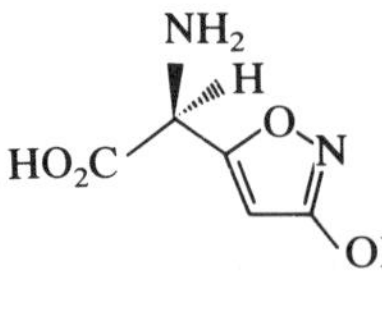

(1.64) 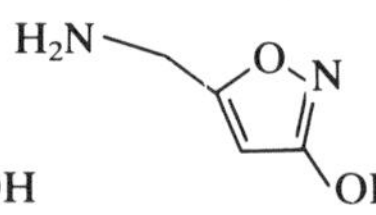   (1.65) 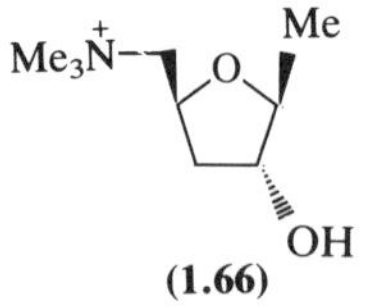   (1.66) 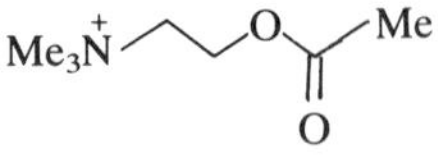   (1.67)

increased salivation, heart rate slowing, stimulation of stomach secretions, etc. It thus mimics the actions of acetylcholine (1.67), a natural neurotransmitter, and has always been of great interest to biologists who seek to delineate the structure and functions of acetylcholine receptors.

The synthesis illustrated in Fig. 1.29, by Brown and Mubarak, utilized D-mannitol, and a possible retro-synthetic analysis is shown in Fig. 1.28. Acid-catalysed dehydration of mannitol yields 2,5-anhydro-D-glucitol as a major product, and this was subjected to a series of protection and deprotection steps, from which an epoxide was eventually obtained. Reduction with Red-Al resulted in regioselective epoxide opening and also reductive removal of the tosylate. A possible mechanism for part of the sequence is shown in equation 1.30:

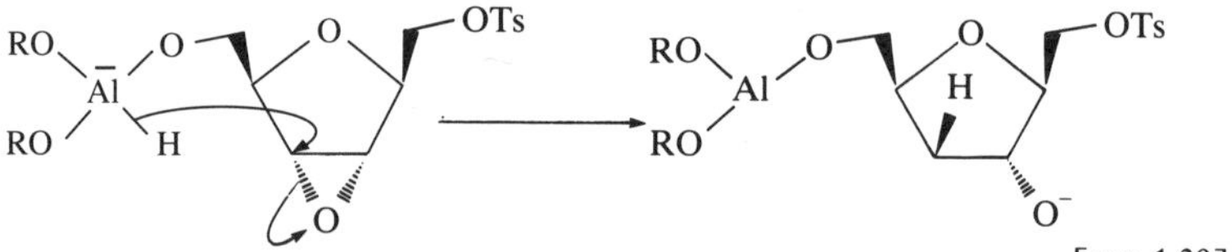

[eqn 1.30]

Finally, nucleophilic displacement of the primary tosylate by trimethylamine produced (+)-muscarine, isolated as its hydrochloride salt.

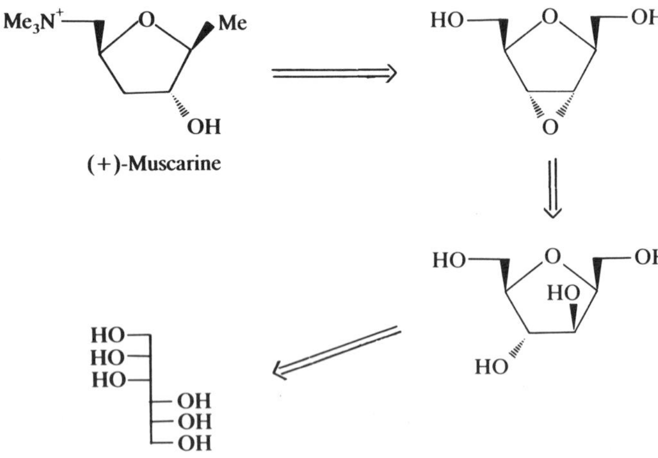

(+)-Muscarine

D-Mannitol

Fig. 1.28.

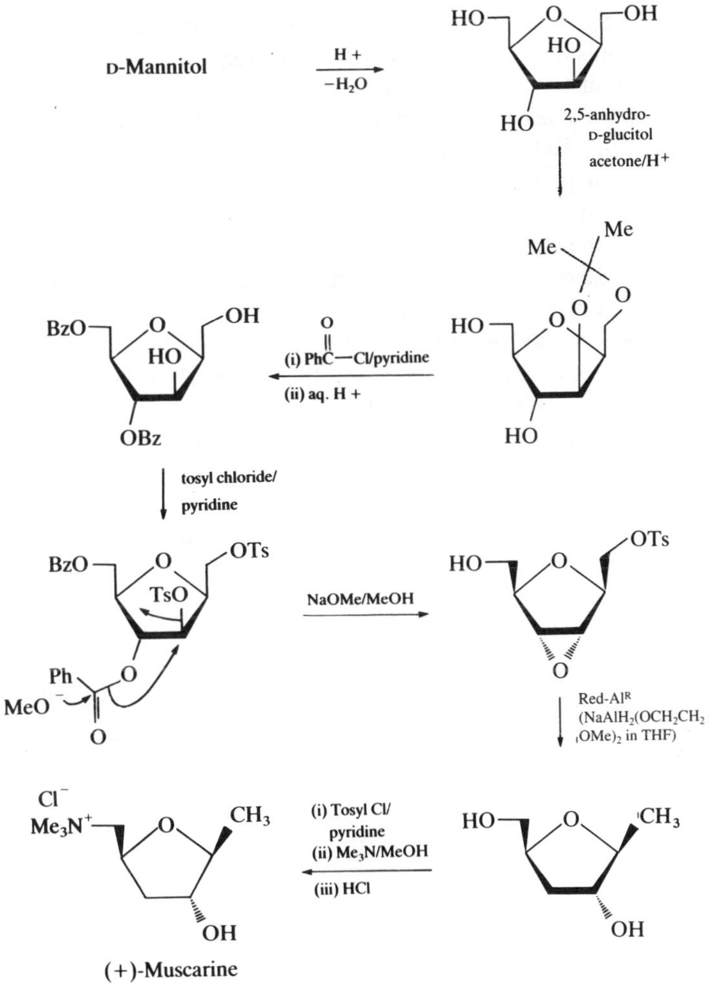

Fig. 1.29.

1.7 Synthesis of carbohydrates

All of the syntheses described so far have utilized carbohydrates as starting materials for the construction of other compounds. However, there is also a demand for novel or uncommon carbohydrates, and this section is concerned with some examples of the preparation of these compounds.

1.7.1 Synthesis of L-(−)-daunosamine

L-(−)-Daunosamine is the carbohydrate portion of the broad spectrum anti-cancer agent adriamycin (**1.14**), and several syntheses of this sugar have been reported using, *inter alia*, D-glucose, D-mannose, L-rhamnose, and L-fucose as starting materials. One particularly simple route due to Horton and Weckerle, is shown in Fig. 1.30. This commenced with methyl-α-D-mannopyranoside and proceeded in nine steps to daunosamine with an overall yield of 40 per cent. Initial formation of the bis-benzylidene derivative was followed by selective acetal fragmentation initiated by addition of butyl lithium. The base apparently abstracts the axial hydrogen at C-3 with subsequent elimination of benzaldehyde. The resultant ketone was converted into its oxime and thence after reduction and acetylation into (primarily) the 2,3-dideoxy 3-amino sugar derivative (**1.68**). Cleavage of the second acetal was achieved through the use of N-bromosuccinimide, with obtention of the bromide (**1.69**). This method is a very effective means of selectively brominating the primary hydroxyl of carbohydrates. Elimination of HBr, removal of the benzoyl group, and catalytic hydrogenation yielded 3-acetamido-2,3,6-trideoxy-β-L-lyxo-hexopyranoside (**1.70**). Subsequent reaction with barium hydroxide removed the acetyl group to yield daunosamine, isolated as its hydrochloride salt. Numerous more recent syntheses have been accomplished, but this one is notable for its brevity, its use of simple reagents, and the fact that it was accomplished on the multigram scale without the need for any chromatographic purification.

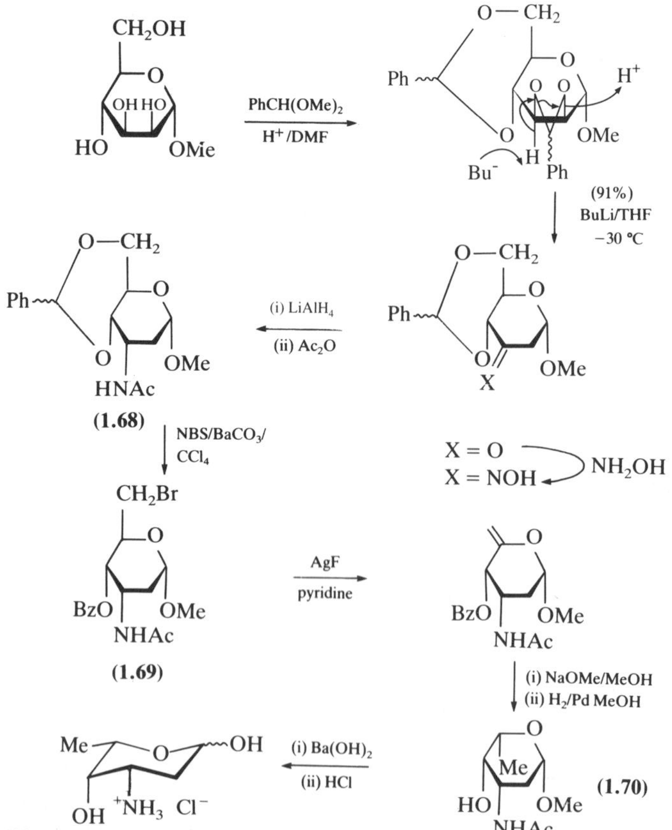

Fig. 1.30.

1.7.2 Synthesis of methyl-L-ribofuranoside

This synthesis is remarkable in its use of an enzymatic reaction to provide an efficient resolution of a racemic intermediate producing one major stereoisomer for further elaboration. Thus the Diels–Alder cycloadduct from furan and dimethyl maleate was converted into the isopropylidene derivative (**1.71**) (Fig. 1.31) by standard methods; and this was then partially hydrolysed using pig liver esterase. The half-ester thus obtained (**1.72**) had an optical purity of around 75 per cent, that is a 75 per cent excess of that particular stereoisomer over the alternative enantiomer. Ozonolysis in methanol provided keto-diester (**1.73**) (a possible mechanism is shown in

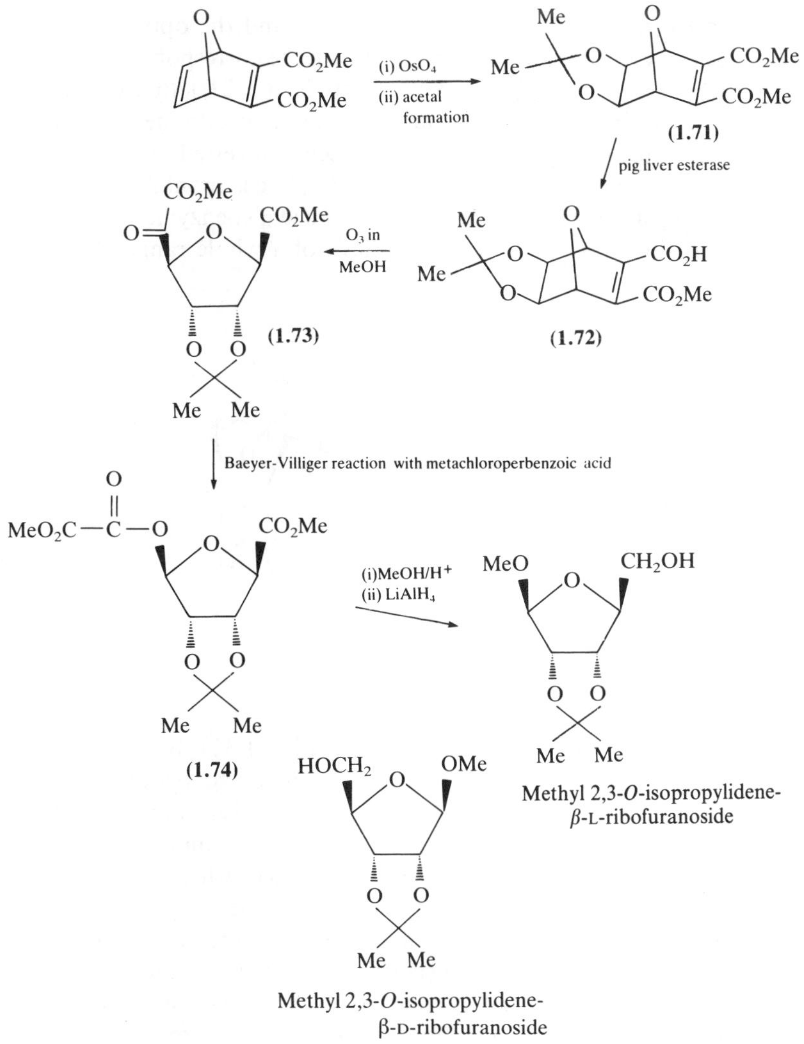

Methyl 2,3-*O*-isopropylidene-
β-L-ribofuranoside

Methyl 2,3-*O*-isopropylidene-
β-D-ribofuranoside

Fig. 1.31.

equation 1.31), and a subsequent Baeyer–Villiger oxidation yielded the oxalate derivative (**1.74**). Methanolysis with methanolic HCl and reduction of the ester group yielded methyl-β-L-riboside as its 2,3-isopropylidene derivative:

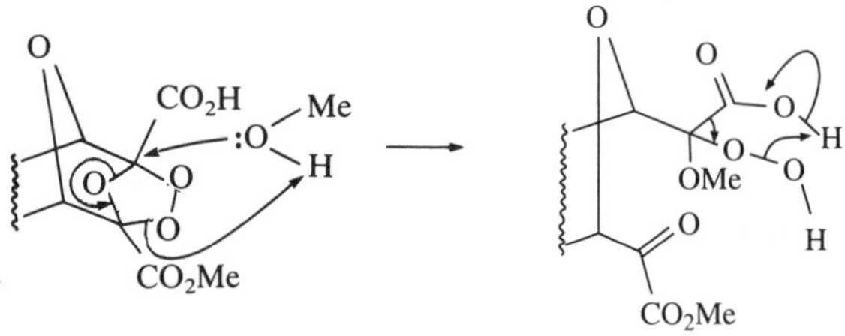

[eqn 1.31]

This had an $[\alpha]_D^{20}$ of $+63.2°$ (CHCl$_3$, $c = 1.5$), and the optical purity of this product was established by comparison with the corresponding D-ribose derivative which had an $[\alpha]_D^{20}$ of $-82.2°$ (CHCl$_3$, $c = 2.0$), thus confirming that the L-form had been formed in around 75 per cent enantiomeric excess.

The key intermediate (**1.74**) has also been converted into D-ribose, ($+$)-showdomycin (**1.75**), and ($-$)-cordycepin (**1.76**), an inhibitor of RNA synthesis, amongst other compounds; and this chemicoenzymatic approach promises to be of great value in the synthesis of a whole range of natural products.

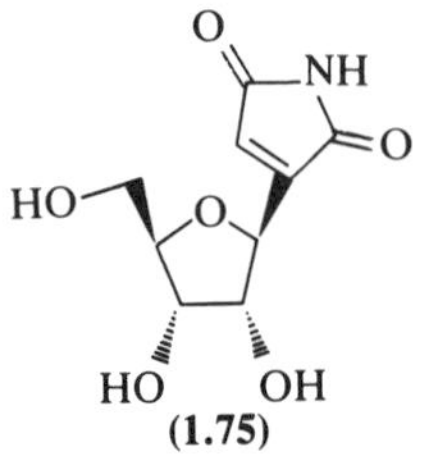

(1.75)

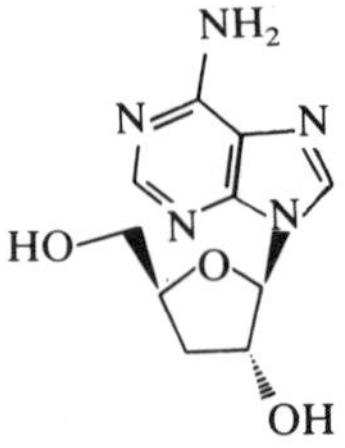

(1.76)

1.7.3   Synthesis of lincosamine

The synthesis of racemic lincosamine (see Fig. 1.32) by Danishefsky and coworkers is representative of a novel approach to pyranose sugars, which involved Diels–Alder cycloadditions between dienes and aldehydes. Lincosamine is of interest since it is the saccharide portion of the clinically useful antibiotic lincomycin. The synthesis commences with a cycloaddition of butenal and 1-benzoyloxy,2-trimethylsiloxy,4-methoxybuta-1,3-diene under Lewis acid catalysis. Hydrolysis of the enol silyl ether and double bond rearrangement provided the major cycloadduct shown in the figure (67 per cent yield). This possesses the correct stereochemistry and appropriate functionality for elaboration into lincosamine. Selective reduction of the ketone (hydride addition from the least hindered face), followed by

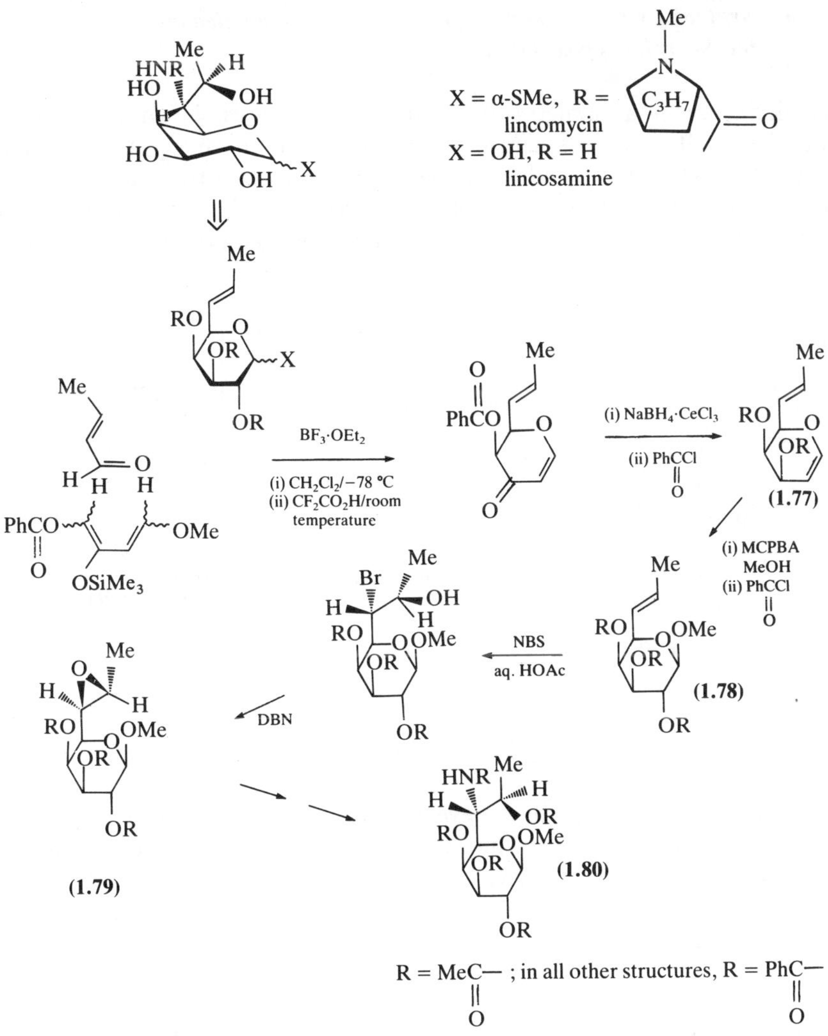

Fig. 1.32.

benzoylation, yielded a dibenzoate (**1.77**). Epoxidation with metachloro-peroxybenzoic acid (a bulky peracid) was site-selective, and presumably produced the α-epoxide which suffered cleavage to yield the methyl-β-galactopyranoside (**1.78**) after benzoylation. Bromohydrin formation was also regio- and stereoselective, and reaction with 1,5-diazabicyclo[4.3.0]non-5-ene (DBN) provided the key epoxide (**1.79**). The remaining steps involved cleavage of the oxirane with azide, reduction, and manipulation of the protecting groups, with obtention of the aminooctose (**1.80**), which can be converted into lincosamine.

The main importance of this synthesis is that it allows access to aminooctoses with a defined stereochemistry at six contiguous chiral centres.

1.7.4   Synthesis via asymmetric epoxidation of allylic alcohols: the Sharpless epoxidation

The induction of asymmetry observed in the synthesis shown in Fig. 1.31 was the result of stereoselective processing by an enzyme. An alternative method of asymmetric induction has been developed by Sharpless, and involves the chiral epoxidation of allylic alcohols. The mechanism of the process is not completely understood, but the methods and results are depicted in equation 1.32:

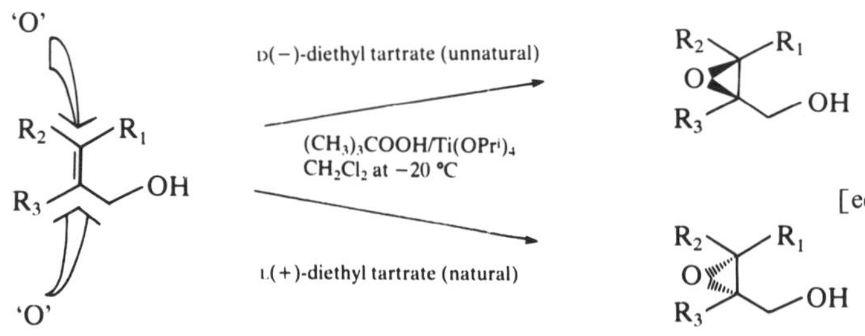

When the alkenol is chiral itself, one essentially pure diastereoisomeric epoxide is obtained from each antipode of diethyl tartrate. Using this method, Sharpless and Masamune have developed routes to a wide range of natural products, including carbohydrates, and a synthesis of D-ribose (as its 2,3:4,5-di-isopropylidene derivative) is shown in Fig. 1.33.

The optical purities of the final products obtained by these methods are usually in excess of 90 per cent, and often closer to 98 or 99 per cent, which means that they compare very favourably with enzyme-catalysed processes. The starting materials are also relatively inexpensive, and this chemistry has been much exploited when chiral molecules are to be synthesized without recourse to the use of carbohydrates.

1.7.5   Synthesis of glycopeptides

Earlier in the chapter the importance of glycopeptides and proteins was mentioned. Almost all of these compounds contain either a β-*N*-glycosidic linkage or a β-*O*-glycosidic linkage, as shown in the general structures (**1.81**) and (**1.82**). With the increasing understanding of the roles of these compounds in molecular recognition has come the need to synthesize complex glycopeptides. A typical synthesis is illustrated in Fig. 1.34.

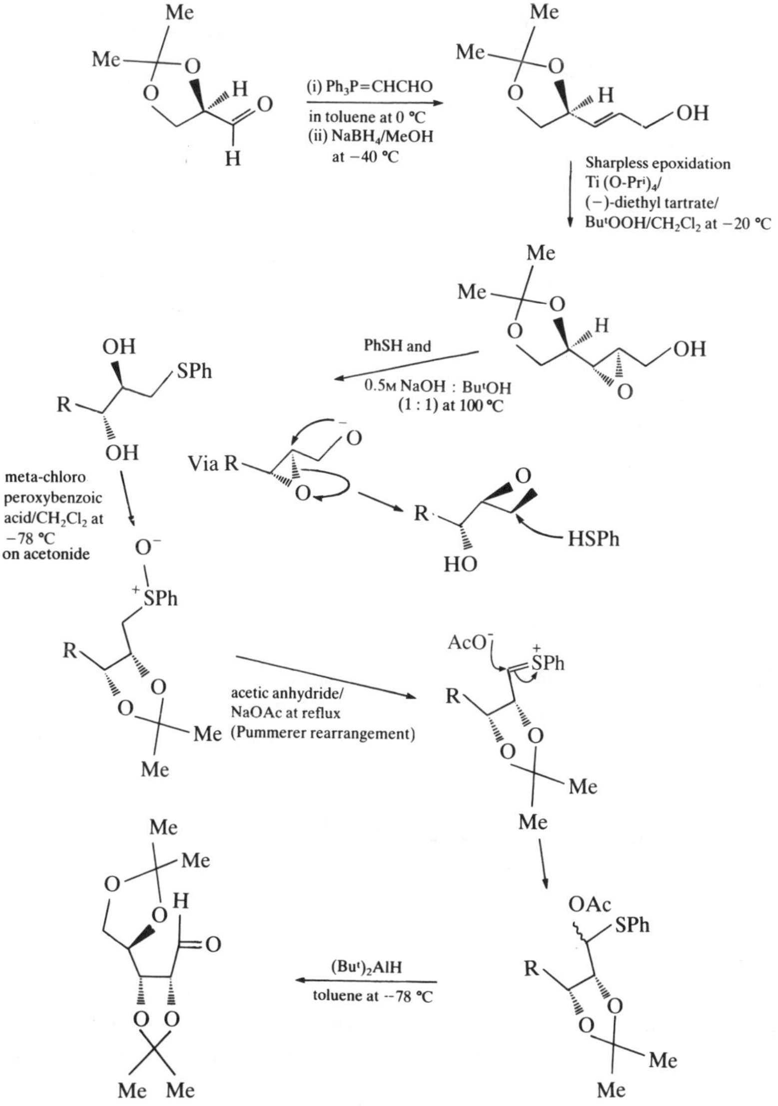

Fig. 1.33.

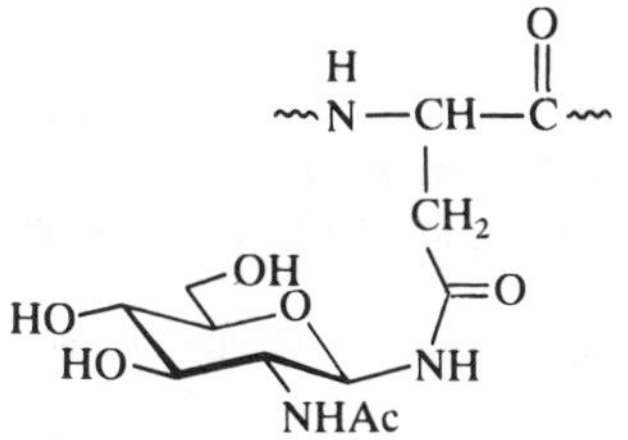

(1.81)

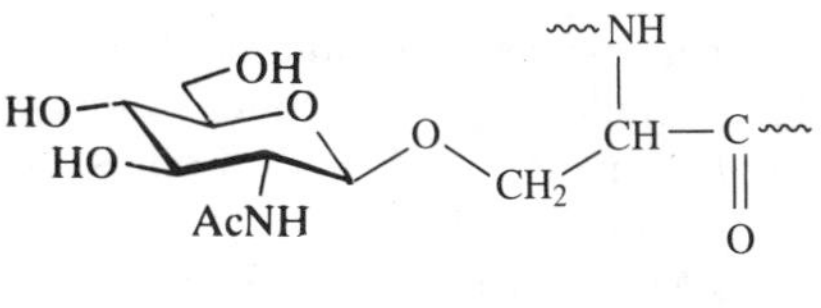

(1.82)

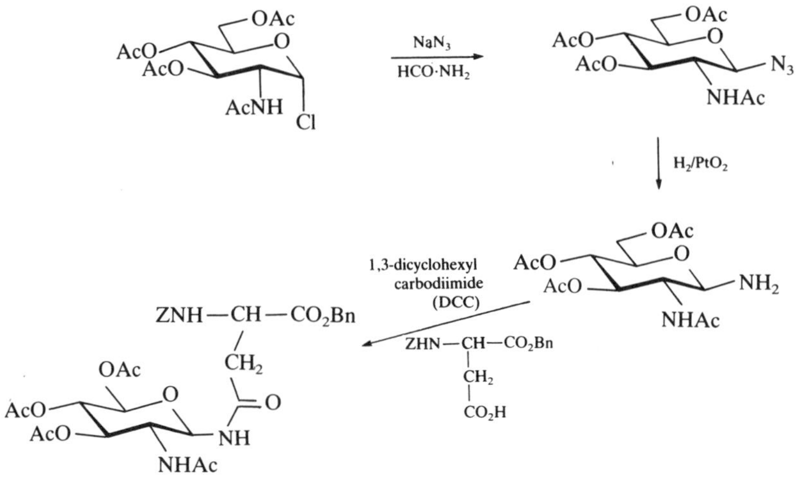

Fig. 1.34.

1.7.6   *Synthesis of an artificial antigen*

As mentioned earlier, the oligosaccharides are of even greater importance in molecular recognition. They are present on the surface of most cells, and contribute very significantly to the immunological 'signature' of the cell. They participate in molecular interactions with oligosaccharides and glycopeptides on other host cells and with those of invading bacteria or viruses. The synthesis of these oligosaccharides is thus of great contemporary interest as a means of understanding their molecular interactions, and for the production of highly specific vaccines and other immunologically significant species.

A typical oligosaccharide synthesis is shown in Fig. 1.35, and depicts the preparation of an artificial antigen precursor for the pathogenic bacterium *Shigella flexneri*. The extensive use of silver triflate for the modern variant of the classical Koenigs–Knorr glycoside synthesis is characteristic of much recent oligosaccharide synthesis.

Further reading

Most textbooks of organic chemistry contain chapters on carbohydrate chemistry, but the following books contain more comprehensive accounts:

1. R.J. Ferrier and P.M. Collins, *Monosaccharide Chemistry*, Penguin Books, London, 1972.
2. J.M. Tedder, A. Nechvatal, A.W. Murray, and J. Carnduff, *Basic Organic Chemistry*, Part IV, Wiley, Chichester, 1972.
3. L. D. Hall, W. Pigman, and D. Horton, *The Carbohydrates*, vols 1A, 1B, 2A, and 2B, 2nd ed., Academic Press, New York, 1980.

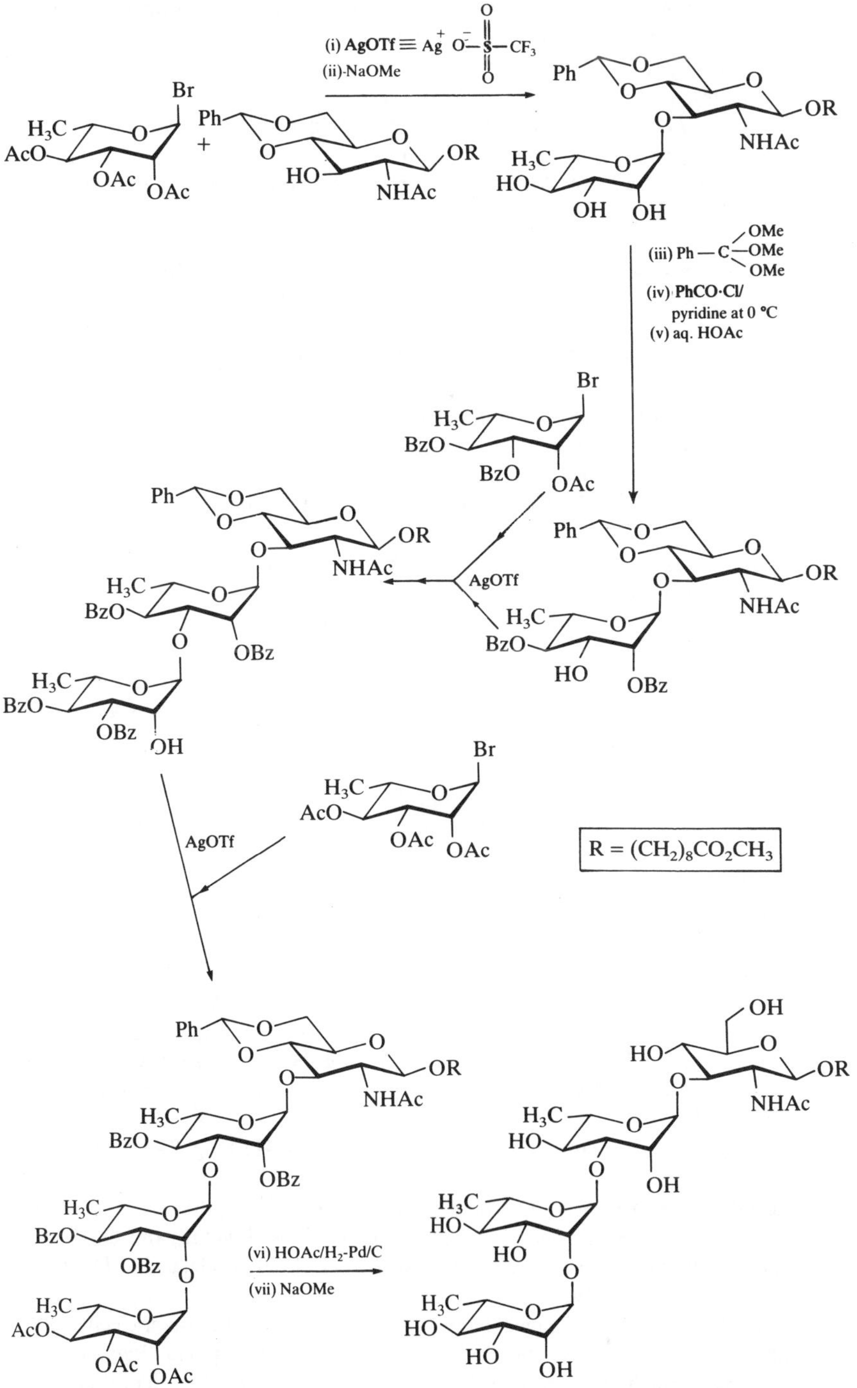

Fig. 1.35 Synthesis of an artificial *Shigella flexneri* type Y antigen precursor.
(Tf ≡ triflate ≡ CF₃—SO₂—.)

4. R.D. Guthrie, *Guthrie and Honeyman's Introduction to Carbohydrate Chemistry*, 4th ed., Oxford University Press, Oxford, 1974.

5. R.L. Whistler, M.L. Wolfrom, and J. N. BeMiller, *Methods in Carbohydrate Chemistry*, vols 1–8, Academic Press, New York, 1962–80.

6. S. Coffey, *Rodd's Chemistry of Carbon Compounds*, vol. 1F, 1967; M.F. Ansell, supplement to 1F, Elsevier, Amsterdam, 1983.

7. J.F. Kennedy and C.A. White, *Bioactive Carbohydrates*, Wiley, London, 1983.

8. R.H. Thomson, *The Chemistry of Natural Products*, chapter 1, Blackie, Glasgow, 1985.

9. Excellent reviews of recent work are given in: *Specialist Periodical Reports, Carbohydrates*, from 1967, the Royal Society of Chemistry, London; *Advances in Carbohydrate Chemistry and Biochemistry*, from 1945, Academic Press, New York; *Carbohydrate Research*, from 1965, Elsevier, Amsterdam.

10. Conformations of carbohydrates: S.J. Angyal, *Angewandte Chemie*, 1969, **8**, 157; S.J. Angyal, *Adv. Carb. Chem. Biol.*, 1984, **42**, 15; J.F. Stoddart, *Stereochemistry of Carbohydrates*, Wiley-Interscience, New York, 1971; R. Barker and A.S. Serianni, *Accounts Chem. Research*, 1986, **19**, 307.

11. Pathway of carbon fixation in photosynthesis: I. Zelitch, *Ann. Rev. Biochem.*, 1975, **44**, 123.

12. Amino sugars in antibiotics: J.F. Kennedy and C.A. White, *Bioactive Carbohydrates*, chapter 12, Wiley, London, 1983.

13. Sucrose chemistry: R. Khan, *Adv. Carb. Chem. Bio.*, 1976, **33**, 235; *Pure Appl. Chem.*, 1984, **56**, 833.

14. Synthetic sweeteners: B. Crammer and R. Ikan, *Chem. Soc. Rev.*, 1977, **6**, 431.

15. Polysaccharide chemistry: G.O. Aspinall, *The Polysaccharides*, vol. 1 (1982), vol. 2 (1983), and vol. 3 (1985), Academic Press, New York.

16. Blood group determinants: R.U. Lemieux, *Chem. Soc. Rev.*, 1978, **7**, 423.

17. Biological significance of oligosaccharides: J. Montreuil, *Adv. Carb. Chem. Bio.*, 1980, **37**, 157; S. Hakamori, *Ann. Rev. Biochem.*, 1981, **50**, 733; Y.-T. Li and S.-C. Li, *Adv. Carb. Chem. Biochem.*, 1982, **40**, 235.

18. Synthesis of glycosides and oligosaccharides: R.R. Schmidt, *Angewandte Chemie*, 1986, **25**, 212; H. Paulsen, *Chem. Soc. Rev.*, 1984, **13**, 15.

19. Reaction of D-galactose with methanol: C.T. Bishop and F.P. Cooper, *Canad. J. Chem.*, 1962, **40**, 224.

20. Protection of hydroxyl groups (and other functionalities): T.W. Greene and P.G.M. Wuts, *Protective Groups in Organic Synthesis*, 2nd edn, Wiley, New York, 1991.

21. The anomeric effect: N.S. Zefirov, *Tetrahedron*, 1977, **33**, 3193; A.J. Kirby, *The Anomeric Effect and Related Stereoelectronic Effects at Oxygen*, Springer-Verlag, New York, 1983.

22. Deoxygenation via xanthates and related species: W. Hartwig, *Tetrahedron*, 1983, **39**, 2609.

23. Acetal formation: A.N. De Belder, *Adv. Carb. Chem. Bio.*, 1977, **34**, 179; A.B. Foster in *The Carbohydrates* (eds L.D. Hall, W. Pigman, and D. Horton), vol. 1A, chapter 11, Academic Press, New York, 1980; J. Gelas and D. Horton, *Heterocycles*, 1981, **16**, 1587.

24. Reactions of Grignard reagents and Wittig reagents at the anomeric centre: Y.A. Zhdanov, Y.E. Alexeev, and V.G. Alexeeva, *Adv. Carb. Chem. Biol.*, 1972, **27**, 227; J.G. Buchanan, A.R. Edgar, M.J. Power, and P.D. Theaker, *Carb. Res.*, 1974, **38**, C22; H. Ohrui, G.H. Jones, J.G. Moffatt, M.L. Maddox, A.T. Christensen, and S.K. Bryam, *J. Amer. Chem. Soc.*, 1975, **97**, 4602.

25. Synthesis of formycin: J.G. Buchanan, A.R. Edgar, R.J. Hutchison, A. Stobie, and R.H. Wightman, *Chem. Commun.*, 1980, 237.
26. Structure elucidation of carbohydrates: B. Lindberg, J. Lonngren, and S. Svensson, *Adv. Carb. Chem. Biochem.*, 1975, **31**, 185; H. Rauvala, J. Finne, T. Krusius, J. Karkkainen, and J. Jarnefelt, *Adv. Carb. Chem. Bio.*, 1981, **38**, 389; L.D. Hall in *The Carbohydrates* (eds W. Pigman, D. Horton, and J.D. Wander), vol. 1B, 1300, Academic Press, New York, 1980; K. Bock and C. Pedersen, *Adv. Carb. Chem. Bio.*, 1983, **41**, 27; *Adv. Carb. Chem. Bio.*, 1984, **42**, 193; J.F.G. Vliegenthart, H. Van Halbeck, and L. Dorland, *Pure Appl. Chem.*, 1981, **53**, 45.
27. Structure elucidation of wilforibiose: S. Tsukamoto, K. Hayashi, and H. Mitsuhashi, *Tetrahedron Lett.*, 1984, 3595.
28. Synthesis using carbohydrates (general): B. Fraser-Reid and R.C. Anderson, *Fortschr. Org. Naturst.*, 1980, **39**, 47; S. Hanessian, *Total Synthesis of Natural Products: The Chiron Approach*, Pergamon Press, Oxford, 1983; T.D. Inch, *Tetrahedron*, 1984, **40**, 3161; K.K. Bhat, S.-Y. Chen, and M.M. Joullie, *Heterocycles*, 1985, **23**, 691.
29. Synthesis of thromboxane B$_2$: E.J. Corey, M. Shibasaki, and J. Knolle, *Tetrahedron Lett.*, 1977, 1625; A.G. Kelly and J.S. Roberts, *Chem. Commun.*, 1980, 228.
30. Synthesis of antibiotic X-14547A (indanomycin): M.P. Edwards, S.V. Ley, S.M. Lister, B.D. Palmer, and D.J. Williams, *J. Org. Chem.*, 1984, **49**, 3503.
31. Synthesis of (−)-shikimic acid: G.W.J. Fleet and T.K.M. Shing, *Chem. Commun.*, 1983, 849.
32. Synthesis of (+)-showdomycin: J. Mann and P.D. Kane, *Perkin Trans. I*, 1984, 657.
33. Synthesis of (+)-*exo*-brevicomin: R.J. Ferrier and P. Prasit, *Perkin Trans. I*, 1983, 1645; A.E. Sherk and B. Fraser-Reid, *J. Org. Chem.*, 1982, **47**, 932.
34. Synthesis of (+)-muscarine: A.M. Mubarak and D.M. Brown, *Perkin Trans. I.*, 1982, 809.
35. Synthesis of macrolide antibiotics carbomycin B and leucomycin A$_3$: K.C. Nicolaou, M.R. Pavia, and S.P. Seitz, *J. Amer. Chem. Soc.*, 1981, **103**, 1222, 1224.
36. Synthesis of (+)-thienamycin: S. Hanessian, D. Desilets, G. Rancourt, and R. Fortin, *Canad. J. Chem.*, 1982, **60**, 2292; T. Kametani, T. Nagahara, and M. Ihara, *Perkin Trans. I*, 1981, 3048.
37. Synthesis of sugars from non-carbohydrate precursors: A. Zamojski, A. Banaszek, and G. Grynkiewicz, *Adv. Carb. Chem. Bio.*, 1982, **40**, 1.
38. Synthesis of (−)-daunosamine: T. Mukaiyama, Y. Goto, and S.-I. Shoda, *Chem. Lett.*, 1983, 671; D. Horton and W. Weckerle, *Carbohydrate Res.*, 1975, **44**, 227.
39. Chemicoenzymatic syntheses: M. Ohno, Y. Ito, M. Arita, T. Shibata, K. Adashi, and H. Sawai, *Tetrahedron*, 1984, **40**, 145.
40. Synthesis of lincosamine: E.R. Larsen and S. Danishevsky, *J. Amer. Chem. Soc.*, 1983, **105**, 6715.
41. Sharpless epoxidation and related chemistry: A. Pfenninger, *Synthesis*, 1986, 89; K.B. Sharpless, *Chem. in Britain*, 1986, **22**, 38; K.B. Sharpless, C.H. Behrens, T. Katsuki, S.W.M. Lee, V.S. Martin, M. Takatani, S.M. Viti, F.J. Walker, and S.S. Woodward, *Pure Appl. Chem.*, 1983, **55**, 589.
42. Synthesis of artificial antigens: P.J. Garegg, *Chem. in Britain*, 1990, **26**, 669.
43. Recent developments in the synthesis of C-glycosides: M.H.D. Postema, *Tetrahedron*, 1992, **48**, 8545.

2 Nucleosides, nucleotides, and polynucleotides

J. Hobbs

2.1 Introduction

In 1869, the Swiss researcher, Friedrich Miescher, isolated a phosphorus-containing substance, which gave an acidic reaction, from cell nuclei and called it *Nucleinsäure* (nucleic acid). For some 70 years investigations of this substance proceeded slowly, and the identities of the bases, the sugars, and of phosphoric acid in the nucleic acids were elucidated, without their true significance being appreciated. Then, at the beginning of the 1940s, Avery and his coworkers demonstrated that nucleic acids carried genetic information, and work on the chemical structures of the naturally occurring polynucleotides was intensified. Largely as a result of the work of Todd, Levene and their schools, the covalent structures of nucleosides (general structures (**2.1**) or (**2.2**), R = H), nucleotides ((**2.1**) or (**2.2**), R = PO_3^{2-}) and the nucleic acids (polymers derived from nucleotides) were established. In 1953 Watson and Crick formulated the 'double helix' structure of deoxyribonucleic acid (DNA), and the elegant rationale for the storage, transcription, and reproduction of genetic information inherent in the model was instantly appreciated and subsequently established by experiment. The science of molecular biology was born. During the 1960s the decipherment of the genetic code by Nirenberg, Khorana, and coworkers rendered it possible to associate a given base sequence in DNA with a sequence of amino acids in an encoded protein; the establishment of sequencing methods, albeit slow and time-consuming, allowed the elucidation of the primary structures of ribonucleic acids of low molecular weight, such as transfer ribonucleic acid (tRNA), and the researches of Khorana laid the basis for the synthesis of longer oligodeoxyribonucleotides, culminating in the synthesis of whole genes coding for tRNA molecules in the 1970s. The last 20 years have seen quite astonishing strides in the chemical and biochemical manipulation of the nucleic acids and their subcomponents. New, efficient methods of nucleoside synthesis have been described, and huge numbers of nucleoside analogues prepared both for biochemical research, and for investigation of their potential therapeutic properties, especially as anti-viral and anti-neoplastic agents. Novel efficient methods for the phosphorylation of nucleosides, including unprotected nucleosides, have been established, and these methods adapted to provide nucleoside chiral phosphates, which have been used to elucidate the stereochemistry of reaction at phosphorus in both enzymic and non-enzymic reactions involving nucleotides. The discovery of the 'restriction endonucleases', which cut DNA at specific sequences, permitting long DNA strands to be cut into pieces of size amenable to sequencing, and the development of 'gel sequencing' techniques of enormous power, have added hugely to our knowledge of gene structure. Finally, the development and refinement of the 'phosphotriester' approach to oligo-nucleotide synthesis, and in particular its adaptation as a solid-phase technique and recent automation, have facilitated enormously the production of oligonucleotides of defined sequence, and their use for gene synthesis, gene manipulation, and site-directed mutagenesis. Genetic engineering on a large scale passed from being a molecular biologist's pipe-dream to becoming reality in well under a decade. This chapter will attempt to summarize briefly some of the most important and widely used chemical and biochemical techniques developed in this field during the last 20 years.

2.2 Nucleosides

The most commonly occurring nucleosides consist of a nitrogenous base linked by a glycosidic bond to C-1 of D-ribofuranose (**2.1**) or 2-deoxy-D-ribofuranose (**2.2**), specifically as the β-anomers. The nitrogenous bases are derivatives of 9*H*-purine, specifically the 6-amino derivative (**2.3**: adenine, Ade) and the 2-amino-6-keto derivative (**2.4**: guanine, Gua), or of pyrimidine, specifically the 2,4-diketo derivative (**2.5**: uracil, Ura) and its 5-methyl analogue (**2.6**: thymine, Thy) and the 2-keto-4-amino derivative (**2.7**: cytosine, Cyt). The common ribonucleosides, formed when (**2.3**) and (**2.4**) are linked at N-9, and (**2.5**) and (**2.7**) at N-1, to C-1 of ribofuranose are thus adenosine, guanosine, uridine, and cytidine (Ado, Guo, Urd, and Cyd, respectively), and the common 2'-deoxyribonucleosides, formed when (**2.3**), (**2.4**), (**2.6**) and (**2.7**) are linked analogously to 2-deoxyribose, are 2'-deoxyadenosine, 2'-deoxyguanosine, 2'-deoxythymidine, and 2'-deoxy-cytidine (dAdo, dGuo, dThd, dCyd, respectively). 2'-Deoxythymidine is, however, usually named 'thymidine'. These are the nucleosides which, as their phosphate esters, compose the common nucleotides and the poly-nucleotides RNA and DNA. Usually, uracil (in the form of uridine) is found in RNA, and thymine (in the form of 2'-deoxythymidine) in DNA: the other bases are found in both types of nucleic acid.

However, many other nucleosides are known to occur naturally, modified in either the base or the sugar moiety, or in both. Viral DNA may

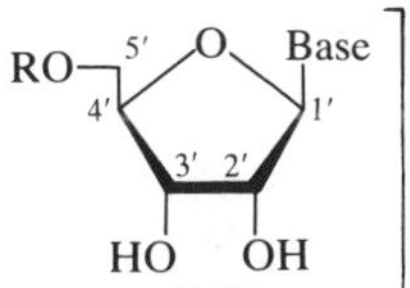

(2.1) 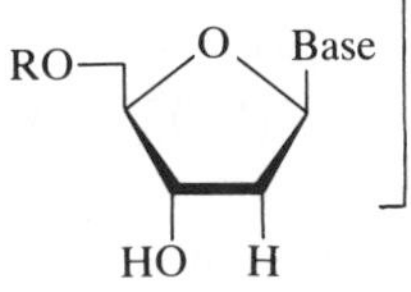 (2.2) 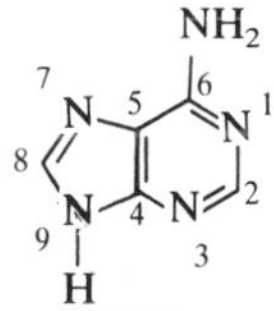 (2.3) 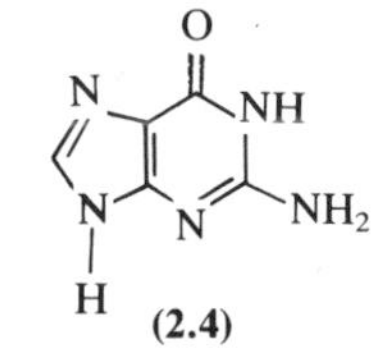 (2.4) 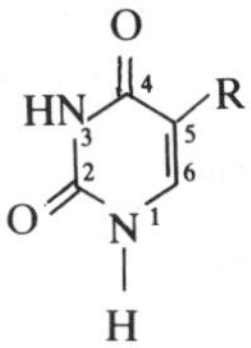 (2.5) R = H (2.6) R = CH$_3$ (2.7)

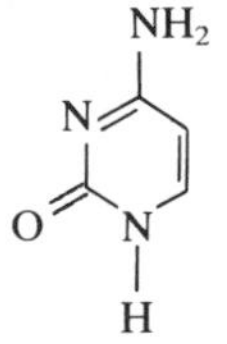

contain 5-methylcytosine or 5-hydroxymethylcytosine in place of cytosine. 6-Methyladenine occurs sometimes in DNA in place of adenine, and 5-hydroxymethyluracil and 5-dihydroxypentyluracil in place of thymine. Different types of RNA may contain modified nucleosides, particularly tRNA which acts as a carrier of amino acids to the ribosome for insertion into proteins in protein biosynthesis. Examples are 4-thiouridine, 5,6-dihydro-uridine, ribothymidine and the C-nucleoside pseudouridine (**2.8**), in which a carbon–carbon bond replaces the normal C—N glycosidic link. N^6-(Δ^2-Isopentenyl)adenosine (**2.9**) is also found in some tRNA molecules and, with its 2-methylthio analogue and the *cis*- and *trans* isomers of N^6-(3-hydroxymethylbut-2-enyl) adenosine (**2.10**), belongs to the group of plant cell division factors known as the cytokinins. Modified sugar derivatives are also found in RNA, chiefly as the 2′-*O*-methylated nucleosides. Generally, such modified nucleosides are formed by post-biosynthetic modification of the nucleic acids by specific enzymes.

Many other nucleosides which are not found in the nucleic acids occur naturally as secondary metabolites, and these include a number of antibiotics

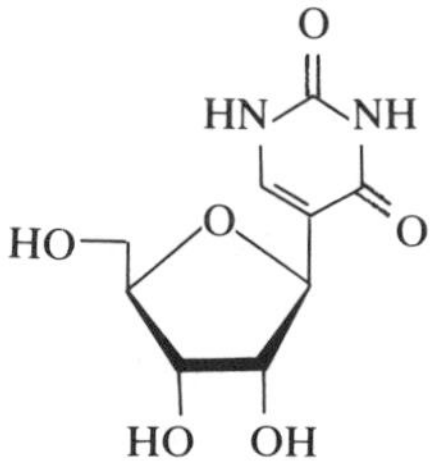

(2.8) 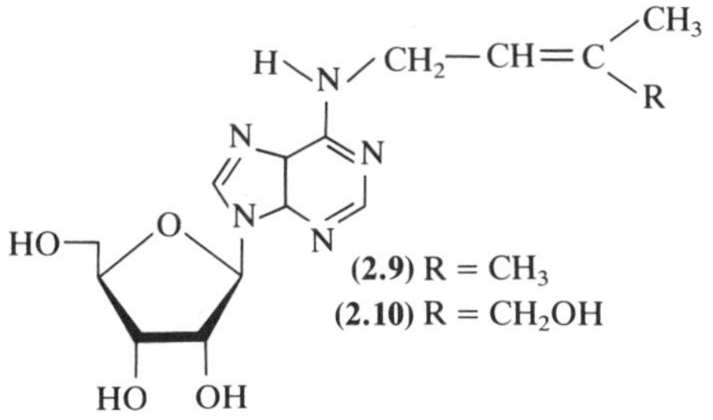 (2.9) R = CH$_3$ (2.10) R = CH$_2$OH

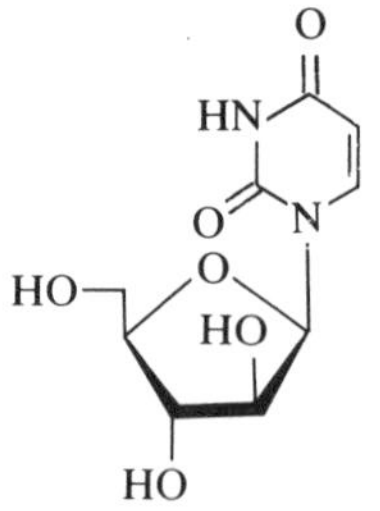

(2.11)

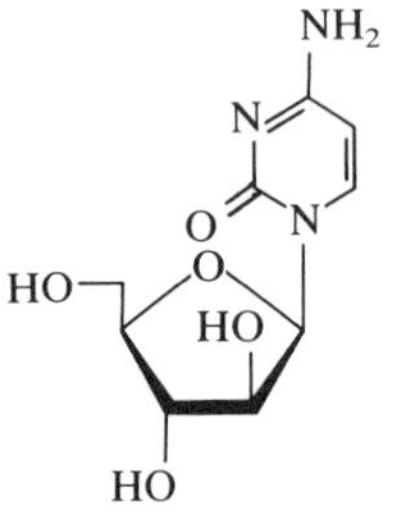

(2.12)

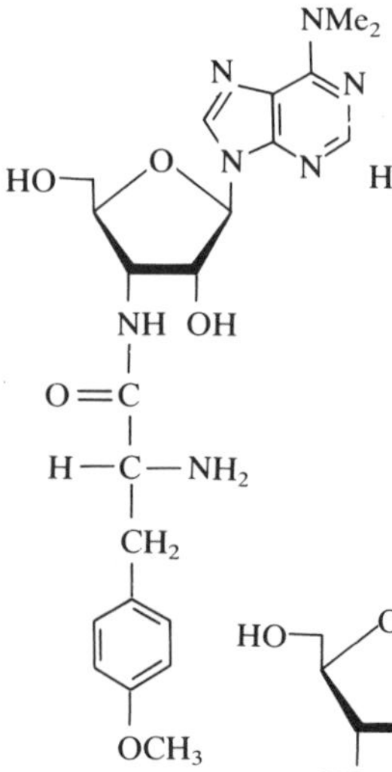

(2.13)

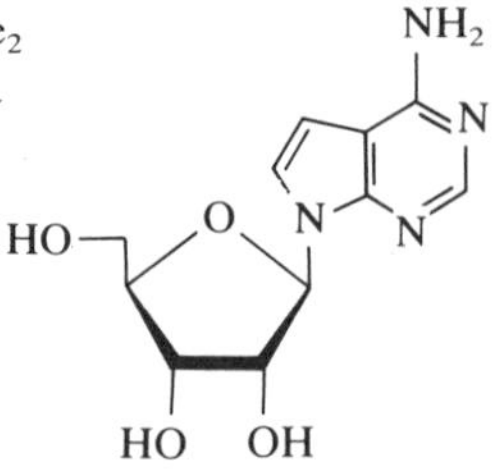

(2.14)

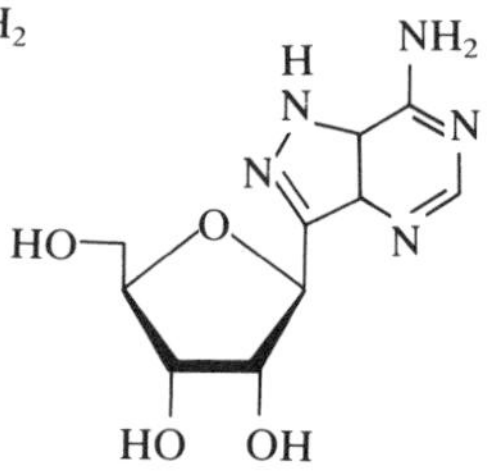

(2.15)

and compounds with antitumour properties. For instance, the arabino-analogue of uridine, *ara*-uridine (**2.11**), occurs naturally in sponges, and the related *ara*-cytidine (**2.12**) and *ara*-adenosine, synthesized chemically, are valuable anti-tumour drugs. 3′-Deoxyadenosine occurs naturally as the antibiotic cordycepin. Aminosugars may also be found, as in the anti-trypanosomal agent puromycin (**2.13**), an inhibitor of protein synthesis. Many base-modified antibiotics are known, such as 7-deazaadenosine (tubercidin, **2.14**) and 8-aza-9-deazaadenosine (formycin, **2.15**), another C-nucleoside. It would be superfluous to attempt further listing of the considerable array of nucleoside antibiotics and other nucleosides of significant biological activity, but excellent texts are available.

2.2.1 *Nucleoside conformation*

The conformation of nucleosides is dependent on their atomic structure and on a number of external influences such as the states of solvation, ionization, and temperature. Three aspects of conformation are of particular interest. In most common nucleosides, the base can rotate freely about the glycosidic bond, although there is a preference for adopting a specific conformation. Definition of these conformations in terms of torsional angles may be confusing, since different conventions for defining the torsional angles have been used in the past. Consider, however, (**2.16**), which represents a bird's-eye view down the glycosidic bond towards the furanose ring, with

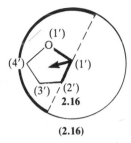

(2.16)

C-1′ of the nucleoside at the centre of a circle. The arrow represents the N(1)—C(2) bond in the pyrimidine nucleosides, or the N(9)—C(4) bond in purine nucleosides. If the arrow points towards the semicircle of thickened perimeter (as shown) the conformation is *syn*, otherwise it is *anti*. The *anti*-conformation is the one usually found in nucleosides, nucleotides, and nucleic acids. However, substitution at C-8 of the purine ring often leads to adoption of *syn*-conformation, as in 8-bromo-guanosine-5′-triphosphate, for instance, which is an inhibitor of DNA-directed RNA polymerase, while guanosine 5′-triphosphate (GTP) is a substrate. Restriction of the ability to rotate about the glycosidic bond, as in the arabinonucleosides, for instance, may be largely responsible for their anti-metabolic activity. The techniques of circular dichroism (CD) and optical rotary dispersion (ORD), and more recently of NMR spectroscopy, have been particularly valuable in determining nucleoside conformation. The 5′-hydroxymethyl group of the furanose sugar can assume three staggered conformations. If the arrow in (**2.17**)–(**2.19**) represents the C(5′)—O(5′) bond, these staggered conformations are *gauche–gauche*, *gauche–trans*, and *trans–gauche* respectively,

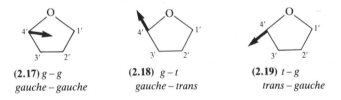

(2.17) *g – g*
gauche – gauche

(2.18) *g – t*
gauche – trans

(2.19) *t – g*
trans – gauche

referring to the direction of the arrow firstly with respect to the C(4′)—O(1′) bond, and secondly with respect to the C(4′)—C(3′) bond, in each case. The nucleosides, nucleotides, and nucleic acids most commonly adopt the *gauche–gauche* conformation. Finally, the sugar ring itself is puckered, like a cyclopentane ring, in which a series of 'twist' and 'envelope' conformations can occur, the first when two atoms are out of the plane of the ring, and the latter when only one atom is. If an atom is 'above' the plane of the ring, i.e. on the same side as the base, in a β-nucleoside, it is said to be '*endo*', and if below the plane of the ring, '*exo*'. The most important conformations of the sugar moiety in nucleosides are shown in (**2.20**) (known as the '*N*'-family of conformations) and (**2.21**) (the '*S*'-family). The forms illustrated all exhibit varying degrees of twist.

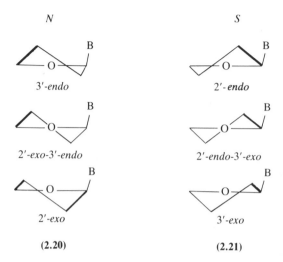

(2.20)　　　　　　　　　　　(2.21)

Most nucleosides and nucleotides in solution are in a flexible state, with *N* and *S* conformations present in equilibrium, the position of equilibrium depending on the substituents on the sugar ring. As deduced from NMR data, adenosine analogues in which the 2′-hydroxy group is replaced by another group have been found to incline more towards a 2′-*exo* conformation as the electronegativity of the substituent increases. In double-helical RNA, and in the 'A' form of DNA, 3′-*endo* puckering is the rule, but the 'B' form of DNA displays a 3′-*exo* conformation.

2.2.2　Nucleoside synthesis

Nucleoside synthesis may be performed either by glycosylation of a base, or by preparing a cyclic sugar derivative bearing a substituent at C-1 which is subsequently transformed to become part of the ring system of the base. For N-glycosides the former method is most commonly used, and for the synthesis of C-nucleosides the latter is almost obligatory.

2.2.2.1　Base glycosylation

Early methods of nucleoside synthesis generally sought to condense a heavy metal salt of a pyrimidine or purine base – usually the silver or mercuric salt – with a glycosyl halide, such as 1-alloxy-1-bromo-2,3,4,6-tetra-*O*-acetyl-α-D-glucose (**2.22**). This procedure suffers from two major disadvantages, however; firstly, the salts are rather insoluble, resulting in incomplete reaction, and secondly, there is a tendency to form kinetically-controlled products which are unable to rearrange to the (desired) thermodynamically controlled products under these reaction conditions. For instance, condensation of the silver salt of N^4-acetylcytosine (**2.23**) with (**2.22**) affords the

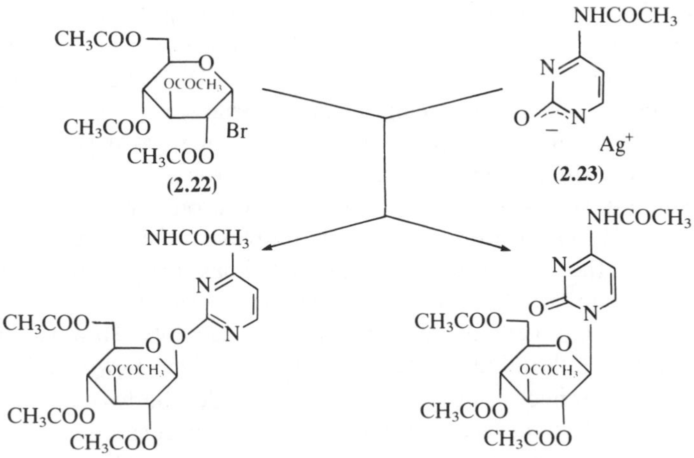

Scheme 1

kinetic product (**2.24**) and the thermodynamic product (**2.25**) in a ratio of roughly two to one (Scheme 1). However, some conversion of (**2.24**) to (**2.25**) can be achieved by heating (**2.24**) with stannic chloride – a Lewis acid – in xylene. A logical progression was, therefore, to solubilize the base as far as possible and to perform the condensation in the presence of a Lewis acid, and it was found that N^6-octanoyladenine, for instance, upon heating with 1,2,3,5-tetra-O-acetyl-D-ribofuranose in 1,2-dichloroethane under reflux in the presence of tin(IV) chloride, gave a fully protected nucleosidic product which, upon deacylation using methoxide, gave adenosine in good yield. Using N^2-palmitoylguanine, instead, guanosine could be prepared similarly, in moderate yield. As we shall see, the strength of the Lewis acid is critically important.

A convenient alternative method for solubilizing the rather insoluble nucleic acid bases consists in silylating them using hexamethyldisilazane
$$\text{H}$$
(HMDS; $\text{Me}_3\text{SiNSiMe}_3$) with a small quantity of trimethylchlorosilane catalyst. Thus, uracil is converted to (**2.26**), and N^6-benzoyladenine to (**2.27**), for instance. While nucleoside synthesis may be performed by heating such silylated bases with peracylated sugar halides such as (**2.22**), with or without

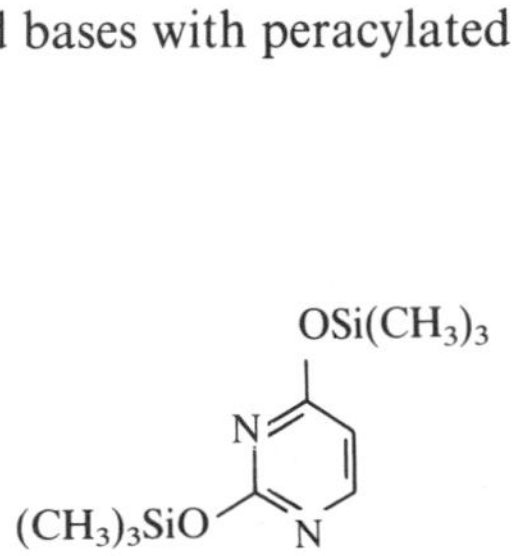

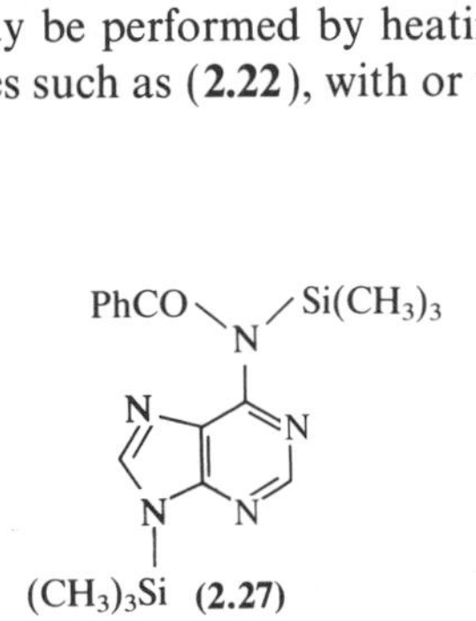

the presence of silver perchlorate or mercuric bromide as catalyst, the best version of this 'Silyl–Hilbert–Johnson Method' consists in condensing the bases with peracylated sugar acetates, e.g. (**2.28**), in the presence of Lewis acids, and in 1,2-dichloroethane or acetonitrile (or a mixture of these) as solvents. Vorbrüggen, in particular, has sought to delineate the stages of the reaction, and distinguishes three separate processes as occurring. Firstly, the catalyst activates the sugar moiety by promoting formation of the electrophilic sugar cation (**2.29**), and complexing the acetate which is lost from C-1 of the sugar. Secondly, the silylated base reacts with (**2.29**) to form the glycosidic bond, with the Lewis acid–acetate complex aiding in desilylation to potentiate the base as a nucleophile. However, the third process – complexation of the silylated base by the Lewis acid catalyst – does not favour nucleoside synthesis, since it renders both a reactant molecule and the catalyst ineffective for promoting or participating in the other processes. The greater the basicity of the silylated base, the more this process will occur, and the more catalyst is required for the reaction. Moreover, the Lewis acid, if not actually forming a complex, will be associated in the region of the nitrogen atom of highest basicity, which in the pyrimidine species is

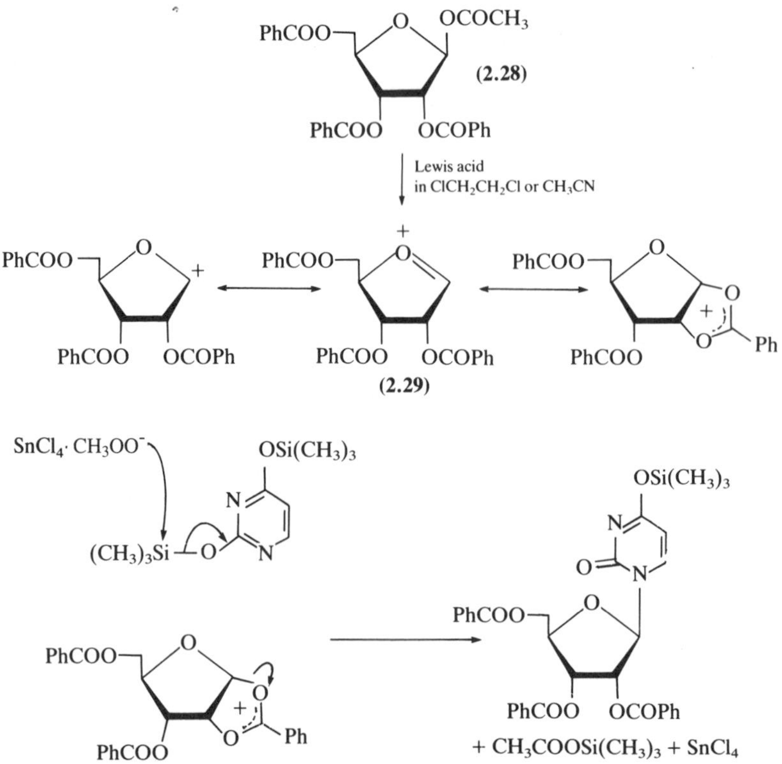

Scheme 2

usually N-1, and glycosylation at N-1 is inhibited in consequence. If glycosylation occurs, it will then occur at the unhindered N-3, to give the unwanted kinetic product. Thus a Lewis acid is required which is strong enough to effect the first process without interacting too strongly with silylated base. Trimethylsilyl (TMS) triflate (trifluoromethanesulphonate), trimethylsilyl nonaflate (nonafluorobutanesulphonate), and trimethylsilyl perchlorate have all been found to act as valuable mild Lewis acid catalysts, trapping the acetate displaced from the sugar during cation formation as trimethylsilyl acetate, and releasing triflate, nonaflate, and perchlorate respectively, any of which can promote the desilylation effected in Scheme 2 by the tin(IV) chloride–acetate complex. The interaction of these Lewis acids (which are weak Friedel–Crafts catalysts) with silylated pyrimidine bases has been quantified by measuring shifts in the ^{13}C NMR spectra of the bases on complex formation, and the TMS-triflate, -nonaflate, and -perchlorate all found to be much weaker Lewis acids than tin(IV) chloride, and consequently more useful when condensing the more basic silylated pyrimidine species with sugar derivatives.

While the 'Silyl–Hilbert–Johnson' method has produced particularly spectacular results in the synthesis of pyrimidine nucleosides, silylation coupled with the use of the weak Lewis acids is also efficacious in purine nucleoside synthesis. Thus, treatment of (**2.27**) with (**2.28**) in acetonitrile in the presence of TMS-triflate followed by deacylation of the product, gives a 70 per cent yield of adenosine. Moreover, the method may conveniently be reduced to a 'one-pot' reaction. For instance, if a mixture of the desired base (*N*-acylated, if necessary), the protected sugar derivative (e.g. (**2.28**)), and trifluoromethanesulphonic acid is treated with HMDS and trimethyl-chlorosilane in acetonitrile under reflux, silylation of the reagents and subsequent condensation occur to give good yields of the protected nucleosides. For many nucleoside syntheses, this is the method of choice. However, there are some exceptions. Guanine nucleosides are generally difficult to prepare in good yield, and seem to be most easily accessible by condensing silylated 2-amino-6-chloropurine (**2.30**) or 2-acetamido-6-chloropurine (**2.31**) with a halosugar derivative in the presence of mercuric cyanide. Deacylation of the product and hydrolysis of the reactive 6-chloro function then gives guanine nucleosides.

It will be seen that in the right-hand illustration of canonical forms of

(**2.30**) R = H
(**2.31**) R = CH$_3$CO

(**2.29**), and in Scheme 2 the formation of benzoyloxonium ion sterically hinders attack from the α-side by the incoming base, and consequently only the β-nucleosides are produced. Indeed, this is also observed if a 2-acetyl, 2-acetamido-, and, in general, any 2-acyl- or aroyl-function is present on the ribose sugar, and led B.R. Baker to formulate the generalization known as 'Baker's rule': the incoming base enters at C-1 *trans* to any acyloxy or aroyloxy function at C-2 of the sugar. In the synthesis of 2'-deoxyribo-nucleosides, no such stereoselective influence is present, and mixtures of the α- and β-anomers are formed.

Supposing one wishes to prepare the β-anomer of an arabinonucleoside, or the α-anomer of a ribonucleoside, by base glycosylation; how may Baker's rule be circumvented? Firstly, one must use a reactive sugar derivative protected at the 2-hydroxy group by a function other than the acyl group. For *ara*-nucleoside synthesis, a condensation utilizing 2,3,5-tri-*O*-benzyl-arabinofuranosyl halides in a two-phase system has been found effective. The sugar halide and the required base, acylated to protect amino groups if necessary, are dissolved in a system of dichloromethane or 1,2-dimethoxy-ethane:50 per cent aqueous sodium hydroxide in the presence of tetra-*n*-butylammonium bisulphate or benzyltriethylammonium chloride as phase-transfer catalyst, and shaken hard, to yield a mixture of (**2.32**) and

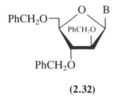

(**2.32**)

its α-anomer. The ratio of the anomeric products formed is dependent on the catalyst used, and the quantity added. It is thought that the catalyst may alter the anomeric ratio in the sugar halide reactant, and the S_n2 displacement of the halide then follows. In a recent stereoselective synthesis of α-ribonucleosides (Scheme 3), 5-*O*-benzoyl-2,3-*O*-isopropylideneribofuranose was treated with 2-fluoro-1-methylpyridinium tosylate in the presence of a tertiary amine, to afford the β-anomer of the 2-hydroxy-*N*-methylpyridinium ribofuranoside (**2.33**) apparently as sole product, and treatment of (**2.33**) with a silylated base gave good yields of the protected α-nucleoside (**2.34**), probably by S_n2 displacement, although some of the β-anomer was also formed, possibly via a competing S_n1 process. Also, treatment of methyl D-ribofuranoside with boron trichloride and N^6-octanoyladenine in chloro-form in the presence of pyridine gives a fair yield of α-adenosine. In this case the boron trichloride is probably acting as a Lewis acid to promote the methoxy group at C-1 of the sugar as a leaving group, and may also become coordinated to the hydroxy groups at C-2 and C-3 of the sugar, to form a cyclic complex.

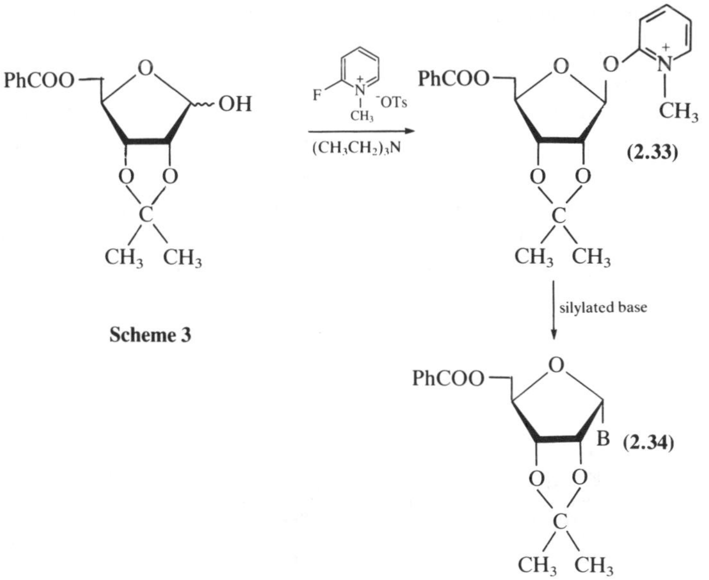

Scheme 3

Nucleosides may also be prepared by introduction of a suitable nitrogenous group at C-1 of the sugar ring, followed by condensation reactions to build up the ring. For instance, treatment of D-ribose with methanolic ammonia affords D-ribopyranosylamine which, when condensed with 2,2-dimethoxy-propane using *p*-toluenesulphonic acid, gives 2,3-*O*-isopropylidene-D-ribofuranosylamine (**2.35**), which condenses with appropriately substituted

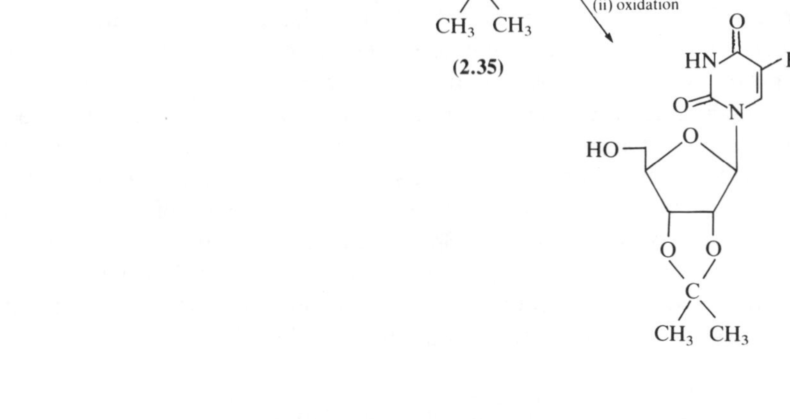

N-ethoxycarbonylacrylamide species to afford 5-substituted uridine derivatives. Again, D-arabinose condenses efficiently with cyanamide to afford (**2.36**), which condenses with ethyl propiolate to afford $O^2,2'$-anhydrouridine (**2.37**). Hydrolysis of (**2.37**) with weak mineral acid then affords *ara*-uridine (**2.11**). In point of fact, most *ara*-nucleosides have been prepared by modification of the sugar moiety of ribonucleosides, rather than by direct base glycosylation or base assembly, as instanced here, while nucleoside syntheses in which the base is constructed *in situ* from species such as (**2.35**) and (**2.36**) have largely been directed to preparing nucleosides containing analogues of the nucleic acid bases.

2.2.2.2   *Nucleoside transformation*

In nucleoside transformation, a naturally occurring nucleoside is transformed chemically into an analogue, a target substance which may be evaluated as a drug or designed to effect a particular process or reaction in chemistry or biochemistry. Huge numbers of nucleoside analogues, modified in sugar and/or base moieties, have been prepared by nucleoside transformation, and a couple of examples must serve to illustrate this important area of nucleoside synthesis. For instance, when cytidine is treated with diphenylcarbonate in DMF at around 150°C in the presence of an equivalent of sodium bicarbonate or water as a proton source, the $O^2,2'$-anhydrocytidine species (**2.38**) is formed, presumably via transesterification with the diphenylcarbonate to give a 2',3'-cyclic carbonate (**2.39**), which undergoes ring closure and loss of carbon dioxide. The cyclonucleoside (**2.38**) has a high pK_a value for protonation on the base, and in the presence of a proton source and water undergoes ready autohydrolysis to form *ara*-cytidine (**2.12**), already mentioned as an anti-tumour agent. A similar set of reactions from uridine via (**2.37**) affords *ara*-uridine (**2.11**). Cyclonucleosides with oxygen (or other heteroatom) bridges between the base and sugar moieties of the nucleoside can be made to the 2'- (as in (**2.38**)), 3'-, or 5'-positions of the sugar ring and have proved valuable in nucleoside transformations. Formation of the cyclonucleoside is often effected by displacement of a mesylate (methanesulphonate) or tosylate (toluenesulphonate) substituent on the sugar ring

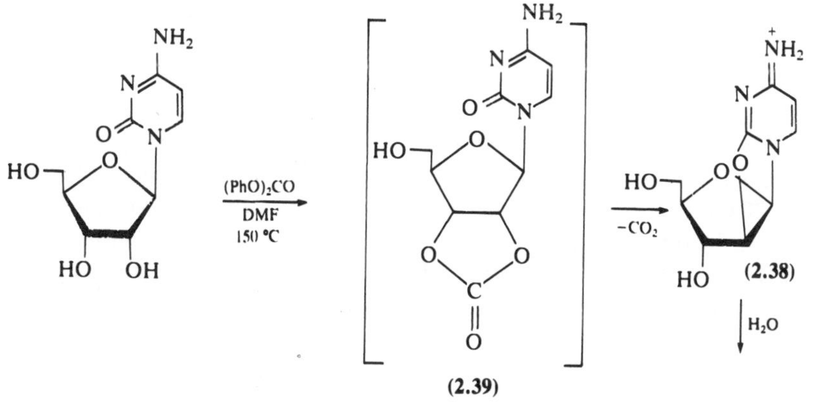

(2.39)

by an oxo-substituent on the pyrimidine ring, giving intramolecular cyclization as in (2.39) → (2.38): an unusual reagent which has proved effective is diethyl 1,1,2-trifluoro-2-chloroethylamine which converts 2'-deoxythymidine (2.40) to O^2,3'-anhydro-2'-deoxythymidine (2.41) rather efficiently. Treatment of (2.41) with lithium azide in hot DMF then affords 3'-azido-2',3'-dideoxythymidine (2.42), widely known as AZT, a drug which shows valuable, if limited, effect in controlling the progress of AIDS (acquired immunodeficiency syndrome).

Space does not permit description of the huge numbers of nucleoside analogues containing modified bases and sugars which have been prepared, but some excellent reviews are available.

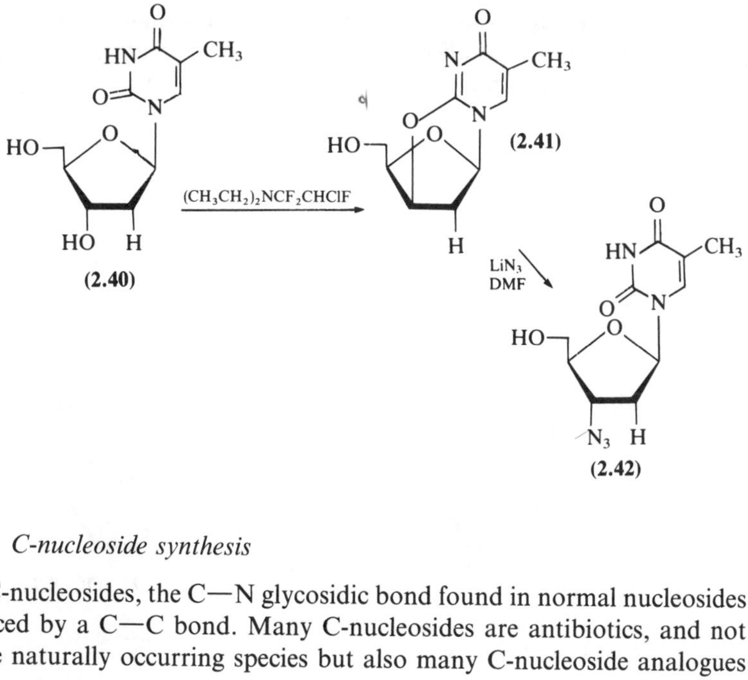

2.2.2.3 C-nucleoside synthesis

In the C-nucleosides, the C—N glycosidic bond found in normal nucleosides is replaced by a C—C bond. Many C-nucleosides are antibiotics, and not only the naturally occurring species but also many C-nucleoside analogues have been synthesized with an eye to their possible medicinal value. Most C-nucleoside syntheses involve the attachment of a functionalized moiety, containing one to four carbon atoms, at C-1 of the sugar ring, after which the base is elaborated by condensation and cyclization. To illustrate this, let us consider several synthetic routes to showdomycin (**2.43**), a maleimide C-nucleoside antibiotic from *Streptomyces showdoensis* (see Chapter 1, p. 52). If (**2.28**) is treated with trimethylsilyl cyanide and tin(IV) chloride in acetonitrile, the nitrile (**2.44**) is formed, which upon hydrolysis with aqueous acid, esterification to form the methyl ester, alkaline hydrolysis of all the ester groups, acetylation of the sugar hydroxy groups using acetic anhydride, and treatment with thionyl chloride, gives (**2.45**). When (**2.45**) is treated with methoxycarbonylmethylenetriphenylphosphorane in the presence of cyanide, the cyanoacrylate (**2.46**) is formed, which upon hydrolysis to the amide and cyclization using acetic anhydride, followed by deacetylation of the sugar ring, affords showdomycin. Alternatively, if 2,3,5-tri-*O*-benzylribose is treated with ethynylmagnesium bromide in THF, and the resulting diol ring closed by treatment with tosyl chloride in pyridine, the ethynyl-ribofuranoside (**2.47**) is formed, and subsequent treatment with carbon monoxide in methanol in the presence of palladium chloride and mercuric chloride gives the dimethyl maleate derivative (**2.48**). Hydrolysis of the ester groups, followed by cyclization to the anhydride using acetic anhydride, amidation with ammonia and re-cyclization with acetyl chloride gives 2′,3′,5′-tri-*O*-benzylshowdomycin, and the benzyl groups may be removed with boron trichloride to give (**2.43**).

In another synthesis, condensation of the Wittig reagent ethoxycarbonyl-ethylidenetriphenylphosphorane with 5-*O*-*t*-butyldimethylsilyl-2,3-*O*-iso-propylideneribofuranose affords (**2.49**), which on treatment with phenyl-selenyl chloride undergoes stereoselective cyclization to give the β-anomer

(**2.43**)

(**2.44**) R¹ = PhCO ; R² = CN
(**2.45**) R¹ = CH₃CO ; R² = COCl

(**2.46**)

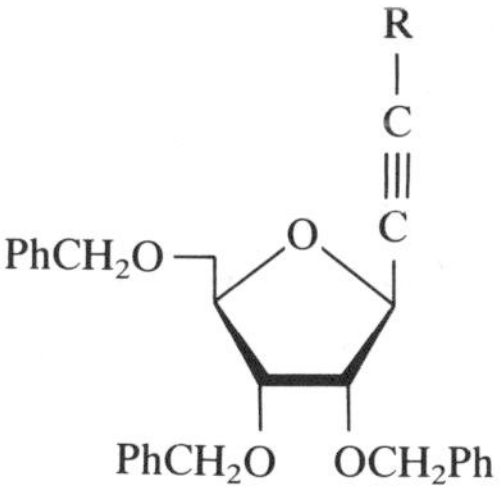

(**2.47**) R = H
(**2.56**) R = CH(OCH₂CH₃)₂

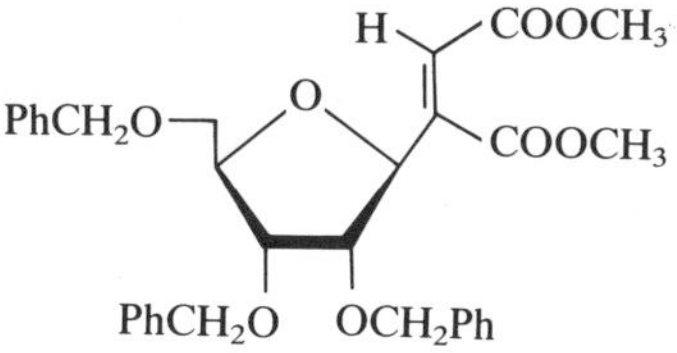

(**2.48**)

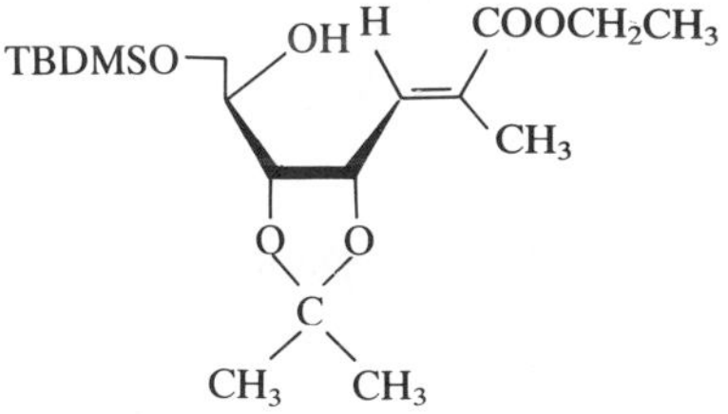

(**2.49**) 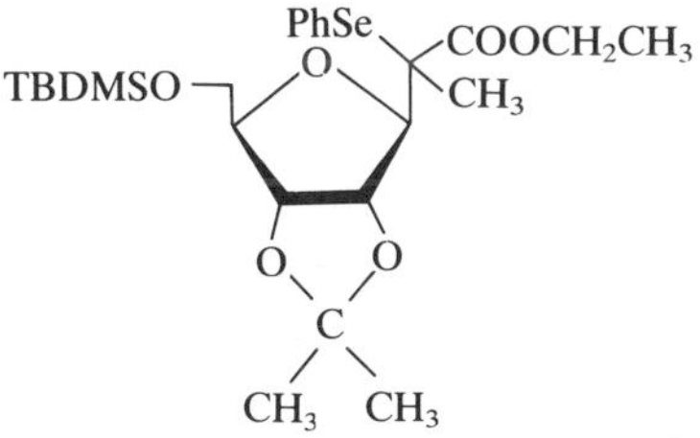   TBDMS = (CH₃)₃C(CH₃)₂Si−   (**2.50**)

(**2.50**), probably as a result of steric hindrance by the isopropylidene group to formation of the α-anomer. Oxidation with hydrogen peroxide is followed by loss of phenyl selenoxide to give (**2.51**), which upon ozonolysis and condensation of the resulting α-ketoester with carbamoylmethylenetriphenylphosphorane affords (**2.52**). Treatment with pyridinium tosylate then effects cyclization, and removal of the protecting groups then affords (**2.43**) (see Chapter 1, p. 51). It is not intended to imply that showdomycin is the sole synthetic aim of C-nucleoside synthesis! However, sugar nitriles such as (**2.44**), alkynyl sugars such as (**2.47**), and the products of Wittig reaction with sugars such as (**2.49**) and its cyclized products are probably the most commonly used intermediates in C-nucleoside synthesis, and these syntheses of showdomycin illustrate the use of all three approaches. An alternative, rather elegant approach to the same compound consists in condensing (**2.28**) with 1,2-bis(trimethylsilyloxy)cyclobut-1-ene using tin(IV) chloride in

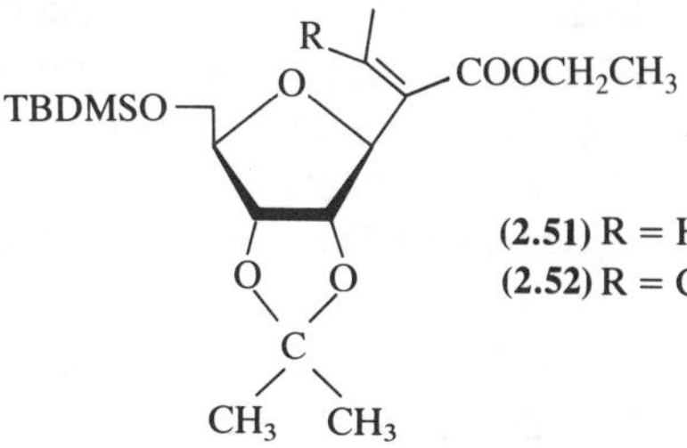

(**2.51**) R = H
(**2.52**) R = CONH₂

methylene chloride to give (2.53), thus constructing a C-glycoside by a process analogous to the Silyl–Hilbert–Johnson route. Treatment of (2.53) with lithium bis(trimethylsilyl)amide affords the corresponding silyl enolate, which undergoes electrophilic attack by nitrosyl chloride to form the oxime (2.54) which undergoes ring-opening on standing in methylene chloride to give the nitrile (2.55). Treatment of (2.55) first with ammonia, and then with trifluoroacetic anhydride, then gives (2.43).

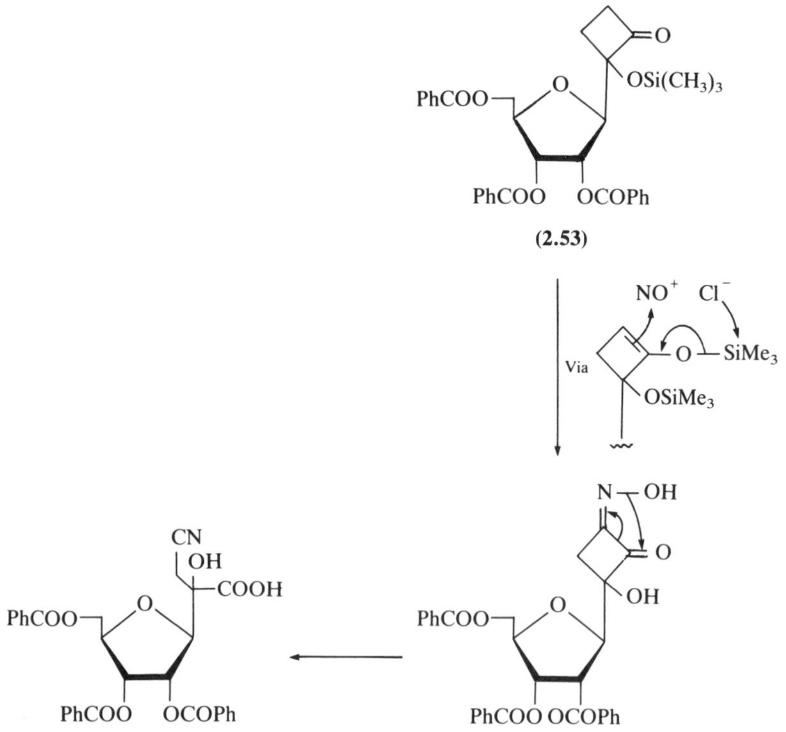

(2.53)

(2.55)   (2.54)

The functionalized ribose derivatives prepared initially during C-nucleoside synthesis may lend themselves to elaboration to form several different species. For instance, condensation of 2,3,5-tri-O-benzylribose with 3,3-diethoxy-prop-1-ynylmagnesium bromide affords predominantly the β-anomer (2.56) (cf. (2.47)) which condenses with hydrazine to form the pyrazole nucleoside (2.57). On treatment with 1-fluoro-2,4-dinitrobenzene and triethylamine the N^1-(2,4-dinitrophenyl)pyrazole (2.58) is formed, and subsequent debenzylation with boron trichloride followed by acetylation then gives the corresponding acetylated sugar derivative (2.59). Nitration of this using copper(II) nitrate in acetic anhydride yields the nitrated derivative (2.60), which is treated with sodium methoxide to remove the dinitrophenyl group (and, simultaneously, the sugar acetate groups), re-acetylated and re-nitrated

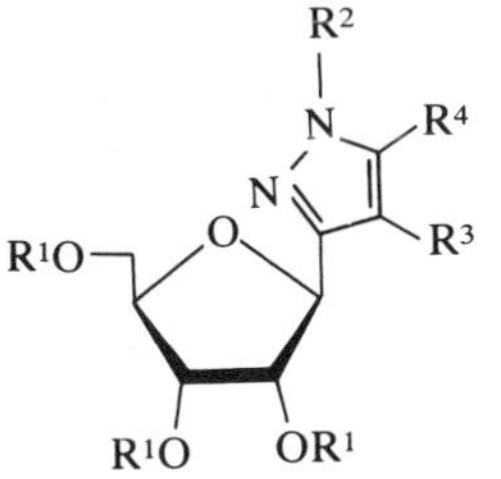

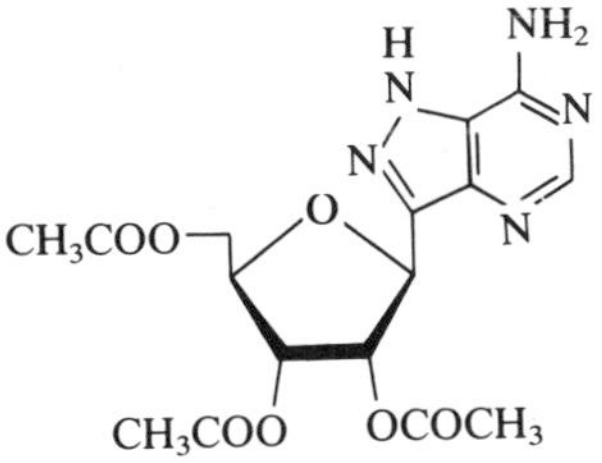

(2.57) $R^1 = PhCH_2$; $R^2 = R^3 = R^4 = H$
(2.58) $R^1 = PhCH_2$; $R^2 = DNP$; $R^3 = R^4 = H$
(2.59) $R^1 = COCH_3$; $R^2 = DNP$; $R^3 = R^4 = H$
(2.60) $R^1 = CH_3CO$; $R^2 = DNP$; $R^3 = NO_2$; $R^4 = H$
(2.61) $R^1 = CH_3CO$; $R^2 = R^3 = NO_2$; $R^4 = H$
(2.62) $R^1 = CH_3CO$; $R^2 = H$; $R^3 = NO_2$; $R^4 = CN$
(2.63) $R^1 = CH_3CO$; $R^2 = H$; $R^3 = NH_2$; $R^4 = CN$
(2.66) $R^1 = CH_3CO$; $R^2 = H$; $R^3 = OH$; $R^4 = CN$
(2.67) $R^1 = CH_3CO$; $R^2 = H$; $R^3 = OH$; $R^4 = CONH_2$
(2.68) $R^1 = R^2 = H$; $R^3 = OH$; $R^4 = CONH_2$

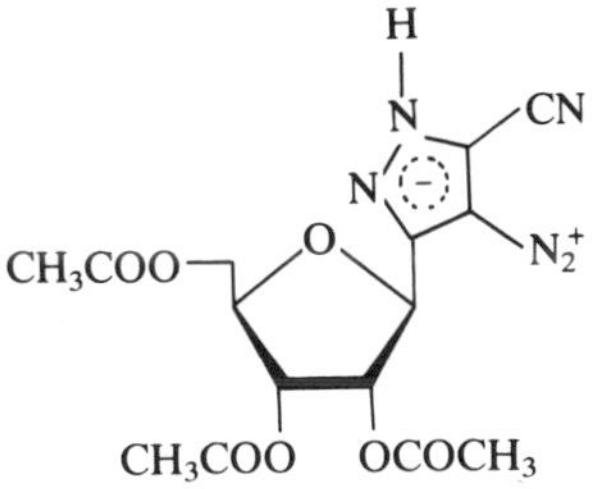

as before to give (2.61). When (2.61) is treated with potassium cyanide, *cine*-substitution occurs, via nucleophilic attack of cyanide at C-5 and loss of the N^1-nitro group as nitrite, affording (2.62) which upon catalytic hydrogenation gives (2.63). Condensation of (2.63) with formamidine acetate gives (2.64), after which deacetylation with methoxide yields the C-nucleoside adenosine analogue formycin (2.15). If, instead, (2.63) is diazotized, the rather stable diazonium species (2.65) is formed, but this, on irradiation in aqueous acetone, yields (2.66) which is hydrolysed to the amide (2.67) by nickel acetate in acetic acid. Deacetylation with methanolic ammonia then affords the C-nucleoside antibiotic pyrazofurin (2.68).

Occasionally, the complete base ring system may be attached to the sugar moiety in a single stage. For instance, 5-lithiated 2,4-bis(*t*-butyloxy) pyrimidine, generated by treating the corresponding 5-bromo derivative with butyllithium, condenses with 2,3,5-tri-*O*-benzylribose to give (2.69) and its C-1′ epimer. Cyclization and debutylation of (2.69) using ethanolic hydrogen chloride, followed by debenzylation with boron trichloride, then affords pseudouridine (2.8).

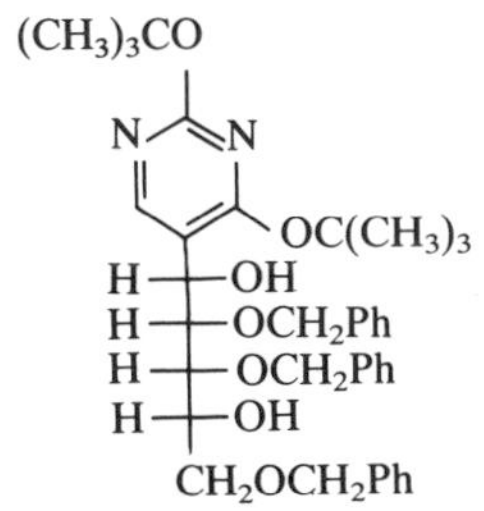

Some intriguing routes to C-nucleosides which start from non-carbohydrate precursors have been devised. In the method which has been most thoroughly developed and exploited, $\alpha,\alpha,\alpha',\alpha'$-tetrabromoacetone has been coupled to furan in the presence of diiron nonacarbonyl $(Fe_2(CO)_9)$ or zinc–silver couple, after which reduction of the dibrominated product with a zinc–copper couple affords the bicyclic ketone (**2.70**). *Cis*-hydroxylation using osmium tetroxide occurs on the least hindered side of the double bond, which after formation of the acetonide and Baeyer–Villiger oxidation affords a mixture of (**2.71**) and its enantiomer. Condensation of (**2.71**) with *t*-butoxybis(dimethylamino)methane then affords (**2.72**), which is functionally closely related to such species as (**2.48**) and (**2.51**), and can be used to synthesize a number of naturally occurring C-nucleosides, or, by starting from substituted furan or tetrabromoacetone species, to prepare a good range of analogues. Another versatile approach starts from the Diels–Alder adducts formed from furan or cyclopentadiene and dimethyl acetylenedicarboxylate, (**2.73**) and (**2.74**) respectively, which are *cis*-hydroxylated and converted to the acetonides, (**2.75**) and (**2.76**), as above, and then subjected to the action of pig liver esterase which specifically hydrolyses only one ester group, giving (**2.77**) and (**2.78**) respectively. Using a number of standard procedures (**2.77**) has been converted to showdomycin, (**2.78**) to the antibiotic aristeromycin (**2.79**), and epoxidized (**2.73**), after similar treatment with pig liver esterase, has been converted to the antibiotic cordycepin (3′-deoxyadenosine). Together these present an excellent example of the use of the stereoselectivity of an enzyme species as a synthetic tool in organic synthesis.

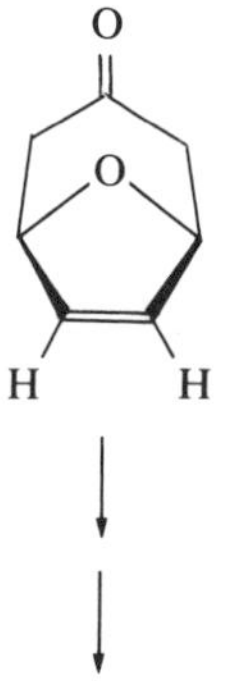

(**2.70**)

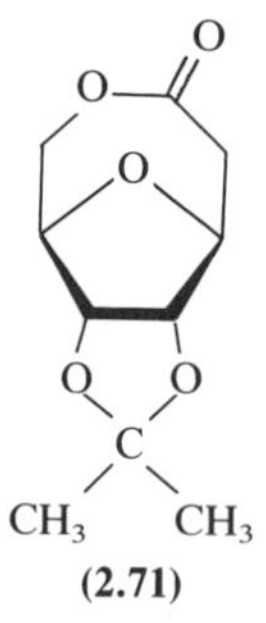

(**2.71**) 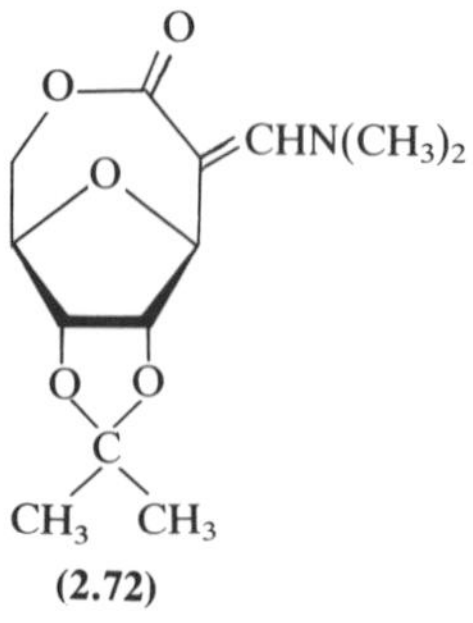   (**2.72**)

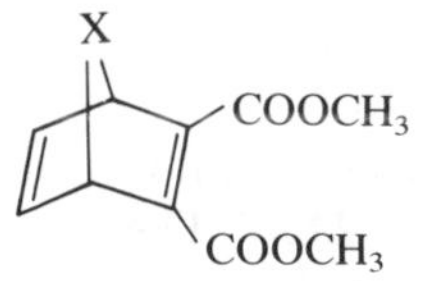

(2.73) X = O
(2.74) X = CH₂

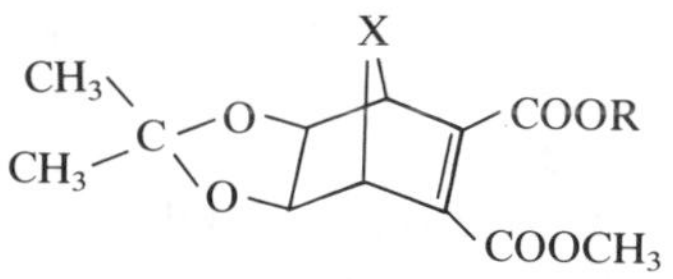

(2.75) X = O; R = CH₃
(2.76) X = CH₂; R = CH₃
(2.77) X = O; R = H
(2.78) X = CH₂; R = H

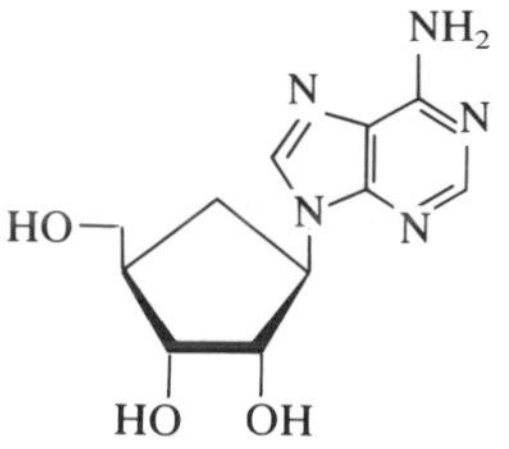

(2.79)

2.2.2.4 Some useful general reactions in nucleoside synthesis

While space does not permit extensive discussion of the many reactions which have been used to prepare nucleoside analogues, several novel and efficient methods of modifying the structures of naturally occurring nucleosides to form other naturally occurring nucleosides or their analogues deserve mention. Appropriately protected ribonucleosides may often be reduced to the corresponding 2'- or 3'-deoxyribonucleosides by treatment with thiocarbonyldiimidazole, phenoxythiocarbonyl chloride, or a similar species to give 2'- and 3'-thiocarbonyl esters which are then treated with tri-*n*-butyltin hydride in the presence of azobisisobutyronitrile (AIBN; a radical initiator). Thus, for instance, (**2.80**), treated in this way and then deprotected, affords cordycepin, and 3',5'-O-protected ribonucleosides may be reduced to 2'-deoxynucleoside species in a similar manner. In a recent extension of this procedure, 5'-O-silyl protected ribonucleosides have been treated with carbon

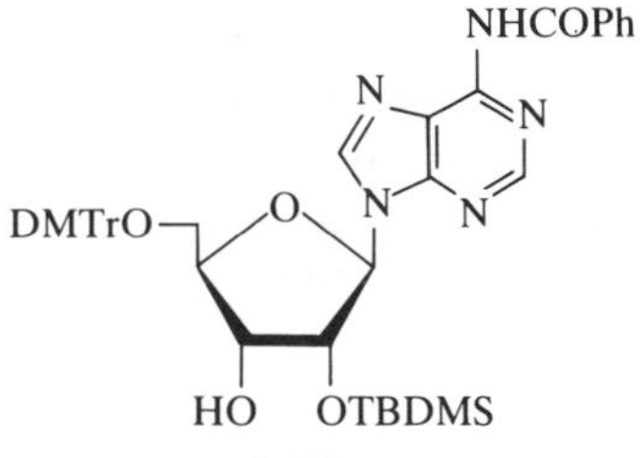

(DMTr =)

('Dimethoxytrityl')

(2.80)

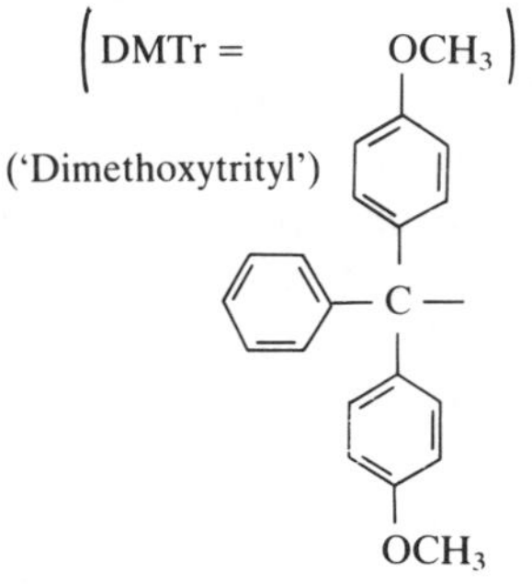

disulphide in the presence of alkali, and then methyliodide, to give their 2′,3′-bis(xanthates) (**2.81**), which when reduced as above gave the corresponding 2′,3′-didehydro-2′,3′-dideoxy species (**2.82**). Further catalytic reduction gave the 2′,3′-dideoxyribonucleosides (**2.83**). The silyl and base protecting groups are readily removed with methanolic ammonia to give the unprotected nucleosides corresponding to (**2.82**) and (**2.83**). A number of other routes to these compounds have also been described, and they are of considerable current interest since some show substantial activity against human immunodeficiency virus (HIV), and are in clinical trial. Also, the 5′-triphosphates of the 2′,3′-dideoxyribonucleosides are the essential

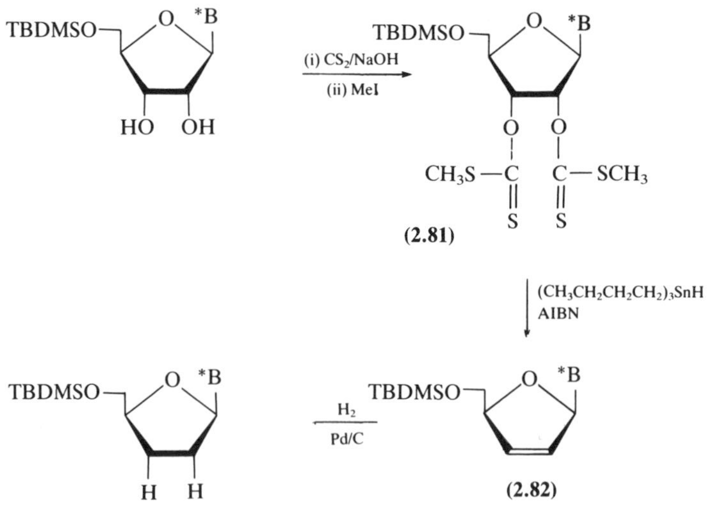

ingredients of the 'dideoxy terminator' methods of DNA sequencing. Ribonucleosides may be monomethylated on the sugar ring to give a mixture of the 2′- and 3′-O-methyl isomers without methylation of the base by slow addition of dilute diazomethane to the unprotected nucleoside in methanol in the presence of tin(II) chloride dihydrate. Stannous methoxide is formed, which undergoes ligand exchange with the *cis*-glycol group of the ribonucleosides giving (**2.84**). This bidentate tin complex is effectively in equilibrium with the two isomers in which tin is monodentate, with the now unliganded oxygen function reacting, formally, as an alkoxide group and becoming methylated. Formation of the tin complex thus promotes the reactivity of the oxygen functions to which it is bonded. Purine nucleosides give predominantly the 3′-O-methylated products, and pyrimidine nucleosides the 2′-O-methylated isomers. Unprotected uracil nucleosides may conveniently be methylated at the 5-position to give the corresponding thymine nucleosides via a Mannich reaction. Thus, 2′-deoxyuridine, treated with formaldehyde and pyrrolidine, affords (**2.85**). On treatment with toluene-4-thiol, the

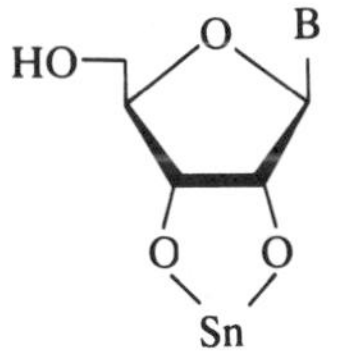

(2.84)

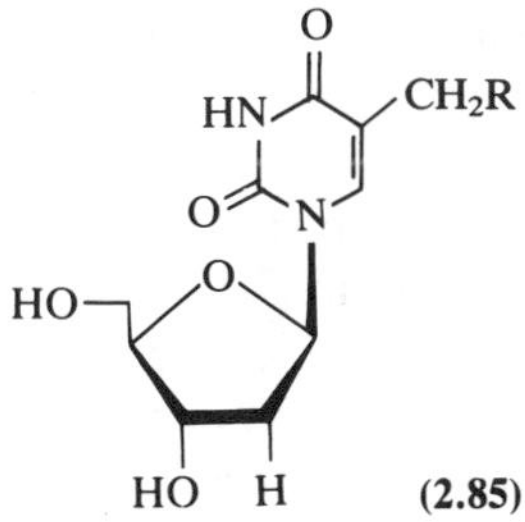

(2.85) R = —N⟨pyrrolidine⟩

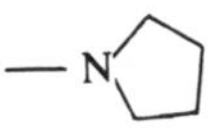

(2.86) R = —S—⟨C₆H₄⟩—CH₃

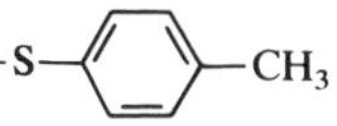

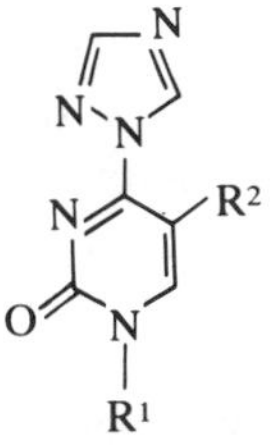

(2.87) R¹ = blocked sugar moiety
R² = H or Me

pyrrolidine group in (**2.85**) is displaced to give 5-(4-tolylthiomethyl)-2′-deoxyuridine (**2.86**), which on reduction with Raney nickel affords 2′-deoxythymidine. In a convenient procedure for converting a keto function at the 4-position of the pyrimidine ring to an amino function, uridine or thymidine nucleosides containing protected sugar residues may be treated with phosphorylchloride, or 2- or 4-chlorophenylphosphorodichloridate, and 1H-1,2,4-triazole in the presence of a tertiary amine to give a 4-(1,2,4-triazol-1-yl)pyrimidinone of type (**2.87**), which, when treated with ammonia, affords the corresponding cytidine species. Alternatively, primary or secondary amines may be used, leading to the formation of N^4-alkylated cytidine species.

2.3 Nucleotides

Nucleotides are phosphoric acid esters of nucleosides. The nucleoside monophosphates are simple primary or secondary alkyl phosphates, the acid functions presenting dissociations at pK_a values of c. 2 and 6.7 respectively. The so-called nucleoside 5′-diphosphates (**2.88**) and nucleoside 5′-triphosphates (**2.89**) are monoalkyl esters of pyrophosphoric acid and tripolyphosphoric acid respectively. These are of central importance in

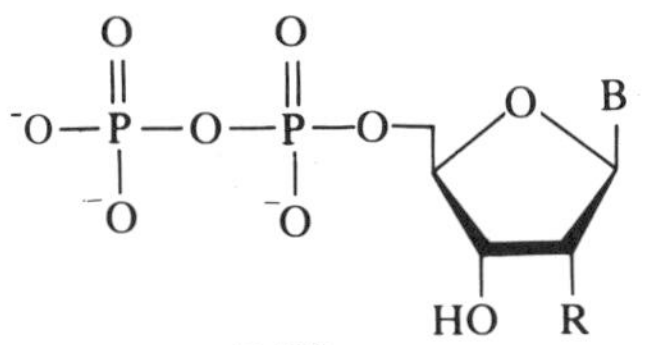

(2.88)   R = H or OH; B = base   (2.89)

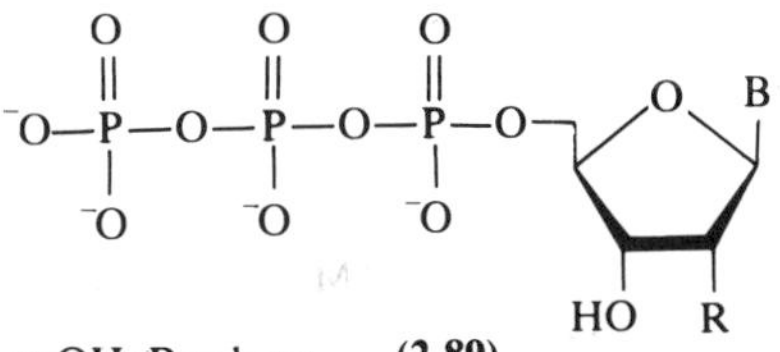

metabolism. The standard free energy of hydrolysis (ΔG^0) is around $-35\ kJ\ mol^{-1}$ for the hydrolysis of the phosphoanhydride bonds in these species, and the phosphate (or, in the nucleoside triphosphates, pyrophosphate) groups have substantial group transfer potential. The terminal phosphate groups of nucleoside 5'-triphosphates may thus be used to phosphorylate other species, such as to prime sugars to enter metabolic pathways. Alternatively, the energy in the phosphoanhydride bond may be transduced to drive active transport processes, or to do mechanical work, as in muscle. The nucleoside 5'-polyphosphates are the energy currency of the cell for general metabolic purposes, and are also the substrates from which the nucleic acids are synthesized. Generally, the polyphosphate substituents are attached at the 5'-hydroxy group of nucleotides: nucleoside polyphosphates containing a 3'-pyrophosphate substituent do occur naturally, but in comparatively meagre quantities, and only under special conditions. Unsymmetrical P^1-(5'-nucleosidyl)-P^2-alkylated pyrophosphates are of central importance in biosynthetic pathways (uridine 5'-diphosphate (UDP)–glucose, etc., in sugar metabolism; cytidine 5'-diphosphate (CDP)–choline, etc., in lipid metabolism). Certain enzyme cofactors are nucleotidic in nature, such as the unsymmetrical pyrophosphates nicotinamide adenine dinucleotide (**2.90**: NAD$^+$) and the analogue which bears an extra phosphate group at the 2'-position of the adenosyl moiety, NADP$^+$ (**2.91**). These, along with flavin–adenine dinucleotide (**2.92**: FAD) and flavin mononucleotide (**2.93**: FMN) are involved in many redox processes of cell metabolism. Note that FAD and FMN are in fact misnomers: the riboflavin component is, in fact, [7,8-dimethyl-10-(1'-D-ribityl)]-isoalloxazine, and thus a derivative of the sugar alcohol ribitol rather than ribose, and not a true nucleoside.

(**2.90**) R = H
(**2.91**) R = PO$_3{}^{2-}$

(**2.92**) R = 5'-AMP
(**2.93**) R = H

(**2.94**) B = adenine-9
(**2.95**) B = guanine-9

Coenzyme A (see the Introduction), which is another unsymmetrical pyrophosphate containing an adenosine 3′,5′-bisphosphate component, is a further nucleotide of central significance in metabolism. Finally, the cyclic dialkyl phosphates adenosine-3′,5′-monophosphate (**2.94**: 'cyclic AMP, cAMP') and guanosine-3′,5′-monophosphate (**2.95**: 'cyclic GMP, cGMP') are of great importance as intracellular 'messengers' which mediate events within cells which are set in train by the binding of hormones externally to the cell membranes, and are also important in the biochemistry of vision and of olfaction.

2.3.1 *Nucleotide biosynthesis*

The biosynthesis of the nucleic acid components effectively occurs to afford the finished component at the nucleotide level, but the routes to the pyrimidine and purine nucleotides are very different. The biosynthesis of pyrimidines begins with condensation of aspartate with carbamoyl phosphate to give *N*-carbamoylaspartate (**2.96**), which loses water to afford L-dihydroorotate (**2.97**). This is then oxidized by a process involving NAD^+ to afford uracil-6-carboxylic acid ('orotic acid') (**2.98**) which is condensed with 5-phospho-α-D-ribofuranosyl-1-pyrophosphate (PRPP) to give orotidine-5′-phosphate (**2.99**: 'orotidylic acid') which is then decarboxylated to give uridine-5′-phosphate (**2.100**: uridylic acid, UMP). Successive transfers of phosphoryl groups from two molecules of adenosine 5′-triphosphate

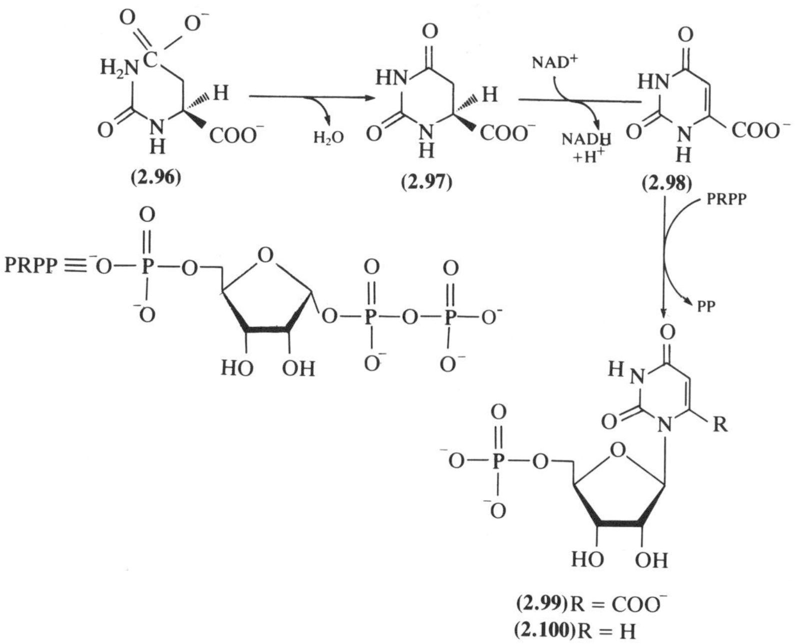

(**2.96**)   (**2.97**)   (**2.98**)

(**2.99**) R = COO^-
(**2.100**) R = H

(ATP) convert UMP to UTP, which is then aminated to cytidine 5'-triphosphate, CTP, by nitrogen transfer from the amide group of glutamine, the energy required for this reaction being supplied by ATP. Cytosine nucleotides are thus formed at the triphosphate level.

The biosynthesis of purines, in contrast, begins with the displacement of pyrophosphate from PRPP by an amino group, again from the amide moiety of glutamine, to give 5-phospho-β-D-ribofuranosylamine (**2.101**), which is then condensed with glycine to afford the phosphoribosylglycinamide (**2.102**). This is then *N*-formylated to (**2.103**) via transfer of the one-carbon fragment from 5,10-methenyltetrahydrofolic acid, and subsequently aminated, again by transfer of an amino group from glutamine, to give the *N*-(phosphoribosyl)amidine (**2.104**). Ring-closure then occurs to afford the 5-aminoimidazole nucleotide (**2.105**), which is carboxylated to give (**2.106**). Condensation of the carboxylate group of (**2.106**) with the primary amino group of aspartate, and subsequent elimination of fumarate, serves

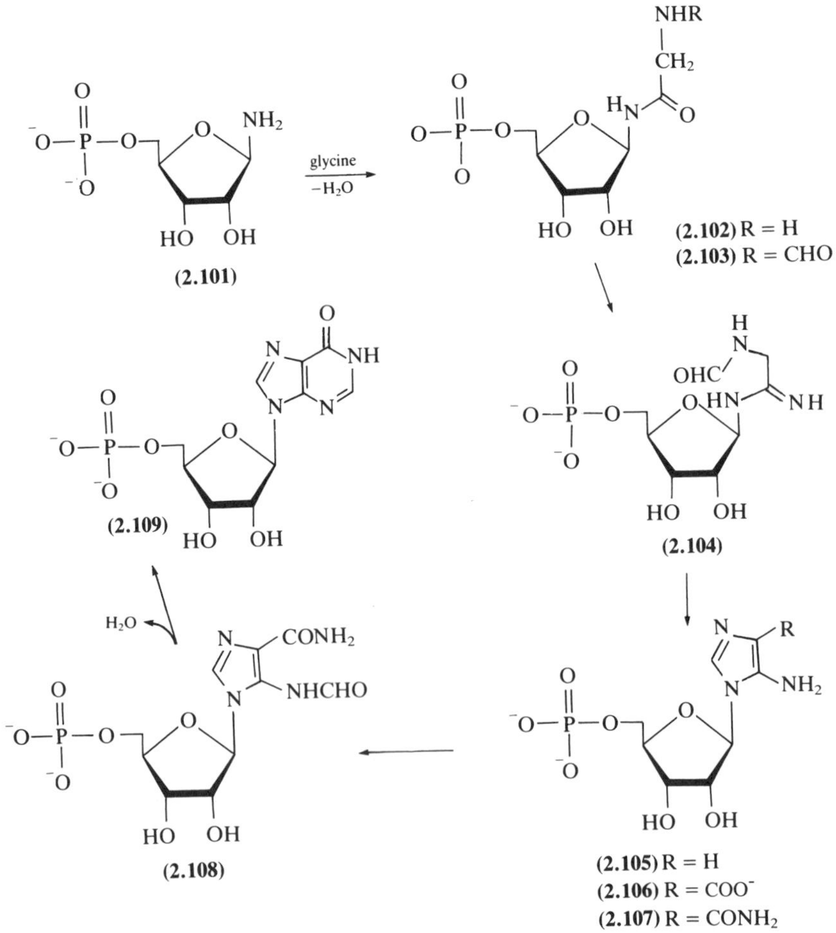

(**2.101**)

(**2.102**) R = H
(**2.103**) R = CHO

(**2.109**)

(**2.104**)

(**2.108**)

(**2.105**) R = H
(**2.106**) R = COO⁻
(**2.107**) R = CONH₂

to convert (**2.106**) to its amide (**2.107**) which is then formylated to (**2.108**) by transfer of a formyl group from 10-formyltetrahydrofolic acid. Ring-closure then affords (**2.109**), in which 6-oxopurine(hypoxanthine) forms a glycosidic bond to ribose-5-phosphate. 9-(β-D-Ribofuranosyl) hypoxanthine is called inosine, and the trivial name of (**2.109**) is inosine-5′-phosphate (IMP, or inosinic acid). IMP may then be aminated, again by the device of condensation with aspartate followed by elimination of fumarate, to give adenosine 5′-phosphate (AMP or adenylic acid) or, alternatively, it may be oxidized in a process involving NAD^+ to give the corresponding 2,6-dioxopurine (xanthine) nucleotide, which is then aminated by transfer of nitrogen from glutamine to give guanosine 5-phosphate (GMP, or guanylic acid). The amination processes, and most of the condensation processes, in these pathways require input of phosphoanhydride bond energy supplied by ATP, or occasionally GTP, and thus illustrate the involvement of nucleotides as energetic molecules and redox coenzymes in the *de novo* biosynthesis of other nucleotides. 2′-Deoxyribonucleotides are generally formed by reduction of ribonucleoside 5′-diphosphates by the action of the enzyme ribonucleotide reductase in a process which involves reduced $NADP^+$ and flavoproteins, although in some bacteria the reduction occurs at the ribonucleoside triphosphate level. The process appears to involve free radicals, but the detailed mechanism is not yet fully known.

2.3.2   Nucleotide synthesis

2.3.2.1   Nucleoside monophosphates

The phosphorylation of nucleosides at the 5′-position is most commonly performed using a phosphoryl halide species. Typically, in early procedures, a nucleoside protected at the 3′- and, if present, 2′-hydroxy group would be treated with, say, dibenzyl phosphorochloridate to afford a phosphotriester such as (**2.110**). Hydrogenolysis of the benzyl groups, and deprotection of the sugar ring with dilute acid, then afforded the nucleoside 5′-mono-phosphate. However, unprotected nucleosides may be phosphorylated predominantly, and frequently exclusively, at the 5′-position, using phos-phoryl chloride in trimethyl or triethyl phosphate at ice temperature. This procedure is effective with ribonucleosides, deoxyribonucleosides, and many sugar-modified nucleoside analogues. The actual phosphorylating agent involved is thought to be the species (**2.111**). The initial product formed is a phosphorodichloridate of type (**2.112**), which is hydrolysed in an aqueous work-up to yield the 5′-monophosphate. If thiophosphoryl chloride is used instead, the corresponding 5′-phosphorothioate is formed. A related procedure, of similar efficiency and regioselectivity, entails treating the unprotected nucleoside with phosphoryl chloride in the presence of water and pyridine or with pyrophosphoryl chloride and pyridinium chloride in acetonitrile as solvent. The phosphorylating agent is here thought to be

(2.110)

(2.111) R = CH$_3$ or CH$_3$CH$_2$

(2.112) R = H or OH

(2.113)

(2.113), and, again, aqueous work-up affords the 5'-monophosphate. It is thought that the protons released in these reactions may confer regioselectivity by increasing the conductivity of the solution and suppressing the reactivity of the secondary alcoholic functions. These are nowadays the most widely used methods for the preparation of nucleoside 5'-monophosphates, although not necessarily the most efficient. A number of rather bulky phosphorylating agents such as bis(2-*t*-butylphenyl) phosphorochloridate, whose regioselectivity for the 5'-position is conferred by steric hindrance, have been described, but, while efficient as phosphorylating agents, they are seldom used: the bulk of the blocking groups which confers regioselectivity tends to suppress their easy removal. However, enzymic phosphorylation of the 5'-position using 4-nitrophenyl phosphate as phosphate donor and a phosphotransferase enzyme from carrots or wheat shoots is sometimes performed, since it offers phosphorylation under rather mild conditions, enzymic specificity, and the capability of recovering unreacted nucleoside quantitatively, even if yields of the nucleotide are not always high. An older method which is still used consists of coupling the 5'-hydroxy group of an otherwise sugar- and base-protected nucleoside with β-cyanoethyl phosphate, using a condensing agent such as DCC or trichloroacetonitrile. Treatment with base then removes the cyanoethyl group as acrylonitrile to afford the corresponding nucleoside 5'-phosphate species. Note that this method avoids exposure to acidic conditions.

Phosphorylation of a nucleoside at any hydroxy group is most efficiently performed by blocking all hydroxy groups other than the one to be phosphorylated, and then using one of the many phosphorylating or phosphitylating species which have been developed for use in oligonucleotide synthesis (see below). Protection of base amino groups may also be required. Care must be taken to ensure that the phosphate group is fully unblocked before the other sugar hydroxy groups are unmasked, since otherwise transesterification leading to the formation of cyclic phosphate species is likely to occur in basic media. Moreover, ribonucleoside 2'- and 3'-phosphates, once synthesized, may not be exposed to acid, since this will result in their mutual interconversion via acid-catalysed intramolecular transphosphorylation.

The most widely used method of forming the 3',5'-cyclophosphate ring found in cAMP and similar species is to treat a nucleoside 5'-monophosphate bearing a free 3-hydroxy group with DCC. Essentially, however, any good leaving group attached to the 5'-phosphate which can be expelled by intramolecular base-assisted attack by the 3'-hydroxy group will lead to formation of the nucleoside 3',5'-monophosphate. For instance, treatment of 2'-deoxythymidine-5'-phosphate with 2,4-dinitrofluorobenzene furnishes initially the 2,4-dinitrophenyl ester of the nucleotide, and the 2,4-dinitrophenyl group is then expelled by fluoride ion to afford the phosphorofluoridate (**2.114**). Treatment with potassium *t*-butoxide then results in formation of the 3',5'-monophosphate with expulsion of fluoride ion. In another useful method, an unprotected nucleoside is treated firstly with trichloromethylphosphonyl chloride in phosphate to afford a trichloro-methylphosphonate of type (**2.115**), which on treatment with potassium *t*-butoxide loses the trichloromethyl anion with ring-closure to form the cyclophosphate.

(**2.114**)   (**2.115**) R = H or OH

Ribonucleoside 2',3'-monophosphates are easily formed by cyclization of a ribonucleoside 2'- (or 3'-) monophosphate with a condensing agent, as described above for the 3',5'-monophosphates.

2.3.2.2 *Nucleoside polyphosphates*

Three methods are commonly used to convert a nucleoside 5'-monophosphate to its 5'-di- or triphosphate. In the first, the monophosphate is treated with morpholine and DCC to afford the corresponding nucleoside-5'-phosphoro-morpholidate (**2.116**). While this is the simplest method of making phosphoromorpholidates, they can be made more efficiently via a 'phospho-triester' approach using a phosphoromorpholidite reagent, for instance, but the simpler (and cheaper!) route remains the one most commonly used. Compound (**2.116**) is, of course, a phosphoramidate, and it is characteristic of this class of compounds that upon protonation of the nitrogen atom the amine component becomes a good leaving group, and the phosphoryl component is consequently transferred to an attacking nucleophile. Thus,

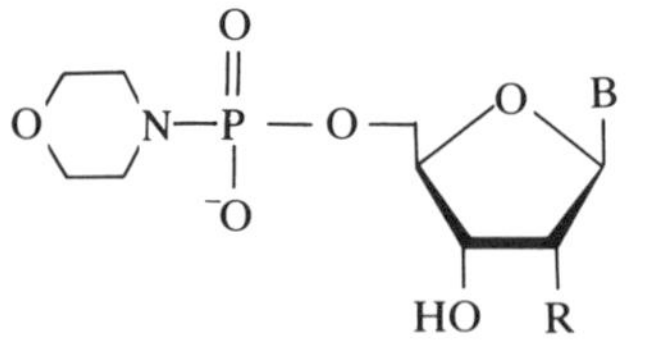

(2.116) R = H or OH 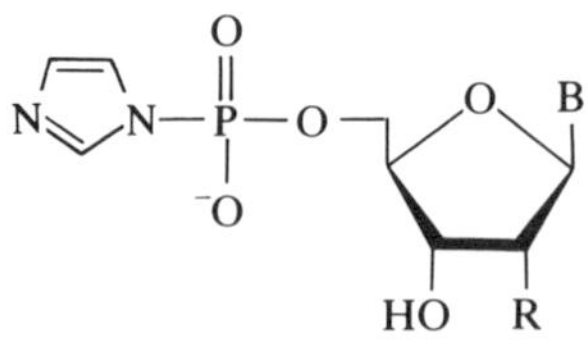   (2.117) R = H or OH

treatment of (**2.116**) with orthophosphate or pyrophosphate in the presence
of a proton source results in the formation of the corresponding nucleoside
5'-di- or triphosphate, respectively. In the second method, the nucleoside
5'-monophosphate is treated with carbonylbis(imidazole) to afford the
5'-phosphorimidazolidate (**2.117**), which possesses high reactivity analogous
to that of the phosphoromorpholidate, and may be converted to the di- and
triphosphate under similar, but milder, conditions. If a ribonucleoside
5'-monophosphate is used with excess carbonylbis(imidazole), the 2'-,3'-
cyclic carbonate of the 5'-phosphorimidazolidate may be formed as a
by-product, but after formation of the di- or triphosphate, the carbonate
ring is easily destroyed at pH 10.5. In the third method, the nucleoside
5'-monophosphate is treated with diphenylphosphorochloridate to afford the
pyrophosphate triester (**2.118**), which on treatment with orthophosphate or
pyrophosphate affords the corresponding di- or triphosphate.

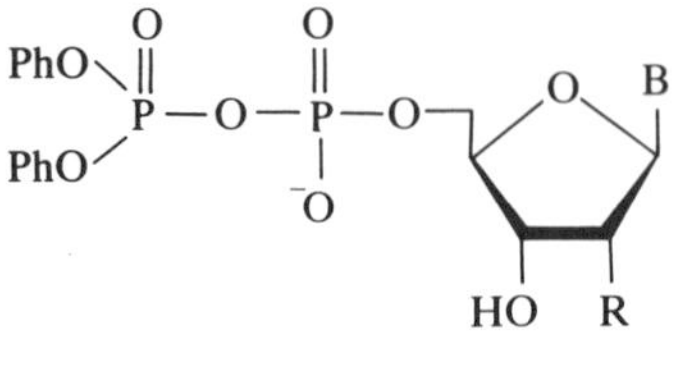

(**2.118**) R = H or OH

The nucleophilic species used to attack (**2.116**)–(**2.118**) need not be
restricted to orthophosphate and pyrophosphate, of course! For instance,
treatment of adenosine 5'-phosphoromorpholidate with ATP affords
P^1,P^4-bis(adenosine-5') tetraphosphate, a compound of great interest as a
cellular growth regulator, and analogues of this and many other P^1,P^n-
bis(aikyl)polyphosphates, both symmetrical and unsymmetrical, have been
made similarly.

It must be pointed out that these are by no means the only methods used
to prepare nucleoside polyphosphates. Nucleoside 5'-(S-methyl or S-ethyl)-
phosphorothioates may also be used, for instance. In the presence of a suitable
oxidizing agent (the choice is determined to some degree by the exact
compound used) oxidation at sulphur and displacement of the oxidized
alkylthio moiety by phosphate or pyrophosphate affords the nucleoside

5′-di- or triphosphate, as appropriate. Again, di-*n*-butylphosphinothioyl bromide reacts with nucleoside 5′-monophosphates to form species of type (**2.119**) which, on treatment with phosphate or pyrophosphate in the presence of silver ions, yield the appropriate 5′-di- or triphosphates. The biochemist may make use of nucleoside monophosphate or diphosphate kinases to achieve the same ends. However, the three methods described at the start of this section have found the most consistent and successful application during the last 20 years.

(**2.119**) R = H or OH

One further point: it is also possible to introduce a polyphosphate chain directly by displacement of, say, tosylate from a nucleoside 5′-*O*-toluenesulphonate species by the polyphosphate. This method offers no great advantage for preparing regular nucleoside 5′-polyphosphates, but is useful for introducing diphosphate analogues, such as imidodiphosphate or methylene (or substituted methylene) diphosphonates at the 5′-position, to give nucleotide analogues modified in the phosphate chain.

2.3.3 *P-chiral nucleotides*

A number of efficient methods for the synthesis of nucleotides chiral at the phosphorus atom have been described recently and used, for the most part, to define the stereochemical course of enzyme-catalysed reactions. Starting from the pioneering work of Eckstein on P-chiral nucleoside phosphoro-thioates, the developments in synthetic methods for these and for P-chiral phosphates have been indivisible, and form an elegant and stimulating area of research. We shall consider a few representative examples of the many available.

When 5′-*O*-acetyluridine is treated with thiophosphoryl tris(imidazolide), the corresponding 2′,3′-*O*,*O*-thiophosphorimidazolidate is formed, which on hydrolysis with water and deacetylation yields 96 per cent of the *endo*-isomer of uridine-2′,3′-monophosphorothioate (**2.120**) and 4 per cent of its *exo*-diastereoisomer. Such compounds were vital in defining the stereo-chemical course of hydrolysis in reactions catalysed by ribonucleases.

(*S*)-Benzoin (**2.121**), prepared by treating (*S*)-mandelic acid with phenyllithium, may be converted to its ethylene ketal and then hydrolysed

(2.120)

in $H_2{}^{18}O$ to afford $({}^{18}O)$-(S)-benzoin (**2.122**) which, on reduction with lithium aluminium hydride, yields $({}^{18}O)$-*meso*-hydrobenzoin (**2.123**). If this is treated with thiophosphoryl bromide in the presence of pyridine, and then with 2′,3′-*O,O*-diacetyladenosine, either (**2.124**) or (**2.125**) is formed, depending on the amount of pyridine present. Reduction and deacetylation of these compounds using sodium in liquid ammonia then furnishes the (Sp) (**2.126**) and (Rp) (**2.127**) diastereoisomers of adenosine 5′-$({}^{18}O)$ phosphorothioate, respectively. (Note that in accordance with usage, the double bond formally drawn between phosphorus and one of its ligand atoms in $P(V)$ species is omitted in these structures, and is neglected when assigning stereochemistry. To save confusion, the negative charges on chiral phosphate and thiophosphate groups have been omitted from (**2.120**) and (**2.126**)– (**2.134**).) This is a useful and versatile procedure: if $({}^{17}O)$ phosphoryl chloride is used instead of thiophosphoryl bromide, a single stereoisomer of the phosphotriester intermediate (**2.128**) is formed and affords adenosine 5′-$[(S)$-${}^{16}O,{}^{17}O,{}^{18}O]$ monophosphate (**2.129**) upon reduction and deacetylation. The (Rp) diastereoisomer (**2.130**) may be obtained either by using $({}^{17}O)$ *meso*-hydrobenzoin instead of (**2.123**), and $({}^{18}O)$ phosphoryl chloride, or by starting from (R)-mandelic acid. If adenosine-5′-diphosphate, ADP, is used in place of 2′,3′-*O,O*-diacetyladenosine, the P_γ-chiral diastereoisomers of ATP, adenosine 5′-$[(S)$-${}^{16}O,{}^{17}O,{}^{18}O]$ triphosphate (**2.131**) and its (Rp) diastereoisomer (**2.132**) may be obtained. Chiral nucleotides such as (**2.131**) and (**2.132**) have been used to study processes of phosphoryl transfer in ATP-requiring enzymes.

One especially useful property of the ${}^{17}O$ nucleus is its nuclear spin of $5/2$. The associated nuclear quadrupole moment shortens considerably the relaxation time of any ${}^{31}P$ nucleus to which it is attached, resulting in line-broadening of the phosphorus signal affected to the point at which it virtually disappears from the ${}^{31}P$ NMR spectrum. If (**2.129**) is treated with diphenylphosphorochloridate, the diphenylphosphoryl group becomes attached to ${}^{16}O$, ${}^{17}O$, or ${}^{18}O$ atoms with equal probability (cf. (**2.118**)). Treatment with potassium *t*-butoxide then effects ring-closure with loss of diphenylphosphate, including the oxygen isotope to which the diphenylphosphoryl group was bound. Only the isotopomers of the chiral cAMP

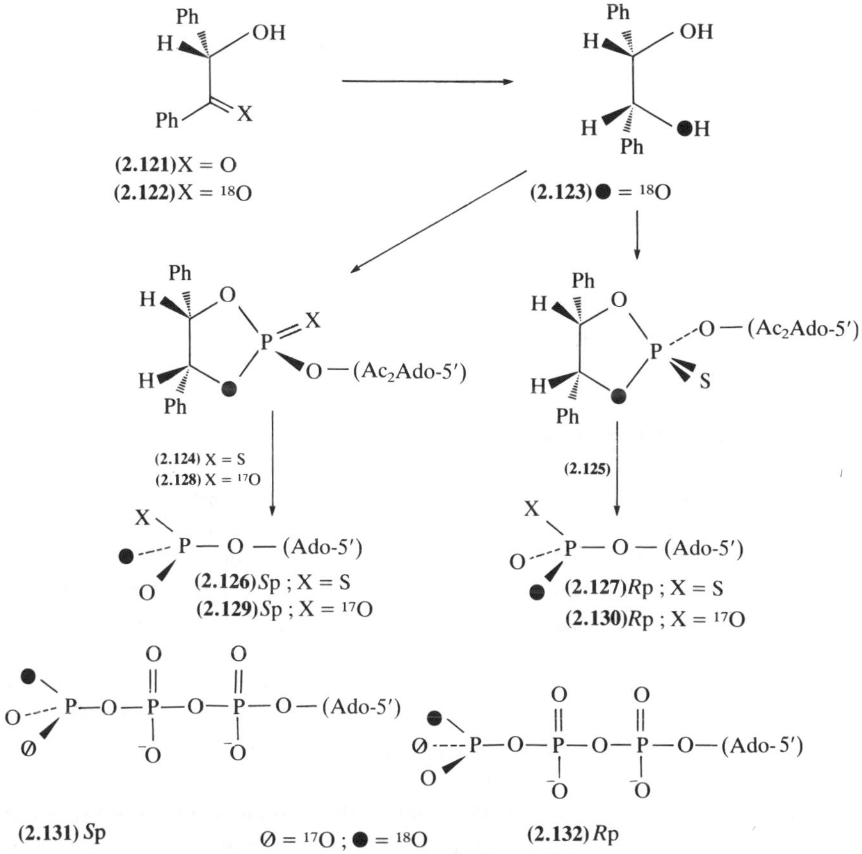

(2.121) X = O
(2.122) X = ^{18}O

(2.123) ● = ^{18}O

(2.124) X = S
(2.128) X = ^{17}O

(2.125)

(2.126) Sp ; X = S
(2.129) Sp ; X = ^{17}O

(2.127) Rp ; X = S
(2.130) Rp ; X = ^{17}O

(2.131) Sp

Ø = ^{17}O ; ● = ^{18}O

(2.132) Rp

formed which have lost the ^{17}O isotope give strong ^{31}P NMR signals, which must therefore be due to (2.133) or (2.134) or a mixture of the two.

Upon esterification by diazoethane, these yield mixtures of the axial and equatorial ethyl esters, which differ in their ^{31}P NMR chemical shifts, the isomers with the esterified oxygen axial exhibiting resonance at higher field. Moreover, the presence of the ^{18}O atom bonded to phosphorus confers a small upfield shift, which is larger when the ^{18}O is doubly bonded to phosphorus (i.e. is not the esterified oxygen) than when it is singly bonded (i.e. the esterified oxygen). Inspection of the ^{31}P NMR spectrum thus permits

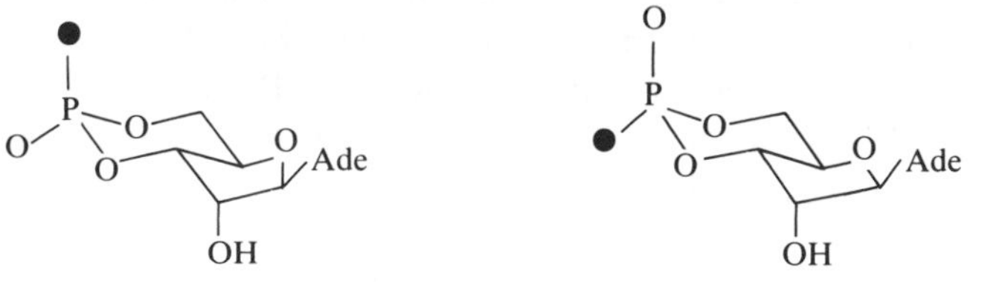

Sp   (2.133)

● = ^{18}O

Rp   (2.134)

the stereochemistry of the adenosine-3',5'-(^{18}O) monophosphate species to be assigned as (Sp) (**2.133**) or (Rp) (**2.134**). In this case, (**2.134**) is found to be formed exclusively from (**2.129**), showing that ring-closure occurs stereospecifically with inversion of configuration at the phosphorus atom. In fact, the situation is complicated by (^{17}O) H$_2$O being available in only *c.* 50 per cent enrichment, but allowance may be made for this when calculating expected peak intensities, and the technique has proved very successful.

Regrettably the lack of space precludes description of the use of P-chiral compounds to investigate the stereochemical course of enzyme catalysis, but the reader is recommended to several excellent reviews.

2.3.4 Some applications of ^{31}P NMR in nucleotide research

The ^{31}P isotope has nuclear spin $I = 1/2$, and the signals of individual phosphorus atoms in polyphosphate chains up to the triphosphate are usually clearly differentiable. ^{31}P NMR spectroscopy is currently of great interest as a technique for investigation of metabolic disturbances *in vivo* as a diagnostic aid and, using surface coils techniques, whole organisms can be examined, the current state-of-the-art being represented by the whole body scanner. In nucleotides the positions of ^{31}P resonances in NMR spectra are dependent on the state of charge of the polyphosphate chain, and consequently on the pH. Thus intracellular pH values *in vivo* may be estimated from the ^{31}P NMR spectra of nucleotides, although it appears that ^{17}O NMR spectroscopy of ^{17}O-labelled nucleotides may provide a more sensitive and accurate method. Self-aggregation of nucleotides in solution, formation of the nucleotide–metal complexes which are the normal substrates for most nucleotide-requiring enzymes and their interaction with the active site, and enzyme kinetics (using a spin-flipping technique) have all been investigated by ^{31}P NMR spectroscopy. Here, too, labelling of the polyphosphate chain with oxygen isotopes has been of considerable value. The uses of ^{17}O and ^{18}O have already been mentioned. The presence of an ^{18}O atom bonded to phosphorus in the polyphosphate chain results in a small upfield shift of the ^{31}P signal, the shift being larger for non-bridging oxygen than for bridging oxygen atoms, and moreover being proportional to the number of ^{18}O atoms bonded. This may be utilized as follows: suppose ADP is treated first with carbonyl bis(imidazole) and then with (^{18}O$_4$) orthophosphate, affording ($\beta\gamma$-^{18}O,γ^{18}O$_3$) ATP (**2.135**), and that this is incubated with an enzyme, a process during which the terminal phosphoryl group is reversibly removed and replaced. If there is free rotation of P$_\beta$ of the intermediate (β-^{18}O$_1$) ADP (**2.136**), randomization of the ^{18}O isotope between the three non-bridging oxygen atoms bonded to P$_\beta$ will occur, and on reformation of the ATP species (now (**2.135**) and (**2.137**)) the P$_\beta$ and P$_\gamma$ resonance position will alter slightly. This and similar experiments on isotopic 'scrambling' and 'washout' have thrown much light on the interactions of nucleotides with

$$\overset{\displaystyle \bullet}{\underset{\displaystyle \bullet}{\overset{\displaystyle \|}{{}^{-}\bullet - \text{P}}}} - \bullet - \overset{\displaystyle O}{\underset{\displaystyle {}^{-}O}{\overset{\displaystyle \|}{\text{P}}}} - O - \overset{\displaystyle O}{\underset{\displaystyle {}^{-}O}{\overset{\displaystyle \|}{\text{P}}}} - (\text{Ado-5}')$$

(2.135)

↓ +enzyme

$$\left[\text{enzyme} - \overset{\displaystyle \bullet}{\underset{\displaystyle \bullet}{\overset{\displaystyle \|}{\text{P}}}} - \bullet \right] + \quad {}^{-}\bullet - \overset{\displaystyle O}{\underset{\displaystyle {}^{-}O}{\overset{\displaystyle \|}{\text{P}}}} - O - \overset{\displaystyle O}{\underset{\displaystyle {}^{-}O}{\overset{\displaystyle \|}{\text{P}}}} - O - (\text{Ado-5}')$$

(2.136)

↓ − enzyme

● = ^{18}O

$$\overset{\displaystyle \bullet}{\underset{\displaystyle \bullet}{\overset{\displaystyle \|}{{}^{-}\bullet - \text{P}}}} - O - \overset{\displaystyle O}{\underset{\displaystyle \bullet}{\overset{\displaystyle \|}{\text{P}}}} - O - \overset{\displaystyle O}{\underset{\displaystyle {}^{-}O}{\overset{\displaystyle \|}{\text{P}}}} - O - (\text{Ado-5}')$$

(2.137)

enzymes, and also on the processes involved in simple chemical hydrolysis of the polyphosphate chains.

2.4 Oligo- and polynucleotides

As a general rule, the naturally occurring oligo- and polynucleotides consist of strings of nucleoside units connected by phosphodiester bridges formed between the 3'-hydroxy group of one unit and the 5'-hydroxy group of the next (**2.138**). An oligonucleotide chain thus has polarity, with a 5'-terminus and a 3'-terminus, and its sequence is conventionally written from 5'-end to 3'-end. Thus adenylyl-(3'-5')uridylyl-(3'-5')-guanosine is represented by ApUpG or (more correctly, by current convention) by A-U-G. Unless it is otherwise specified, the 3'-5'-internucleotidic link is assumed. The presence of a 5'-terminal phosphate residue is indicated by a small 'p', as in pA-U-G, and the same sequence but with a 3'-terminal phosphate residue would be A-U-Gp. The presence of deoxyribonucleoside residues is shown by the prefix 'd', which may precede each individual residue, e.g. dA-dT-dG, or the whole sequence d(A-T-G). An exception to the general rule of 3'-5'-internucleotidic links is found in a 2'-5'-linked oligoadenylate species, (2'-5')pppA-A-A, which is synthesized in cells exposed to one of the class of proteins known as interferons. Among other activities, this curious species binds to and activates

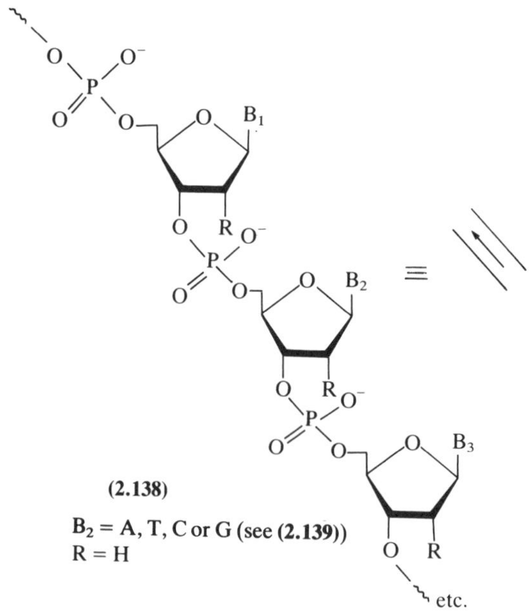

(2.138)

B_2 = A, T, C or G (see **(2.139)**))
R = H

etc.

a specific cellular endoribonuclease which digests messenger RNA (mRNA), thereby preventing its translation in protein synthesis. As a result, the replication of infecting virus may be suppressed in a cell in which interferons are released.

DNA exists most commonly as a right-handed double-stranded helical structure ('B'-DNA) containing 10 base pairs per complete turn of length (or 'pitch') 3.4 nm (**2.139**). The two strands are of opposite polarity, the bases are arranged according to complementarity rules: adenine on one strand forms a base pair with thymine on the other, involving two hydrogen bonds, (**2.140**), and guanine on one strand base pairs with cytosine on the other, forming three hydrogen bonds (**2.141**). These are 'Watson–Crick' base pairs. Alternative arrangements involving hydrogen-bonding to N-7 of the purine bases ('reverse Hoogsteen pairs') are possible, and occur in certain polyribonucleotide triple helices, and possibly also in special triple-stranded structures in DNA. The 'B' form of DNA occurs under conditions of > 66 per cent humidity and high salt concentration and is thought to be the form normally present in cells. Under conditions of lower humidity and salt concentration, a different form of the double helix is seen, with 11 base pairs per turn and a helical pitch of 2.815 nm. This 'A' form is also the characteristic helical arrangement of RNA (**2.138**, R = OH; B_n = A, C, G or uracil (U)), which is normally single-stranded but may curl round on itself to form 'hairpin loops' in which internally self-complementary sequences base-pair, to give a double-helical form. In this case adenine forms a base pair with uracil. It has also been found that certain sequences rich in guanine and

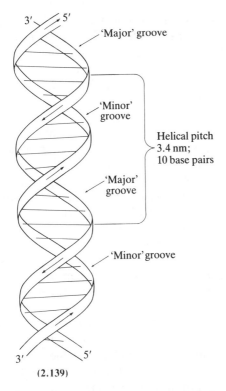

(2.139)

B-DNA: the strand polarity and designation of 'Major' and 'Minor' grooves are shown. Each quasi-horizontal line (actually tilted at 6° to a plane perpendicular to the helical axis) represents an adenine–thymine (A–T; **2.140**) or guanine–cytosine (G–C; **2.141**) base pair.

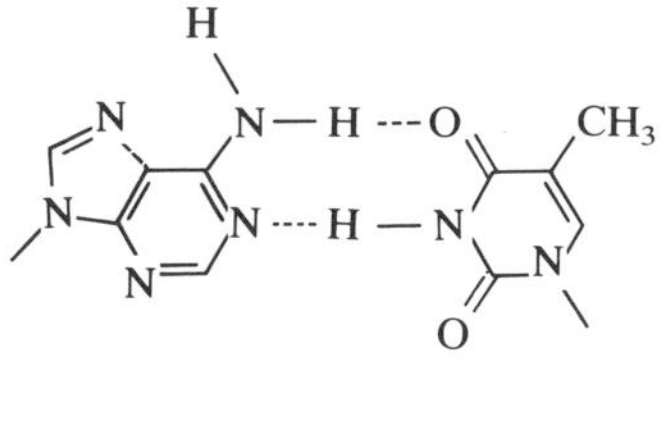

A   T   G   C

(2.140) 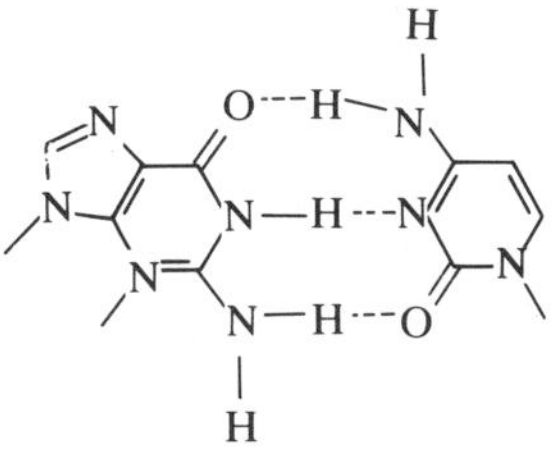   (2.141)

cytosine residues, such as the self-complementary alternating copolymer poly d(G-C), can assume a left-handed helical arrangement in high salt concentrations, in which the backbone has a characteristic 'zig-zag' arrangement which has been styled 'Z'-DNA. This helix contains 12 base pairs per turn, with a pitch of about 4.35 nm. Evidence using specific antibodies raised against 'Z'-DNA suggests that this helical form occurs naturally *in vivo*, and its possible significance is a source of great interest and conjecture. Most recent work has revealed further subtleties in nucleic acid structures, and the reader is advised to seek up-to-date reviews (see 'further reading').

2.4.1   Biosynthesis

The biosynthesis of DNA involves semi-conservative replication of the double helix. Specific proteins bind to 'initiation points' on the double helix, resulting in partial 'melting', or separation of the strands, to form replication 'forks' (**2.142**). The parting of the double helix to bare the complementary sequences of the bases and, thus, the genetic information on its constituent strands makes available the template sequences required for replication of the DNA. Essentially an enzyme, DNA polymerase, synthesizes a new strand from its 5'-end, elongating the chain in a 3'-direction. The enzyme uses 2'-deoxynucleoside-5'-triphosphates as substrates, and requires a 3'-hydroxy group on a primer oligonucleotide which is complementarily base-paired to the single-stranded template in accordance with the Watson–Crick base-pairing rules (adenine to thymine, cytosine to guanine) before synthesis can commence. Then, the 3'-hydroxy group is thought to displace pyrophosphate from the 5'-triphosphate moiety of the next residue added, to form the internucleotidic link, this being the simplest interpretation of the experimental observation that inversion occurs, as a rule, at the phosphorus atom. The newly added residue in turn forms a complementary base pair with the residue lying opposite on the template strand. Accidental misincorporations at this new terminus are excised by 3' → 5' exonuclease 'proofreading' enzyme activities: it is essential that the

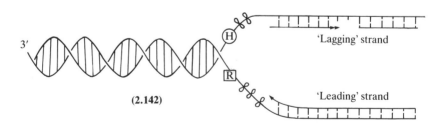

(**2.142**)

DNA replication: DNA is unwound by the combined effects of helicase H and *Rep* protein R with tetrameric single-strand binding protein: maintaining the single strands apart while replication by DNA polymerase takes place on short RNA primers complementary to the parental strands (see text). Removal of the RNA primers from the 5'-termini of the 'Okazaki fragments' thus formed is followed by gap-filling and ligation to give continuous strands.

fidelity of DNA replication should be very high, in order that the genetic information be preserved intact. The DNA polymerases of *Escherichia coli* have been studied very carefully. The main work of DNA replication *in vivo* is thought to be performed by a complex enzyme, DNA polymerase III*, together with an auxiliary protein, copolymerase III*. However, a more useful and easily handled enzyme, DNA polymerase I ('pol I') is normally used for *in vitro* work, and seems to function primarily to fill gaps in DNA sequences during repair processes *in vivo*. In a single polypeptide chain of molecular weight 109 000, it contains not only the polymerase activity, but also $3' \rightarrow 5'$ and $5' \rightarrow 3'$ exonuclease activities. This chain can be fragmented, to give two portions, and the larger portion, known as 'Klenow fragment', which contains the polymerase and $3' \rightarrow 5'$ exonuclease functions, is much used in recombinant DNA research. The more complex eukaryotic cells contain nuclear DNA polymerases α, β, and γ, which perform equivalent functions to the polymerases I, II, and III of prokaryotes.

DNA is, in fact, made discontinuously, as a number of small pieces called 'Okazaki fragments' after their discoverer. In each case a short RNA primer complementary to part of the template strand sequence is first synthesized, and the DNA fragment is then built up, one residue at a time, at the 3'-end of this primer, as indicated above. Then the RNA primers are digested away by specific ribonucleases, or by the $5' \rightarrow 3'$ exonuclease function of pol I, leaving a number of short DNA fragments forming base-paired duplexes with the original, unbroken, template strand. The gaps between these fragments are then filled in by the polymerase functions of pol I and, for small gaps, pol III, until eventually the 3'-OH terminus of each augmented fragment lies adjacent to the 5-phosphate terminus of the next fragment, the bases of the two terminal nucleotides being pair complementary with adjacent bases on the template strand. Each two adjacent fragments are now joined by the enzyme DNA ligase (Latin, *ligare*, to bind). The enzyme isolated from *E. coli* uses NAD^+ as substrate, from which the 5'-phosphate group of one DNA fragment displaces nicotinamide mononucleotide to form a P^1-(5'-adenosyl)-P^2-(5'-oligonucleotidyl) pyrophosphate. This unsymmetrical pyrophosphate is then attacked, probably directly, by the 3'-hydroxy group of the other DNA fragment, with displacement of AMP, to form the internucleotidic link (**2.143**). Not all DNA ligases use NAD^+ as substrate: the enzymes from eukaryotes, and also the form encoded by bacteriophage T4, use ATP. However, in summary, a DNA strand is replicated by DNA polymerase, and its nicks sealed by DNA ligase.

RNA is transcribed from a DNA template by the enzyme RNA polymerase, and again the enzyme from *E. coli* has been studied most thoroughly. Synthesis occurs from the 5'-terminus in the direction of the 3'-terminus, with ribonucleoside 5'-triphosphates being used as substrates. Synthesis begins at an initiation site called a 'promoter' site without a primer being required, and commences with insertion of a purine ribonucleoside 5'-triphosphate as the 5'-terminus, followed by a pyrimidine residue. The nascent RNA molecule thus starts ppp(Purine) p (Pyrimidine) .... The chemistry of the process is

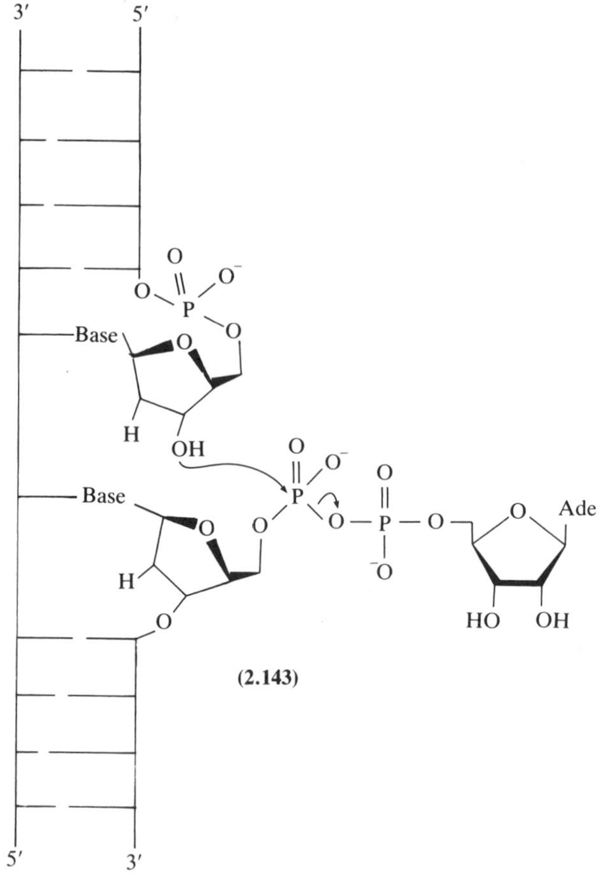

(2.143)

DNA ligation: AMP is displaced to seal the gap in the DNA backbone. For diagrammatic purposes, the base-pair spacing and the glycosidic bond have been distorted.

essentially the same as that seen in DNA replication, with successive additions of the ribonucleotides with bases complementary to those on the (antiparallel) DNA template strand, excepting that uracil here base-pairs with adenine. RNA synthesis is continuous, although strand elongation may be subject to pauses, the significance of which is not clear. The RNA transcripts formed are then processed to form the functional RNA molecules. For transfer RNA (tRNA), which carries amino acids to the ribosome in protein synthesis and contains a high proportion of modified nucleosides, this process involves trimming by ribonucleases and modification of specific nucleoside residues by specialized enzymes. Ribosomal RNA (rRNA) molecules occur in a number of discrete sizes, formed by the processing by nucleases of a much larger precursor molecule. Finally, messenger RNA (mRNA), which carries the code to direct protein synthesis, is formed in eukaryotes by the removal of a (usually) large number of non-translated intervening sequences (or 'introns') from the sequence originally transcribed. The remaining, expressed sequences ('exons') are joined together ('RNA splicing') to form the coding mRNA product. A recent remarkable discovery is that in some cases, RNA can perform self-splicing, resulting in the excision of an intron without the agency of an enzyme.

This means that RNA can possess catalytic properties which are normally regarded as the special function of the enzymes, and the name 'ribozyme' has been coined for catalytically active RNA. The ribozymes which have been characterized up to now are rather specific RNA sequences catalysing either one cleavage reaction, or a closely related set of reactions. However, the finding that the nucleic acids are by no means chemically inert lumps of information has promoted interested speculation as to whether their activity has been significant in chemical evolution.

Certain animal viruses, the retroviruses, contain only single-stranded RNA molecules, and their mode of replication involves in some cases the formation of a complementary DNA (cDNA) strand which is synthesized from 5'-end to 3'-end by using the single-stranded RNA as template, a short primer, and deoxynucleoside triphosphates as substrates. The enzyme which performs this process is called RNA-directed DNA polymerase or, most commonly, reverse transcriptase, and the enzyme encoded by avian myeloblastosis virus, in particular, is of value in genetic engineering, since it can be used to form double-stranded DNA from a single-stranded mRNA template.

The retroviruses have recently come under acute scrutiny since HIV (the causative agent of AIDS) belongs to this class, and several of the drugs showing significant activity against HIV, including AZT (**2.42**) and some 2',3'-dideoxy- and 2',3'-dideoxy-2',3'-didehydronucleosides (cf. (**2.82**) and (**2.83**)), appear to function by being converted *in vivo* into their 5'-triphosphates which are powerful inhibitors of HIV reverse transcriptase. Moreover, if misincorporated at the end of a growing DNA chain, these nucleotides necessarily terminate the chain since no 3'-hydroxy group is available for further chain extension.

Terminal deoxynucleotidyl transferase is an unusual DNA polymerase, in that when supplied with a deoxynucleoside 5'-triphosphate, as substrate, it will progressively add a homopolymer tail to a 3'-hydroxy group located at a non-base-paired 3'-terminus of a DNA chain or duplex. No template strand is necessary. Thus, if dTTP is supplied, an oligo (dT) tail is added; if dATP, then an oligo (dA) tail is formed; and so on.

It will be noted that in all these polymerization reactions, pyrophosphate is displaced from the nucleoside triphosphate substrates. This is subsequently degraded to orthophosphate by the enzyme pyrophosphatase. Were this not so, the polymerization reactions could be reversed by pyrophosphorolysis of the polynucleotides, a process which would have disastrous consequences *in vivo*. The action of pyrophosphatase thus renders the polymerization reactions irreversible.

2.4.2   Oligonucleotide synthesis

Oligonucleotide synthesis is most commonly a matter of stepwise chain elongation, in which a phosphodiester bridge is introduced, usually as a $3' \rightarrow 5'$ link, between two protected nucleosidic or nucleotidic fragments, and the product is used in its turn as one component of a further similar reaction of chain elongation. Since construction of the internucleotidic link involves specific phosphorylation of a 3'- or 5'-OH group, any other similarly reactive sites in the molecule must be masked by protecting groups, which must be capable of subsequent specific removal, either to permit the next stage of chain elongation, or during final deblocking to afford the unprotected oligonucleotide, under conditions in which neither the nucleoside units nor the phosphate links are affected. The correct choice of protecting groups, and their manipulation, is therefore essential.

2.4.2.1   Protection of sugars and bases

The amino functions of the nucleoside bases are usually protected as amide derivatives: the N^6-function of adenine by the benzoyl group, the N^4-function of cytosine by the anisoyl group (or often, also, the benzoyl group), and the N^2-function of guanine by the isobutyryl group. These may be introduced as follows: 2'-deoxyadenosine, on treatment with benzoic anhydride in the presence of pyridine, affords $N^6,O^{3'},O^{5'}$-tribenzoyl-2'-deoxyadenosine (**2.144**). On treatment with 2 M sodium hydroxide, the comparatively labile ester groups are hydrolysed to leave N^6-benzoyl-2'-deoxyadenosine (**2.145**). Similar procedures may be used to introduce the other base-protecting groups, all of which may be removed using concentrated ammonia.

However, the introduction of the acyl functions at the exocyclic amino groups heightens the risk of depurination during oligodeoxyribonucleotide synthesis, while the use of concentrated ammonia to remove the protecting acyl groups at the conclusion of oligonucleotide synthesis may lead to cleavage of the internucleotidic phosphodiester bonds, particularly in oligoribonucleotides. There is presently an increasing trend to using the phenoxyacetyl group to protect the amino functions on the bases (e.g. as in (**2.146**)), since it is removed using ammonia in mild conditions and depurination and internucleotidic cleavage associated with its use are reportedly minimal.

The 5'-OH group of the sugar residue is most commonly protected by the acid-labile triphenylmethyl (trityl) functions, more specifically the (4-methoxyphenyl)diphenylmethyl ('monomethoxytrityl') and bis(4-methoxyphenyl)phenylmethyl ('dimethoxytrityl' DMT) groups. Each substitution of a 4-methoxy group of a phenyl ring of the trityl group increases the lability of trityl alkoxides to acid by about tenfold. The dimethoxytrityl group is the most widely used compromise between efficiency as a protecting group during conditions of oligonucleotide synthesis and ease of removal. It is introduced by treatment of the base-protected nucleoside with dimethoxytrityl chloride.

(2.144) R¹ = R² = R³ = PhCO
(2.145) R¹ = R² = H; R³ = PhCO
(2.146) R¹ = R² = H; R³ = PhOCH₂CO
(2.147) R¹ = DMTr; R² = H; R³ = PhCO

(2.148) *B = Protected base

Thus, treatment of (**2.145**) with one equivalent of this reagent in the presence of pyridine results in alkylation of the primary (and sterically more available) 5'-OH group to give the 5'-*O*-dimethoxytrityl species (**2.147**), and the same procedure is applied when preparing other 5'-*O*-tritylated species. The trityl groups are removed by acidic reagents, such as 80 per cent acetic acid, or 2 per cent benzenesulphonic acid, or di- or trihaloacetic acids. However, these carry some attendant risk of depurination where deoxynucleoside species are concerned, and much work has been focused on determining conditions which allow facile detritylation without depurination. Zinc bromide in methylene chloride–isopropanol can be used for this purpose, although the use of 3 per cent dichloroacetic acid in dichloromethane, or 3 per cent trichloroacetic acid in nitromethane–methanol, or trifluoroacetic acid in 1,2-dichloroethane have all been described as giving rapid detritylation without depurination and are more often employed.

The 9-fluorenylmethoxycarbonyl group has found particular application for protecting the 5'-OH group of ribonucleoside units for oligoribonucleotide syntheses in which, in addition, an acid-labile protecting group is present at the 2'-OH function (e.g. as in (**2.148**)). It is readily removed using DBU (1,8-diazabicyclo(5.4.0)undec-7-ene) in acetonitrile, conditions which are highly convenient for use in solid-phase oligonucleotide synthesis by the 'phosphite' and 'phosphoramidite' approach, but side-reactions prevent its use in the 'H-phosphonate' procedure (see below).

Oligoribonucleotide synthesis requires that the 2'-OH group of the nucleoside should also be protected. Acid-labile functions such as the tetrahydropyranyl (as in (**2.148**)), 4-methoxytetrahydropyranyl, and tetrahydrofuranyl groups have been used for this purpose. Firstly, a base- and 3',5'-protected ribonucleoside must be constructed. For instance, treatment of adenosine with triethyl orthoacetate and *p*-toluenesulphonic acid yields the 2',3'-ethoxyethylidene derivative (**2.149**), which upon benzoylation followed by treatment with dilute acetic acid affords (**2.150**) together with its 2'-*O*-acetyl positional isomer, as a separable mixture. The 2'-OH protecting group is then introduced using 2,3-dihydro-4*H*-pyran (**2.151**), or

NH$_2$

HO—

(2.149)

H$_3$C   OCH$_2$CH$_3$

NHCOPh

PhCOO—

CH$_3$COO   OH

(2.150)

OCH$_3$

(2.151)

(2.152)

(2.153)

4-methoxy-5,6-dihydro-2H-pyran (**2.152**), or 2,3-dihydrofuran (**2.153**), in the presence of an acid catalyst such as *p*-toluenesulphonic acid. The 3′- and 5′-OH groups may then be unblocked and re-protected as desired. Alternatively, the ribonucleoside may be subjected to limited silylation, using *t*-butyldimethylsilyl (TBDMS) chloride, and the isomeric 2′,5′- and 3′,5′-*O*,*O*-bis (TBDMS) derivatives separated. The 3′,5′-isomer may then be protected at the 2′-OH, as above. The silyl groups may then be removed by treatment with fluoride ion. Alternatively, hydrolysis of the 2′,5′-*O*,*O*-bis (TBDMS) isomer with 80 per cent acetic acid results in selective removal of the 5′-silyl protection, leaving the 2′-OH group protected by TBDMS. The TBDMS group itself is a valuable protecting group in oligonucleotide synthesis, and may be used at any position. During oligonucleotide synthesis it is conveniently removed using tetra-*n*-butylammonium fluoride (TBAF) in tetrahydrofuran. Its use as a protecting group for the 2′-OH function is preferred by some to that of the acid-labile groups described above, a drawback being that its bulk may sterically hinder efficient coupling during the construction of the internucleotidic link.

The 2-nitrobenzyl group has also been used to protect the 2′-OH position, being introduced using 2-nitrobenzyl bromide and sodium hydride, and best removed by irradiation at pH 3.5 with light of wavelength > 300 nm.

2.4.2.2   *The phosphodiester method*

The procedure for oligonucleotide synthesis known as the 'phosphodiester method' is so called because the intermediates separated following each coupling stage are phosphodiesters. For instance, 5′-*O*-monomethoxytrityl-2′-deoxythymidine may be coupled with 3′-*O*-acetyl-2′-deoxythymidylic acid (formed by acetylating 2′-deoxythymidine 5′-phosphate (dTMP)) using 2,4,6-triisopropylbenzenesulphonyl chloride (TPS-Cl) to afford (**2.154**). The process is thought, from ^{31}P NMR evidence, to involve formation of a mixed sulphonic–phosphoric anhydride which reacts with pyridine present in the mixture to form a pyridinium phosphoryl species (**2.155**). Pyridine is then displaced upon attack of the free 3'-OH group to form (**2.154**). Note that the phosphate must be activated with the coupling agent prior to addition of the 3'-OH component, since otherwise sulphonation of the latter will occur as a competing reaction. Once (**2.154**) is formed, treatment with dilute ammonia will remove the acetyl group to give (**2.156**), which can be coupled with another 3'-*O*-acetylated nucleoside-5'-phosphate, and the chain thus extended. Finally, the sugar and base-protecting acyl groups and the monomethoxytrityl group may be removed as described previously.

In this example the chain was extended in the 3'-direction using 5'-nucleotide units blocked at the 3'-position. This is more useful for oligodeoxyribonucleotide synthesis, since in oligoribonucleotide synthesis a bulky group protecting the 2'-OH function may hinder coupling. The chain could alternatively be extended in the 5'-direction using 3'-nucleotide units blocked at the 5'-position. In such a case, (**2.154**) would arise from coupling 5'-*O*-monomethoxytrityl-2'-deoxythymidine-3'-phosphate with 3'-*O*-acetyl-2'-deoxythymidine, and the chain would be extended by detritylation of (**2.154**), followed by coupling to the next 5'-*O*-monomethoxytritylnucleoside-3'-phosphate.

The phosphodiester method suffers from several major disadvantages. Firstly, the products from each stage of the sequence are ionized species,

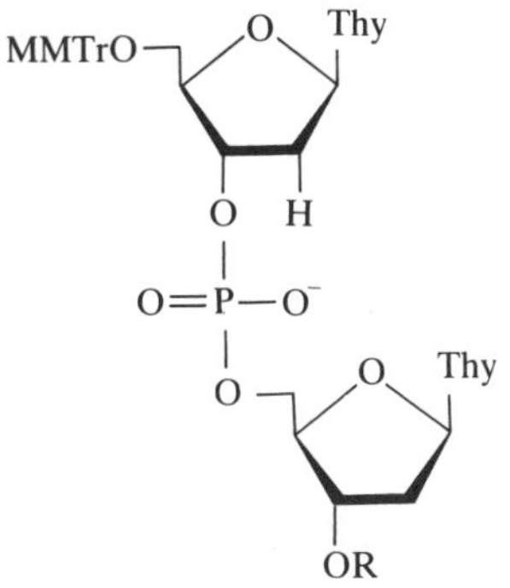

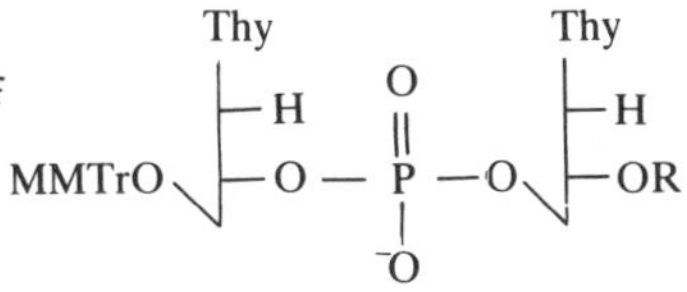

(**2.154**) R = CH$_3$CO
(**2.156**) R = H

MMTr (monomethoxytrityl) ≡

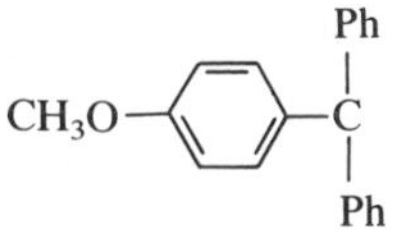

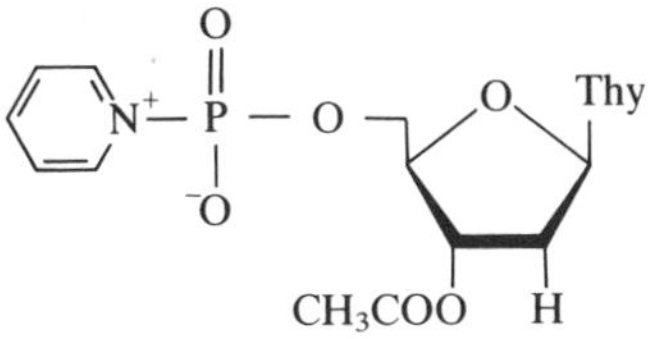

(**2.155**)

and commonly require to be purified using ion-exchange chromatography. This is not only laborious, but also effectively limits the quantity of material which can conveniently be handled. Secondly, the ionized species require to be manipulated in aqueous buffers, and the protecting groups may be lost in these conditions. Thirdly, and most seriously, the nucleophilic oxyanions of the phosphodiesters formed, and also the protecting groups, may react with the condensing agents employed to form the phosphodiester links, leading to condensed by-products such as pyrophosphates and phosphoramidates, which may in turn give rise to unwanted modifications at the phosphodiester links, chain cleavage, deprotection at the protecting groups, etc. The consequent losses in yield, added to manipulative difficulty, render the phosphodiester method an unrealistic approach for lengthy iterative syntheses, and it is nowadays rarely used, and only for very short oligomers such as dinucleoside monophosphates. The disadvantages associated with its use are largely eliminated in the phosphotriester method, which, in its solid-phase methodology, particularly using phosphoramidite intermediates, represents the current state-of-the-art.

Notwithstanding its disadvantages, the earliest syntheses of genes were performed using oligonucleotides prepared by the phosphodiester method, and its historical importance is beyond dispute.

2.4.2.3 *The phosphotriester method*

In the phosphotriester method, all oligonucleotidic intermediates are isolated as phosphotriesters which, being electroneutral, may be purified readily using silica gel chromatography, and particularly high performance liquid chromatography (HPLC). This permits faster purification, and the handling of larger quantities, than can normally be realized using ion-exchange methods.

Firstly, a base-protected 5'-*O*-dimethoxytrityl-2'-deoxynucleoside such as (**2.147**) is treated with a monofunctional phosphorylating agent such as (**2.157**), in the presence of a base such as *N*-methylimidazole, to give the phosphotriester (**2.158**). Species such as (**2.157**) are prepared by successive displacements of chloride ion from phosphoryl chloride with one equivalent of each of the appropriate phenols or alcohols, in the presence of base. Now, the dimethoxytrityl group may be removed from (**2.158**) using one of the reagents described above, to give (**2.159**), and the cyanoethyl group removed from (**2.158**) using triethylamine in pyridine to give (**2.160**). The protected base is essentially unaffected by these processes. Then, (**2.159**) and (**2.160**)

(**2.157**)

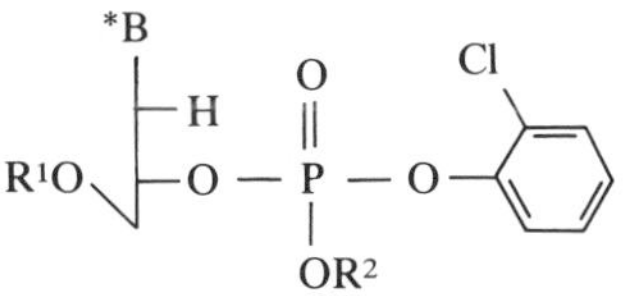

(2.158) R^1 = DMTr; R^2 = CH_2CH_2CN
(2.159) R^1 = H; R^2 = CH_2CH_2CN
(2.160) R^1 = DMTr; R^2 = H

In these and subsequent formulae, *B is a protected base.

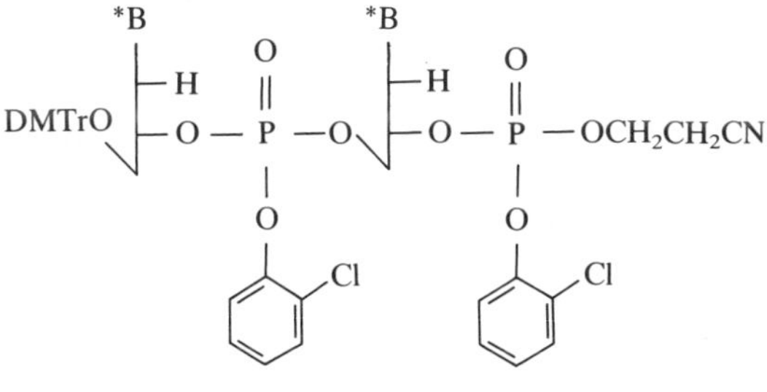

(2.161)

are coupled together using a condensing agent, to give (2.161), which is, of course, electrically neutral and consequently easily purified as indicated above. It may, in its turn, be deblocked at the 5′-end and coupled to a further unit of type (2.160), or at its 3′-end and coupled to a further unit of type (2.159). Alternatively, 5′- or 3′-deprotected dimer blocks of type (2.161) may be used as the chain-augmenting units, thus permitting the chain to be lengthened two residues at a time, and larger blocks may also be coupled. There is good reason for this: as the oligonucleotides grow longer, chromatographic separation is facilitated by using oligomer blocks to extend the chain, since the differential between R_F values of reagents and products is thereby increased.

The condensing agents used to couple units such as (2.159) and (2.160) are almost always arylsulphonyl species. Mesitylenesulphonyl-$1H$-3-nitro-1,2,4-triazolide (2.162) or the corresponding tetrazolides are preferred, since they give rapid condensation, high yields, and few side-reactions (such as sulphonation of the uracil or thymine bases at O^4, or guanine at O^2, or of the 5′-OH group of (2.143) or its equivalent in the condensation mixture). These reagents are prepared by treatment of the corresponding arylsulphonyl chloride with the appropriate nitrogenous base. The coupling reactions are

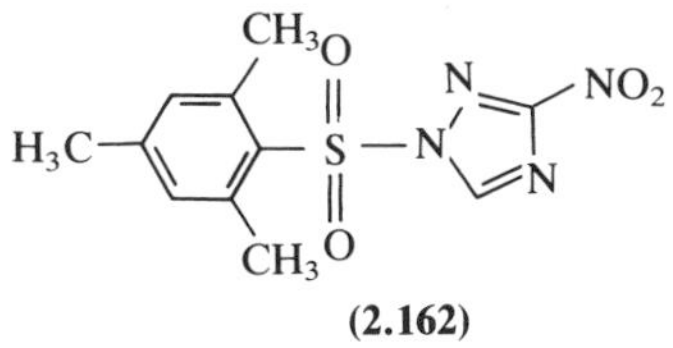

(2.162)

also catalysed by bases such as 4-dimethylaminopyridine (DMAP), N-methylimidazole, and tetrazole. On the basis of ^{31}P NMR investigations, it has been suggested that (**2.160**) reacts with the condensing agent to form a mixed sulphonic–phosphoric anhydride, which is attacked by base (pyridine, DMAP, N-methylimidazole, or tetrazole) to form the corresponding phosphorylated base (in analogy to (**2.155**)) with displacement of the sulphonate. The base moiety is in turn displaced by attack of (**2.159**) to complete the esterification and form the internucleotidic link. Variants of this sequence have also been suggested.

The above procedure describes a typical phosphotriester condensation, but many agents similar to (**2.157**) have been synthesized and used. Often the aryl component in these reagents is the 2- or 4-chlorophenyl group, but the nature of the other group, which is removed prior to each coupling stage in order to unmask the phosphate dissociation, may vary widely. The 2,2,2-trichloroethyl and 2,2,2-tribromoethyl groups have been used in place of cyanoethyl: both are removed specifically using a zinc–copper couple, or zinc in acetylacetone. Irrespective of the protecting groups used, however, the fundamental approach is the same in each case. An alternative route to species such as (**2.158**) lies in treating the base-protected 5'-O-dimethoxytrityl-2'-deoxynucleoside with one equivalent of the aryl phosphorodichloridate, or arylphosphorobis(triazolidate) and then treating the aryl nucleosidyl phosphorochloridate (or phosphorotriazolidate) thus formed with the desired protecting alcohol or amine: e.g. with β-cyanoethanol, to form (**2.158**).

2.4.2.4   The 'phosphite' procedure

During the 1980s, a rapid, efficient, and versatile variant of the phosphotriester method evolved and proved readily amenable to automation for solid-phase synthesis. In this form it has become the procedure of choice for making small quantities of oligodeoxyribonucleotides such as are widely used in sequencing, site-directed mutagenesis, genetic engineering and other aspects of research. Essentially the oligonucleotide chain is elongated to form each internucleotidic link initially as a phosphite triester, which is oxidized to phosphate during each elongation cycle.

The solid phase is generally porous silica or controlled pore glass (CPG), derivatized to carry a spacer arm with a primary amino group at its terminus, as, for example, in aminopropylsilica. This is treated with succinic anhydride to form succinamidopropylsilica, and the free carboxy group is then condensed with the 3'-hydroxyl group of a base-protected 5'-O-dimethyltrityl-2'-deoxynucleoside, typically using DCC, to afford (**2.163**). After detritylation as described previously to give (**2.164**), synthesis proper is performed, using base-protected 5'-O-dimethoxytrityl-2'-deoxynucleoside-3'-O-methyl (or 2-cyanoethyl)-N,N-diisopropylphosphoramidite (**2.165**) units for each elongation step, in the presence of a base, usually tetrazole. Model reactions suggest that the tetrazole behaves initially as a proton donor, protonating

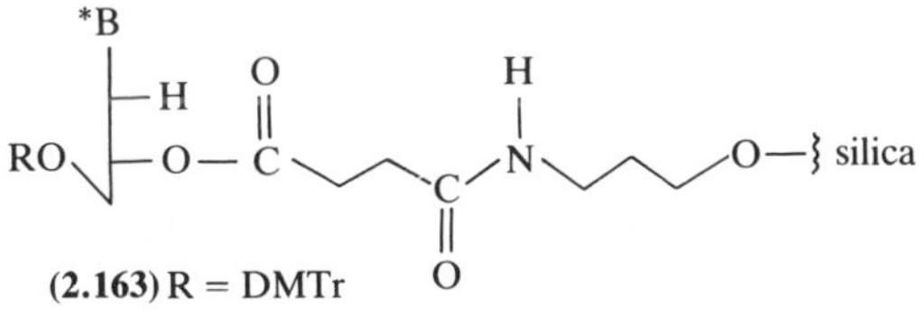

(2.163) R = DMTr
(2.164) R = H

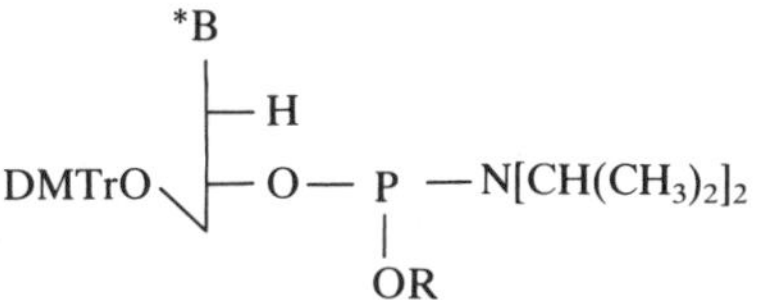

(2.165) R = CH$_3$ or CH$_2$CH$_2$CN

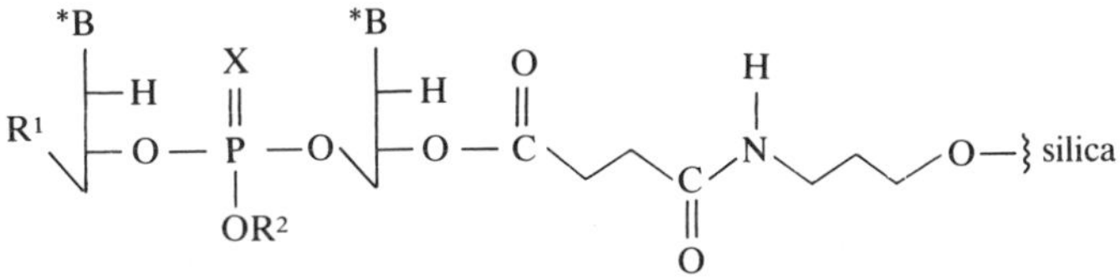

(2.166) R^1 = DMTr; R^2 = CH$_3$ or CH$_2$CH$_2$CN; X is absent
(2.167) R^1 = DMTr; R^2 = CH$_3$ or CH$_2$CH$_2$CN; X = O
(2.168) R^1 = H; R^2 = CH$_3$ or CH$_2$CH$_2$CN; X = O
(2.169) R^1 = DMTr; R^2 = CH$_3$ or CH$_2$CH$_2$CN; X = S

(2.165) and then displacing diisopropylamine to form a phosphotetrazolidite which is the actual reactive species. Coupling is rapid and, in the presence of excess (2.165), near-quantitative, to give (2.166). Following coupling, treatment with acetic anhydride is performed in a 'capping' step, to acylate any unreacted (2.164) which might otherwise become elongated in subsequent elongation cycles, giving rise, eventually, to spurious 'failure' sequences contaminating the desired product sequence. After flushing away excess reagents, oxidation of (2.166) to the corresponding phosphate (2.167) is performed, commonly using aqueous iodine, after which flushing with dry reagents, and detritylation with, for instance, dichloroacetic acid, is performed to give (2.168), ready for the next elongation cycle. The procedure is very rapid, an elongation cycle being completed in only a few minutes, and highly efficient, with coupling yields of 98 per cent or better. An added versatility consists in the possibility of using different agents in the oxidation step: if aqueous iodine is replaced by sulphur in pyridine, phosphorothioate links are formed (e.g. in (2.169)) and may thus be introduced *ad lib.* in oligonucleotide sequences. Alternatively, if water containing an oxygen isotope other than oxygen-16 is used with iodine, the oxygen isotope may be introduced at a specific phosphodiester link as desired. Oligonucleotides containing such specifically modified links have afforded much useful information on structure and enzymic degradation of oligonucleotides.

2.4.2.5 The 'H-phosphonate' procedure

Introduced in the mid-1980s, this variant route to oligonucleotide synthesis offers speed and efficiency comparable with that of the phosphite procedure and has found use in lengthy syntheses. The monomer building blocks are *H*-phosphonates of general formula (**2.170**). These can be formed in several ways: treatment of a base-protected 5'-*O*-dimethoxytrityl-2'-deoxynucleoside with 2-chloro-5,6-benzo-1,2,3-dioxaphosphorin-4-one ('salicyl chlorophosphite') (**2.171**) gives (**2.172**), which forms (**2.170**) on hydrolysis in aqueous pyridine. Tris(triazolyl)phosphite or tris(imidazolyl)phosphite may be used as alternatives to salicyl chlorophosphite. Then, treatment of (**2.170**) with pivaloyl chloride (Bu^{t}CO·Cl) in the presence of an immobilized nucleosidic component such as (**2.164**) affords the coupled *H*-phosphonate product (**2.173**), which is detritylated using dichloroacetic acid to give (**2.174**), ready for the next coupling step. A 'capping' step to block any

(**2.170**)   (**2.171**)   (**2.172**)

(**2.173**) R = DMTr
(**2.174**) R = H

unreacted 5'-OH groups may be performed using pivaloyl chloride and 2-cyanoethyl-*H*-phosphonate, for instance, although some practitioners claim that this is unnecessary: the pivaloyl chloride serves as a 'capping' reagent. The hydrogen atom bonded to phosphorus serves, in a sense, as its protecting group. Coupling yields are commonly of the order of 95 per cent, and reported elongation cycles have ranged between 4 and 15 minutes in length. At the end of elongation, a single oxidation step using iodine in aqueous pyridine serves to oxidize all the *H*-phosphonate links to phosphodiester links, and alternative oxidizing protocols with other agents may be used to generate all-thiophosphate links or all-phosphoramidate links if oligonucleotide analogues are to be studied. Despite the predominant utilization of the phosphoramidite methodology, the *H*-phosphonate method is finding some use and affords an elegant supplement to the oligonucleotide chemist's now formidable array of efficient synthetic methods.

2.4.2.6 Other strategies in oligonucleotide synthesis

The previous sections have concentrated principally on oligodeoxyribo-nucleotide synthesis, but the methods are equally applicable to oligo-ribonucleotide synthesis, in which protected nucleosides of type (2.175) are prepared and used in either conventional, phosphite, or *H*-phosphonate phosphotriester syntheses. 2'-5'-Linked oligonucleotides such as '2-5A' have been synthesized similarly, starting from base- and 3',5'-protected ribonucleosides.

Oligoribonucleotide synthesis obviously offers extra complications due to the necessity to block the 2'-hydroxy function with a group which will remain stable during deblocking of the 5'-hydroxy function in elongation cycles, but which is removable at terminal deblocking under conditions which will not cause isomerization of the internucleotidic link. Thus, for instance, it is often risky to use acid-labile protection at both 2'- and 5'-positions: unless the media used are perfectly dry, detritylation or depixylation (pixyl = 9-phenylxanthen-9-yl) of the 5'-hydroxy group under acidic (zinc bromide in 1,2-dichloroethane) conditions can lead to substantial loss of a tetrahydro-pyranyl group from the 2'-position. The consequences of accidental loss of the 2'-protecting group are potential isomerization or cleavage at the internucleotide link. While some remarkably elegant protecting groups have been developed for protecting the 2'-position, the use of the TBDMS (*t*-butyldimethylsilyl) group for this purpose has proved reliable: it is stable under acidic deprotection conditions, but readily removed using fluoride ion at the conclusion of synthesis. While all the 'phosphotriester'-related procedures have been used successfully in oligoribonucleotide syntheses, the stratagem presently favoured employs the solid-phase phosphoramidite procedure using subunits of type (2.176) for elongation. Also, use of the phenoxyacetyl group to protect the exocyclic amino groups of the nucleic

(2.175) R^1 = MMTr or DMTr or pixyl
R^2 = TBDMS or tetrahydropyranyl (thp)
or methoxytetrahydropyranyl (Mthp)
or 2-nitrobenzyl

Pixyl (9-phenylxanthan-9-yl) ≡ ; thp as in (2.148)

Mthp ≡

2-nitrobenzyl ≡

acid bases in oligoribonucleotide synthesis permits efficient base deacylation during the deblocking sequence without the premature removal of TBDMS groups from the 2′-positions and possible chain cleavage. A good alternative to subunits of type (**2.176**) is (**2.177**), employing the Fmoc (9-fluorenyl-methoxycarbonyl) group for protection of the 5′-position, and the acid-labile methoxytetrahydropyranyl group at the 2′-position. After each coupling step, the Fmoc group is removed using DBU in acetonitrile.

(**2.176**) R^1 = pixyl or DMTr   (**2.177**)

It must be emphasized that all steps in oligonucleotide synthesis proceed in good to excellent yields, as they must if a product of any length is to be prepared in reasonable yield. Two side-reactions may present serious complications and depress yields, however. The first is the already mentioned depurination, particularly deadenylation, which may occur when the growing oligonucleotide chain is detritylated at the 5′-terminus with an acidic reagent. Effective recent efforts to prevent this have included protecting the N^6-amino group of adenine as an amidine derivative, and this may eventually become standard practice. Secondly, the coupling agent may sulphonate uracil or thymine at O^4, or guanine at O^2, and protecting groups for these base substituents have also been developed and shown to be effective by improving coupling yields. The 4-nitrophenylethyl group, which has also been used to protect the internucleotidic link and which is removed by β-elimination using DBU, affords a versatile example. However, these synthetic refinements do not yet seem to have become widely used in laboratories concerned with gene synthesis, and only time and experience will tell whether the improved yields justify the extra synthetic complications.

In solid-phase syntheses, oligonucleotides are generally lengthened in the 5′-direction from an immobilized 3′-terminus. However, synthesis in the 3′-direction from an immobilized 5′-terminus does offer certain advantages: firstly, the solid-phase-bound oligomer need not be subjected to acidic reagents at any point during the extension cycle, and secondly, any by-product resulting from sulphonation of the 5′-OH component by the coupling agent is washed off the column. On the other hand, each extension cycle then involves deblocking a phosphate dissociation, which may be awkward and time-consuming.

Despite its recent spectacular successes, oligonucleotide synthesis remains a rapidly evolving area, and many more excellent synthetic procedures and strategies have evolved than this chapter can hope to cover. The interested reader is encouraged to seek recent reviews in this area for further information.

2.4.3 Assembly of longer oligonucleotides and genes

A long oligodeoxyribonucleotide duplex, such as a synthetic gene, is prepared by first synthesizing a number of shorter oligonucleotides by solid-phase phosphotriester methods. For instance, preparation of a human leucocyte interferon gene (514 base pairs) involved the synthesis of 67 shorter oligomer segments. The sequences of these single-stranded segments are chosen in such a way that when two of the segments (which have their sequences complementary and on opposite strands in the final gene sequence) are mixed in solution, they anneal according to the Watson–Crick base-pairing rules to leave a short single-stranded 'tail' or 'sticky-end' at either end of the resulting duplex. Now, when two such duplex fragments bearing 'sticky ends', which lie contiguously in the complete gene sequence, are mixed in solution, one 'sticky end' on one fragment will be complementary to a 'sticky end' on the other fragment, and the two ends therefore anneal, to give a longer duplex with a single nick, i.e. an internucleotidic link missing, in each strand. Then, treatment with a suitable DNA ligase, such as T4 DNA ligase, and ATP, seals the nicks to form a longer duplex which still has a 'sticky end', complementary to the end of the next piece of gene sequence, at each end. By repetition of this process the entire gene is assembled.

2.4.4 Nucleic acid sequencing

In the mid-1970s, two conceptually similar techniques for sequencing nucleic acids were described, the significance of which has subsequently led to their originators sharing a Nobel prize. Sanger's 'plus-and-minus' method, later refined as the 'dideoxy-sequencing method', relies essentially on enzymes for its application, while the 'Maxam–Gilbert sequencing' procedure is essentially 'chemical'. We will concentrate here on the latter method.

The treatment of DNA with dimethyl sulphate results in alkylation to form 3-methyladenine and 7-methylguanine residues (as in (**2.178**)), *inter alia*, and the glycosidic bonds attached to these bases become labile. On heating at pH7, the alkylated bases are lost, but in dilute acid, 3-methyladenine is lost preferentially. Subsequent treatment with piperidine then cleaves the chain at the points of depurination ((**2.179**) → (**2.181**)). The treatment of DNA with hydrazine, on the other hand, causes the pyrimidine bases to be attacked at C-6, and the bases to fragment leaving, ultimately, ribosylurea (**2.182**) and ribosylhydrazones. In 2 M sodium chloride, however, cytosine is attacked in preference to thymine. Again, subsequent treatment with piperidine leads to cleavage of the chain, this time where the pyrimidine base has been fragmented.

These reactions are utilized as follows: a single-stranded fragment of DNA, up to several hundred nucleotides long, is isolated. Any phosphate residue at the 5′-terminus is removed enzymatically (alkaline phosphatase) to leave

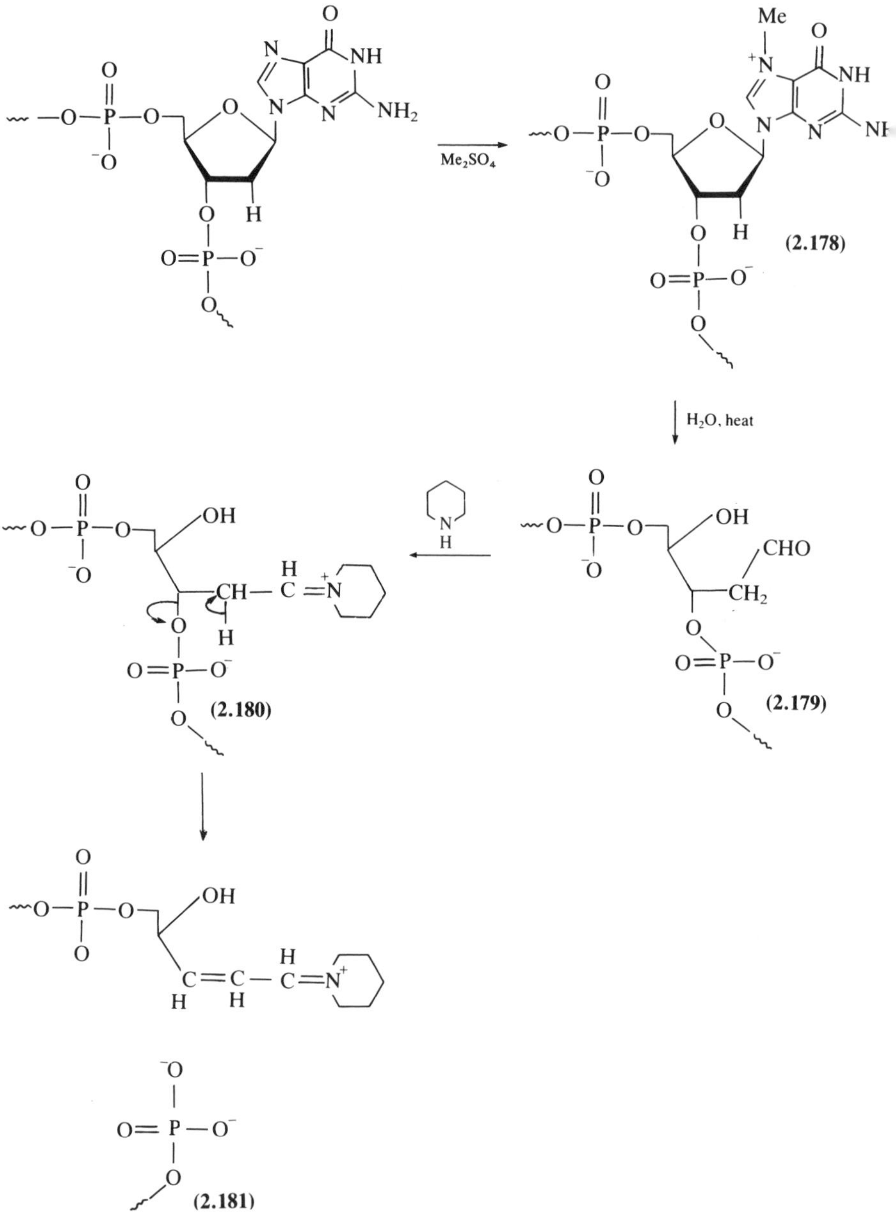
Me₂SO₄
H₂O, heat
(2.178)
(2.179)
(2.180)
(2.181)

a 5′-OH group. A radiolabel is then introduced by phosphorylation of this terminus using polynucleotide kinase and [γ-^{32}P] ATP, the radioactive γ-phosphoryl group of the ATP being transferred to the polynucleotide, which is then divided into four portions. Two portions are then treated with dimethyl sulphate. One of these (A) is then treated with piperidine; the other (B) is incubated in dilute acid, then treated with piperidine. The other two

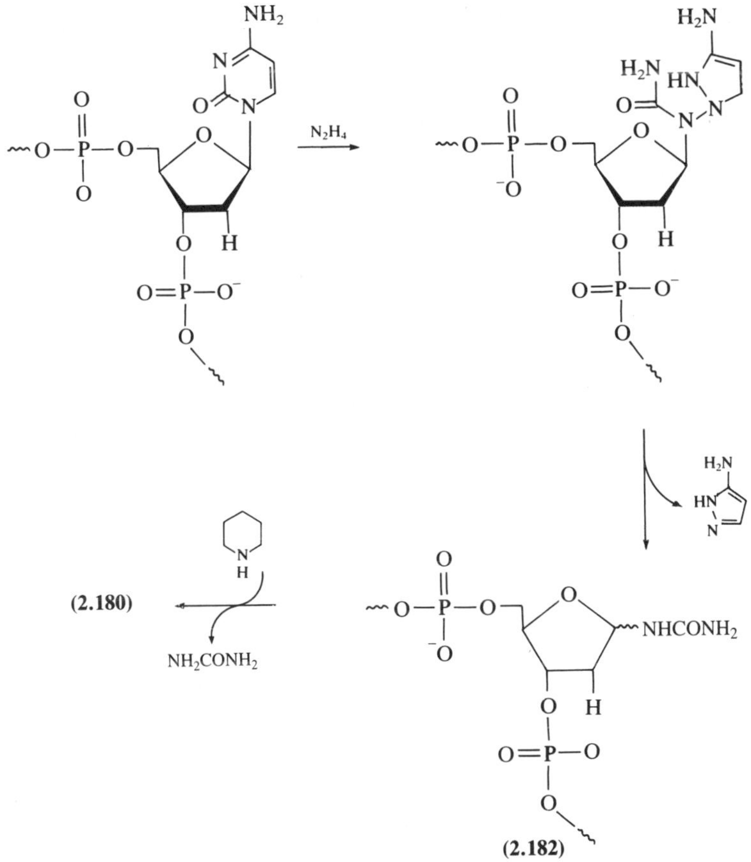

(2.180)

(2.182)

portions are treated with hydrazine, with added 2 M NaCl (C) or without NaCl (D), and subsequently with base. The trick consists in using reagents for methylation or hydrazinolysis at concentrations so low that only *one statistical hit per polynucleotide chain* takes place. Each portion of the original DNA, treated as above, has now given rise to a different population of *5′-end-labelled* polynucleotide fragments:

(i)   In (A),   *all possible chain lengths of the DNA* which terminate where guanine was originally present are found, since one guanine was alkylated at random in each molecule of the polynucleotide.

(ii) In (B),   all possible chain lengths of the DNA terminating where adenine or guanine were originally present are found.

(iii) In (C),   ditto, terminating where cytosine was originally present.

(iv) In (D),   ditto, terminating where cytosine or thymine were originally present.

The components in the four mixtures are then separated electrophoretically side-by-side in denaturing polyacrylamide gels, a procedure which separates the components according to chain length, with the shortest species migrating furthest, and autoradiography performed to reveal the positions of labelled components as a pattern of bands in each gel, from which the base sequence of the DNA is read by inspection. If bands are seen at the same position in gels from (A) and (B), guanine was present; if in (B), but not in (A), adenine was present; if in (C) and (D), cytosine was present; if in (D) but not in (C), thymine was present.

This is the 'Maxam–Gilbert' method. A similar sequencing method for RNA has been devised, in which diethyl pyrocarbonate is used for modifications of the adenine bases. Moreover, these methods have applicability far beyond simple sequencing. If, for instance, a polynucleotide interacts with another polynucleotide, or a protein, in such a way as to cause hindrance to attack on the polynucleotide by base-specific reagents such as the above, comparison of the pattern of bands generated by the sterically

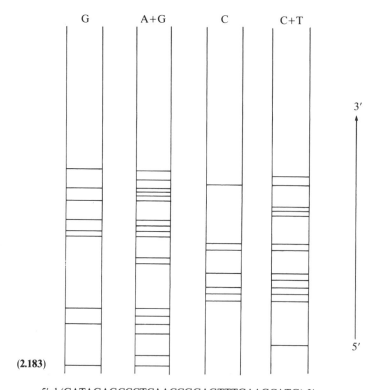

(2.183)

5'-d (GATAGAGCCCTCAACCGGAGTTTGAAGCATG)-3'

A representation of the sequencing gels generated in Maxam– Gilbert DNA sequencing. The shortest fragments, at the bottom, have run furthest and have the greatest spacing between them. These correspond to the 5'-end of the sequence. Reading upwards, the sequence is read 5' → 3'.

protected polynucleotide with that of a control sample affords information on tertiary interactions of polynucleotides with each other, or with proteins. This technique, now dubbed 'footprinting', has found much application in molecular biology.

2.4.5 Recombinant DNA

In answer to the question 'why synthesize genes?', one must cite their use in recombinant DNA work. To explain the significance of this area of research, the reader must first be aware of the function of its principal tools.

Restriction endonucleases are enzymes which cleave double-stranded DNA at specific base sequences, generally four to six nucleotides long, in both strands of the DNA duplex, generating 3'-OH and 5'-phosphate termini. The sites recognized for cleavage generally have sequences of diad symmetry, and cleavage may take place at the centre of symmetry (to generate 'flush' or 'blunt' ends) or symmetrically about the centre of symmetry, generating 'cohesive' or 'sticky' ends. More than 500 of these enzymes have now been described, and the recognition sites, centre of symmetry and mode of cut of a few of the most important are shown in Table 2.1.

Plasmids are extrachromosomal circular DNA molecules found in bacterial cells. Their presence is not normally essential to their host cell, but the genes which they carry may be essential to it under certain conditions. For instance, they may carry genes conferring resistance to antibiotics, without which the host cell is vulnerable. Their molecular weights range from about 10^6 to 10^8, i.e. *c.* 1600–160 000 base pairs. They are reproduced by the apparatus of the host organism during its own reproductive cycle, and the genes which they carry are expressed by the host cell.

A completely random base sequence in DNA could be expected to contain a restriction site for an endonuclease recognizing a six-base-sequence once every 4^6, or 4096, base pairs. Thus, a small plasmid may contain one, or only a few, cleavage sites for a given enzyme. On treatment with the enzyme a plasmid containing a single 'restriction site' is linearized, but bears two self-complementary 'sticky ends', or else 'blunt ends'.

Let us suppose that one wishes to obtain a large quantity of a protein from, say, a higher animal, which is normally only available in small quantities from natural sources, such as interferon, insulin, or somatostatin. Direct chemical protein synthesis is at present not a realistic method for obtaining large quantities of long proteins. If, however, one can transfer the genetic material directing the synthesis of the protein into a bacterium, and the bacterium will then make functional protein, one has in essence an infinitely variable protein factory at one's disposal. However, there is a snag. The mRNA directing the synthesis of the desired protein, if from a eukaryotic source, would have had its introns removed in the maturation process described earlier (p. 107), but the DNA from which it was transcribed contained all the intervening sequences. If this DNA were to be grafted into

Table 2.1 Some important restriction endonucleases and their restriction sequences.

Designation	Source organism	Sequence and mode of cleavage					
Cohesive ends				Diad line			
Eco RI	*Escherichia coli*	G	A A	T T	C	
		C	T T	A A	G	
Bam HI	*Bacillus amyloliquefaciens H*	G	G A	T C	C	
		C	C T	A G	G	
Hae II	*Haemophilus aegyptius*	Pu	G C	G C	Py	
		Py	C G	C G	Pu	
Hind III	*Haemophilus influenzae*	A	A G	C T	T	
		T	T C	G A	A	
Pst I	*Providencia stuartii*	C	T G	C A	G	
		G	A C	G T	C	
Taq I	*Thermus aquaticus*		T C	G A		
			A G	C T		
Blunt ends							
Hae I	*Haemophilus aegyptius*	(A)	G G	C C	(T)	
		(T)	C C	G C	(A)	
Sma I	*Serratia marcescens*	C	C C	G G	G	
		G	G G	C C	C	

a plasmid, no mechanism would exist in the bacterium to direct the maturational splicing of the RNA transcribed from it prior to its translation, so functional protein could not be formed.

However, it may be feasible to isolate the mRNA from which the protein of interest is derived, normally by a method involving hybridizing (i.e. annealing) the mRNA to a DNA sequence complementary to part of the molecule (another use for oligonucleotide synthesis!). Then, reverse transcriptase is used to prepare the cDNA molecule from the mRNA. The mRNA is digested by a ribonuclease activity associated with the enzyme, and another complementary DNA strand is then made, so that one eventually obtains blunt-ended double-stranded DNA containing an uninterrupted sequence coding for the desired protein. If the mRNA is not available, knowledge of the protein sequence permits the sequence of a DNA strand which would encode the protein to be deduced using the genetic code, and a completely synthetic gene can be made.

How is the foreign DNA introduced into the plasmid? Several methods are possible. Suppose a plasmid has been cleaved so as to produce blunt ends, and the DNA to be introduced also contains blunt ends. The enzyme DNA ligase encoded by *E. coli* bacteriophage T4 is able to catalyse blunt-end joining, but requires high concentrations of DNA for the process to proceed efficiently. Nevertheless, this is a workable method. A better one is to use 'homopolymer tail joining'. The blunt-ended linearized plasmid is treated with terminal deoxynucleotidyl transferase and, say, dATP, to add poly(dA) tails to the 3'-OH termini, and the foreign DNA to be introduced is treated with the same enzyme and dTTP, to add poly(dT) tails at its 3'-OH termini. On mixing the plasmid and foreign DNA, the complementary tails become annealed. Any gaps may be filled in using Klenow fragment and sealed using DNA ligase.

Suppose one wishes to insert blunt-ended foreign DNA at a restriction site on the plasmid having cohesive ends. This can be performed if the blunt-ended DNA is first blunt-end-ligated at each end, using T4 DNA ligase, to an 'adaptor', an oligonucleotide which is often about 10 residues long and which has been synthesized to contain the same restriction sequence. The foreign DNA, now carrying the adaptor sequence at each end, is treated with the restriction endonuclease to generate the same cohesive ends. The cleaved plasmid and foreign DNA are then mixed, and the cohesive ends anneal. Ligation completes the insertion process.

The adapted plasmid 'vector' must then be introduced into the host organism. For many bacterial species, this can be accomplished by suspending the cells in cold calcium chloride solution in the presence of the plasmid, which typically becomes stably established in *c.* 0.1 per cent of the cells. How, then, can one identify plasmids containing the foreign DNA from those in which the insertion process has failed?

Consider the *E. coli* plasmid pBR 322, which contains genes conferring resistance to the antibiotics ampicillin and tetracycline. The *Bam* HI cleavage site is located within the gene for tetracycline resistance, and if foreign DNA is inserted at this point, the ability of the plasmid to confer resistance to tetracycline on its host organism is lost. Suppose, now, that wild-type *E. coli* is treated with a population of pBR 322 plasmids, some of which contain an insert at the *Bam* HI site, and then left to grow in a medium containing tetracycline and cycloserine. Tetracycline prevents the growth of the wild-type *E. coli* (which lacks resistance to it) and also of the *E. coli* containing the altered plasmid in which the gene conferring resistance has been inactivated. The cells containing the unaltered pBR 322 grow – and are destroyed, since cycloserine is cytotoxic to growing cells. The surviving cells, which lacked resistance to tetracycline, are harvested and grown on agar plates containing ampicillin. The wild type, again lacking resistance, cannot grow, but the cells containing the altered plasmid now form colonies. Many plasmids have been engineered to afford sophisticated 'cloning vehicles' in which these and similar procedures are applied.

This is not the end of the story: there are other considerations to be taken into account before an engineered gene produces the desired functional protein. However, the reader who has absorbed this will have little difficulty in understanding the uses of synthetic oligonucleotides in site-directed mutagenesis, and can only be encouraged to seek further, in books and journals devoted to molecular biology, for more detailed accounts of this active and fascinating field of research.

2.4.6   *Copying DNA: the polymerase chain reaction*

In the late 1980s, an elegant procedure was developed which permits a DNA sequence, which may be several thousand bases in length, to be copied (or 'amplified') *ad lib.*, without the need to insert it into a microorganism, grow the microorganism, and subsequently harvest and isolate the specific DNA sequence required. In theory, billions of copies of a DNA sequence can be generated from a single molecule. The procedure is known as the 'polymerase chain reaction', or PCR, and depends on the activity of a thermostable DNA polymerase enzyme such as that from the thermophilic bacterium, *Thermus aquaticus*. However, the chemist has an essential role to play in the procedure since the ends of the DNA sequence to be amplified are defined by two oligodeoxyribonucleotides, normally prepared on a DNA synthesizer.

The process is illustrated in (**2.184**);the double-stranded DNA to be amplified is heated in the presence of an excess of two oligonucleotides, each complementary to a sequence on one of the strands, and then cooled: heating causes the DNA strands to separate, and on cooling the 'primer' oligonucleotides anneal to the longer DNA strands. The *Taq* DNA polymerase is then used, with dATP, dTTP, dCTP, and dGTP, to elongate the primer oligodeoxyribonucleotides to form complementary strands on the longer DNA templates. The heating–cooling/annealing–polymerization cycle is repeated again and again: the enzyme is adequately thermostable to retain its activity through repeated cycles. The span of the original DNA sequence between the positions of annealing of the two oligonucleotide primers is copied specifically, and the number of copies formed increases geometrically with the number of PCR cycles performed. Thus, 30 cycles – by no means an unrealistic number – can generate 2^{30}, or around 10^9 copies of a tract of DNA.

This formidable procedure is already finding wide application in diverse fields: forensic science, DNA sequencing, the insertion and amplification of mutations into DNA, and diagnostics, to name but a few.

It is instructive, and satisfying, to reflect how much of the current progress and excitement in cell and molecular biology and biotechnology, and hopes for the future, rest on the solid underpinning of the work of organic chemists. The past two decades have seen unimaginable progress, and it is entertaining (if a waste of time) to speculate where molecular biology would be today if, for instance, gel sequencing and solid-phase oligonucleotide synthesis had not been developed. For the present and future, we can expect to learn much more about the different physical shapes nucleic acid molecules can adopt, and their significance in cellular processes such as control and differentiation. We are also on the verge of understanding more about catalytic processes in RNA molecules, and their roles in systems which have little to do with the functions normally associated with tRNA, mRNA, and rRNA. And we may anticipate that in answering the problems which arise in these and other areas, the skills, insights, and technology developed by the chemical community will be required, and not found lacking.

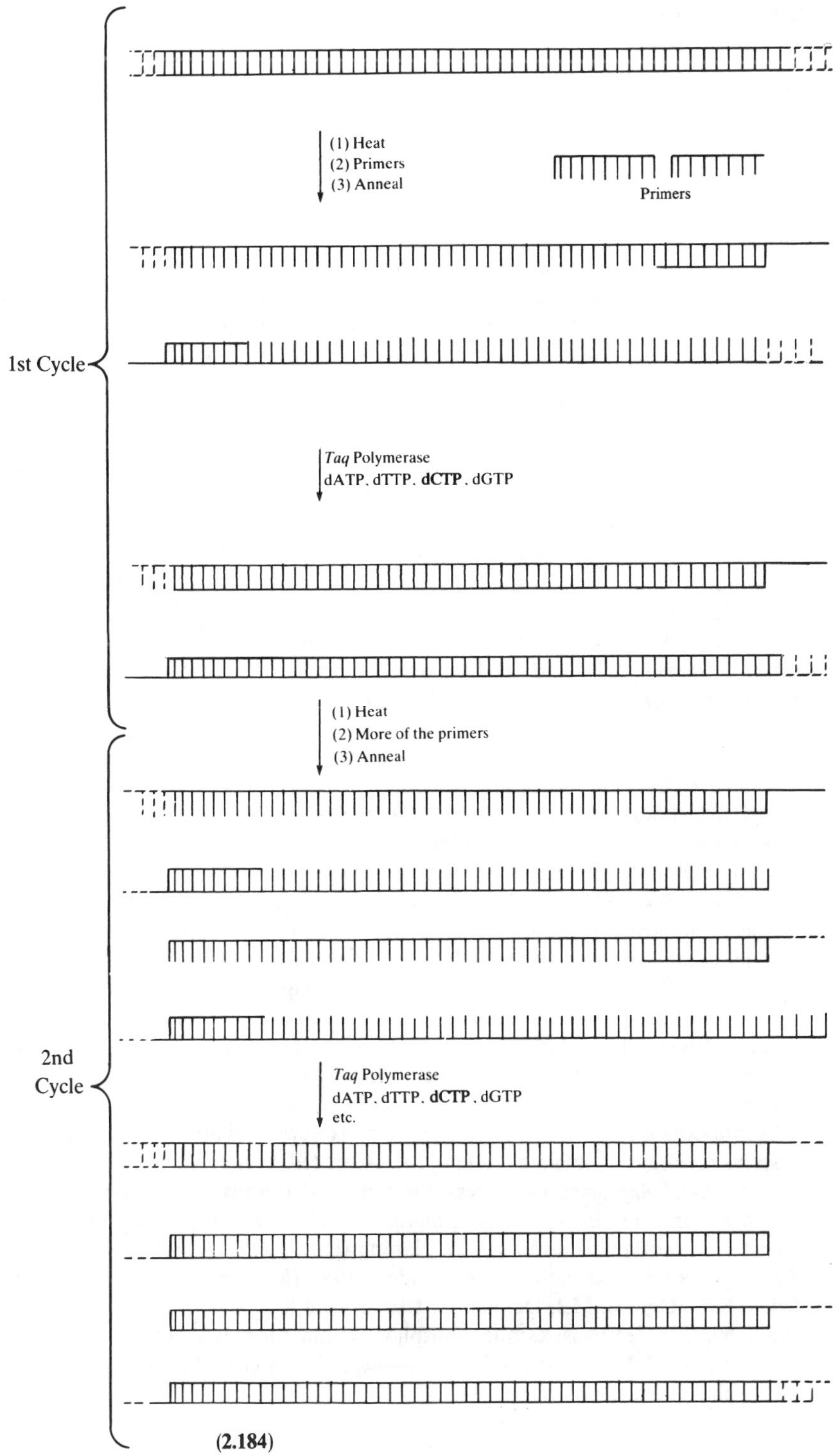

1st Cycle

(1) Heat
(2) Primers
(3) Anneal

Primers

Taq Polymerase
dATP, dTTP, **dCTP**, dGTP

(1) Heat
(2) More of the primers
(3) Anneal

2nd
Cycle

Taq Polymerase
dATP, dTTP, **dCTP**, dGTP
etc.

(2.184)

Further reading

1. P.O.P. Ts'o (ed.), *Basic Principles in Nucleic Acid Chemistry*, vols 1 and 2, Academic Press, London, 1974.

2. G.M. Blackburn and J.B. Hobbs, 'Nucleic Acids' in *Annual Reports B* of the Royal Society of Chemistry, RSC Press, 1985, vol. 82, pp. 311–351 and references therein.

3. The yearly reviews by D.W. Hutchinson, and later J.B. Hobbs and R. Costick, on 'Nucleotides and Nucleic Acids' in *Organophosphorus Chemistry* (Specialist Periodical Reports of the Royal Society of Chemistry), RSC Press, 1970–1992, vols 1–23, provide a comprehensive and continuing survey of progress in this field.

4. Nucleosides: R.J. Suhadolnik, *Nucleoside Antibiotics*, Wiley, Chichester, 1970; *Nucleosides as Biological Probes*, Wiley, Chichester, 1979; H. Vorbrüggen in *Nucleoside Analogues: Chemistry, Biology and Medical Application* (eds R.T. Walker, E. de Clercq, and F. Eckstein), Plenum Press, London, 1979 (a most valuable book which deserves to be better known: a mine of information); E. de Clercq and R.T. Walker (eds), *Targets for the Design of Antiviral Agents*, Plenum Press, London, 1984; N.R. Williams *et al.*, *Carbohydr. Chem.*, 1985, **17**, 186; J.G. Buchanan and R.H. Wightman, *Top. Antibiot. Chem.*, 1982, **6**, 229.

5. Showdomycin has been used to illustrate some roots to C-nucleosides; more detailed accounts of some syntheses are to be found in: L. Kalvoda, *J. Carbohydr., Nucleosides, Nucleotides*, 1976, **3**, 47; T. Inoue and I. Kuwajima, *J. Chem. Soc., Chem. Commun.*, 1980, 251, and references therein; S.R. James, *J. Carbohydr., Nucleosides, Nucleotides*, 1979, **6**, 417 (for a review of methods of C-nucleoside synthesis).

6. Nucleotides: K.H. Scheit, *Nucleotide Analogues: Synthesis and Biological Function*, Wiley, Chichester, 1980.

7. Chiral nucleotides and their applications (the references cited in these reviews give comprehensive access to the primary literature): F. Eckstein, *Angew. Chem., Int. Ed. Engl.*, 1983, **22**, 423; G. Lowe, *Acc. Chem. Res.*, 1983, **16**, 244; M. Cohn, *Acc. Chem. Res.*, 1982, **15**, 326; W.J. Stec, *Acc. Chem. Res.*, 1983, **16**, 411.

8. Nucleic acid structure: much valuable information has been elucidated in this area in recent years, and a thorough grounding may be obtained in: W. Saenger, *Principles of Nucleic Acid Structure*, Springer-Verlag, Berlin, 1984; F.A. Jurnak and A. McPherson (eds), *Biological Macromolecules and Assemblies*, vol. 2, Ch. 1–4, Wiley, Chichester, 1985; J.A. McCammon and S.C. Harvey, *Dynamics of Proteins and Nucleic Acids*, Cambridge University Press, Cambridge, 1987 (an appreciation of the dynamics of nucleic acid structures).

9. Oligonucleotide synthesis: there are so many reviews available in this area that it seems invidious to single out a few: M.J. Gait (ed.), *Oligonucleotide Synthesis – A Practical Approach*, IRL Press, Oxford, 1984 (an invaluable book for the bench researcher); E. Sonveaux, *Bioorg. Chem.*, 1986, **14**, 274 (an excellent, recent, and substantial review); S.A. Narang (ed.), *Synthesis and Application of DNA and RNA*, Academic Press, London, 1987 (a recent book covering the field); K.K. Ogilvie, M.J. Damha, N. Usman, and R.T. Pon, *Pure Appl. Chem.*, 1987, **59**, 325 (emphasizes the phosphoramidite procedure and its broader applications); R. Stroemberg, *Chem. Commun., Univ. Stockholm*, 1987, 1 (reviews the *H*-phosphonate procedure).

10. Nucleic acid sequencing: A.J.H. Smith, A.M. Maxam, and W. Gilbert, in *Methods in Enzymology* (eds L. Grossman and K. Moldave), vol. 65, Academic Press, London, 1980, pp. 499–580; S.M. Weissman, *Methods of DNA and RNA Sequencing*, Preager, New York, 1983; F. Sanger, *Ann. Rev. Biochem.*, 1988, **57**, 1.
11. The polymerase chain reaction: R.K. Saiki *et al.*, *Science*, 1988, **239**, 487; J.L. Marx, *Science*, 1988, **240**, 1408 (offers an enjoyable descriptive introduction to the procedure).

3 Amino acids and peptides

R.S. Davidson and J. B. Hobbs

3.1 Introduction

Proteins are essential to life. They play a variety of important roles, e.g. as enzymes, hormones, and participants in ion transport and the immune defence system. The proteins are complex molecules, the basic building blocks being the α-amino acids. These acids are not only linked in a particular sequence for a particular protein but because of their chirality, and the nature (e.g. polarity) of side-chains, they also have a well-defined three-dimensional structure. It is this particular structure which often defines the particular special reactivity of the proteins.

In this chapter some of the fundamental properties of proteins are dealt with, and special methods for their synthesis are then considered. It has taken some of the most brilliant chemists to develop simple syntheses for nature's basic materials. Similarly, the building up of polypeptides from constituent amino acids has led to the application of most elegant chemistry, with practical applications always in mind. The determination of the structure of polypeptides has to be done with great care if artefacts are to be avoided and results obtained with tiny amounts of material.

The efforts of chemists in the field of protein chemistry are complemented by those of the biochemists who have unravelled the natural processes for production of proteins, their complex interrelationships, and how they react with other species. Such has been the progress that natural processes are being harnessed to produce not only natural proteins but also proteins in which specific amino acids are replaced by others in order that properties may be altered in an advantageous manner.

3.2 Synthesis of α-amino acids

There are many well-established chemical routes for the synthesis of α-amino acids. In most cases a racemic mixture is obtained, and consequently this has to be resolved if pure enantiomers are to be obtained. A few enantioselective routes exist. This section deals with:

(1) classical synthetic routes;
(2) some newer synthetic routes;
(3) syntheses based on α-amino acids as chiral building blocks;
(4) resolution of racemic mixtures of α-amino acids.

In this chapter absolute configuration is described, using both the D,L and the *R*,*S* nomenclature. The reason for retaining the old D,L nomenclature is that this is commonly used by biochemists and features in current chemical catalogues. Thus the student needs to be familiar with this system and also to be able to deduce absolute configuration on the basis of the *R*,*S* system, e.g.

Cysteine 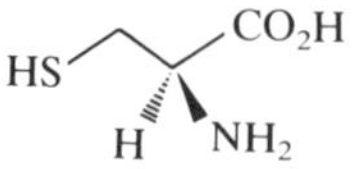 L *R*

Leucine 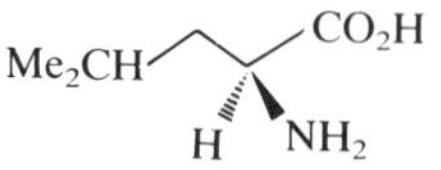 L *S*

Usually L-amino acids have the *S*-configuration, the most important exception being cysteine.

3.2.1 *Classical synthetic routes*

Details of these routes may be found in most good undergraduate student textbooks.

Route A:

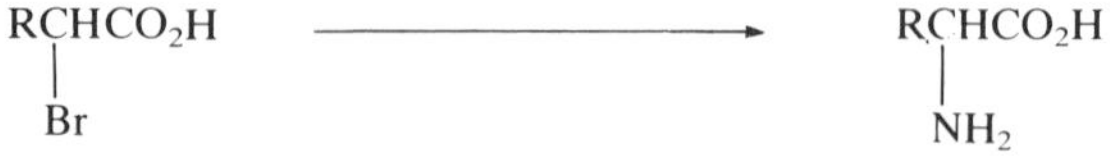

The C—Br bond is susceptible towards nucleophilic attack by an ammonia equivalent. The classical example of this strategy is the Gabriel synthesis (Fig. 3.1). Initial reaction of the α-bromo acid with the potassium salt of phthalimide is followed by hydrazinolysis to yield the amino acid and the cyclic hydrazide. More recent approaches employ phthalide replacements which are more easily removed, e.g. Boc_2NK.

Route B:

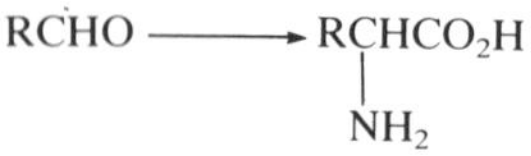

The Strecker synthesis (Fig. 3.2) is the classical method for amino acid synthesis and was developed in the 1930s. The major limitation is the harsh conditions of the hydrolysis.

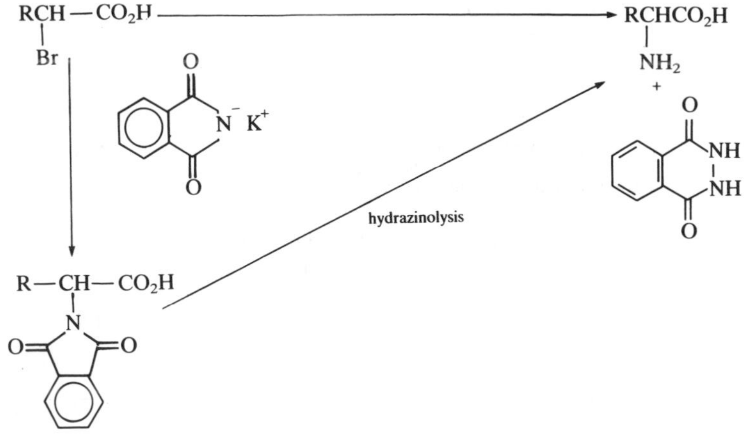

Fig. 3.1 Gabriel synthesis.

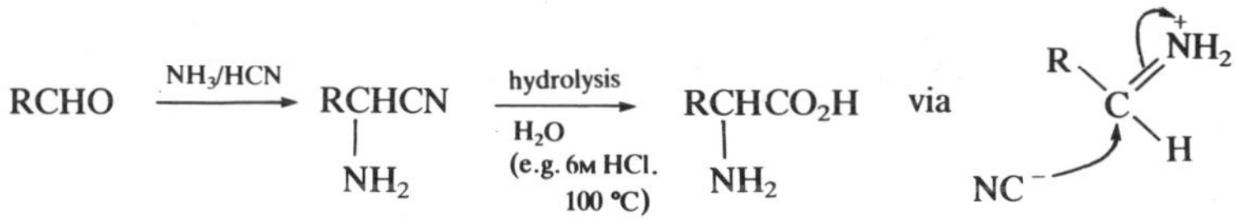

Fig. 3.2 Strecker synthesis.

Route C: ArCHO ⟶ ArCH₂CHCO₂H
                                    |
                                   NH₂

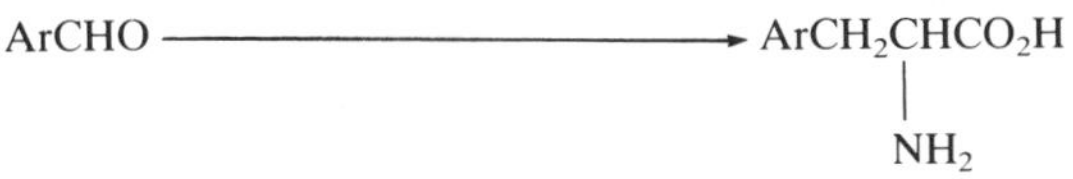

The method shown in Fig. 3.3 is specific for the preparation of the aryl amino acids and utilizes cheap and readily available starting materials (i.e. glycine and ethanoic acid anhydride).

Route D: R-Br ⟶ RCHCO₂H
                              |
                             NH₂

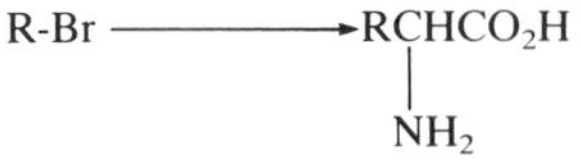

The species R—Br needs to be a primary or benzyl/allylic bromide since the crucial step involves nucleophilic attack upon the C—Br bond by a carbanion, viz. as shown in Fig. 3.5. The acetamidomalonate derivative shown in Fig. 3.4 is a versatile intermediate. It contains the

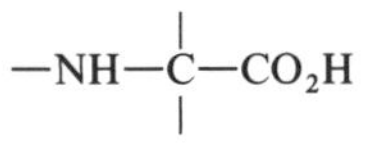

—NH—C—CO₂H
        |

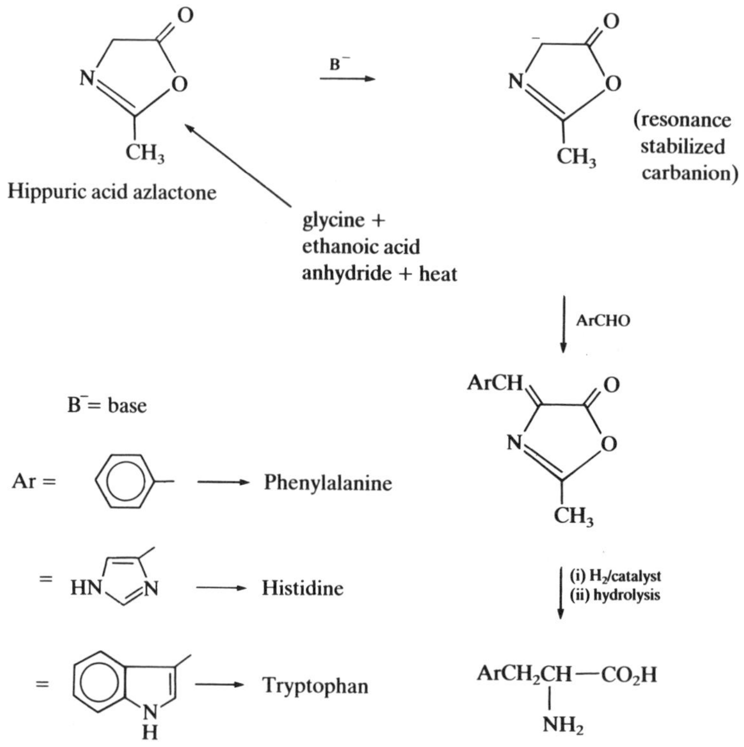

Fig. 3.3 Azlactone method: oxazolone.

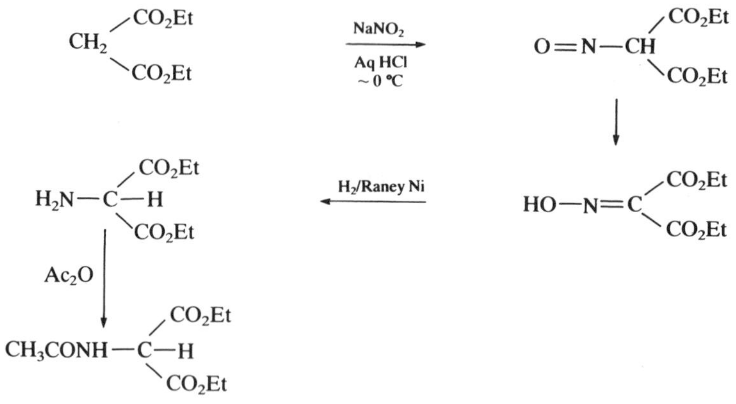

Fig. 3.4.

unit and furthermore the C—H bond is easily activated by base to give a carbanion which may then be used in a nucleophilic displacement reaction or a Michael addition reaction (Fig. 3.5).

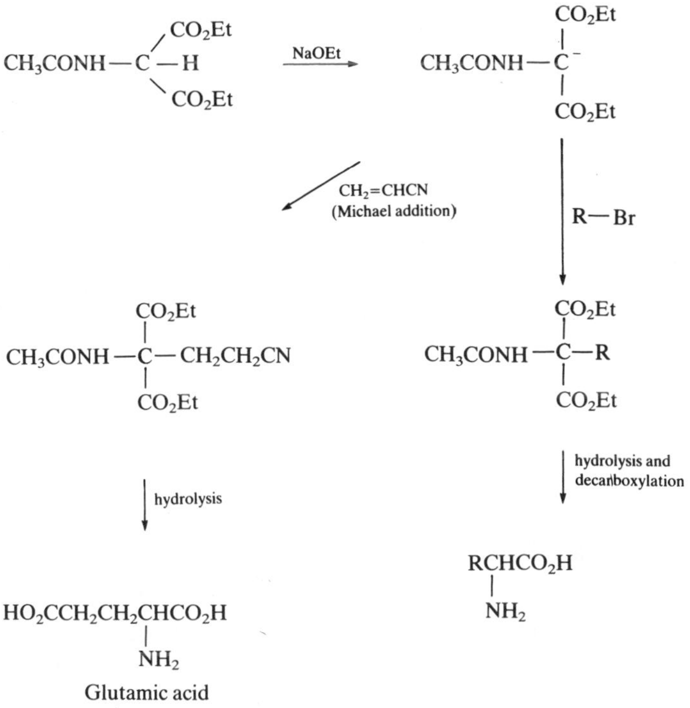

The reaction scheme shows:

$$CH_3CONH-\underset{\underset{CO_2Et}{|}}{\overset{\overset{CO_2Et}{\diagup}}{C}}-H \xrightarrow{\text{NaOEt}} CH_3CONH-\underset{\underset{CO_2Et}{|}}{\overset{\overset{CO_2Et}{|}}{C^-}}$$

With CH₂=CHCN (Michael addition) to the left and R—Br to the right.

$$CH_3CONH-\underset{\underset{CO_2Et}{|}}{\overset{\overset{CO_2Et}{|}}{C}}-CH_2CH_2CN \qquad CH_3CONH-\underset{\underset{CO_2Et}{|}}{\overset{\overset{CO_2Et}{|}}{C}}-R$$

hydrolysis (left) ; hydrolysis and decarboxylation (right)

$$HO_2CCH_2CH_2\underset{\underset{NH_2}{|}}{C}HCO_2H \qquad\qquad R\underset{\underset{NH_2}{|}}{C}HCO_2H$$

Glutamic acid

Fig. 3.5.

3.2.2 *Newer synthetic routes*

3.2.2.1 *Enantioselective synthesis*

This synthesis is based on the fact that in many biological systems, α-ketocarboxylic acids are transformed into α-amino acids by a process known as transamination which involves vitamin B_6 (pyridoxine, (**3.1**)).

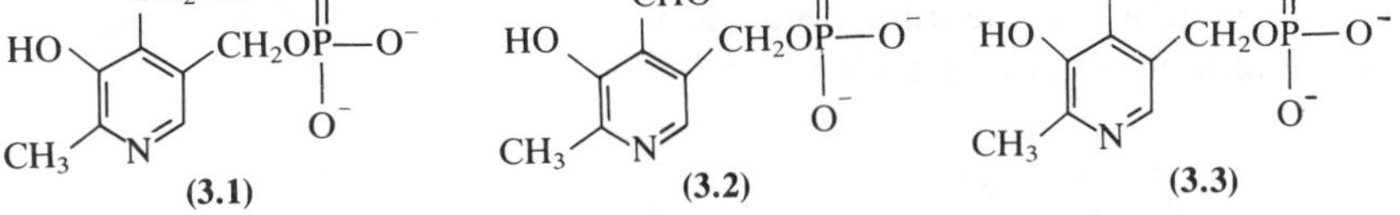

(**3.1**) Pyridoxine (vitamin B_6)

(**3.2**) Pyridoxal

(**3.3**) Pyridoxamine

The process of interconversion of α-ketocarboxylic and α-amino acids is shown in Fig. 3.6 using partial structures for the vitamin and its derivatives.

In the enantioselective synthesis, developed by Corey, which mimics the biological transamination process, the transaminating agent is X, and the

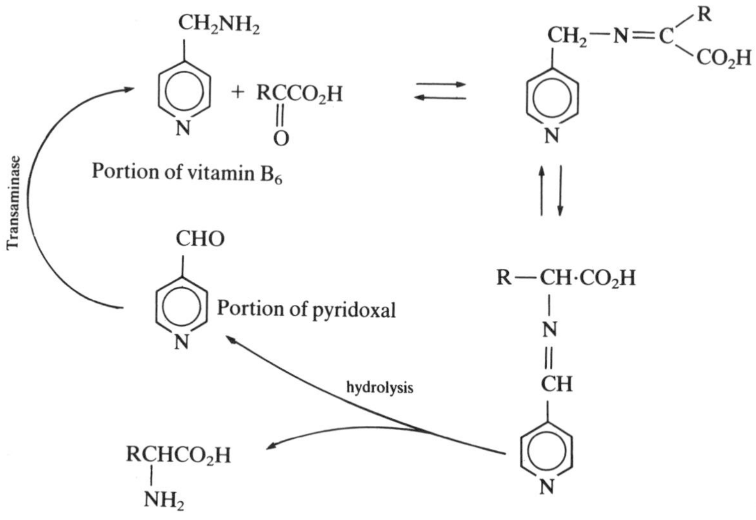

Fig. 3.6.

process proceeds as shown in Fig. 3.8. Initial condensation between X and an α-keto-ester yields Y, which is reduced stereoselectively and the product hydrolysed to regenerate the amino alcohol X and an amino acid.

To improve the efficiency of the asymmetric induction step further, the following reagent was used:

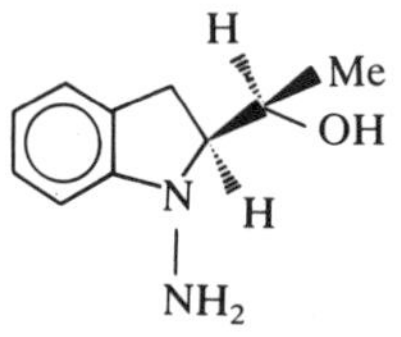

Fig. 3.7.

The use of this secondary alcohol further decreases the possibility of reduction of the amine group occurring from the top face.

Use of (**3.4**) as the aminating agent led to the formation of the other enantiomeric series of α-amino acids.

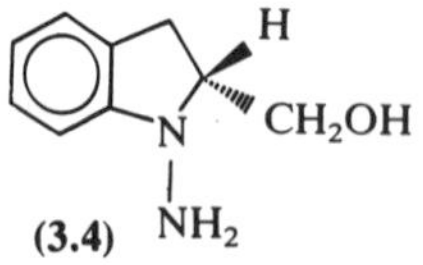

Transamination has also been accomplished using an immobilized enzyme (commercially available) in the synthesis of γ-hydroxy-L-glutamic acid (Fig. 3.9).

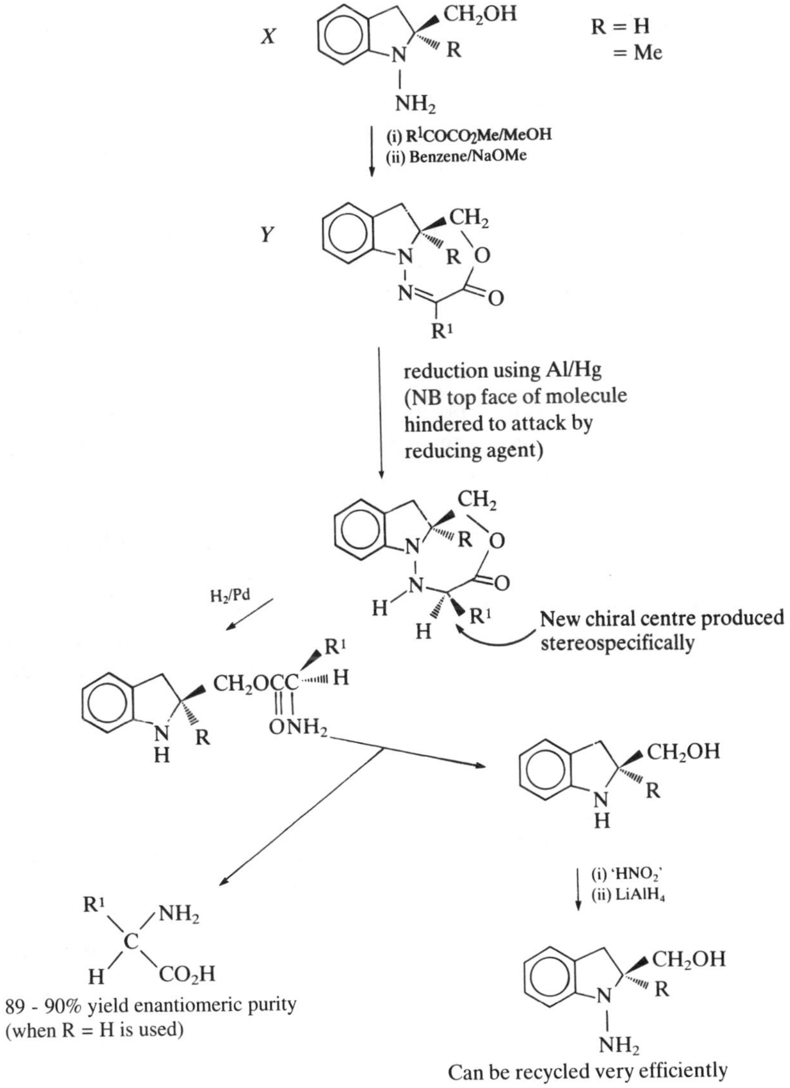
X
CH₂OH
R
N
NH₂
R = H
= Me
(i) R¹COCO₂Me/MeOH
(ii) Benzene/NaOMe
Y
CH₂
R O
N
N
O
R¹
reduction using Al/Hg
(NB top face of molecule
hindered to attack by
reducing agent)
H₂/Pd
New chiral centre produced
stereospecifically
CH₂OCC
ONH₂
(i) 'HNO₂'
(ii) LiAlH₄
89 - 90% yield enantiomeric purity
(when R = H is used)
Can be recycled very efficiently

Fig. 3.8.

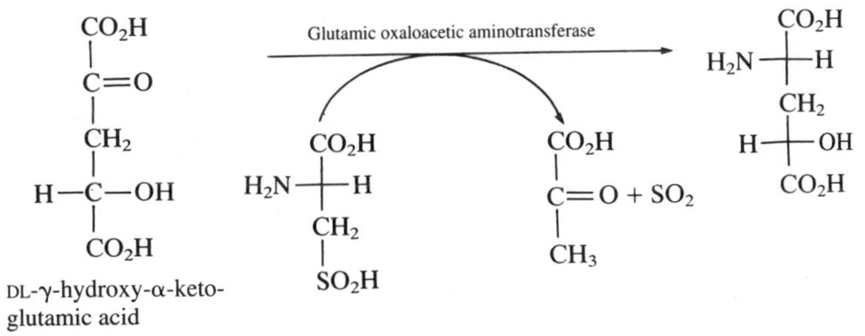
Glutamic oxaloacetic aminotransferase
CO₂H
C=O
CH₂
H—C—OH
CO₂H
H₂N
H
SO₂H
C=O + SO₂
CH₃
OH
DL-γ-hydroxy-α-keto-
glutamic acid

Fig. 3.9.

3.2.2.2   Synthesis of unusual α-amino acids

Nikkomycin B.   Nikkomycin B (D) is a γ-hydroxy-α-amino acid. Some such compounds act as inhibitors for chitin synthetase. The nikkomycins and the group as a whole are thus potential insecticides since chitin is the main structural polysaccharide of insect skeletons (see Chapter 1).

2-Ketobutanoic acid was converted (Fig. 3.10) into the amide oxime which formed the trianion upon reaction with Bu^{t}Li in *N,N,N',N'*-tetramethyl-ethylenediamine (TMEDA). Reaction of this with 4-methylbenzaldehyde followed by addition of trifluoroacetic acid yielded the dihydroisoxazole, which could be converted in nikkomycin B by reduction and hydrolysis.

Cyclic amino acids.   Whilst proline and its derivatives are well known, there is interest in other cyclic amino acids. Aminocyclopropanecarboxylic acid, for instance, is the biogenetic precursor of ethene, which in turn is involved in such processes as the ripening of fruit.

Figure 3.11 shows a synthesis of coronamic acid which utilizes a classical reaction for formation of the cyclopropane ring.

The first step of this reaction sequence is the formation of the carbanion derived from the malonate ester. The quaternary ammonium salt having a hydroxyl anion has a finite solubility in the organic phase, and, furthermore,

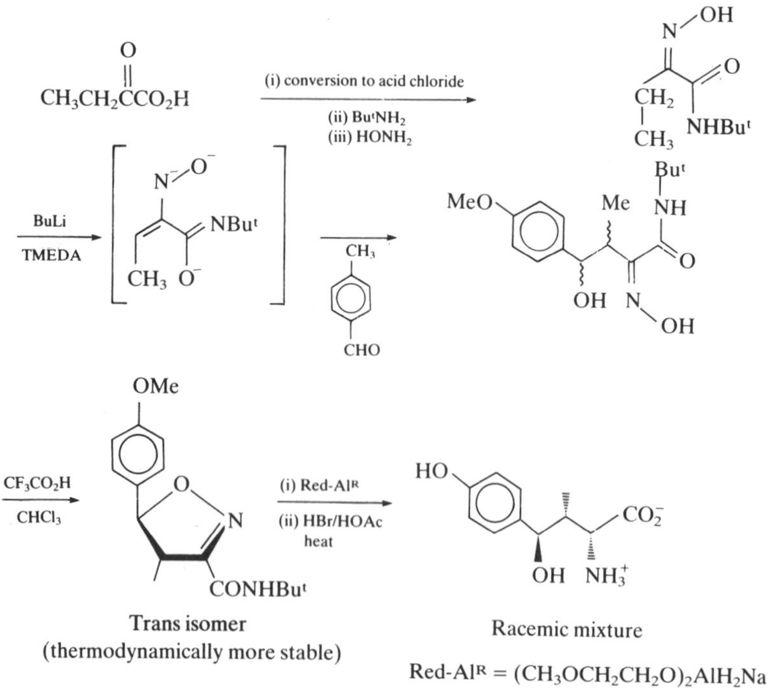

Fig. 3.10.

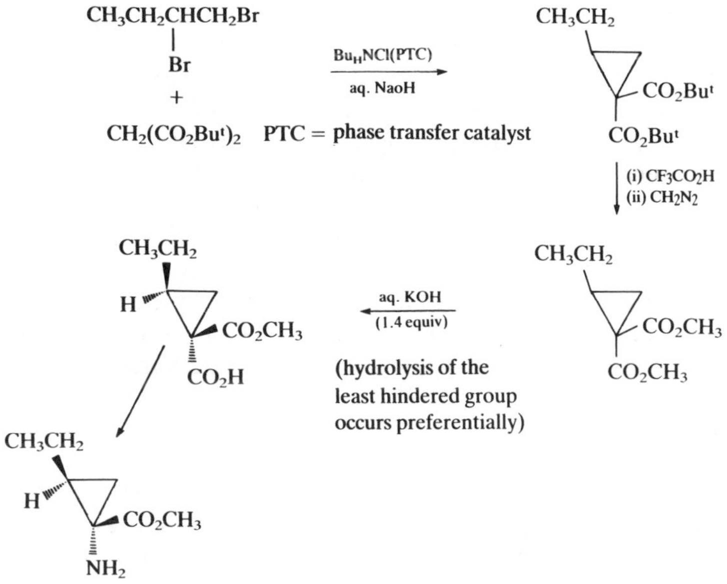

CH₃CH₂CHCH₂Br
|
Br

+

CH₂(CO₂Buᵗ)₂   PTC = phase transfer catalyst

Bu₄NCl(PTC)
aq. NaOH

CH₃CH₂
⟍
CO₂Buᵗ
|
CO₂Buᵗ

(i) CF₃CO₂H
(ii) CH₂N₂

CH₃CH₂
⟍
CO₂CH₃
|
CO₂CH₃

aq. KOH
(1.4 equiv)

(hydrolysis of the
least hindered group
occurs preferentially)

CH₃CH₂
⟍
H
CO₂CH₃
CO₂H

CH₃CH₂
⟍
H
CO₂CH₃
NH₂

Fig. 3.11.

since the hydroxyl anion will not be solvated by water molecules, it will act as a very strong base:

$$Bu_4NCl_{aq.} \overset{H_2O}{\rightleftharpoons} Bu_4NOH_{aq.} \qquad \text{(aq. = aqueous phase)}$$

$$Bu_4NOH_{aq.} \rightleftharpoons Bu_4NOH_{org.} \qquad \text{(org. = organic phase)}$$

The reactions which ensue are typical nucleophilic substitution reactions exhibited by carbanions.

Conversion of the CO_2H group to NH_2 was accomplished as follows:

$$RCO_2H \longrightarrow RCOCl \longrightarrow RCON_3 \overset{heat}{\longrightarrow} RN{=}C{=}O \overset{hydrolysis}{\longrightarrow} RNH_2$$

i.e. the Curtius rearrangement reaction.

3.2.3 Syntheses based on α-amino acids as chiral building blocks

The synthesis of polypeptides requires the use of enantiomerically pure α-amino acids and for all stages of the assemblage of these building blocks to occur without racemization. This is just a small facet of the topic of enantioselective synthesis since there is a rich variety of compounds to be synthesized in optically pure form. The classical approach to the synthesis of compounds containing more than one chiral centre is to identify an intermediate which possesses a chiral centre and also groups which will help in the design and prosecution of the optical resolution process. More recently, with the growth of knowledge concerned with creating optically active compounds via asymmetric induction, the necessity for carrying out optical resolutions has decreased, but has not been obviated. In the last few years it has been recognized that a vast array of optically pure compounds occurs naturally, and many of these compounds contain more than one chiral centre. It is not surprising, therefore, that chemists are seeking to use such compounds as their synthetic base, and methods available for manufacturing naturally occurring building blocks such as carbohydrates and α-amino acids are increasing. The use of optically active α-amino acids demands that none of the synthetic methods employed will cause racemization.

3.2.3.1   L-Vinylglycine

L-Vinylglycine which has been isolated from mushrooms and is implicated in the mechanism of action for threonine synthetase, in which *O*-phosphohomoserine is converted into threonine, has been synthesized from optically active α-amino acids which are readily available, e.g. L-methionine acid and L-glutamic acid. Synthesis from the latter is shown in Fig. 3.12.

The α-amino group of L-glutamic acid is protected by standard methodology utilized in peptide synthesis, i.e. it is reacted with benzyloxy-carbonyl chloride. Reaction of this product with formaldehyde produces an intermediate *N*-hydroxymethyl compound which undergoes intramolecular cyclization to give (A). Thus in (A) we have a substituted α-amino acid in which the glycine residue is protected. Subjection of (A) to lead tetra-acetate

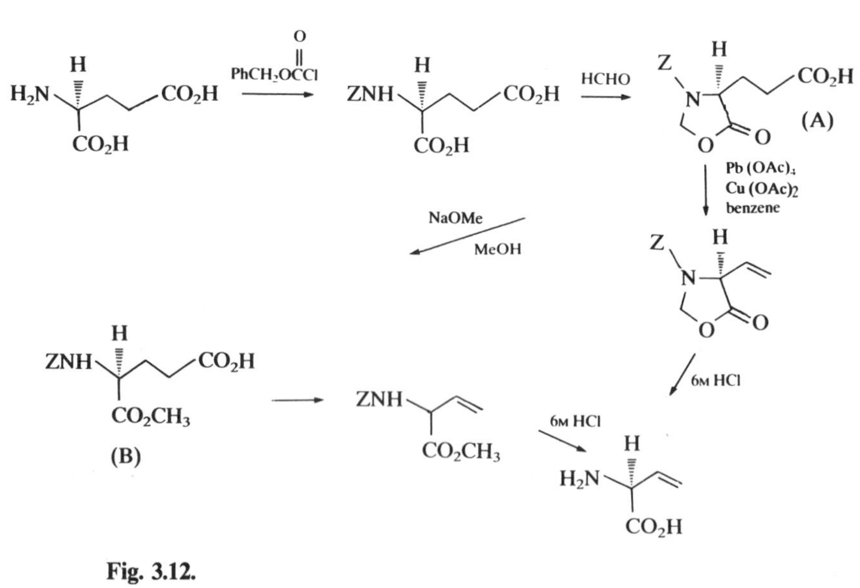

Fig. 3.12.

leads to decarboxylation (a reaction in which lead tetra-acetate is commonly involved) to give the protected L-vinylglycine. Alternatively, (A) can be transformed in (B) by the action of sodium methoxide in methanol. Compound (B) is particularly valuable since it is a derivative of L-glutamic acid in which the two carboxyl groups are chemically distinguishable. Thus compound (B) can now be selectively decarboxylated using lead tetra-acetate to give the N^α-protected L-vinylglycine. The chiral centre in the target molecule is of course directly derived from the chiral centre in the starting material.

3.2.3.2 *Alkylation at centres other than the chiral centre*

Unless special precautions are taken (see Section 3.2.3.1), there is the danger that, because of the acidity of the α-hydrogen atom, α-amino acids will undergo racemization if they are subjected to basic reaction conditions. Many alkylation reactions require such conditions. In Fig. 3.13, alkylation of an amino acid is carried out with an alkyl cuprate which obviates the necessity for carrying out the alkylation process in a strongly basic medium.

3.2.3.3 *Side-chain manipulation*

Many important compounds are derived from α-amino acids via modification of their side-chains, e.g. partial synthesis of ACE (angiotensin-converting enzyme) inhibitor. The function of the ACE is to transform angiotensin I into angiotensin II which is a powerful pressor agent, i.e. it raises the blood pressure by constricting the arterioles and stimulates the secretion of aldosterone. Thus a possible way of controlling high blood pressure is to administer an ACE inhibitor (Fig. 3.14).

C-protected alanine is reacted with ethyl 4-oxo-4-phenyl-2-butenoate in a Michael reaction. A new chiral centre is created, and hence, since there is no asymmetric induction, a diastereomeric mixture results. Fortunately, the diastereoisomers can be separated by recrystallization. The $(S)(S)$ diastereo-isomer is reduced and can then be converted by esterification into an ACE inhibitor.

There is an increasing interest in selective and efficient iron sequestering agents for therapeutic, diagnostic, and agricultural applications. Such a compound, occurring naturally, is enterobactin. This macrocyclic trilactone

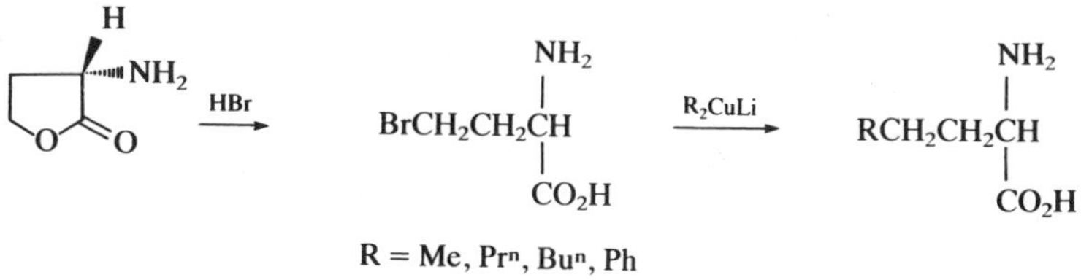

R = Me, Pr^n, Bu^n, Ph

Fig. 3.13.

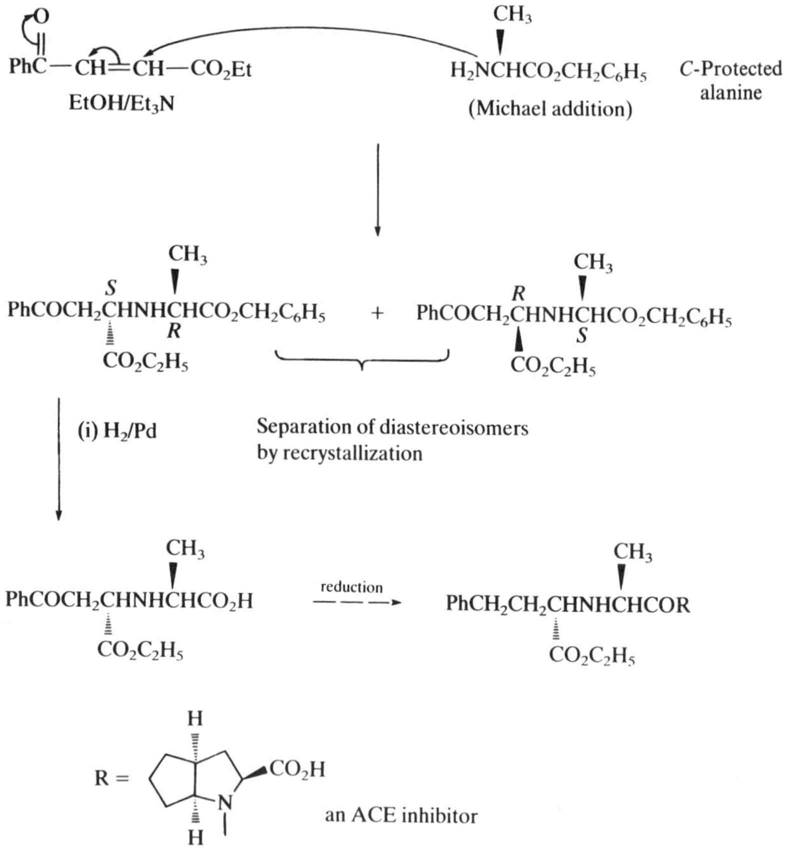

PhC$-$CH$=$CH$-$CO$_2$Et $\qquad\qquad$ H$_2$NCHCO$_2$CH$_2$C$_6$H$_5$ *C*-Protected alanine

$\overset{O}{\underset{}{\parallel}}$ (with CH$_3$ above H$_2$NCHCO$_2$CH$_2$C$_6$H$_5$)

EtOH/Et$_3$N $\qquad\qquad\qquad\qquad$ (Michael addition)

PhCOCH$_2$CHNHCHCO$_2$CH$_2$C$_6$H$_5$ +   PhCOCH$_2$CHNHCHCO$_2$CH$_2$C$_6$H$_5$

(left: CH$_3$ above, *S* / *R* labels, CO$_2$C$_2$H$_5$ below) (right: CH$_3$ above, *R* / *S* labels, CO$_2$C$_2$H$_5$ below)

(i) H$_2$/Pd $\qquad$ Separation of diastereoisomers by recrystallization

PhCOCH$_2$CHNHCHCO$_2$H $\xrightarrow{\text{reduction}}$ PhCH$_2$CH$_2$CHNHCHCOR

(left: CH$_3$ above, CO$_2$C$_2$H$_5$ below) (right: CH$_3$ above, CO$_2$C$_2$H$_5$ below)

R = (bicyclic structure with H, CO$_2$H, N, H)   an ACE inhibitor

Fig. 3.14.

bearing sideophores (catechol residues that chelate to iron) has been synthesized via a short, elegant route which relies upon tin to act as a template thereby directing the cyclization process (Fig. 3.15).

In this synthesis N$^\alpha$-protection is effected by means of the trityl group, (Ph$_3$C$-$) $\equiv$ Trt. Intramolecular cyclization to give the β-lactone (C) is induced by the powerful and well-used esterification reagent – a mixture of DCCI and 4-dimethylaminopyridine. The β-lactone (C) is reacted with the stannoxane (D). It is possible that this reaction proceeds via multiple insertions of the β-lactone (C) into (D) followed by expulsion of the cyclized product (E) and regeneration of the tin template. Detritylation is easily accomplished by treatment with acid (owing to the ease of formation of the triphenylmethyl cation). Acylation of the amino groups is carried out by standard peptide bond formation, namely by reaction of the amino groups with an active ester (formation of the 4-nitrophenate anion being the driving force; Fig. 3.16).

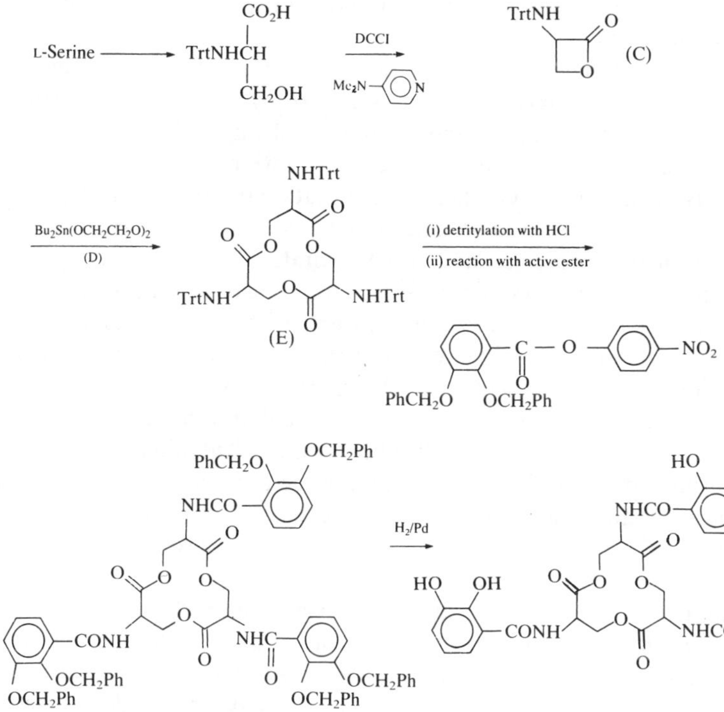

Fig. 3.15.

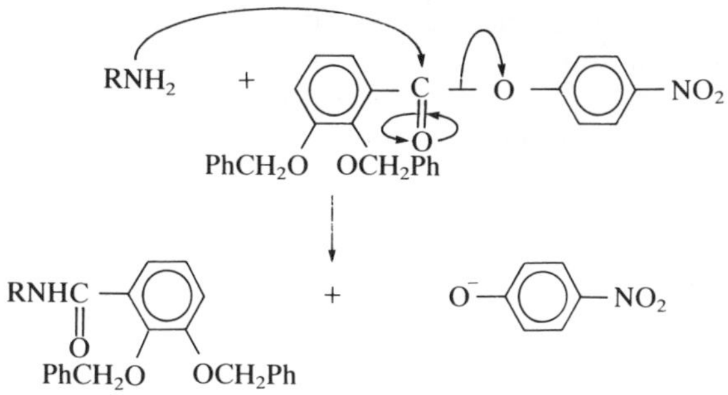

Fig. 3.16.

3.2.4 Resolution of racemic mixtures of α-amino acids

The following methods have been employed:

(i) Formation of diastereoisomeric salts with optically active bases (often alkaloids such as brucine) or acids, e.g. camphorsulphonic acid. The diastereoisomers are separated by crystallization or chromatography and the amino acid regenerated by addition of whichever is appropriate – either acid or base.

(ii) Formation of diastereoisomeric amides by reaction with optically active acids, e.g. L-menthoxyacetic acid followed by separation (crystallization or chromatography) and regeneration by hydrolysis – avoiding alkaline conditions which may cause racemization.

(iii) Chiral HPLC. Several optically active stationary phases have been developed which are based on natural optically active compounds, e.g. α-amino acids, proteins, and carbohydrates. By using supports functionalized with L-proline, racemic amino acids can be separated. The L-proline-carrying phase is loaded first with copper ions and the resulting immobilized proline–copper complex can reversibly complex with additional ligands present in the mobile phase. When a mixture of D and L amino acids is eluted through the column the retention time for the D and L enantiomers will differ owing to the difference in the equilibrium constants for the formation of the D amino acid-(Cu^{2+})L-proline and L-amino acid-(Cu^{2+})L-proline diastereoisomeric pairs. The amino acids are usually eluted as the copper complexes. Figure 3.50 shows the resolution of racemic alanine, serine, leucine (A), proline, aminobutyric acid and threonine (B) on a column derivatized with N'-benzyl-(R)-1,2-diaminopropane. In this case the amino acids were eluted as their copper(II) complexes. Such separations can be scaled up, e.g. a 1 litre column containing 300 g L-hydroxyproline derivatized resin was found to resolve completely up to 20 g racemic proline in one elution cycle.

(iv) Resolution has also been accomplished by transforming the α-amino acids as their esters into diastereoisomeric mixtures of Schiff bases by their reaction with optically active ketones, e.g. 2-hydroxypinan-3-one (**3.5**). The diastereoisomers are separated by chromatography and the amino acids regenerated by hydrolysis.

(**3.5**)

(v) Resolution using enzymes. The proteases are enzymes capable of hydrolysing the peptide bond. If a racemic mixture of an acetylated amino acid is subjected to such enzymes, deacetylation of L-amino acid derivative occurs preferentially. Separation of acylated and non-acylated amino acid can be accomplished by standard means. Aminoacylases may be used to accomplish deacetylation. These enzymes can be immobilized on inert supports with retention of their activity and this has important commercial consequences. Flow

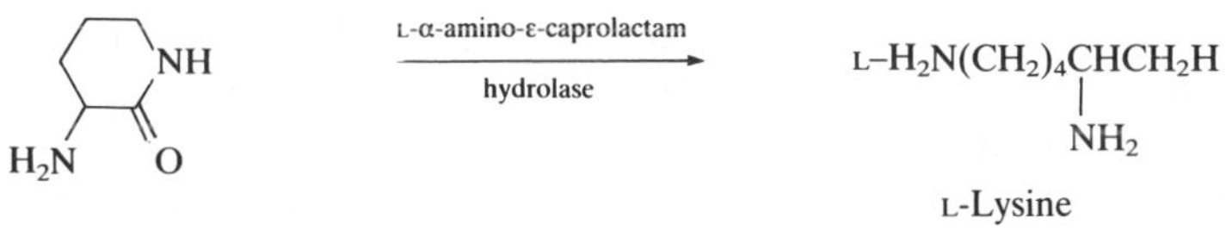

Fig. 3.17.

reactors have been designed in which the acetylated amino acid is passed through a bed of the immobilized enzyme and the products separated via simple means. Using a 1000 litre capacity column containing the immobilized enzyme, it has proved possible to obtain $200-700 \, kg \, l^{-1}$ of enantiomerically pure amino acid, and an added bonus is that, although the immobilized enzyme does become deactivated, regeneration of the columns is a relatively simple process.

Enantiomerically pure α-amino acids have also been produced using enzymic processes, for example production of L-lysine from DL α-amino-ε-caprolactam (Fig. 3.17). The D-α-amino-ε-caprolactam need not be wasted. Reaction of this material with D-α-amino-ε-caprolactam racemase regenerates the racemic starting material which can be subjected to the resolution process.

In another process, the natural process for producing α-amino acids via transamination has been imitated. If a mixture of a keto acid ammonia and pyridoxal phosphase is passed over immobilized lyase, transamination takes place. Using this process, optically active tryptophan and tyrosine have been produced as have some pharmaceutically important compounds such as L-dopa and serotonin (see Chapter 7).

3.3　Biodegradation of the amino acids

Amino acids which are ingested in higher animals via the breakdown of proteinaceous food in the gut are commonly used in turn for the biosynthesis of protein, or may become precursors of hormones, the nucleic acid bases, the porphyrins, cell membrane components, or some vitamins. However, if they are ingested in excess of the quantities required to maintain body protein, they may be degraded oxidatively and used ultimately to provide energy. For most of the amino acids, the catabolic process begins with trans-amination, usually to α-ketoglutarate, to form L-glutamate and a new α-keto acid, which may be degraded further. If, on administration to a starving rat, an amino acid is degraded to give rise to increased levels of glucose in the circulation, it is termed 'glucogenic', while if it gives rise to acetyl CoA (or acetyl CoA and acetoacetate) it is called 'ketogenic'. Of the common amino acids, only leucine is purely ketogenic, while phenylalanine, isoleucine, lysine, and tyrosine are simultaneously ketogenic and glucogenic, the others being only glucogenic. Each amino acid has an individual breakdown pathway, although in several cases amino acids display similar modes of degradation, or channel into a degradative pathway common to a number of amino acids. The topic is dealt with at length in good biochemistry texts; here we shall confine discussion to basic principles and a few illustrative examples of chemical interest.

3.3.1 Transamination

The transaminases, or aminotransferases, contain pyridoxal phosphate (**3.6**) tightly but non-covalently bound as a prosthetic group. This condenses with the α-amino group of the amino acid to form an aldimine (**3.7**), which can tautomerize to the ketimine (**3.8**), which is hydrolysed to form the α-ketoacid and pyridoxamine (**3.9**). The pyridoxamine then condenses (usually) with the α-ketoglutarate to form a ketimine and the above reaction sequence reverses to regenerate pyridoxal phosphate and form L-glutamate. The equilibrium constants for the reactions catalysed by the transaminases are close to unity.

Since, however, the glutamate formed is oxidized apparently via α-ketoiminoglutarate (**3.10**) which undergoes hydrolysis to release ammonia and reform α-ketoglutarate, the transaminase-catalysed reactions serve to collect the α-amino groups of at least 12 amino acids (alanine, arginine, asparagine, aspartic acid, cysteine, isoleucine, leucine, lysine, phenylalanine, tryptophan, tyrosine, and valine) in the form of glutamic acid which in turn gives rise to ammonia. The oxidative degradation of glutamate affords a

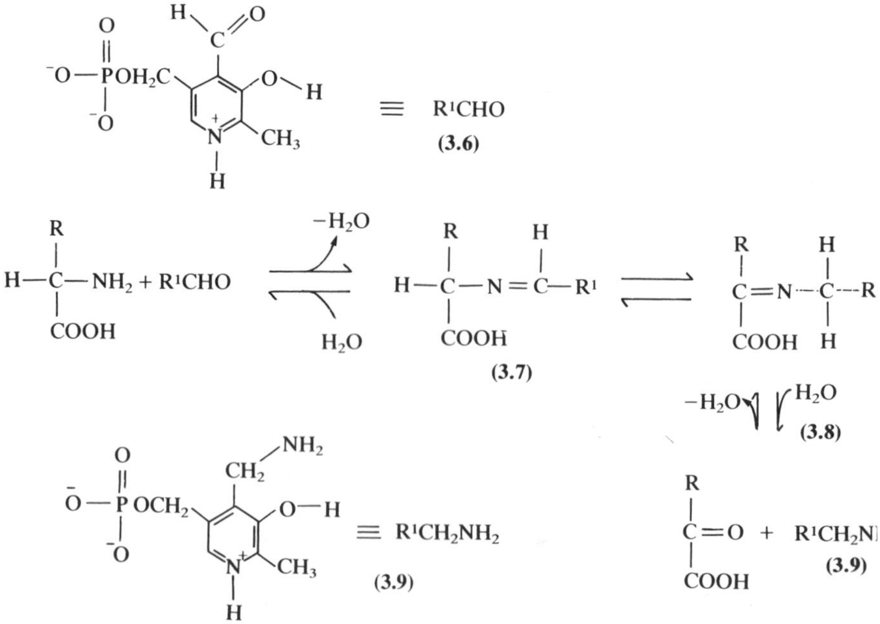

(3.11)

(3.14)

NH_2
|
COOH CH_2 CHO
| | |
CH_2 CH_2 CH_2
| | |
CH_2 CH_2 CH_2
| | |
$C = NH$ $HCNH_2$ $H—C—NH_2$
| | |
COOH   COOH   COOH

(3.10)   (3.12)   (3.13)

molecule of NADH, which is oxidized by the respiratory chain with concomitant formation of three molecules of ATP, and thus oxidative deamination gives rise to useful biosynthetic energy. In the intermediate (3.11) formed during transamination, the phenolic oxygen and the nitrogen of the Schiff base are thought to be coordinated via a proton which serves to hold the π-conjugated system in a planar configuration. The protonated pyridine ring nitrogen can then act as an electron sink, to accommodate the electrons present in bonds a, b, or c. Consequently, pyridoxal phosphate is not a prosthetic group exclusive to the transaminases (in which bond a is broken during the aldimine → ketimine step) but is also found in the amino acid racemases (bond a is broken and re-formed with racemization at C_α), the decarboxylases (bond b is broken), and in enzymes catalysing side-chain cleavage (bond c is broken if R is hydroxymethyl or a similar methylol).

3.3.2 *The metabolic fate of the α-ketoacids*

The α-ketoacids formed by transamination of the α-amino acids are, in some cases, members of metabolic pathways. Pyruvate (formed from alanine), oxaloacetate (formed from aspartate), and α-ketoglutarate (formed from glutamate) can all feed into the citric acid cycle, and from there form glucose via glucogenesis; hence their classification as glucogenic species. Valine, leucine, and isoleucine are all transaminated and the resulting α-ketoacids acted upon by an α-ketoacid dehydrogenase, α-ketoisovalerate dehydrogenase, to give branched acyl CoA derivatives which undergo, wholly or partially, the β-oxidation sequence seen in fatty acid catabolism to afford (eventually) succinyl CoA (from valine) and acetyl CoA (from leucine). In one rare genetic condition, α-ketoisovalerate dehydrogenase is inactive and the α-ketoacids accumulate in the bloodstream, causing mental retardation and imparting the characteristic odour to the urine of sufferers from which the disorder takes its name, maple syrup urine disease.

Arginine is hydrolysed by arginase in the course of the urea cycle, affording urea (the principal nitrogenous excretion product in primates and terrestrial animals) and ornithine (**3.12**) which affords glutamic semialdehyde (**3.13**) on transamination. Hydrolysis of the oxidation product (**3.14**) of proline also gives rise to (**3.13**), which is then oxidized to glutamate and degraded as described above, thus illustrating a second commonality of pathway.

Phenylalanine suffers a particularly interesting degradative fate. In the first step of its catabolism it is oxidized to tyrosine using molecular oxygen by the enzyme phenylalanine 4-monooxygenase. If $^{18}O_2$ is used, one atom of oxygen-18 is incorporated in the tyrosine, the other being reduced to (^{18}O) water. Moreover, if the phenylalanine is tritiated at the 4-position, it is found that the resultant tyrosine contains the tritium label at the 3-position. This remarkable migration was first discovered by researchers in a National Institute of Health laboratory in the USA, and consequently dubbed the NIH shift. The enzyme contains iron(II) ions, and requires NADPH and dihydrobiopterin as cofactors. The mechanism remains a matter of some conjecture, but a plausible possibility is shown in Fig. 3.18. The tetrahydrobiopterin may form a peroxide (**3.15**) which oxidizes phenylalanine, being functionally equivalent to a hydroxonium ion, HO^+. The

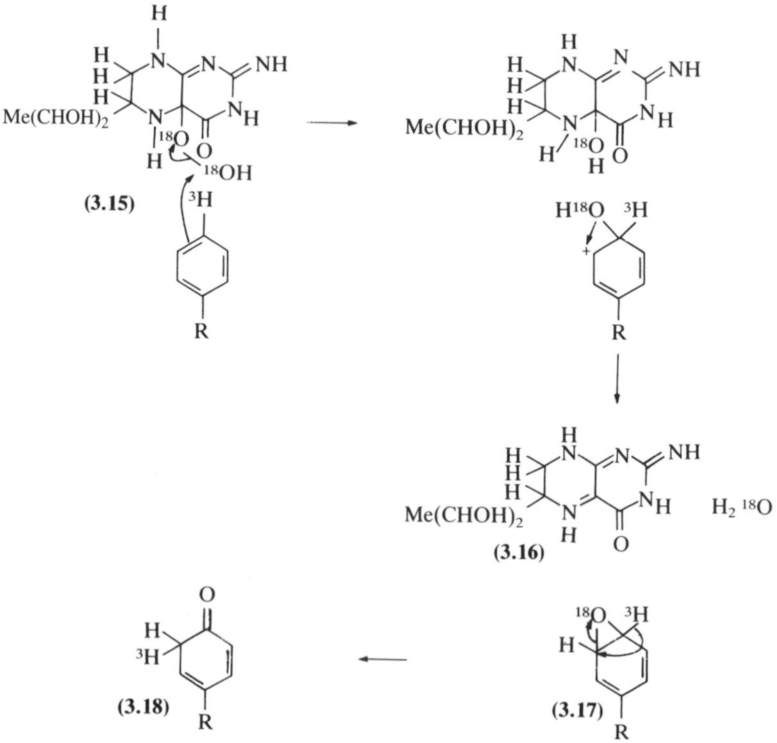

Fig. 3.18.

resultant hydroxylated tetrahydrobiopterin loses water to become dihydro-biopterin (**3.16**), and the arene oxide (**3.17**) of phenylalanine is formed. Ring-opening of the epoxide with a concomitant 1,2-shift of tritium then affords (**3.18**) which aromatizes to give tyrosine. In model reactions, arene oxides have been prepared and shown to exhibit the NIH shift. Due to a recessive mutant gene, about one person in 40 000 lacks the functional phenylalanine-4-monooxygenase enzyme, and is unable to form tyrosine in this way. Instead, phenylalanine undergoes transamination to form phenylpyruvic acid, which is excreted in the urine, a condition known as phenylketonuria. Infants born with this disease rapidly suffer mental retardation unless reared on a low phenylalanine diet, and consequently the test for phenylketonuria is one of the first metabolic checks applied to new-born children. Since the increasing use of the non-nutritive sweetener aspartame (L-aspartyl-L-phenylalanine methyl ester) in prepared foods is likely to increase the ingestion of phenylalanine in the diet, sufferers from this disease need to maintain constant awareness of the composition of the foods they eat, together with their additives.

Tyrosine undergoes transamination to afford *p*-hydroxyphenylpyruvic acid (**3.19**), which is oxidized by molecular oxygen and simultaneously decarboxylated to afford homogentisic acid (**3.20**), in a reaction catalysed by *p*-hydroxyphenylpyruvate dioxygenase. If $^{18}O_2$ is used, the labelling pattern is found to be as indicated (**3.20**), and it is seen that an NIH-type shift of the methylenecarboxy function has occurred. The mechanism of the reaction is uncertain, although it is tempting to speculate that an organic peroxide, possibly similar to (**3.15**), may be involved (see Fig. 3.18). Oxidation by homogenistic acid 1,2-dioxygenase then opens the aromatic ring to afford maleylacetoacetate (**3.21**) which is then isomerized enzymically to fumarylacetoacetate and hydrolysed to afford fumarate, a component of the citric acid cycle, and acetoacetate which can be converted to two molecules of acetyl CoA. Thus phenylalanine (and tyrosine) are simultaneously glucogenic and ketogenic. It will be noted that homogentisic acid is a substituted hydroquinone. In another rare 'inborn error of metabolism', the 1,2-dioxygenase is non-functional, and homogenistic acid is excreted in the urine, which blackens upon standing, particularly rapidly in alkaline conditions, in the same way that alkaline hydroquinone solutions blacken, and for the same reasons of aerial oxidation. Sufferers from this disease, alkaptonuria, also display dark pigmentation in their connective tissue, but the consequences are less dire than those of phenylketonuria or maple syrup urine disease.

3.3.3   Biosynthesis of the amino acids

There is considerable variation in the capacity of different organisms to synthesize amino acids. All higher plants and many bacteria can synthesize

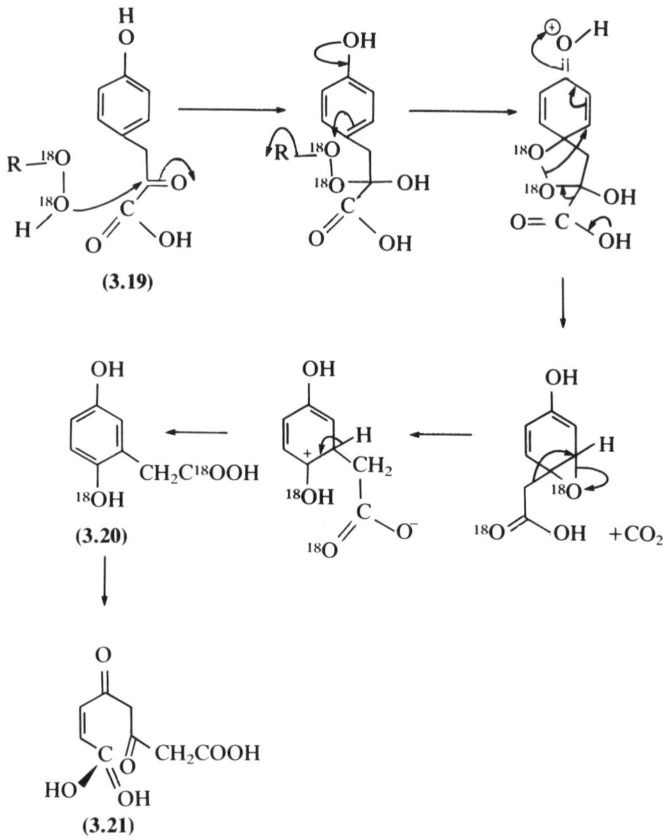

(3.19)

(3.20)

(3.21)

Fig. 3.19.

all 20 of the common amino acids, but other bacteria must be supplied with the amino acids that they cannot synthesize and therefore these become essential nutrients for growth. Man and the albino rat are able to synthesize 10 of the 20 common amino acids: glutamic acid, glutamine and proline, alanine, aspartic acid and asparagine, tyrosine, serine, glycine, and cysteine. The remaining 10 are thus the essential amino acids that must be supplied in the diet.

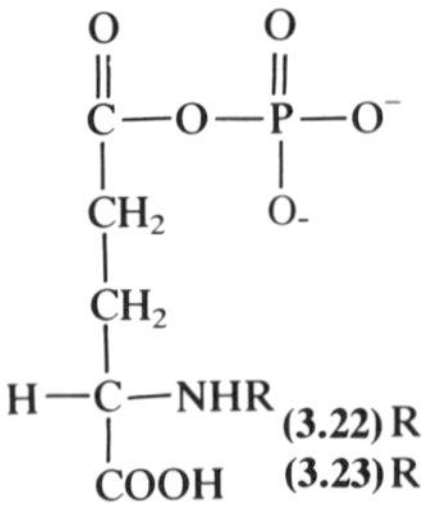

(3.22) R = H
(3.23) R = COCH$_3$

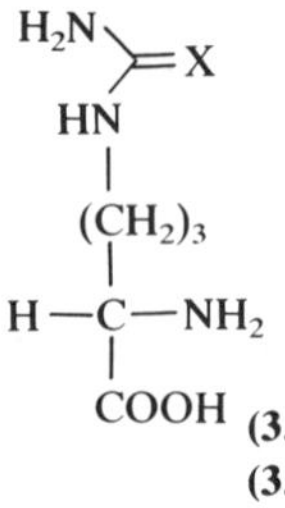

(3.24) X = O
(3.25) X = NH

Each of the 20 different amino acids is synthesized by a different multi-enzyme sequence. In some cases these are short, and are virtually the reverse of the degradative pathways described in the previous section. For instance, glutamate is formed by the reductive amination of α-ketoglutarate, alanine and aspartic acid via transamination from glutamate to pyruvate and oxaloacetate respectively, and proline via the intramolecular cyclization of glutamic semialdehyde (**3.13**) to give (**3.14**) which is then reduced by NADPH. Glutamic semialdehyde appears to be formed via the reaction of glutamate with ATP to form γ-glutamyl phosphate (**3.22**). The phosphate, which is of course a good leaving group, is displaced upon reduction by NADPH or NADH to afford (**3.13**). A similar process involving the formation and reduction of N-acetyl-γ-glutamyl phosphate is involved in the biosynthetic route to L-ornithine (**3.12**), via transamination of N-acetyl-glutamic γ-semialdehyde. The ornithine then condenses with carbamoyl phosphate giving L-citrulline (**3.24**) which is then aminated via condensation of the ureido group with L-aspartate, followed by β-elimination of fumaric acid to leave arginine (**3.25**). The processes from ornithine to arginine constitute the greater part of the urea cycle, which (as mentioned previously) is completed by the hydrolysis of arginine to afford ornithine and urea. If the phosphoryl group of (**3.22**) is displaced by ammonia, glutamine is formed, and the biosynthesis of asparagine takes place similarly, via β-aspartyl phosphate. Acyl phosphates are thus versatile intermediates in the biosynthesis of amino acids.

Tyrosine is formed by the hydroxylation of phenylalanine, as detailed previously, and phenylalanine and tryptophan both arise via the shikimate pathway, which is described elsewhere (see Section 1.6.2). It would be superfluous to consider in detail the biosyntheses of all the 20 common amino acids, which may be found in good biochemical texts, so, after the simple examples described above, we shall consider only one complex pathway in any detail: the biosynthesis of histidine.

In the first step of this remarkable sequence of reactions, ATP is glycosylated at the N-1 position by 5-phospho-α-D-ribofuranosyl-1-pyro-phosphate, a metabolite more commonly associated with the biosynthesis of pyrimidine nucleotides, with displacement of pyrophosphate to form N'-(5-phosphororibosyl)-ATP (**3.26**). The quaternization at N-1 facilitates hydrolytic attack at C-6 with ring-opening to afford (**3.27**), which is converted via an isomerase into the phosphoribulose derivative (**3.28**). The next step may be rationalized as ammonolysis of (**3.28**) using ammonia derived from the carboxamide group of glutamine to release 5-amino-imidazole-4-carboxamide ribonucleotide (**3.29**) and afford 1-formimino-ribulose-5-phosphate which cyclizes spontaneously to afford imidazole glycerol phosphate (**3.30**). This readily loses water with subsequent ketonization to give rise to imidazole acetol phosphate (**3.31**), in a process not unlike the conversion of (**3.27**) to (**3.28**). Transamination of (**3.31**) using glutamate as donor, followed by hydrolytic release of phosphate by a phosphatase, affords L-histidinol

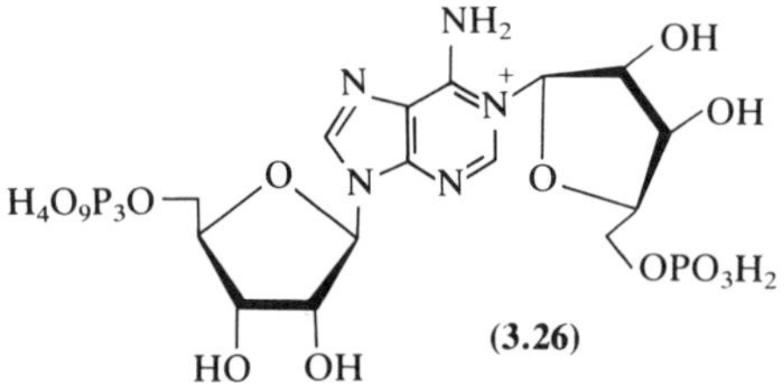

NH$_2$

OH

OH

H$_4$O$_9$P$_3$O

OPO$_3$H$_2$

HO   OH

(3.26)

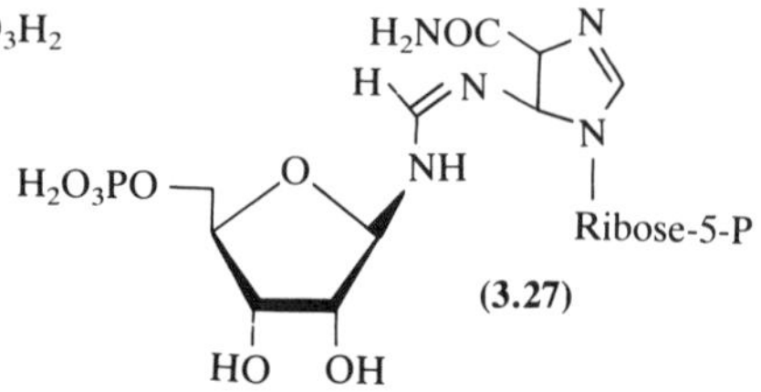

H$_2$NOC   N

H   N

H$_2$O$_3$PO   NH

Ribose-5-P

HO   OH

(3.27)

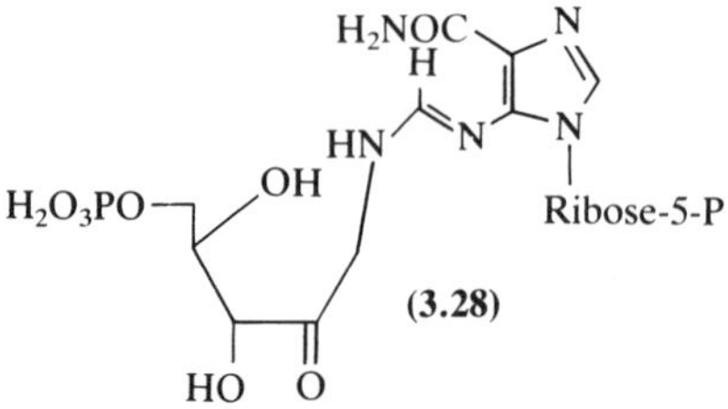

H$_2$NOC   N

H

HN   N

H$_2$O$_3$PO   OH   Ribose-5-P

(3.28)

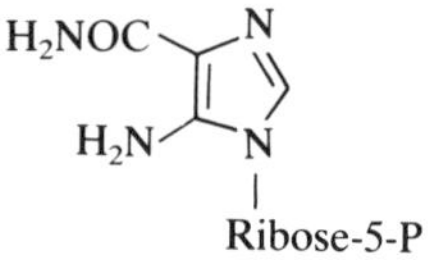

HO   O

H$_2$NOC   N

H$_2$N   N

Ribose-5-P

(3.29)

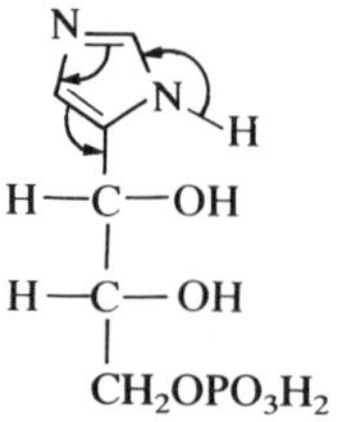

N

N

H

H—C—OH

H—C—OH

CH$_2$OPO$_3$H$_2$

(3.30)

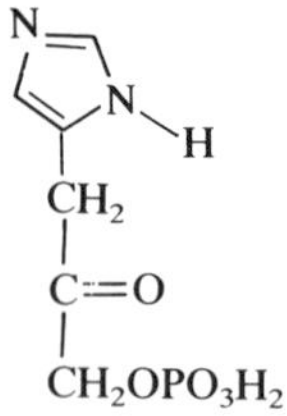

N

N

H

CH$_2$

C=O

CH$_2$OPO$_3$H$_2$

(3.31)

N

N

H

CH$_2$

H—C—NH$_2$

CH$_2$OH

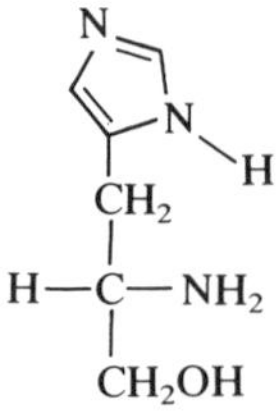

(3.32)

(**3.32**), after which two rounds of oxidation by the NAD$^+$-linked histidinol dehydrogenase affords L-histidine.

This biosynthetic sequence was eludicated in an elegant series of investigations by Ames, using mutants of *Salmonella typhimurium* and *Escherichia coli*. The ten enzymes involved in the sequence in *S. typhimurium* are encoded by nine contiguous structural genes which are under the control of a single operator gene, the whole constituting the *his* operon. If histidine is absent from the growth medium, the bacterium is able to synthesize it, unless there is a mutation in the *his* operon which prevents one or more of these functional enzymes being formed. If, however, the activity of a mutagen corrects the mutated gene to restore it, and thus the full biosynthetic pathway, to functionality, the resulting 'revertant' is able to grow on a histidine-free medium. This is the basis of the 'Ames test', in which strains of *S. typhimurium* which contain mutations in the *his* operon, and are thus unable to grow in the absence of histidine, are exposed to potential mutagens, and the number of revertants – revealed as colonies now able to grow despite the absence of histidine – are counted. The test thus affords a quantitative assessment of mutagenicity, and, while it has some shortcomings, it is one of the most widely used methods for assaying compounds of unknown mutagenicity. We thus see that the biosynthetic sequence of at least one amino acid offers a valuable practical application, albeit in rather a roundabout way!

3.4   Chemical synthesis of peptides

The synthesis of polypeptides requires methods for assembling an array of amino acids in a predetermined unique way in high yield, and under conditions that do not cause racemization. Thus, optically active amino acids (which are bifunctional compounds) have to be coupled in a specific manner and this requires that one of the functional groups be *protected* or *blocked*. The basic concepts for peptide synthesis are as shown in Fig. 3.20, i.e. protection, activation, coupling, and specific deprotection.

If the amino acids contain reactive groups in their side-chains (e.g. NH_2 as in lysine, CO_2H as in aspartic acid, SH as in cysteine) these will also require protection.

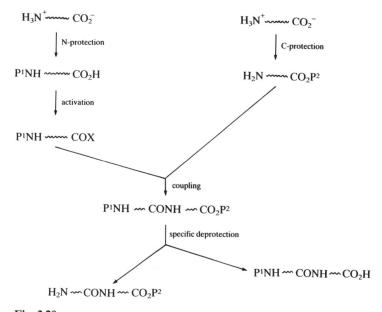

Fig. 3.20.

The challenges faced by peptide chemists are daunting. From the outlined strategy it can be seen that protecting groups are required which can be applied under mild conditions. Mild conditions are stipulated so that racemization is kept to the absolute minimum. A further requirement for protecting groups is that they can be removed specifically so that the group required in the next step can be uncovered specifically. Since a protected peptide may sometimes contain two or more protecting groups, this is a matter of some importance. The design of activation and coupling procedures also requires that the processes occur in high chemical yields with a total lack of racemization. Why is the lack of racemization stressed? If a peptide is assembled with some racemization occurring at every step, the yield of the optically pure product will be low and it will be contaminated with diastereoisomers from which it can be separated only with a great deal of difficulty.

Considerations must also be given to strategy employed for constructing a simple linear polypeptide containing many amino acid residues. The options are as follows:

Option 1   A ⟶ (A-B) ⟶ (A-B-C) ⟶ (A-Z)   i.e. growth from the C-terminus

(linear synthesis)

Option 2   Z ⟶ Y-Z ⟶ X-Y-Z ⟶ (A-Z)   i.e. growth from the N-terminus

(linear synthesis)

Option 3

L -----➤ (A-----L)
Z -----➤ (M-----Z)
➤ (A——Z)

Growth from the N-terminus to give smaller peptides which are then coupled to give the products

(convergent synthesis)

Option 4

A -----➤ (A-L)
M -----➤ (M-Z)
➤ A——Z

Growth from the C-terminus followed by coupling of the smaller peptides

(convergent synthesis)

Note that when a sequence of amino acids is written A-B-C or Pro-Ile-Leu, the amino acid on the left carries the free amino group, and the one on the right the free carboxyl group.

In general, growth of the chains is carried out from the N-terminal residue. Problems with racemization are most likely to occur at the activation step. Thus with growth from the carboxyl end, every activation is likely to lead to racemization at each of the peptide links. The activation process on a low molecular weight compound can be carried out under milder conditions than a higher molecular weight compound because of its more favourable solubility characteristics. It is factors such as solubility which also lead to the favouring of construction of polypeptides via synthesis of smaller units and then knitting these together, i.e. a convergent synthesis such as option 3. Another important factor favouring the convergent approach is that higher yields can be attained than with the linear approach.

The growth of the polypeptides can be carried out as a typical solution phase organic synthesis involving reaction followed by purification of the product using conventional procedures or, alternatively, the peptides may be grown on a polymer support (solid or liquid).

3.4.1 N-protecting groups

The object of protecting an amino group is to decrease its nucleophilicity so that it will not participate in the coupling reaction. Acylation of amino groups is known to accomplish this objective, i.e.

$$CH_3COCl + H_2NR \rightarrow CH_3CONHR$$
$$(3.33)$$

However, an important criterion for a protecting group is that it is removed under mild conditions such that cleavage of peptide bonds, modification of side-chain groups and racemization do not occur. Clearly the stringent conditions required for cleaving the amide (3.33) preclude the use of simple acyl groups.

Transformation of the amino groups into urethanes is the prime method of choice since the groups can be introduced with little or no racemization and features can be readily built into the protecting group which will aid its ready removal, e.g. by acid or mild alkali.

3.4.1.1 *Benzyloxycarbonyl group (Z) and related groups*

This group is introduced by reaction of the amino acids with the stable chloroformate (3.34) using a mixture of an organic and aqueous solvent mixture together with the application of vigorous stirring. Reaction with amino acid esters is usually carried out with chloroform as solvent.

$$PhCH_2OCCl + H_2N \sim CO_2P^2 \longrightarrow PhCH_2OCNH \sim CO_2P^2$$
$$(3.34) \quad \overset{\parallel}{O} \qquad\qquad\qquad (3.35) \quad \overset{\parallel}{O}$$

The fact that urethanes (3.35) can be cleaved by at least two different methods adds to the versatility of this group and accounts for much of its popularity.

Since the urethane (3.35) is benzylic, hydrogenolyses can be utilized to remove the group (Fig. 3.21). The hydrogenolysis requires very mild

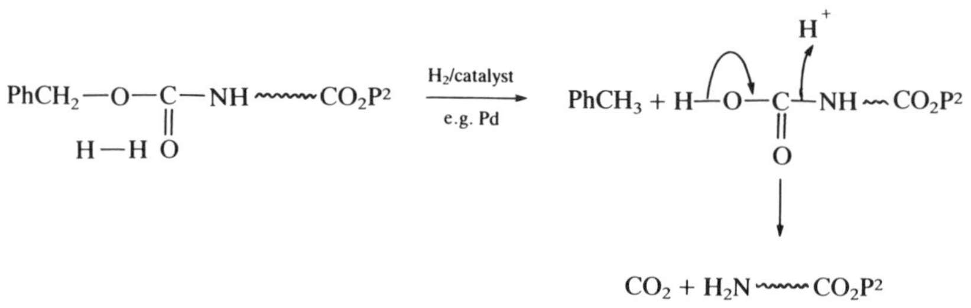

PhCH$_2$—O—C—NH ～～～CO$_2$P^2 $\xrightarrow[\text{e.g. Pd}]{\text{H}_2/\text{catalyst}}$ PhCH$_3$ + H—O—C—NH ～～CO$_2$P^2

$\qquad\qquad$ H—H O $\qquad\qquad\qquad\qquad\qquad\qquad\qquad\qquad$ O

$$\downarrow$$

CO$_2$ + H$_2$N ～～～CO$_2$P^2

Fig. 3.21.

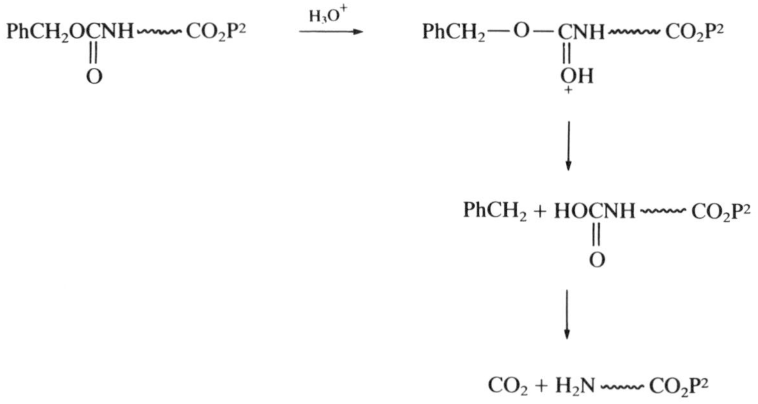

PhCH$_2$OCNH ～～～CO$_2$P^2 $\xrightarrow{\text{H}_3\text{O}^+}$ PhCH$_2$—O—CNH ～～～CO$_2$P^2

$\qquad\quad$ O $\qquad\qquad\qquad\qquad\qquad\qquad\qquad\qquad\quad$ OH

$$\downarrow$$

PhCH$_2$ + HOCNH ～～～CO$_2$P^2

$\qquad\qquad\qquad$ O

$$\downarrow$$

CO$_2$ + H$_2$N ～～～CO$_2$P^2

Fig. 3.22.

conditions. Clearly this method cannot be applied if the peptide has sulphur-containing residues since these lead to poisoning of the catalyst.

Cleavage of the urethane can be carried out with acid and this occurs mainly via an S_N1 mechanism. HBr in acetic (ethanoic) acid is the standard reagent for removal of this protecting group (Fig. 3.22). The formation of benzyl carbocations can lead to problems owing to the occurrence of S-benzylation of methionine residues and C-benzylation of tyrosine and tryptophan.

The urethanes (**3.35**) are stable towards the acidic conditions which are used to remove another popular N-protecting group, the *t*-butoxycarbonyl group. This together with the stability of the urethanes towards alkali makes the use of benzyloxycarbonyl groups very popular.

There is a wide variety of substituted benzyloxycarbonyl groups which can be used, e.g. 4-chloro and 4-nitrobenzyloxycarbonyl. The greater crystallinity of these derivatives compared with the unsubstituted compound has been the main driving force for their employment since this property aids purification of the derivatives.

The boron derivative (**3.36**) is useful since it can be reacted with diols to yield derivatives whose hydrophilicity is determined by the nature of the

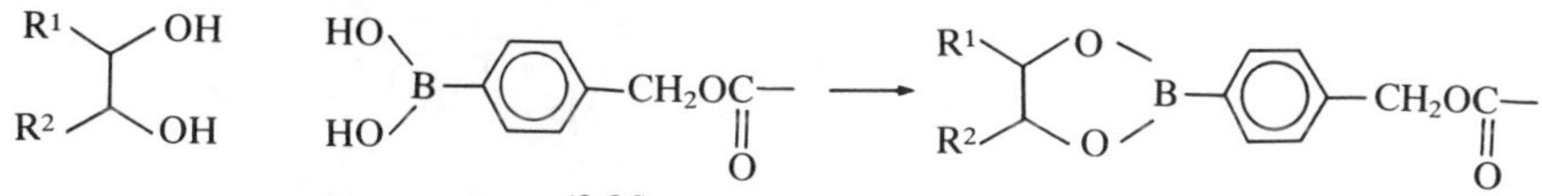

Fig. 3.23.

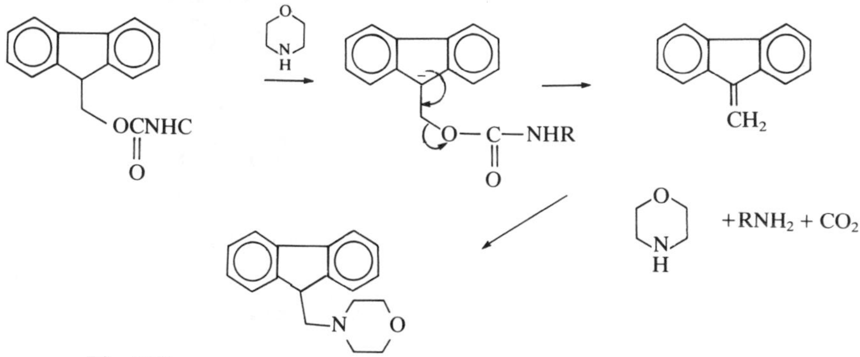

Fig. 3.24.

diol employed (Fig. 3.23). The Dobz group can be removed using conditions which are similar to those employed for the benzyloxycarbonyl group.

A particularly popular group is the fluorenylmethoxycarbonyl (Fmoc) group (**3.37**):

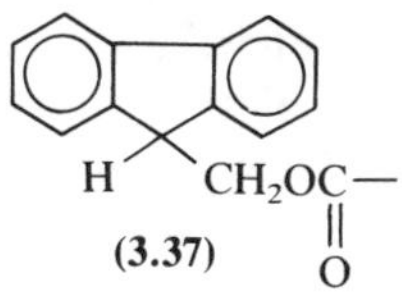

This group is introduced in the usual way using fluorenylmethoxycarbonyl chloride. The derivatives are stable towards acid but are cleaved by mild bases, e.g. piperidine or morpholine (Fig. 3.24). Very usefully, Fmoc can be used in the presence of acid-cleavable protecting groups. This allows it to be removed or retained with a high degree of specificity. It has proved to be particularly useful in solid-phase peptide syntheses.

3.4.1.2 t-Butoxycarbonyl group (Boc) and related groups

This group cannot be introduced in the same way as benzyloxycarbonyl groups due to the relative instability of t-butyl chloroformate (**3.38**)

$$Bu^tO-C-Cl \rightarrow (CH_3)_2C{=}CH_2 + CO_2 + HCl$$
$$\overset{\|}{O}$$

(**3.38**)

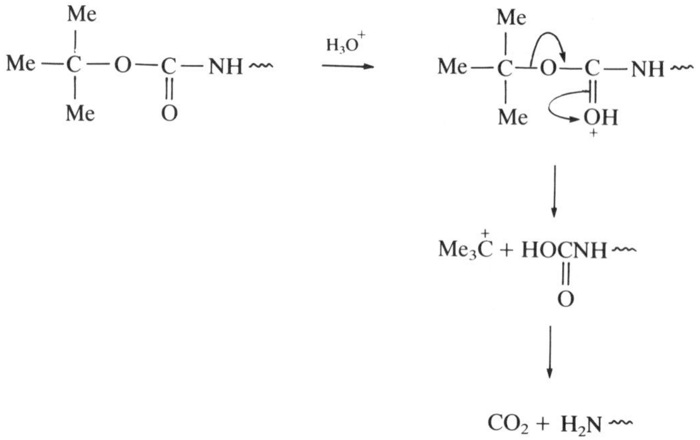

$$\underset{\substack{| \\ Me}}{\overset{\substack{Me \\ |}}{Me-C}}-O-\underset{\substack{\| \\ O}}{C}-NH\sim \quad \xrightarrow{H_3O^+} \quad \underset{\substack{| \\ Me}}{\overset{\substack{Me \\ |}}{Me-C}}-O-\underset{\substack{\| \\ OH \\ +}}{C}-NH\sim$$

$$\downarrow$$

$$Me_3\overset{+}{C} + \underset{\substack{\| \\ O}}{HOC}NH\sim$$

$$\downarrow$$

$$CO_2 + H_2N\sim$$

Fig. 3.25.

This problem has been overcome by the use of anhydride (**3.39**):

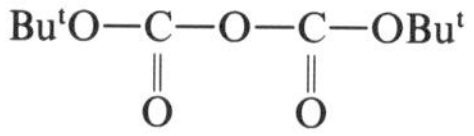

$$\underset{\substack{\| \\ O}}{Bu^tO-C}-O-\underset{\substack{\| \\ O}}{C}-OBu^t$$

(**3.39**)

The *t*-butoxycarbonyl group is stable towards catalytic hydrogenation, sodium in liquid ammonia, alkali and bases such as hydrazine. It is, however, rapidly cleaved under mild acidic conditions mainly via an S_N1 process (e.g. 1 M HCl in acetic acid, HBr in acetic acid, 25 per cent trifluoroacetic acid in dichloromethane), due to the ease of formation of the *t*-butyl cation (Fig. 3.25). The formation of carbocations can lead to the occurrence of unwanted side-reactions, which can be suppressed by the addition of scavengers such as thiophenol. The *t*-butoxycarbonyl group is invaluable since it is cleaved under milder conditions than those required to cleave the benzyloxycarbonyl group. When these two groups are present, either may be selectively removed (*t*-Boc via acid and the Z group by hydrogenolysis). This is termed *orthogonal protection*.

The 2-(4-biphenyl)-isopropyloxycarbonyl group (Bpoc) (**3.40**) is used as a protecting group. The derivatives are remarkably sensitive towards acids, due to the liberation of a highly resonance-stabilized tertiary carbocation. For example, the rate of cleavage of this group by 80 per cent acetic acid at 22–25°C is 3000 times faster than the Boc group. Thus the Bpoc group

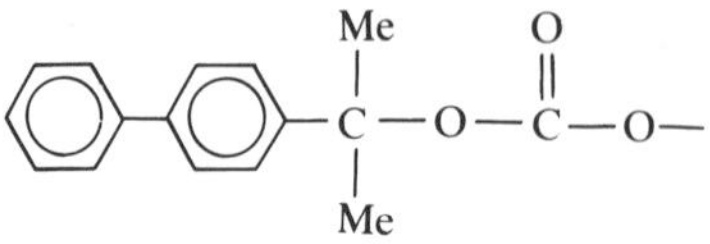

(**3.40**)

can be removed under conditions which leave Boc and *t*-butyl esters unaffected thus making a valuable contribution to the armoury of the peptide chemist. The group can be introduced via either its carbonate or its azide.

3.4.1.3 The trityl group

The triphenylmethyl group (Trityl, Trt) can be introduced by reaction of the amino group with triphenylmethyl chloride (trityl chloride):

$$Ph_3CCl + H_2N \text{---} \quad \xrightarrow[Et_2NH]{} \quad Ph_3 \text{---}$$

Deprotection can be accomplished via either hydrogenolysis or mild acid (aqueous acetic (ethanoic) acid). Under special conditions (90 per cent aqueous trifluoroethanol containing HCl) the group can be cleaved without affecting such groups as Bpoc and Boc.

The lack of nucleophilic character displayed by the tritylated amino group is attributed to steric hindrance by the trityl group. Cleavage by acid is facilitated by the formation of the triphenylmethyl carbocation:

$$Ph_3CNH \text{---} \quad \xrightarrow{H_3O^+} \quad Ph_3\overset{+}{C}NH_2 \text{---} \quad \longrightarrow \quad Ph_3\overset{+}{C} + H_2N \text{---}$$

3.4.2 Selective protection of α,ω-diaminocarboxylic acids

Depending upon the strategy of a polypeptide synthesis, it may be necessary to protect an ω-amino group throughout the synthesis. This requires the use of a protecting group which will remain stable throughout the protection–deprotection sequences carried out on the α-amino group. Clearly the cleavage conditions for the N^α and N^ω-protecting groups have to differ significantly. Thus, if the N^α is acid-labile, the N^ω needs to be either stable to acid or base-labile or to be less acid-labile than the N^α-protecting group (see Table 3.1).

Table 3.1.

Residue for N^α protection	Conditions for cleavage	Suitable N^ω protecting group
PhCH$_2$OCO (Z)	H$_2$/Pd	Boc, etc.
Boc	Cold CF$_3$CO$_2$H	Z, Tos[a], Pht[a]
Bpoc	80% HOAc	Boc, Z, Tos[a], Pht[a]
Trt	HCl in CF$_3$CH$_2$OH pH 4	Boc, Bpoc, Z
Fmoc	Base	Boc, Z, Bpoc, Dpp, etc.

a Tos, Tosyl; Pht, Phthalimido.

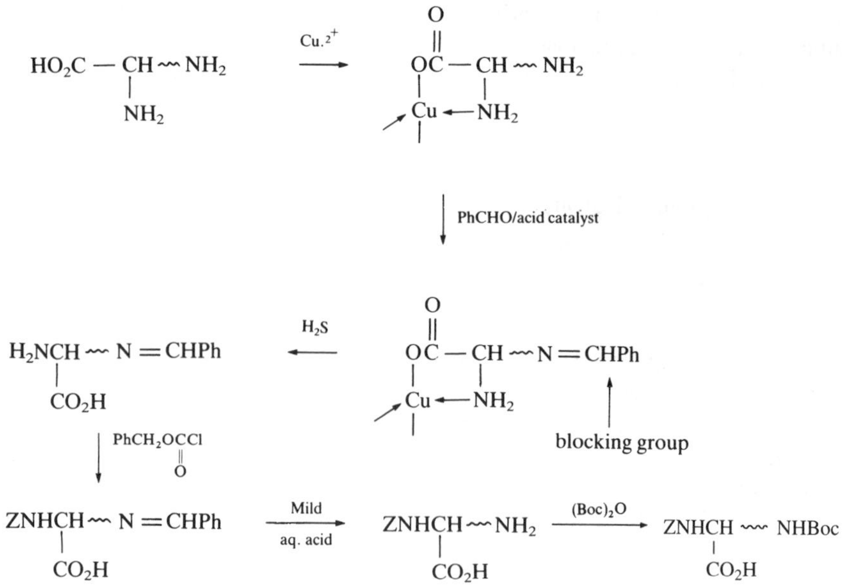

Fig. 3.26.

In order to protect the ω-amino group selectively the N^α amino group is usually protected as its copper complex.

Following introduction of the N^ω-protecting or blocking group, the copper is removed using hydrogen sulphide. Using this strategy Boc-Lys(Z)-OH has been prepared (Fig. 3.26).

N^α- and N^ω-protecting groups are summarized in Table 3.2.

3.4.3   C-protecting groups

Protection of the carboxyl group (summarized in Table 3.3) is necessary in order to prevent it being acylated in the coupling reaction. Furthermore, the protecting groups may also be used to aid solubility in appropriate solvents and to facilitate purification of the products.

The usual method of protection is to form an ester. Methyl esters are easily prepared but suffer from the disadvantage that their removal by alkali can give rise to many side-reactions, including racemization. Phenyl esters have been found to be particularly useful since they can be cleaved by alkaline hydrogen peroxide (pH 10.5):

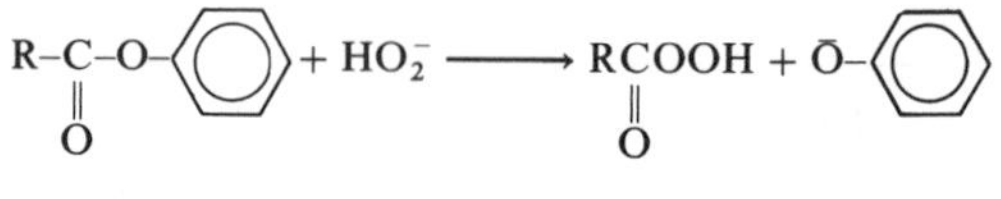

Table 3.2 N^α- and N^ω-protecting groups.

Benzyloxycarbonyl PhCH$_2$OCO—	Z	H$_2$/Pd (unless S present), HBr/HOAc (Boc stable under these conditions).
Fluorenylmethoxycarbonyl 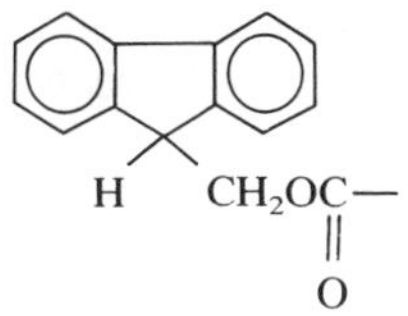 H CH$_2$OC— ‖ O	Fmoc	Stable in acid; mild bases, morpholine and piperidine cleave; can be used in the presence of acid-cleavable group.
t-Butoxycarbonyl Bu^{t}OCO— ‖ O	Boc	Stable towards H$_2$/Pd, Na/liq. NH$_3$, bases. Cleaved by 25% CF$_3$CO$_2$H in CH$_2$Cl$_2$.
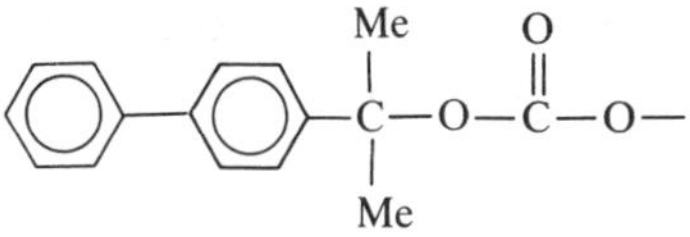	Bpoc	*Very* sensitive to acids. Can be cleaved in presence of Boc.
Trityl Ph$_3$C—	Trt	Removed by H$_2$/Pd or mild acid.

Table 3.3 C-protecting groups.

Phenyl esters	Cleaved by alkaline H$_2$O$_2$ (pH 10.5).
Benzyl esters PhCH$_2$ Ph$_2$CH (use Ph$_2$CN$_2$)	H$_2$/Pd (absence of S) CF$_3$CO$_2$H/HOAc. H$_2$/Pd (absence of S) CF$_3$CO$_2$H/HOAc.
t-Butyl (acid + ⟩=CH$_2$ + acid catalyst)	Stable to acid conditions used for cleaving Bpoc and Trt.

Order of ester reactivity towards acid hydrolysis:

Diphenylmethyl > Bu^t > Trimethylbenzyl > Benzyl

In this reaction the hydroperoxy anion successfully attacks the carbonyl group of the ester since the leaving group is the stable phenolate anion. This reaction produces a peracid which then decomposes to give the desired carboxylic acid. Importantly, little or no racemization occurs under these conditions.

Clearly the use of C-protecting groups which are cleaved by base enables one to operate on the N-terminal residues with protecting groups that are cleaved by acid.

The use of benzyl and related esters is also attractive since, in the absence of sulphur, these groups can be removed by hydrogenolysis. Esterification is carried out by treating the amino acid with benzyl alcohol in the presence of hydrogen chloride or other acid catalysts such as *p*-toluenesulphonic acid. Diphenylmethyl esters (ODpm) are also used and can be prepared by reacting the α-amino acid with diphenyldiazomethane:

$$RCO_2H + Ph_2CN_2 \rightarrow RCO_2CHPh_2 + N_2$$

These derivatives are also cleaved by hydrogenolysis. The benzyl esters can also be cleaved by acid, e.g. trifluoroacetic and/or 0.2 M hydrochloric acid in nitromethane. Many substituted benzyl esters, e.g. 4-nitrobenzyl (ONb) and 4-bromobenzyl, are more stable to acid (e.g. hydrogen bromide in acetic (ethanoic) acid and trifluoroacetic acid) and their use allows a greater degree of flexibility in the N-protecting groups which can be employed; e.g. the acidolysis of benzyloxycarbonyl groups (Z) can be carried out in the presence of 4-nitrobenzyl esters. All these benzyl esters can of course be cleaved by hydrogenolysis.

The 2,4,6-trimethylbenzyl ester (OTmb) as a protecting group is slightly more stable towards acid than the *t*-butyl ester. It can be removed by 2 M HBr in acetic acid and by anhydrous trifluoroacetic acid.

Another advantage of benzyl esters is that they increase the solubility of the amino acid and peptides in organic solvents. Indeed, solubility may become the limiting factor for the size of peptide which can be prepared.

A particularly useful ester protecting group is the *t*-butyl ester. For steric reasons, the esters of the *t*-butanol cannot be prepared by many of the usual methods. They can, however, be prepared by reaction of the amino acid or N-protected amino acid with isobutene in an organic solvent containing an acid catalyst, e.g. *p*-toluenesulphonic acid:

t-Butyl esters are particularly useful in combination with N-protecting groups which are cleaved by hydrogenolysis since they are completely stable under these conditions (orthogonal protection). *t*-Butyl esters are also stable under the acidic conditions used to cleave Bpoc and Trt amine protecting groups. Not surprisingly, *t*-butyl esters are stable towards nucleophilic attack and hence their use in conjunction with protecting groups cleaved by base is particularly attractive. The reactivity of these esters towards acidolysis is diphenylmethyl (ODpm) $\simeq$ *t*-butyl (OtBu) > trimethylbenzyl (OTmb) > benzyl.

Table 3.4 Cleavage conditions for some N- and C-protecting groups.

	H$_2$O room temperature	75% HOAc 1 h 2°C	1 M HCl in HOAc 30 min 25°C	50% TFA in CHCl$_2$ 30–60 min 25°C	2M HBr in HAc 30–60 min 25°C	H$_2$/Pd/C
N-Protecting						
Bpoc	–	+	+	+	+	–
Boc	–	–	+	+	+	–
Z	–	–	–	–	+	+
Trt	–	+	+	+	+	+
C-Protecting						
ODpm	–	–	+	+	+	+
OtBu	–	–	+	+	+	–
OTmb	–	–	±	±	+	+
OPtm	–	–	±	±	+	–
OBzl	–	–	–	–	–	+
ONb	–	–	–	–	–	+
OPh	–	–	–	–	–	–

+ Indicates cleavage; − indicates no reaction; ± partial cleavage.

Other C-protecting groups include phthalimidomethyl esters (OPtm) (**3.41**), which are cleaved by acid under conditions similar to those employed for the trimethylbenzyl group and are stable to hydrogenolysis, but which are rapidly cleaved by sodium thiophenoxide and also by hydrazine.

Table 3.4 contains data on the conditions required to cleave various N- and C-protecting groups. From this table one can determine which is the best combination of N- and C-protecting groups so as to allow one to operate specifically at one terminus.

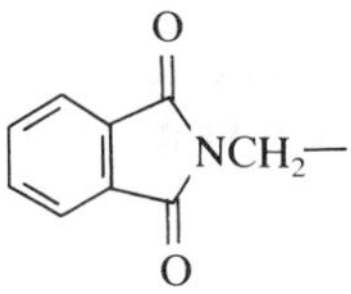

(**3.41**)

3.4.4 Selective protection of carboxyl groups of the mono amino dicarboxylic acids

With aspartic and glutamic acids there is often the need to keep the side-chain carboxyl group protected throughout the construction of the polypeptide. γ-Alkyl esters (methyl ethyl and *t*-butyl) of glutamic acid have been prepared by a transesterification process and used in peptide synthesis. β-Benzyl aspartate has also been prepared from the acid by esterification with benzyl alcohol containing sulphuric acid and used in peptide synthesis. From Section 3.2.3.1 it can be seen that combined protection of the α-amino and carboxyl group can be achieved, leaving the side-chain carboxyl group available for manipulation.

3.4.5   Protection of other functional groups

3.4.5.1   Hydroxyl

The hydroxyl groups of serine, threonine and hydroxyproline, and the phenolic hydroxy group of tyrosine, may need to be protected. Methods include *O*-benzylation and *O*-*t*-butylation. Removal of the benzyl group can be accomplished by hydrogenolysis or Na/liq. NH_3 and by acidic conditions. The *O*-*t*-butyl esters are also cleaved under acidic conditions. However, the presence of these groups in a peptide severely limits the ways in which the peptide may be further extended using acid and sensitive protecting groups, since the cleavage conditions for these groups are so similar to those for many N- and C-protecting groups.

3.4.5.2   Sulphydryl groups

Protection of the SH group of cysteine during synthesis is essential since its anion and itself are both such powerful nucleophiles. Furthermore, oxidation of the thiol group to give cysteine occurs very readily. The most widely used group is the *S*-benzyl group which can be removed by sodium in liquid ammonia. However, this cleavage reagent can also lead to cleavage of peptide materials. More recently the acetamidomethyl group has been introduced:

$$CH_3CONHCH_2OH \rightleftharpoons CH_3CONH\overset{+}{C}H_2 \xrightarrow{\text{RSH}} RSCH_2NHCOMe$$

Cleavage is brought about by Hg(II) ions in 50 per cent acetic (ethanoic) acid. A great advantage of this group is that it is stable under the acidic conditions required to remove Z and Boc groups.

3.4.5.3   Imidazoyl group of histidine

The imidazoyl group possesses a nucleophilic nitrogen and therefore it has to be protected. This is usually done with benzyl chloromethyl ether:

The protecting group can be removed by hydrogenolysis.

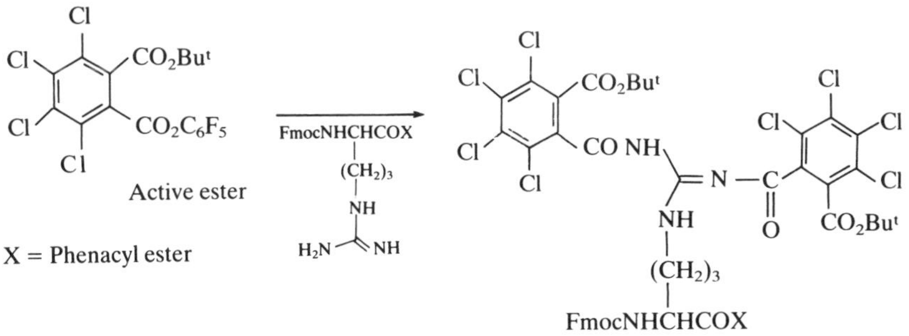

Fig. 3.27.

3.4.5.4 Guanidine group in arginine

The guanidine group is highly nucleophilic and can be protected by introduction of a nitro group which is a highly electron-withdrawing group:

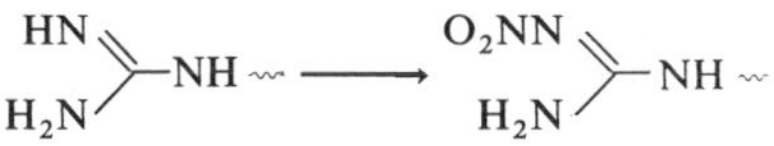

The nitro group can be removed by hydrogenolysis.

With peptide synthesis now being carried out with base-labile N^α-protecting groups there is the need for a good base stable, acid-labile protecting group for guanidine. An example of proven suitability is in Fig. 3.27. This protecting group can be removed with DMF containing 5 per cent 0.01 M aqueous hydrochloric acid (24 h at room temperature).

3.4.6 Activation and coupling

Perhaps the most obvious way of the 'activation and coupling' method involves the use of acyl halides:

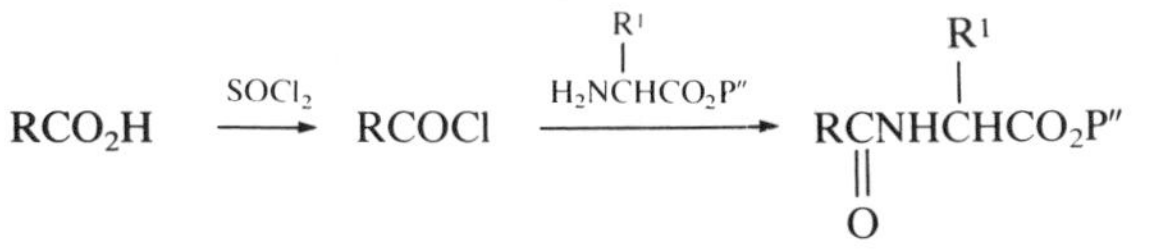

This method has found application, but the possibility of the occurrence of racemization has led to a decline in its use.

Clearly the activation step requires the introduction of a good leaving group (Y):

Examples include:

Y = OAr   Phenolate

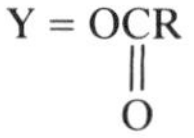

$$Y = \underset{\underset{O}{\overset{\|}{}}}{OCR} \quad \text{Anhydride}$$

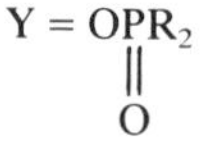

$$Y = \underset{\underset{O}{\overset{\|}{}}}{OPR_2} \quad \text{Anhydride}$$

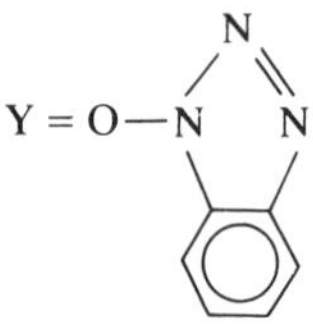

$$Y = O-N \quad \text{Hydroxylamine}$$

$$Y = -N \quad \text{Imidazole}$$

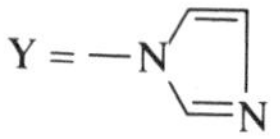

The active ester methods for activation of different groups are now described.

3.4.6.1 Phenolates

The development of this historically important method has concerned the discovery of groups which are particularly good leaving groups. An old favourite is the *p*-nitrophenolate group (Fig. 3.28). Clearly the formation of resonance-stabilized phenolate ion favours the breakdown of the tetrahedral intermediate. The *p*-nitrothiophenolate esters are even more reactive. Other esters of importance include pentachlorophenyl and pentafluorophenyl esters. The aminolysis of these active esters is accelerated by the addition of 1-hydroxybenzotriazole, probably via ester interchange to give the acylated triazole. In the absence of steric effect it appears that the reactivity of the active esters is determined by the acidity of the phenolate component.

3.4.6.2 Active esters as intermediates in coupling reactions

Many coupling agents form covalent compounds with carboxylic acids which are highly reactive towards nucleophiles.

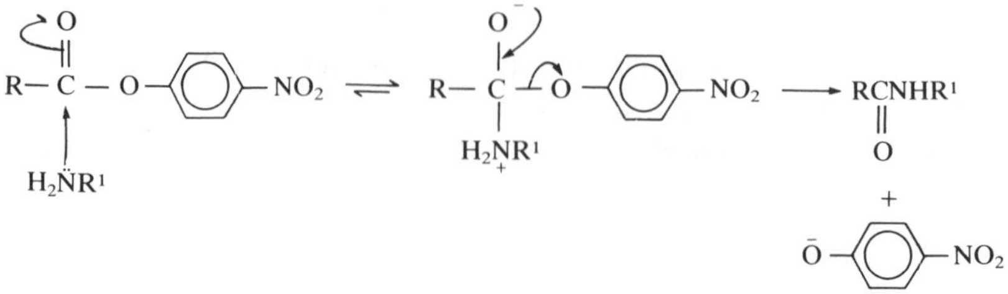

Fig. 3.28.

With dicyclohexylcarbodiimide (DCC), the nucleophile (Nu⁻) may be either the amino group which produces the peptide bond, or another molecule of the acid.

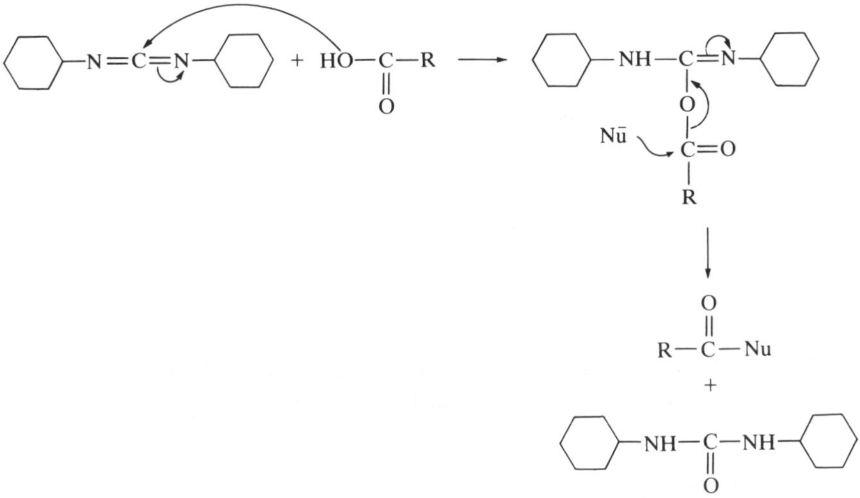

If the latter process occurs, a symmetrical anhydride results which can then react with an amino group to give a peptide bond (see Section 3.4.6.3):

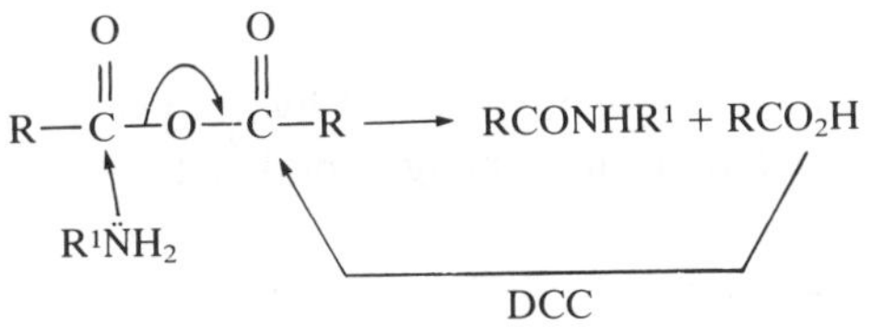

Whether or not anhydrides play a significant part in the coupling reactions with DCC is not known.

DCC is perhaps the most commonly used carbodiimide. The reaction produces a urea which, when organic solvents are used for the reaction, precipitates, so aiding its removal. Coupling can also be carried out in aqueous solution and carbodiimide (**3.42**) is useful for this purpose since the terminal amine group aids solubilization.

$$C_2H_5N=C=N-CH_2CH_2CH_2N(CH_3)_2$$
(**3.42**)

Another class of active ester is the *N*-arylhydroxylamines (sometimes referred to as 'hydroxamic acids'). It was found that DCC coupling reactions were aided (and racemization reduced) when *N*-hydroxyamines (**3.43**) such as *N*-hydroxysuccinimide (**3.44**) and 1-hydroxybenzotriazole (**3.45**) were added (Fig. 3.29).

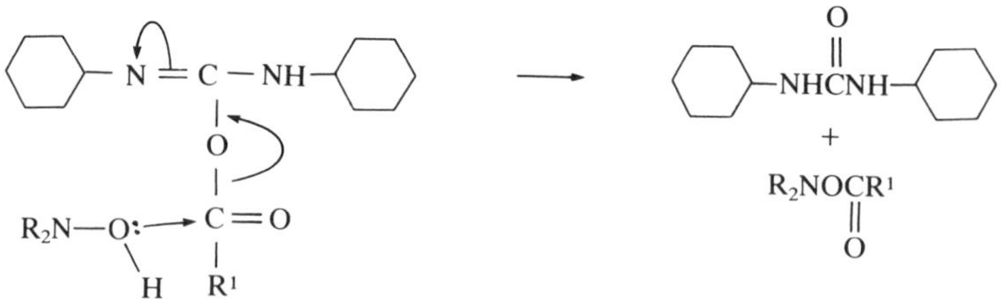

(3.43)

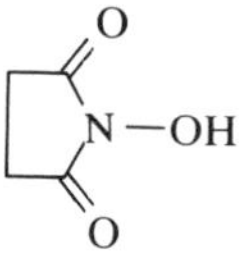

(3.44) 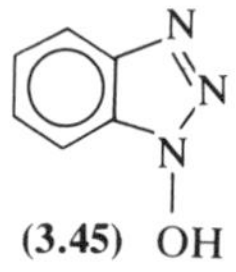   (3.45) OH

Fig. 3.29.

The esters so derived are particularly reactive towards amines. This is thought to be due to the nitrogen of the hydroxylamino group anchomerically assisting aminolysis:

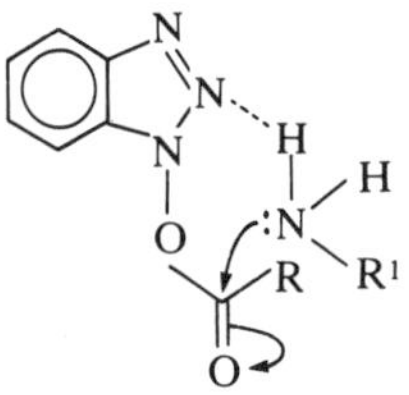

Certainly the use of DCC in conjunction with 1-hydroxybenzotriazole is an attractive and well-used method for forming peptide bonds.

3.4.6.3 *Mixed anhydrides as intermediates in coupling reactions*

Peptide bonds can be formed via reaction of anhydrides with amino groups (Fig. 3.30). Clearly the use of symmetrical anhydrides can only give a maximum yield for the amide of 50 per cent based on the amount of anhydride used. For this reason the use of mixed anhydrides was introduced. These anhydrides were so designed that the nucleophile would preferentially attack the correct carbonyl group (Fig. 3.31).

The mixed anhydride is made by reacting the free acid with a chlorocarbonate (e.g. isobutyl chlorocarbonate) in the presence of a base (e.g. N-methylmorpholine). With R^2 in (**3.46**) as isobutyl group the nucleophilic attack by the amine is directed towards the desired carbonyl group. Nucleophilic attack on carbonyl group (B) in **3.46** is suppressed by

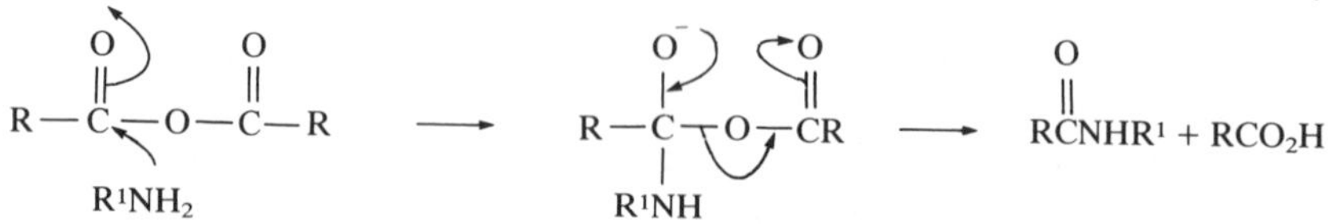

Fig. 3.30.

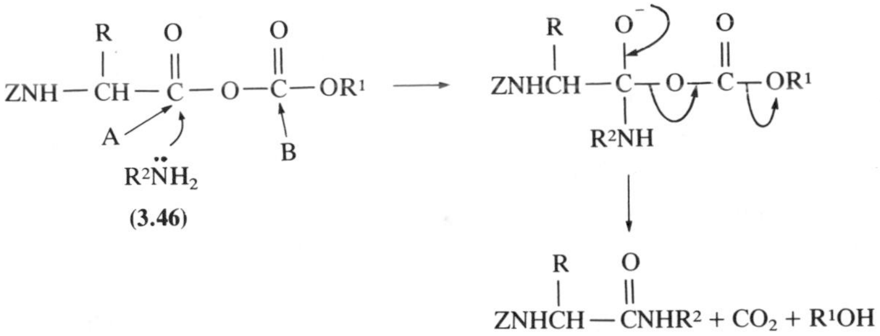

Fig. 3.31.

its conjugation with oxygen and the electron release properties of the alkoxyl group.

Reagents such as (**3.47**) have been devised so as to simplify the mixed anhydride coupling method (Fig. 3.32). Thus using the reagent (**3.47**), the acid and the amine are mixed together in a suitable solvent (e.g. *N,N*-dimethylformamide) at room temperature. After 15–24 hours the reaction mixture is worked-up. The reaction is unfortunately accompanied by some racemization and some side-reaction, e.g. urethane formation.

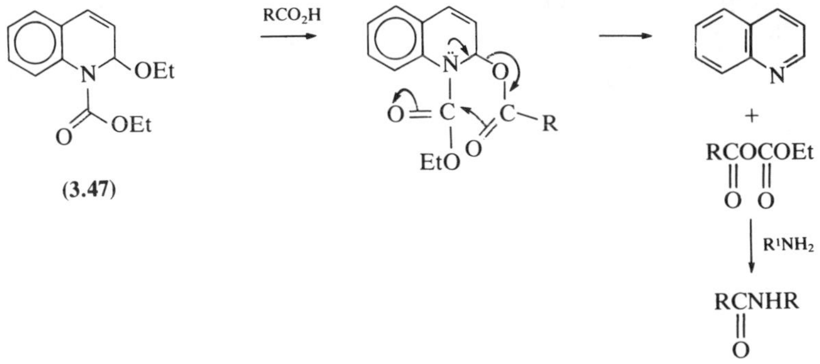

Fig. 3.32.

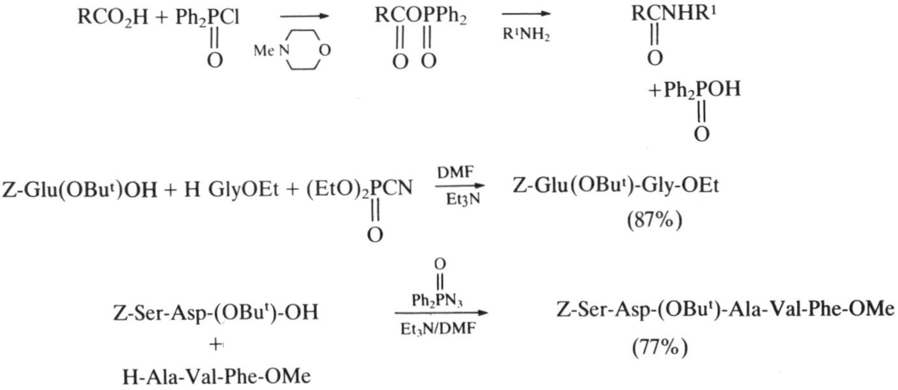

Fig. 3.33.

One of the prime factors affecting the efficiency of the mixed anhydride coupling procedure is ensuring that nucleophilic attack by the amino group is confined to the desired carbonyl group. The use of mixed anhydrides derived from phosphorus-containing acids has a number of advantages over the classical mixed anhydride approach which includes a decreased tendency for the anhydrides to disproportionate and higher regioselectivity in aminolysis. The methodology is shown in Fig. 3.33, where some examples are also given.

The use of the phosphorus azide allows coupling to be carried out in the presence of unprotected hydroxyl groups in side-chains. Furthermore, the coupling appears to be attended by little, if any, racemization.

In recent years there has been a return to the use of symmetrical anhydrides because the reaction produces the amide in a high state of purity.

3.4.7   Summary of strategies and methods available for synthesizing peptides in solution

The multitude of functional groups and coupling methods which have been developed to aid the protein chemist attests to the fact that there is no single clearly defined successful pathway which can be described. Each polypeptide requires the chemist to exercise judgement and skill if the goal is to be attained.

One of the greatest problems is racemization. This usually arises in the activation step as a result of either enolization or oxazolone formation (Fig. 3.34). If the oxazolone forms a carbanion, reprotonation produces racemization (Fig. 3.35).

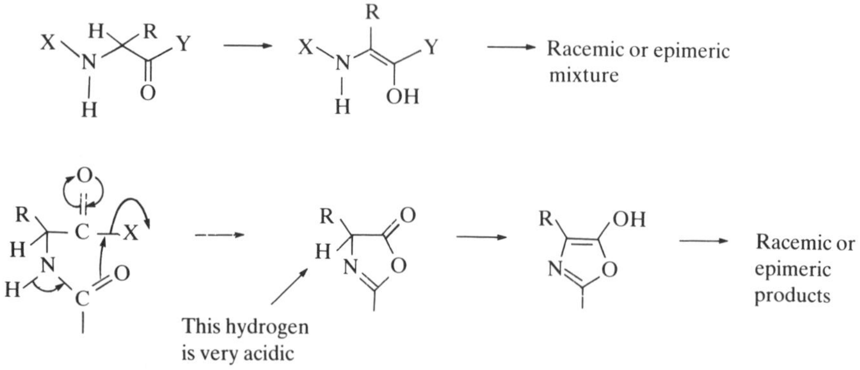

Fig. 3.34.

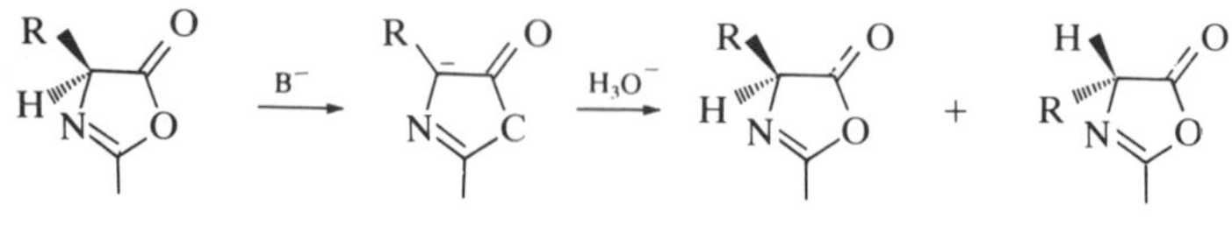

Fig. 3.35.

The former processes can occur anywhere along a polypeptide chain where secondary amide bonds are present together with an appropriate chiral centre. It is for this reason that growth of a polypeptide by elongation from a C-terminal residue is disfavoured. Furthermore, if the strategy requires segmental assembly, activation of the polypeptide may cause racemization. To minimize these problems it is usual to activate a peptide possessing either glycine or proline as the C-terminal residue, thereby minimizing the chance of racemization at the residue most prone to being affected.

The protection of side-chain substituents of α-amino acids presents particular problems. Usually these are protected by groups that are not as easily removed as those used for N^{α}-protection or preferably are removed by a totally different process, e.g. by using acid-sensitive N^{α} groups and groups cleaved by hydrogenolysis in the side-chain. One of the problems associated with protecting every group is that the peptides become insoluble in solvents commonly used for purification. There is also another undesirable aspect of carrying out reactions with high molecular weight compounds which argues in favour of using the segmental approach to synthesis. For high molecular weight compounds the ratio of the reacting group to the other groups within the molecule can be very small. Thus when operating with very small amounts of high molecular weight compounds it becomes necessary to increase the concentration of reagents if reasonable reaction rates are to be attained.

Efforts have been made to speed up the process of solution phase peptide synthesis. In the REMA (*r*espective *e*xcess *m*ixed *a*nhydride) method, excess of the mixed anhydride of the N^{α}-alkyloxycarbonylamino acid is added to peptide carrying the free amino group. When coupling is complete, the excess anhydride is destroyed by the addition of aqueous potassium bicarbonate. The cycle of deprotection followed by acylation with the mixed anhydride is repeated without purification of the intermediates. The success of the method depends upon the use of excess of the mixed anhydride to ensure quantitative coupling and removal of the unreacted anhydride by such a simple process as hydrolysis. Some remarkably complex peptides having full biological activity have been synthesized by this method.

3.4.8 Example of peptide synthesis using the solution method

The synthesis of the C-terminal and seven peptide segment of chicken VP (a vasoactive intestinal peptide) is shown in Fig. 3.36. This synthesis is shown using the 'common shorthand' utilized by peptide chemists. The amino acids to be coupled are set out with the N-terminal acid being on the left-hand side and the C-terminal acid on the right-hand side. A free carboxyl group is denoted by OH on the right-hand side of the line appropriate to the amino acid and a free amino acid group by H on the left-hand side of the line. Side-chain protecting groups are denoted by a short line 45° to the right of the vertical line. Horizontal lines represent bonds.

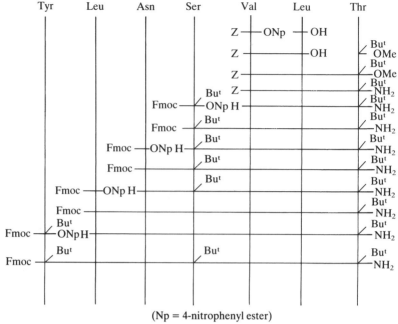

(Np = 4-nitrophenyl ester)

Fig. 3.36.

In the above synthesis the peptide bond is accomplished via the *p*-nitrophenate active ester. Valine, N-protected by a Z group, is coupled to leucine which has its carboxyl group unprotected. This is activated by dicyclohexylcarbodiimide (DCC) and then coupled to threonine methyl ester which has its side-chain hydroxyl group as a *t*-butyl ether. In the next step the methyl ester is converted into the amide with ammonia:

$$RCO \cdot OCH_3 + NH_3 \rightarrow RCONH_2 + CH_3OH$$

Removal of the Z group by hydrogenolysis gives us a tripeptide with a free N^{α}-terminal group. This peptide is coupled to serine in which Fmoc is used as the N^{α}-protecting group and the side-chain hydroxyl group is protected as a butyl ether. The carboxyl group is activated as its *p*-nitrophenate ester. Coupling gives a tetrapeptide in which the N^{α}-terminal group can be deprotected with morpholine. Asparagine and leucine are coupled in a similar way. Tyrosine has a phenolic hydroxyl protected as its *t*-butyl ether and its carboxyl group activated as its *p*-nitrophenate ester. The peptide formed could have all the protecting groups removed by sequential treatment with morpholine and acid to give Tyr-Leu-Asn-Ser-Val-Leu-ThrNH$_2$.

3.4.9   *Use of solid supports in peptide synthesis: the Merrifield approach*

The synthesis of peptides by the solution phase procedure is tedious and time-consuming. The introduction in 1963 by Merrifield of a method whereby polypeptides could be constructed on the surface of a solid inert support revolutionized peptide syntheses. Like all good strategies it is simple. A protected amino acid is covalently attached to a solid resin via its carboxyl group. When coupling is deemed complete the resin is filtered off, thereby simply and efficiently removing excess reagent, etc. Deprotection is carried out using the appropriate reagent and when complete the resin, which now has attached an amino acid carrying a free amino group, is filtered off. An N^{α}-protected amino acid can now be coupled with the resin-supported amino acid. When coupling is complete, the resin is again freed from excess reagents by filtering and washing. In principle this process can be carried out *ad infinitum* to produce a polypeptide of any desired length. The simplicity of the method lends itself to automation which has been done with success. It is, however, usual practice to use the Merrifield method to produce peptides containing about 30 amino acid residues. These large residues are usually coupled together using a solution-phase method. When the peptide has been full grown on the Merrifield resin it is usually cleaved from the resin by exposure to strong acids such as hydrobromic acid in trifluoroacetic acid or by liquid hydrogen fluoride. The method is indeed simple, but for successful execution, attention has to be paid to details.

3.4.9.1   The resin

The first resin to be used, which is still extremely popular, was a chloromethylated copolymer of styrene and divinylbenzene (Fig. 3.37). The

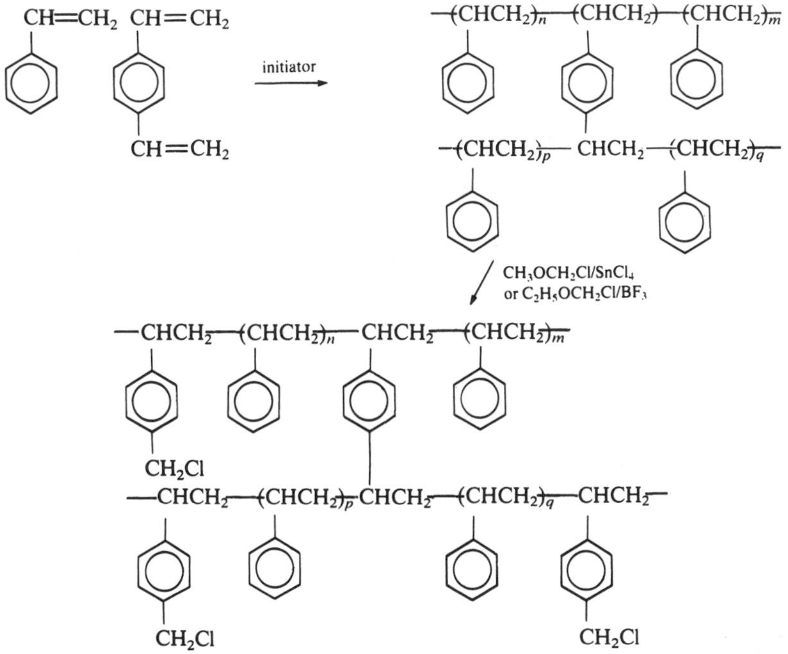

Fig. 3.37.

divinylbenzene is used to create a cross-linked polymer network which imparts mechanical strength. Chloromethylation can be carried out with a variety of reagents. Usually the choice of reagent is determined by the desired extent of chloromethylation. For large peptides, an effective low concentration of chloromethyl groups is apparently necessary. Another necessary feature of the resin is that it should swell in solvents used for carrying out the chemical reaction. Swelling increases the surface area of the resin, thereby increasing the probability of reactions at its surface and also minimizing the possibility of the head group of a growing peptide being buried in the support, so rendering its derivatization difficult. It also, of course, aids penetration of the polymer network by the reagents.

Whilst it is impossible to overemphasize the impact that the Merrifield procedure has had upon the development of protein synthesis, the resin has nevertheless certain defects. Failures occuring in the synthesis of peptides supported on the resin lead, after the final cleavage, to mixtures of peptides that are difficult to purify. Purification is aided by N^α-protection of the growing peptide with (**3.48**) followed by acid cleavage from the resin. Purification of Sulfmoc-protected peptides can be carried out on a weakly basic ion exchange column and then the Sulfmoc residue removed by base.

(**3.48**)

Reverse phase and ion exchange HPLC are now extensively used to purify peptides produced by the Merrifield and other solid-phase methodologies.

Another problem associated with the use of solid supports is that cleavage of the peptide resin ester bond requires the use of strong acid which gives rise to a number of unwanted side-reactions. In order to obviate this problem, resins with spacer groups have been used. These groups not only increase the distance between the polymer and the growing peptide but have a group from which the peptide can be grown and also easily removed. Examples include (**3.49**) where Ⓡ ≡ resin.

(**3.49**)

This system is used in conjuction with amino acids which are N^α-protected with Bpoc group. The mild conditions used for removal of this protecting group do not lead to cleavage of the anchoring group. Cleavage of the anchoring group is accomplished by use of 0.5 per cent trifluoroacetic acid (TFA) in dichloromethane.

If an aminomethylated resin is used, the 4-hydroxymethylphenylacetic acid spacer group (**3.50**) can be used.

$$HOCH_2 - \!\! \bigcirc \!\! - CH_2CONHCH_2 - \!\! \bigcirc \!\! - \boxed{R}$$
(**3.50**)

After the polypeptide has been grown on the resin, cleavage of the anchoring group is carried out with anhydrous hydrofluoric acid at 0°C. If these conditions are too vigorous, related modified supports such as (**3.51**) and (**3.52**) may be used.

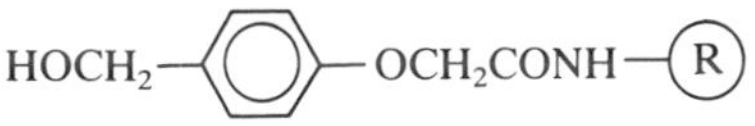

$$HOCH_2 - \!\! \bigcirc \!\! - OCH_2CONH - \boxed{R} \qquad HOCH_2 - \!\! \bigcirc \!\! - CH_2CH_2CONH - \boxed{R}$$

(cleaved by TFA at 25 °C) (**3.51**) 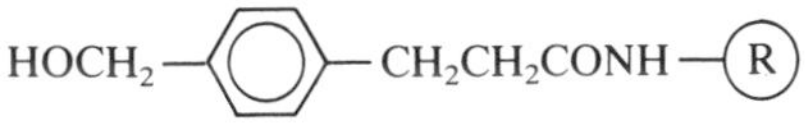   (cleaved by HBr at 25 °C) (**3.52**)

Since a number of peptides end in a C-terminal amide, attention has been given to resins which, when cleaved from the peptide, will produce a C-terminal amide. The earliest and most commonly used residue is the benzhydrylamine residue (**3.53**) (Fig. 3.38).

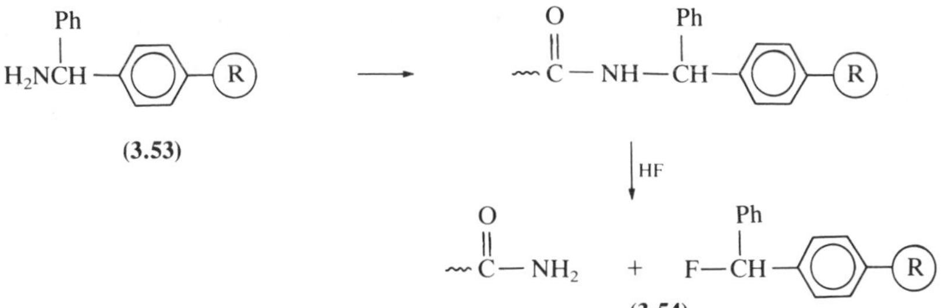

Fig. 3.38.

Use of the derivatized resin (**3.55**) in Fig. 3.39 enables peptides to grow which can be cleaved from the resin by hydrazine to produce hydrazides. Having obtained the hydrazide, these may be converted to acylazides and used to couple to another peptide.

3.4.9.2   Use of protecting groups in the Merrifield solid phase peptide synthesis

Most of the N-protecting groups used in liquid phase synthesis can be used in conjunction with the solid supports. However, removal of groups by catalytic hydrogenation is not practical because of the difficulty of using a heterogeneous catalyst in conjunction with the polymer. In these systems it was also found that the use of Boc protecting groups is preferable to Z groups on account of the acidic conditions used for removing Z groups,

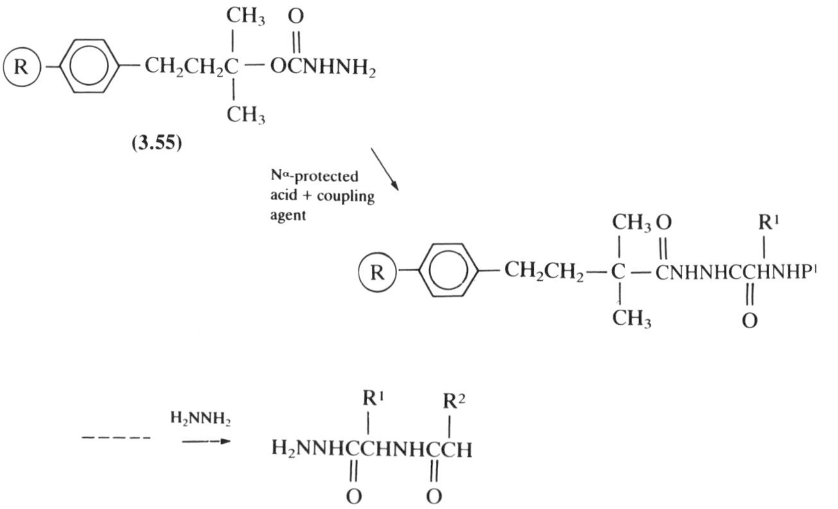

Fig. 3.39.

causing some cleavage of the growing peptide from the support. Removal of the Boc groups can be carried out under mild conditions (0.1–1.0 M hydrochloric acid in acetic (ethanoic) or trifluoroacetic acid) without affecting the anchoring linkage. Other groups, which can be used with advantage, include the Bpoc and the related group (**3.56**).

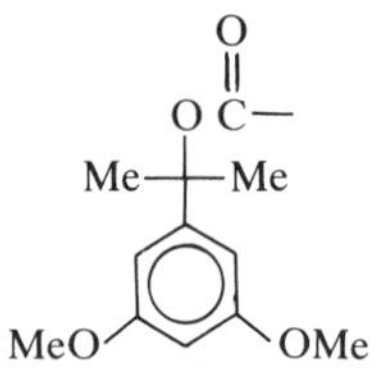

(3.56)

If the Boc or Bpoc group is used for N-protection it follows that side-chain substituents must be protected with groups that are more resistant to acid such as the Z group. With such a combination, partial loss of the Z group during successive protecting and deprotecting of N sequences does not occur to a significant extent. With the Z as a blocking group, its removal is effected when the grown peptide is removed from the polymer support with liquid HF.

3.4.9.3   *Formation of peptide bonds in solid phase peptide synthesis*

The most favoured method of formation of peptide bonds utilizes activation by use of dicyclohexylcarbodiimide in either the presence or absence of

1-hydroxybenzotriazole. The use of excess Boc amino acid and DCC in excess enables coupling to be carried out in a few minutes. This method is acceptable provided all functional groups are protected. Some active esters have been used, but in general these esters are not sufficiently reactive for them to be used extensively.

Synthesis of angiotensin II via the Merrifield solid phase method. Angiotensin II (Fig. 3.40) is an octapeptide produced by the action of ACE upon angiotensin I:

Asp-Arg-Val-Tyr-Ile-His-Pro-Phe-His-Leu

Angiotensin I

|ACE

Asp-Arg-Val-Tyr-Ile-His-Pro-Phe + His-Leu

Angiotensin II

Angiotensin II stimulates a rapid increase in blood pressure as may be required when doing physical exercise. Clearly angiotensin II has the potential to be a therapeutic agent for increasing cardiac activity and ACE inhibitors will be of value in controlling the blood pressure of people suffering from high blood pressure.

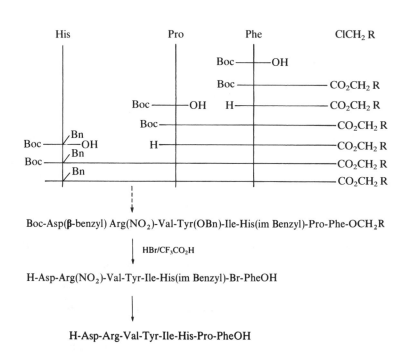

Boc-Asp(β-benzyl) Arg(NO_2)-Val-Tyr(OBn)-Ile-His(im Benzyl)-Pro-Phe-OCH_2R

| HBr/CF_3CO_2H

H-Asp-Arg(NO_2)-Val-Tyr-Ile-His(im Benzyl)-Br-PheOH

H-Asp-Arg-Val-Tyr-Ile-His-Pro-PheOH

Fig. 3.40 Angiotensin.

The synthesis of angiotensin II using the Merrifield solid-phase method was one of the early triumphs of the method. Inspection of the structure of angiotensin II shows that a number of side-chains will need protecting and these protecting groups will have to be stable under the conditions used for manipulating the N-amino groups. Protection was achieved using the following group:

His (im Benzyl)
Tyr (*O*-Benzyl)
Arg (NO_2)
Asp (Benzyl ester)

The Boc group was exclusively used for N^{α}-protection and coupling was carried out with *N*,*N*-dicyclohexylcarbodiimide.

Boc-Phe-OH + $ClCH_2$(R)

↓

Boc-Phe-OCH_2(R)

↓ HCl/HOAc

H-Phe-OCH_2(R)

↓ Boc PrOH/DCC

Boc-Pro-Phe-OCH_2(R)

↓ HCl/HOAc

H-Pro-Phe-OCH_2(R)

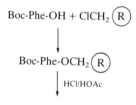

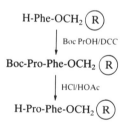

The synthesis was continued in a similar manner to give:

Boc-Asp(β-benzyl)-Arg(NO_2)-Val-Tyr(OBn)-Ile-His(imBn)-Pro-Phe-OCH_2-R

↓ HBr (CF_3CO_2H) (removal of protecting groups and cleavage from the resin)

H-Asp-Arg(NO_2)-Val-Tyr-Ile-His(imBn)-Pro-Phe-OH

↓ H_2/Pd (hydrogenolysis to remove NO_2 from Arg and imBn from His)

H-Asp-Arg-Val-Tyr-Ile-His-Pro-Phe-OH

3.4.10 Use of solid supports in peptide synthesis: the use of polyacrylamide resins

The use of polyacrylamide resins, introduced by R.C. Sheppard in the mid-1970s, is now challenging the well-established Merrifield approach.

The resin is prepared by suspension polymerization from acrylamide, a functionalized acrylamide and a diacrylamide. Figure 3.41 shows a partial structure for the resin, important features being the presence of a few

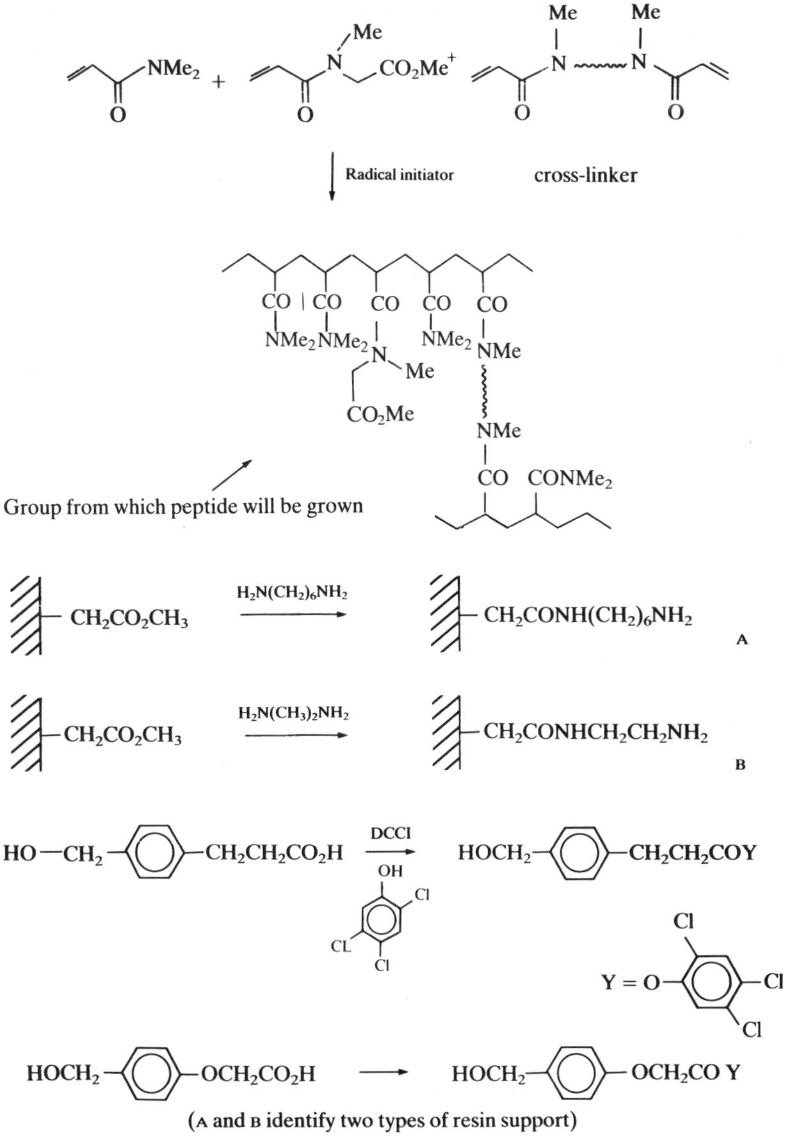

Fig. 3.41.

cross-links and the presence of groups (ester) to which the building blocks for the peptide can be attached. A most useful feature of the resin is that it swells in solvents such as dichloromethane and *N,N*-dimethylformamide which are commonly used in the coupling reactions, etc., in peptide synthesis. Thus if the dry resin having an approximate volume of 2.2 ml is added to excess dichloromethane it swells to a volume of 20 ml. Such an enormous degree of swelling is clearly advantageous in aiding permeation by the reagent. Since the swelling is observed with so many solvents it means that the resin can be used in solvent systems which are optimal for the protection and deprotection processes. Another very useful feature is that the resins can be supported on Kieselguhr (an inorganic clay) by polymerizing the materials in the pores of the support. Such a matrix can be used in a 'continuous flow' peptide synthesizer (i.e. a synthesizer in which the reagents and solvents flow through a column containing the resin).

To attach the C-terminal residue of the peptide to be synthesized requires that a spacer group be attached to the resin. Usually the free ester function on the resin is reacted with α,ω-diamine (e.g. 1,6-diaminohexane or 1,2-diaminoethane) so as to leave a primary amino group on the resin for functionalization. The N^α-protected C-terminal acid of the peptide to be synthesized, as its symmetrical anhydride or pentafluorophenyl ester, is reacted with the 2,4,5-trichlorophenoxy ester of 3-[4-(hydroxymethyl)-phenyl]propionic acid or 4-(hydroxymethyl)phenoxyacetic acid. The amino acid may now be attached to the resin via the active ester as, for example, in Fig. 3.42.

It will be noted that the peptide is linked to the resin via a reversible linkage. Cleavage of the peptide from the support is via the ester group

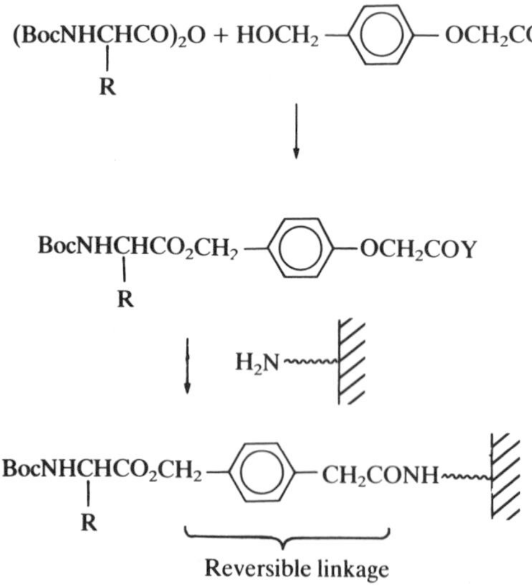

Fig. 3.42.

$$O$$
$$\parallel$$
$$-CNCH_2CONHCH_2CH_2NH_2$$
$$\mid$$
$$Me$$

Fig. 3.43.

$$O$$
$$\parallel$$
$$-CNCH_2CONHCH_2CH_2NHNLeu-H$$
$$\mid$$
$$Me$$

Fig. 3.44.

formed between the C-terminal amino acid and this reversible linkage. When the phenoxyacetic acid group is used, cleavage can be achieved with trifluoroacetic acid, whereas with the 3-phenylpropionic acid derivative, anhydrous hydrogen fluoride is required.

The synthesis of H-Arg-Pro-Lys-Pro-Gln-Gln-Phe-Phe-Gly-Leu-MetNH$_2$ is an example of peptide synthesis using a polyacrylamide solid support. The polyacrylamide shown before and modified with ethylenediamine was utilized (Fig. 3.43). An internal reference spacer group was introduced by reacting the free amino group with N$^\alpha$-Fmoc norleucine as its symmetrical anhydride (Fig. 3.44).

The Fmoc group was removed with piperidine in DMF. The number of active groups on the polyacrylamide support can be determined by measuring the ninhydrin colour yield of the norleucine residues. Knowing the number of active groups enables one to determine how much activated amino acid should be added at each step.

The next step was to introduce the reversible link and then the C-terminal acid of the peptide (i.e. methionine). Figure 3.45 shows the sequence of reactions used. Removal of the Boc group then allowed Fmoc-protected leucine to be coupled as its symmetrical anhydride. The remaining amino acids were coupled in a similar way except for glutamine which was activated via formation of its 4-nitrophenolate ester. The N$^\varepsilon$ group of lysine can be protected via a Boc group or more successfully via a trifluoroacetyl group. Cleavage of the peptide from the reversible linkage was accomplished by ammonolysis and the side-chain groups removed by a single trifluoroacetic acid treatment. Following purification of the peptide by HPLC a 47 per cent yield of pure product was attained.

3.4.11 Application of protease-catalysed peptide bond formation in peptide synthesis

The ability of the proteases to cleave peptide bonds is well known and this reaction is put to good use in determining the primary structure of peptides.

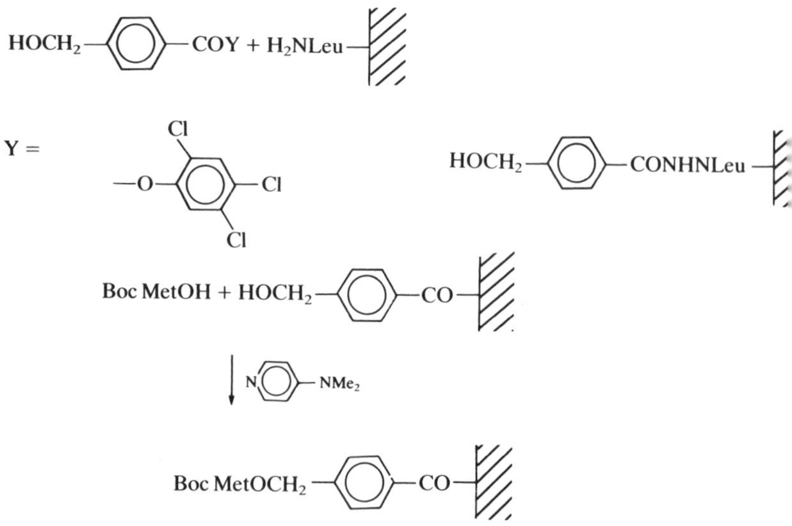

Fig. 3.45.

It should not be forgotten that the function of the enzyme is as a catalyst which brings the reactants together in the appropriate arrangement for reaction to occur. Specificity of the enzyme relates to its ability to recognize the shape of a particular amino acid and it is for this reason that the enzymes catalyse reactions of the naturally occurring amino acids (L series), and these reactions occur without racemization:

$$RCO_2^- + H_3\overset{+}{N}R^1 \rightleftarrows RCO_2H + H_2NR^1 \rightleftarrows RCONHR^1 + H_2O$$

Reaction occurring on the enzyme surface

Peptide bond formation on the enzyme surface is reversible; which process dominates depends upon concentration of reagents, etc. For peptide bond formation to occur the amino group has to be unprotonated and the carboxylic acid group un-ionized. The use of water-miscible organic solvents can be used to achieve, at least in part, these conditions. Such solvent systems often unfortunately reduce the activity of the enzyme. Thermodynamics can be used to great advantage if it is arranged for the product of the reaction to be insoluble, e.g. if the peptide being produced precipitates. Peptide bond formation can also be brought under kinetic control.

In the reaction scheme shown in Fig. 3.46, kinetically controlled accumulation of the product containing the free peptide bond occurs if $k_2 \gg k_3 + k_4$ and $k_4[H_2N-R^1] > k_3[H_2O]$.

A beautiful example of thermodynamic control of peptide formation is afforded by the synthesis of the artificial sweetener aspartame (Asp-Phe-OMe) which is obtained by the coupling of Z-Asp with Phe-OMe. An insoluble salt is formed with the enantiomer of the amino component,

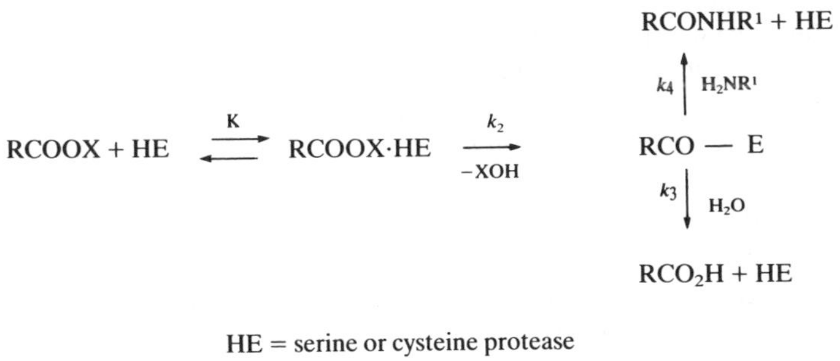

$$\text{RCOOX} + \text{HE} \underset{\text{-XOH}}{\overset{K}{\rightleftharpoons}} \text{RCOOX·HE} \xrightarrow{k_2} \begin{array}{c} \text{RCONHR}^1 + \text{HE} \\ \\ k_4 \uparrow \,\, \text{H}_2\text{NR}^1 \\ \\ \text{RCO} - \text{E} \\ \\ k_3 \downarrow \,\, \text{H}_2\text{O} \\ \\ \text{RCO}_2\text{H} + \text{HE} \end{array}$$

HE = serine or cysteine protease

Fig. 3.46.

Z-Asp-Phe-OMe·Phe-OMe which results in an almost quantitative shift of the equilibrium towards the product.

What does the future hold for the use of enzyme-catalysed coupling reactions in peptide synthesis? Two particular advantages which stand out are (a) the reactions occur without racemization, and (b) the amino acids to be assembled require minimal side-chain protection. Against these has to be weighed the fact that the enzymes show amino acid specificity and therefore there is not a single enzymatic coupling methodology which is universally applicable. However, extensive work has shown the value of using N-terminal-*exopeptidases* (carboxypeptides). *Endopeptidases* are useful for short sequences provided that the peptide does not contain the amino acid for which the enzyme shows specificity except in the position where the peptide bond is to be formed. Assemblage of a polypeptide by use of enzymes requires careful choice of the enzyme and the coupling conditions. With the routine employment of immobilized enzymes growing it should not be long before flow methods for peptide bond formation are developed, thereby simplifying the use of these materials.

3.4.12   *Polypeptide synthesis using recombinant DNA*

In spite of the elegant advances in 'chemical' polypeptide synthesis, it is clear that the synthesis on a large scale of long polypeptides presents enormous practical difficulty and is very costly in time and resources. However, the synthesis of substantial quantities of longer polypeptides is highly desirable, particularly for use in biomedical research and therapy, and in little more than a decade the development of recombinant DNA technology has permitted this to be realized. The processes involved are outlined at the end of Chapter 2, and have been applied with conspicuous success to prepare proteins such as interferons, insulin, somatostatin, and the blood-clotting factor VIII.

Moreover, either by using oligonucleotides containing a mismatching base or sequence which may be annealed to an otherwise complementary gene sequence from a parent organism and extended and ligated to form the complete gene sequence containing a defined point mutation, or via the total synthesis of genes containing defined mutations, application of the same recombinant DNA techniques can result in the synthesis by cells of mutated proteins containing one or more defined amino acid changes from those found in the native protein. Examination of the altered characteristics of the protein, the changes in its enzymic properties (if any), etc., may throw considerable light on the significance of the amino acid residue in determining the function of the protein. 'Protein engineering', as it has been dubbed, is rapidly affording previously unattainable insights into the relationships between primary structure and protein function, to the point where the notion of custom-designed enzymes for industrial uses is by no means far-fetched. The interested reader is encouraged to consult reviews of this exciting and fast-expanding area.

Does this mean that chemical polypeptide synthesis is outdated? By no means, though its use is likely to be of greater value in the synthesis of relatively short oligopeptides. But the recent discovery, development, and commercial value of a compound as simple as the 'non-nutritive' sweetener aspartame (L-aspartyl-L-phenylalanine methyl ester) shows that the skills of the peptide synthetic chemists will continue to be required. Also, those proteins which depend for their function upon post-translational modification by enzymes of the parent organism as the final stage in their biosynthesis may not be obtainable in fully active form using recombinant DNA methods unless the systems responsible for the modification (e.g. to form unusual amino acids such as hydroxyproline or $N^\varepsilon,N^\varepsilon,N^\varepsilon$-trimethyllysine) in the parent organism can be isolated and utilized. Failing this, there seems little alternative to chemical synthesis of the polypeptide.

3.5   Some specific peptides of interest

Between the monomeric amino acids and the large polypeptides which are the essential components of protein there lies a vast range of peptides exhibiting a wide diversity of structure and biological function. Some are typical – that is, they contain only L-amino acids linked by peptide bonds – but many are atypical, consisting of two or more amino acid residues joined by a peptide bond but possessing in addition some structural feature not normally found in protein. It is instructive to examine a number of these compounds as illustrative examples.

3.5.1   *Linear peptides*

A number of important pituitary hormones, such as human oxytocin (**3.57**) and human vasopressin (**3.58**), are simple linear oligopeptides, involved in the release of milk from mammary gland and the regulation of blood volume and pressure, respectively. Note that each contains an intramolecular disulphide bond and glycinamide at the C-terminal, and that despite the

```
     ┌──S─S────────────┐
     Cys-Tyr-Ile-Glu-Asn-Cys-Pro-Leu-Gly NH₂
                    (3.57)
```

```
     ┌──S──S──┐
     Cys-Phe-Glu-Asn-Cys-Pro-Arg-Gly NH₂
                    (3.58)
```

```
                 ┌───────S──S──────────────────┐
     Ala-Gly-Cys-Lys-Asn-Phe-Phe-Trp-Lys-Thr-Phe-Thr-Ser-Cys
                    (3.59)
```

Asp-Arg-Val-Tyr-Ile-His-Pro-Phe
(3.60)

difference in their functions, they differ in only two amino acid residues. Somatostatin (3.59) is secreted by the hypothalamus and is a release-inhibiting factor, preventing the release of the growth hormone somatotropin from the pituitary gland. It also acts on the pancreas, preventing the release of both insulin and glucagon, leading to a lowering of blood glucose concentration, thus stimulating interest as a possible source of treatment for diabetics.

The simple octopeptide angiotensin (3.60), formed via hormonally controlled cleavage of a serum α_2-globulin, is the most powerful hypertensive compound known, stimulating the smooth muscles of blood vessels and reducing blood flow through the kidneys and the excretion of fluid and salts. The enkephalins, methionine enkephalin (3.61) and leucine enkephalin (3.62), are also formed by the cleavage of longer precursors, the endorphins, but seem to function as the body's natural opiates, binding to opiate receptors in the brain to prevent the perception of pain. As oligopeptides they are, of course, prone to rapid degradation *in vivo* by tissue peptidases, and in recent years much synthetic effort has been directed towards the synthesis of compounds which are stable to peptidases but which possess similar opiate agonist activity.

Hypothalamic thyrotropic hormone releasing factor (3.63), pyroglutamyl-histidinyl-prolinamide, is secreted in minute quantities by the hypothalamus, and stimulates the release of thyrotropic hormone from the pituitary gland.

Tyr-Gly-Gly-Phe-Met   Tyr-Gly-Gly-Phe-Leu
(3.61)   (3.62)

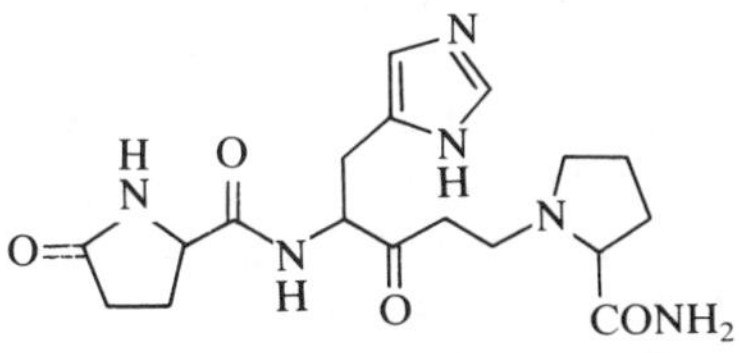

(3.63)

Note that the γ-carboxylic group of the N-terminal glutamic acid has cyclized to the α-amino group to form the pyroglutamyl moiety, a pyrrolidin-2-one species. In other compounds the γ-carboxylic group is involved in straight chain amide bonds, as for instance in glutathione (**3.64**), γ-glutamyl-cysteinyl-glycine, the most widely distributed small peptide and an example of an atypical peptide. One of its chief functions in mammalian tissues is to mediate the transport of amino acids across cell membranes in a process known as the γ-glutamyl cycle. The not unrelated tripeptide (**3.65**), containing a δ-homoglutamyl bond and in addition a D-valine residue, is the common precursor of the penicillin and cephalosporin antibiotics. D-Amino acids, while relatively seldom encountered in higher organisms, are not uncommonly encountered in microorganisms such as the peptidoglycan of the bacterial cell wall.

(**3.64**)

(**3.65**)

3.5.2 Cyclic peptides

Cyclic dipeptides, the 2,5-diketopiperazines, are widespread in nature, and have been isolated as natural products from higher animals, plants, and microorganisms. They are also formed readily from peptides and proteins on thermolysis, and on acid or enzymatic hydrolysis, thus necessitating care to ensure that their presence in natural sources is not artefactual. As an example, *cyclo*-L-His-L-Pro (**3.66**) has been identified in human blood. Note that the two peptide bonds in these species are constrained to having *cis*-geometry. Proline is often encountered as a component of the naturally occurring 2,5-diketopiperazines, possibly because the activation energy required for it to form the *cis* configuration in its *N*-acylated derivatives is lower than that for other peptide bonds. While a number of cyclic dipeptides formed simply from primary L-amino acids are known, the majority of compounds in this class are formed via considerable modification of the diketopiperazine skeleton, frequently involving *N*-methylation and oxidative processes, although it is not clear whether modification occurs before or after cyclization. Examples include (**3.67**), and the microbial metabolite (**3.68**), which contains a β-chloroamine function and, like many nitrogen mustards, possesses useful anti-tumour properties. A more highly modified diketopiperazine is gliotoxin (**3.69**) which represents one of the simpler members of a group of compounds of considerable structural complexity in which the ring is spanned by a disulphide bridge. Another complex group is derived from, or contains, a tryptophan moiety, and these include the vasoconstricting ergot peptides (see Section 7.5.3.2).

Cyclic peptides to which the term 'homodetic' is applied are those in which the ring system is derived from amino acids forming regular amide links, while in 'heterodetic' peptides the ring system is formed from amides and other heteroatom linkages. Despite the widespread occurrence of the diketopiperazines, no homodetic tripeptides and only a few tetrapeptides are known, presumably reflecting the relative stabilities of 6, 9, and 12-membered rings. Larger cyclic peptides are more common and more rigid than their linear counterparts, owing to the conformational constraints of the ring and also to transannular hydrogen bonding. Members of this group are frequently bacterial or fungal metabolites, and include antibiotics such as gramicidin S (**3.70**) and toxins such as malformin A (**3.71**) which induces malformations in germinating beans, the ionophore antamanide (**3.72**) from

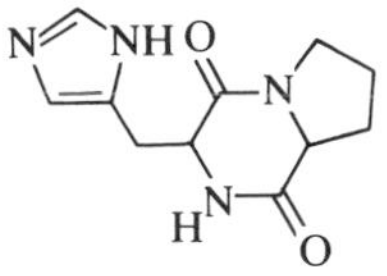

(3.66)

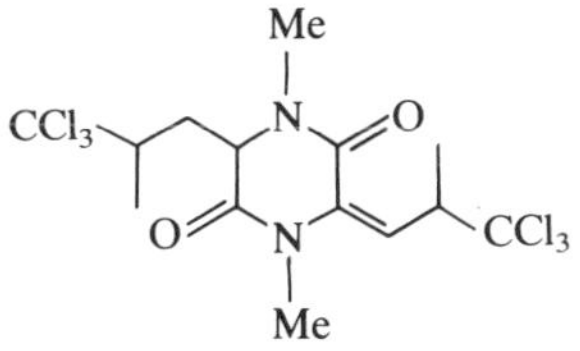

(3.67)

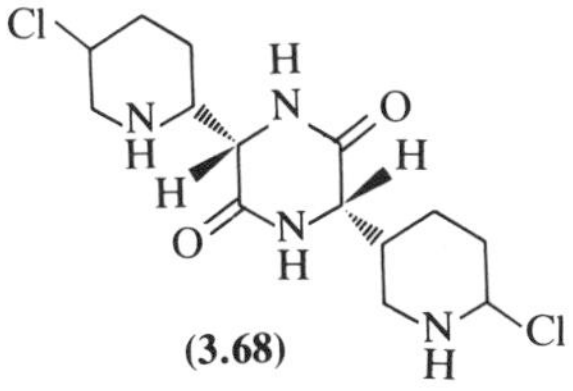

(3.68)

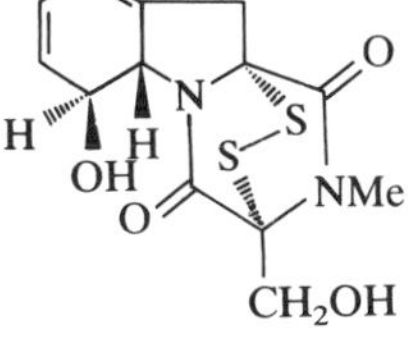

(3.69)

┌─ D-Phe-Pro-Val-Orn-Leu ─┐

└─ Leu-Orn-Val-Pro-D-Phe ─┘

(3.70)

┌─ D-Cys-D-Cys-Val ─┐

└──── Ile-D-Leu ────┘

(3.71)

┌─ Val-Pro-Pro-Ala-Phe ─┐

└─ Phe-Phe-Pro-Pro-Phe ─┘

(3.72)

the death cap mushroom, *Amanita phalloides*, and the highly toxic phallotoxins and amatoxins from the same source, exemplified by α-amanitin (**3.73**), a powerful inhibitor of RNA polymerase. The amatoxins vary in the structures of the side-chains represented in α-amanitin by L-asparagine and L-dihydroxyisoleucine. The phallotoxins are heptapeptides with broadly similar structures to the amatoxins but containing a sulphur bridge in place of the sulphoxide, which damages the membranes of liver cells, releasing potassium ions and enzymes. The cyclosporins form a set of homodetic undecapeptides which have attracted great interest in recent years owing to their immunodepressant properties, which have been widely exploited in modern transplant surgery. The most commonly found basic cyclosporin structure (**3.74**) varies only in the nature of a single side-chain, although other members of the group vary in respect of the *N*-methylation of the peptide bonds, or lack the hydroxy group in the longest side-chain.

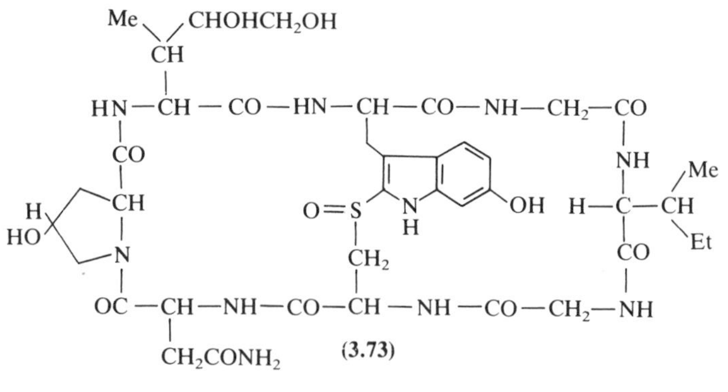

(3.73)

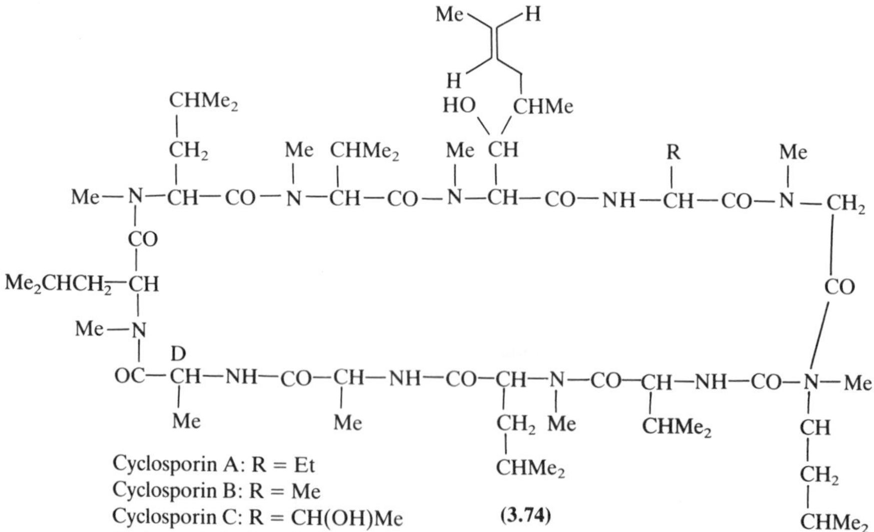

Cyclosporin A: R = Et
Cyclosporin B: R = Me
Cyclosporin C: R = CH(OH)Me
Cyclosporin D: R = CHMe$_2$

(3.74)

Heterodetic peptides in which one or more of the normal bonds is replaced by an ester linkage, are referred to as depsipeptides, or cyclic lactones. Some possess a regular array of alternating amide and ester linkages, while in others the array is irregular. Possibly the best-known member of the former group is the ionophore valinomycin (**3.75**), effectively a cyclic trimer of a basic structural unit each of which contains two ester and two amide links. The ionophores are nature's crown ethers. Valinomycin forms a complex with potassium ions in which the ester groups turn inwards to form a highly polar cavity to accommodate the metal ion, while the hydrophobic side-chains, pointing outwards, confer lipid solubility, and the amide bonds stabilize the complex by intramolecular hydrogen bonding. The hydrophobic nature of the exterior of the complex confers solubility in biological lipid membranes, and thus valinomycin and related compounds act to ferry metal ions across such membranes, tending to collapse any metal ion gradients which a cell may attempt to maintain for vital processes. Valinomycin accommodates potassium and rubidium ions snugly, but the central cavity is too large for sodium and lithium ions, which are poorly complexed. It is thus selective for potassium rather than sodium ions, more so than for any other ionophore known. By contrast, antamanide (**3.72**) is selective for sodium rather than potassium ions. Of the irregular depsipeptides, probably the most celebrated and widely studied example is actinomycin D (**3.76**),

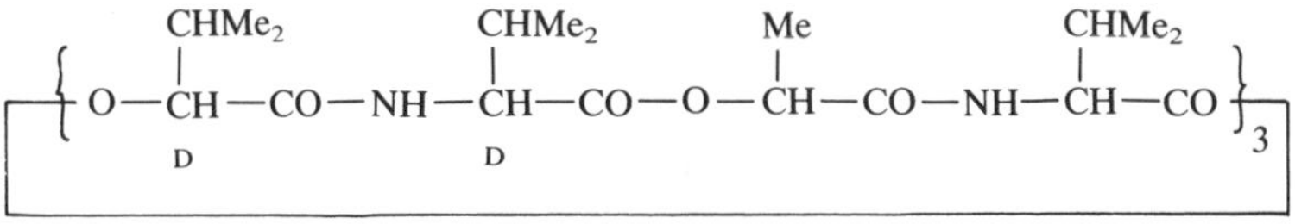

(**3.75**)

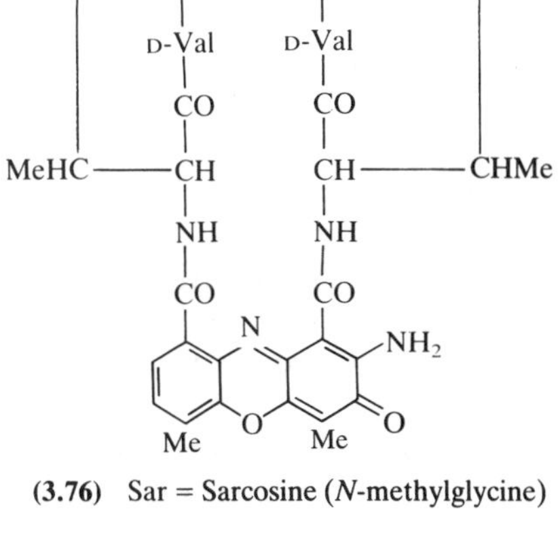

(**3.76**) Sar = Sarcosine (*N*-methylglycine)

which in fact contains two identical depsipeptide rings linked to a phenoxazone chromophore. The two peptide bonds flanking the proline residues are *cis*, in each ring. Actinomycin D is a powerful inhibitor of DNA-dependent RNA synthesis, i.e. of transcription, and thus prevents the biosynthesis of protein. X-ray studies suggest that the phenoxazone ring is intercalated between adjacent G–C base pairs of DNA, with the guanine residues (which are on opposite strands) forming strong hydrogen bonds from the $2\text{-}NH_2$ groups of guanine to the carbonyl groups of the threonine residues of (**3.76**). While the actinomycins are among the most powerful anti-tumour agents known, their high toxicity has largely precluded their use in the treatment of cancer, but they remain valuable tools for the biochemist. The quasi-twofold symmetry of (**3.76**) is also found in two other depsipeptide antibiotics of closely similar structure, echinomycin and triostin A, each containing two quinoxaline rings which are believed to intercalate in similar manner to the phenoxazone ring of (**3.76**).

These few examples must serve to illustrate some of the roles and range of complexity of the smaller peptides of biological significance. The number of peptide antibiotics alone runs into hundreds, and their isolation, structure elucidation, synthesis, mode of action, and exploitation provide a constant challenge to the chemist, the biochemist, the microbiologist, the pharmacologist, and the clinician alike. For more detailed information the reader is referred to specialist volumes and reviews in this area.

3.6   The biosynthesis of proteins

Protein biosynthesis is an extremely complicated and still not fully understood process. For a fuller description, the reader is recommended to a good textbook of biochemistry, but we shall consider certain elements of the process and seek to draw analogies between the chemical and biochemical syntheses of protein.

Synthesis takes place on the ribosomes, which are large ribonucleo-protein conjugates, and the process has been most intensively studied in the *E. coli* system, in which the intact ribosomes sediment in the ultracentrifuge with a sedimentation rate of 70 S (Svedberg units). These 70 S particles are complexes consisting of a 30 S particle and a 50 S particle, the degree of association to form 70 S particles depending on the concentration of magnesium ions present. The 30 S particle contains a 16 S ribosomal RNA (rRNA) molecule, and 21 different proteins, while the 50 S particle contains 5 S rRNA and 23 S rRNA and 34 different proteins. Aided by an 'initiation factor', messenger RNA (mRNA), which contains in its base sequence the information specifying the sequence of the protein to be made, becomes bound to the 30 S subunit.

As in chemical synthesis, in order to form polypeptides the amino acid residues must be activated. This occurs in a two-step process (Fig. 3.47).

First, an aminoacyl-tRNA synthetase condenses the carboxylate group of the amino acid with ATP, to form an aminoacyl adenylate (**3.77**) with displacement of pyrophosphate in a reversible process. There is at least one aminoacyl-tRNA synthetase for each of the 20 common amino acids, for reasons which will become evident. Note that the amino acid has been activated as a mixed anhydride. The same enzyme which catalysed the activation now transfers the aminoacyl moiety of (**3.77**) to the 2'-OH or the 3'-OH (the exact group acylated seems to vary for different synthetases) of the 3'-terminal adenosine moiety of a tRNA molecule, thus forming the carboxylate ester (**3.78**). For each of the common 20 amino acids there exists at least one tRNA species which is cognate, i.e. specific to that amino acid. The tRNA (which always has a 3'-CpCpA terminus) bears a three-base sequence, the 'anticodon', which is complementary according to the base-pairing rules for nucleic acids to a three base 'codon' sequence on the mRNA. The aminoacyl synthetase specifically links each amino acid to its cognate tRNA molecule. The codons on the mRNA, by base-pairing in a complementary, antiparallel sense with the anticodons of different tRNA species with their attached amino acids, specify which amino acids are to be inserted into protein, and in what order. The process of 'charging' the tRNA molecule with its cognate amino acid is freely reversible *in vitro*, implying that the free energy of hydrolysis of the aminoacyl tRNA (**3.78**) is at least comparable to that of the aminoacyl adenylate (**3.77**), and thus that (**3.78**), while ostensibly simply a carboxylate ester of a secondary alcohol, is in fact a highly reactive species. In the cell, the charging reaction is rendered irreversible by the hydrolysis, by pyrophosphatase, of the pyrophosphate released during aminoacyl adenylate formation.

The tRNA molecule thus acts as an 'adaptor', by means of which the encoded amino acid is plugged into the correct position in the sequence specified by the coded information on the mRNA molecule. In order to ensure that proteins of correct sequence are formed, the accuracy of charging the tRNA with their cognate amino acids must be very high, and this in turn

(3.77)

(3.78)

implies that the aminoacyl-tRNA synthetases show very high specificity for the amino acid and tRNA molecules which they bind as substrates. The actual error rate for misinsertion of amino acids into proteins is about 1 in 10^6. Some of the synthetases show very high initial discrimination of the amino acids, especially those with more polar side-chains, but with some of the less polar amino acids an appreciable degree of mischarging occurs. This is normally corrected by 'proofreading' mechanisms in which the aminoacyl-tRNA is hydrolysed before the mischarged tRNA species becomes bound to the ribosome and misincorporation into protein can occur. Several models have been advanced to explain the 'proofreading' mechanism, but at present none seems to be universally valid. Recent results using tRNA species containing specifically mutated sequences suggest that only a few bases in the sequence of tRNA (normally 75–90 nucleotides long) need to be altered in order to change the amino acid with which the molecule is preferentially charged, and thus the tRNA recognition sites on the synthetases appear to be highly sensitive to changes in special short sequences in the molecules.

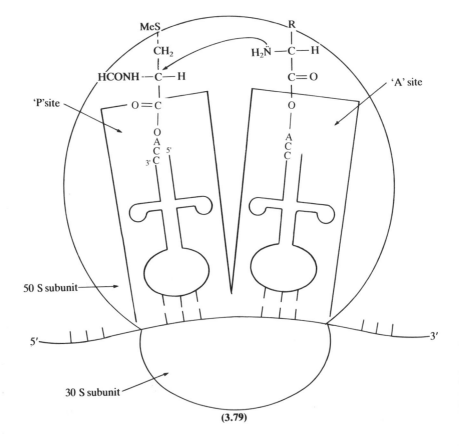

(3.79)

Fig. 3.47.

Aided by further protein initiation factors, and using phosphate anhydride bond energy in the form of GTP, the first amino acid in the encoded protein sequence – which in prokaryotes is N-formylmethionine (fMet) attached to its cognate tRNA molecule ($tRNA^{fMet}$) – becomes bound to the mRNA–30 S complex to give an 'initiation complex', and subsequent binding of the 50 S subunit with displacement of the initiation factors, GDP, and orthophosphate affords the functional 70 S ribosome with fMet–$tRNA^{fMet}$ positioned in the so-called 'P' (for peptidyl) site (cf. (**3.79**)).

The anticodon of the $tRNA^{fMet}$ forms a complementarily base-paired hybrid with the 'initiation codon' AUG on the mRNA sequence, and the reading frame of the mRNA is thus set as a three-base(codon)-at-a-time series of codewords in the 3′ direction along the mRNA sequence. The aminoacyl-tRNA with anticodon complementary to the next codon on the mRNA then binds to the adjacent 'A' (for aminoacyl) site, via a complex involving a soluble protein 'elongation factor' and GTP which is subsequently hydrolysed to give (**3.79**). The peptide bond is then formed by a process of transpeptidization, with the primary amino group of the amino acid in the A site attacking the active ester of N-formylmethionine, resulting in the formation of fMet–$a.a._2$–$tRNA^{a.a.2}$ in the 'A' site and discharging the $tRNA^{fMet}$ in the 'P' site. There then follows a process known as 'translocation', which requires another soluble elongation factor and energy which is again supplied by the hydrolysis of GTP to GDP and orthophosphate. During this process, the discharged tRNA molecule in the 'P' site leaves the ribosome, and the peptidyl-tRNA complexed to the codon on the mRNA is transferred to the 'P' site.

Another way of looking at the process is to consider the ribosome as 'slipping', i.e. being translocated, three bases along the mRNA sequence in the 3′-direction during the process, thus bringing the peptidyl-tRNA into the 'P' site and exposing the next codon on the mRNA sequence in the 'A' site, ready for attachment of the next aminoacyl-tRNA with anticodon complementary to this new codon, and thus for inception of the next round of elongation. These cycles of elongation and translocation continue until a termination codon is reached, when the ester bond attaching the completed polypeptide to the tRNA corresponding to the C-terminal residue is hydrolysed to release the polypeptide from the ribosome. The polypeptide then undergoes post-synthetic modification, with deformylation of the N-terminal methionine, followed, possibly, by removal of a number of amino acid residues from the N-terminus, modification of certain residues, chain cleavage, etc., to afford the final functional form of the protein.

Codon–anticodon mispairing offers another possible source of misinsertion of amino acids during protein synthesis, and it is thought that here, too, a 'proofreading' step intervenes in which mispaired aminoacyl-tRNA species are discharged before misinsertion can occur. This is wasteful of cellular energy since a molecule of GTP is hydrolysed for each new aminoacyl-tRNA species brought to the 'A' site during elongation. A minimum of four 'high energy' phosphoric anhydride bonds are hydrolysed for each new amino acid residue added to the growing polypeptide chain, but in fact the energy bill is even more costly, and the difference is thought to represent the price the cell pays for high fidelity protein synthesis. The processes of protein synthesis occurring in eukaryotes are very similar but differ in some minor details.

Thus, despite the many complex higher order interactions involving protein factors and the ribosome structure, the fundamental chemistry of protein synthesis is simple. A carboxylic-phosphoric anhydride (the aminoacyl adenylate) is first formed, and converted to a highly reactive carboxylate ester (the aminoacyl-tRNA species). Peptide bond formation occurs via attack of the primary amino group of another amino acid residue on this reactive ester, with tRNA acting as a leaving group. Some of the proteins of the 50 S subunit seem to interact to effect the transpeptidization, but no single protein component has been identified as catalysing the process on its own.

3.7 Structure determination of proteins and polypeptides

Of the many challenges that face the protein chemist, obtaining pure material and in sufficient quantity are not the least. Indeed, they are far from trivial since the use of impure material, for example, can completely waste the time required for sequencing. The quantity of available material determines the scale upon which degradations are carried out and consequently can jeopardize the reliability of the results and restricts the techniques which can be applied to sequencing.

Many of the interesting biologically active peptides that are currently being studied are only present at concentrations of microgram per gram of tissue. Because of their activity these peptides can usually be detected by highly sensitive biological assays. Such assays are used to monitor the efficacy of extraction procedures and to determine whether the experimental conditions are denaturing the protein. Thus in the extraction of some opioid peptides from rat pituitaries, the activity of the extract was monitored at the picomole level with a receptor binding assay using neuroblastoma–glioma hybrid cells. Factors which can affect the yield of extracted material from tissues include general protein breakdown during the period between death of the animal and extraction and co-extraction of enzymes which degrade the material under investigation.

In those cases where the peptides are localized within an animal, plant, or microbial cell in one type of subcellular particle, these particles may be separated by differential centrifugation (see later). Following this process the cells have to be disrupted (by mechanical means, ultrasound) and the peptide extracted.

The process of extraction often requires much time in order to achieve optimal conditions. Ideally, the extraction process should also achieve some purification. For those proteins which possess a well-defined three-dimensional structure, extraction is usually carried out with a dilute salt solution where the ionic strength, temperature, pH, and the nature of the ions present are carefully controlled.

Insoluble proteins present formidable problems because they have to be subjected to drastic conditions of extraction. Structural proteins such as collagen (a major component of skin and tendon, and important in the healing of wounds), elastin (present in arterial walls and ligaments), and keratin (chief component of hair, fur, wool, horn) are all difficult to extract and our knowledge of their structure is therefore far from complete.

3.7.1 Purification of peptides

Purification techniques and methods for assessment of purity usually depend upon one or more of the following properties of peptides: molecular weight, density, molecular shape, and ionic charge. Some of the more commonly used techniques are now described.

3.7.1.1 Gel permeation (or exclusion) chromatography

This form of chromatography effects separation on the basis of molecular size and utilizes gels which are inert cross-linked polymers possessing pores and cavities of a well-defined size. When such a gel is packed into a column and a mixture of linear peptides applied at the top, in a suitable buffer solution, the constituents pass down the column and are eluted. Material having the highest molecular size is eluted first and as the elution continues, compounds having smaller and smaller sizes are eluted. Separation is effected according to size: molecules possessing a size greater than that of the pores are not retained, i.e. are excluded. Molecules having a similar or smaller size to the pores enter the interior of the gel particles and are thus retarded or require large volumes of eluting buffer to be washed out of the column.

Gels possessing different pore sizes are available for separating materials whose molecular weight may be in the order of a few thousand to those having a molecular weight of approximately one million. By utilizing several columns packed with gels having different pore sizes, it is possible to obtain extracts containing materials whose molecular weights are within a few thousand of one another. Even purification to these limits is not sufficient to ensure that one is dealing with a single product.

3.7.1.2 Ion exchange chromatography

Separation by this technique relies upon materials differing in their surface charge. A solution of the mixture to be separated contained in a suitable buffer solution is applied to a column packed with a synthetic resin which possesses charged groups. The resins are of two types: cationic and anionic. The cationic resins separate cations and possess negatively charged groups, whilst the anion resins possess positively charged groups. Commonly used ion exchange resins are sulphonated polystyrenes. Other ion exchange supports include diethylamine-cellulose (an anion exchanger), when the amino group is protonated. The progress of a peptide down the column will depend upon the charge of the resin and the charge of the peptide, which is in turn dependent upon the pH of the buffer solution. Clearly by eluting with a buffer solution the pH of which is either slowly increased or decreased, the rate of elution of the peptide can be altered.

In many cases the resolution attained by column chromatography can be considerably bettered by the use of high performance liquid chromatography (HPLC) (which is also known as liquid chromatography, LC). LC employs pumps to force the eluent solution through the columns. Ion exchange (and also molecular sieve) packing materials are available for use with this technique.

Use of both ion exchange LC and gel permeation chromatography in combination for the separation of mixtures is a particularly powerful method. Detection of the proteins is often carried out by derivatizing to yield a fluorescent product. Thus the protein mixture may be derivatized with either fluorescamine or *o*-phthaldehyde and the mixture separated or the stream of eluent from the column split and a portion derivatized with the reagent. With fluorescamine the sensitivity of detection is well below 100 pmol of peptide.

3.7.1.3 Affinity chromatography

If a support material can be derivatized with a material that has a special affinity for the desired polypeptide then purification can be effected by the use of affinity chromatography. Materials which can be purified in this way include enzymes (where the column material is derivatized with a suitable substrate) and antibodies (where the column is derivatized with a suitable antigen). After the impurities have been eluted the column is washed with a solution of the free specific eluent.

3.7.1.4 Electrophoresis

Separation by this technique relies upon the individual polypeptides possessing different surface charges. Basically an electric field is applied to a solution of the mixture which is at a particular pH, and as a consequence the charged polypeptides migrate towards the appropriately charged electrode.

In *zone electrophoresis* the polypeptide mixture is applied to an inert support (e.g. paper) or a gel (e.g. acrylamide gel) – usually across the centre. The voltage is applied for some time. Once the positions of the polypeptides on the paper have been located (e.g. by application of a reagent such as fluorescamine or ninhydrin) the paper can be cut up and the polypeptides eluted with a suitable solvent.

In *isoelectric focusing* a stable pH gradient is set up by applying a voltage across the ends of a column that contains a solution of mixed amphoteric substances possessing different isoelectric points, and either sucrose or glycerol.

These amphoteric substances distribute themselves between the electrodes, the position being dependent upon the charge. Because of the added sucrose or glycerol there is also a density gradient, the highest density being at the bottom of the tube. When a protein is added to the stabilized pH gradient it migrates in the electric field until it reaches the pH region which is the same as that of its isoelectric point. At this pH, it has of course zero mobility. Thus a mixture of polypeptides can be separated on the basis of their isoelectric points.

3.7.1.5 Ultracentrifugation

If centrifugation is carried out using very high rotor speeds, large centrifugal forces are generated which cause dissolved macromolecules to sediment in the direction of the force field. The larger the polypeptide, the faster it moves to the bottom of the centrifuge tube. This method is particularly useful for separating cell extracts from the cell debris.

In zonal density gradient centrifugation a density gradient of a substance such as sucrose or glycerol, in aqueous buffer, is set up in an ultracentrifuge tube prior to centrifugation. The sample is applied to the top of the tube and then the solution centrifuged. The polypeptide moves down the tube at a rate determined by its size, shape, and density.

3.7.2 Evaluation of purity

Methods used for purifying polypeptides can be used for determining purity. Thus the material is checked for homogeneity in size using ultracentrifugation and gel filtration, and homogeneity in charge using electrophoretic methods (employing a number of different supports and over a wide pH range). Clearly if the material has been obtained in a pure state its biological activity should not alter if it is subjected to further purification processes.

3.7.3 Strategy for polypeptide sequencing

The basic strategy for polypeptide sequencing has not changed significantly over the last three decades. However, there have been tremendous advances in increasing the sensitivity of the chemical methods, and greater use of spectroscopic methods (mass spectrometry and NMR) in order to allow the structures to be determined of materials available in minute quantities – often a few nanomoles. The introduction of automated analytical procedures has significantly decreased the time required to carry out analyses.

The process of sequencing allows the primary structure to be determined, i.e. the amino acid sequence. In the case of proteins it will uncover whether they are *simple proteins*, i.e. their structure is an assemblage of α-amino acids, or *conjugated proteins*, i.e. materials made up of an assemblage of α-amino acids attached to a non-polypeptide. The non-polypeptide, known as a *prosthetic group*, may be a sugar (the compound is known as a glycoprotein), a haem (the compound is known as a haemoprotein), or phosphorus-containing residues (compounds known as phosphoprotein).

From a knowledge of the α-amino acid sequence it is possible to gain some idea concerning the polypeptide's *secondary structure*, i.e. which portions of the chain will adopt an α-helical conformation and which will form a β-pleated sheet. Determination of the *tertiary structure*, i.e. the overall folding of the polypeptide chain, and the *quaternary structure* in the case of those molecules composed of more than one protein chain, can only be accomplished by carrying out a full X-ray structural analysis. Such an analysis requires the material to be crystalline and for the crystals to possess particular features.

The strategy follows these steps:

Step 1.   Total hydrolysis to yield a mixture of amino acids which is analysed to give the *total amino acid composition* of the protein. It is this step which also reveals whether other materials, e.g. carbohydrates, are covalently linked to the protein. This step identifies the acids and their relative amounts but *not* their sequence.

Step 2.   End group analysis. The N-terminal and C-terminal acids are identified by appropriate labelling and hydrolysis reactions.

Step 3(4).   Partial acid hydrolysis. The polypeptide is hydrolysed to give a number of peptides smaller than itself. These peptides are sequenced. Provided the smaller peptides contain some common information a sequence of the polypeptides can be established, e.g.

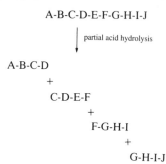

If the total hydrolysis showed that the polypeptide contains 1 mole of each of the amino acids A–J, then the sequence A–J can be established from the partial hydrolysis.

Step 4(3).   Enzymatic cleavage. Enzymes are used to break up the polypeptide into smaller peptides whose structures are then determined. This method is complementary to the partial acid hydrolysis procedure. However, in many cases the enzymes cleave at specific sites and therefore the knowledge concerning the identity of some of the terminal amino acids is acquired. Furthermore, it is unusual for partial acid hydrolysis to give the full structure of the polypeptide since usually the peptides produced by hydrolysis do not contain sufficient common information. The enzymatic cleavage may therefore yield additional information and consequently both acidic and enzymatic cleavages are used.

Step 5. Non-enzymatic specific cleavage of the polypeptide. This is usually carried out with a reagent which attacks a specific residue, e.g. cyanogen bromide attacks methionine. Cleavage in this way yields smaller fragments which may be suitable for sequencing or may have to be broken down further to yield even smaller units.

In addition to these chemical processes, the technique of mass spectrometry is also employed. Using some of the newer methods for ionization, molecular ions of the order of several thousand can be observed and the fragmentation pattern can give information concerning the sequence of the amino acids. This method is particularly useful for N-terminally blocked polypeptides, e.g. polypeptides possessing an N-terminal acetylglycine, since such materials are not easily sequenced using the conventional procedures. At the time of writing the sensitivity of the mass spectrometric method (about 1 nmol) is less than that of the chemical methods (a few picomoles).

3.7.4 Methods involved in sequencing

The pressure to develop methods of sequencing which require smaller and smaller amounts of material has led to the introduction and use of fluorescent labels and radiolabels. Detection of materials using these labels is significantly more sensitive ($> 10^3$) than procedures depending upon light absorption.

Most of the fluorescent labels are used for labelling amino groups and in particular N-terminal residues; see, for examples, Fig. 3.48(a)–(e). A variety of radiolabels has been employed, e.g. tritium for C-terminal determination and ^{35}S for the Edman degradation (see later). Such labels aid the location of materials after chromatography by such techniques as autoradiography.

3.7.4.1 Details of strategy

Step 1: total acid hydrolysis

The commonest method of hydrolysis involves heating the polypeptide with 6 M hydrochloric acid (constant b.p. hydrochloric acid) at 100°C for 24 h under an inert atmosphere. This process leads to the degradation of some amino acids, e.g. tryptophan, and corrections have to be made to the final result. Alkaline hydrolysis is used occasionally even though many acids are destroyed by the treatment (e.g. serine and threonine). However, tryptophan is stable under these conditions. Enzymatic hydrolysis is also used. This method possesses several advantages, e.g. amino acids are not destroyed by the process and many labile side-chain groups such as guanidine residues are not destroyed. The defect of this method is that the hydrolysis rarely

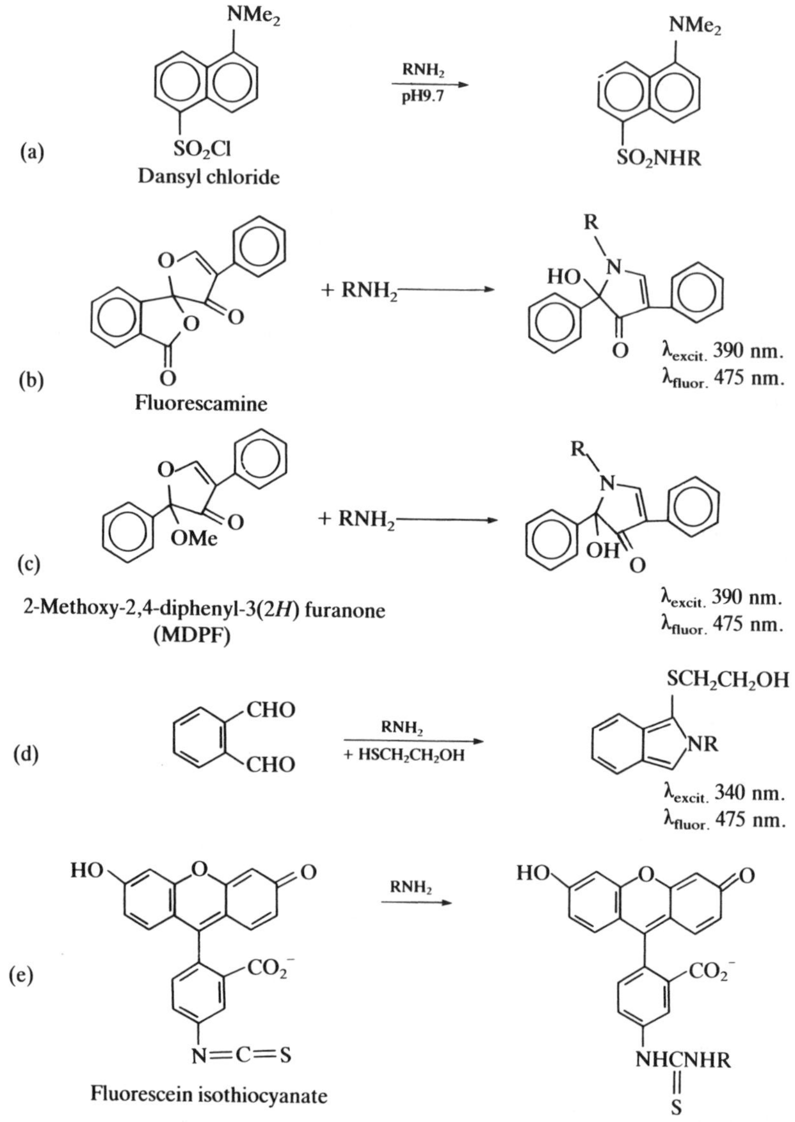

(a) Dansyl chloride

(b) Fluorescamine

$\lambda_{excit.}$ 390 nm.
$\lambda_{fluor.}$ 475 nm.

(c) 2-Methoxy-2,4-diphenyl-3(2*H*) furanone (MDPF)

$\lambda_{excit.}$ 390 nm.
$\lambda_{fluor.}$ 475 nm.

(d)

$\lambda_{excit.}$ 340 nm.
$\lambda_{fluor.}$ 475 nm.

(e) Fluorescein isothiocyanate

Fig. 3.48. Positive ion FABs mass spectrum of methionyl-lysyl-bradykinin.

goes to completion. Nevertheless, this method is particularly useful for identifying acid- and base-labile amino acids.

Having obtained a mixture of the constituent amino acids it is now necessary to separate and identify the acids.

Analysis of mixtures of amino acids. Following the total hydrolysis of a peptide it is necessary to determine which amino acids are present and also to quantify the relative amounts of each.

One of the standard methods is to use ion exchange chromatography. The pK_A and pK_B values of amino acids are dependent upon the nature of the side-chains and this is reflected in their isoelectric points (pH at which the amino acid does not migrate in an electric field; Table 3.5).

If a cationic resin is used (often a sulphonated polystyrene) the acids will be retained via their amino groups. Elution with acid will cause the acid of lowest isoelectric point to elute in preference to the others. To achieve the most efficient separation (in terms of resolution and time), gradient elution is usually employed, i.e. the pH of the eluent is gradually increased throughout the elution period. The eluted amino acids are conventionally detected by their reaction with ninhydrin which produces a blue colour, the intensity of which can be monitored spectrophotometrically (Fig. 3.49).

Most α-amino acids react with *o*-phthaldehyde to give a highly fluorescent product which can be detected (with a spectrofluorometer down to the 100 pmol level). More recently, prederivatization of the amino acid with 9-fluorenylmethyl chloroformate has become common practice. The introduction of the Fmoc group allows the amino acid to be detected by UV spectroscopy and with even greater sensitivity by spectrofluorometry. An example of a separation using this method is shown in Figs 3.50–3.52.

The process of amino acid analysis has now been fully automated, and the use of reliable fluorescence detectors and derivatization methods has increased the sensitivity (80 pmol at the time of writing). Care has to be taken with the method of hydrolysis of the peptides since residues such as tryptophan are destroyed by an acid digest. If alkaline hydrolysis conditions are used, tryptophan is not destroyed, but arginine, cysteine, serine, and threonine are decomposed.

In recent years peptides have been isolated which contain D-amino acids.

Table 3.5.

Amino acid	Formula	Isoelectric point
Glycine	$H_2NCH_2CO_2H$	6.0
Leucine	$H_2NCH_2CHCO_2H$ $\quad\quad\quad\;\; \mid$ $\quad\quad\quad CH(CH_3)_2$	6.0
Serine	H_2NCHCO_2H $\quad\; \mid$ $\quad CH_2OH$	5.7
Aspartic acid	H_2NCHCO_2H $\quad\; \mid$ $\quad CH_2CO_2H$	2.8
Lysine	H_2NCHCO_2H $\quad\; \mid$ $\quad (CH_2)_4NH_2$	10.0

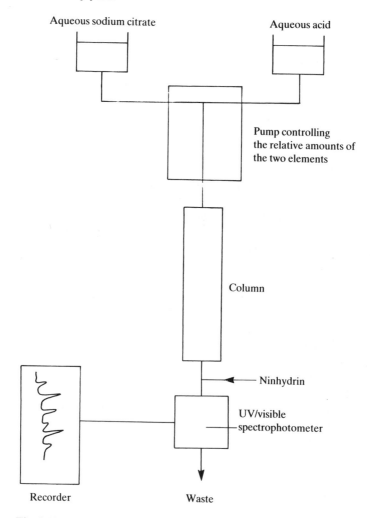

Fig. 3.49.

To ascertain the chirality of each amino acid requires the amino acids to be separated using a chiral support. Dansyl derivatives have been successfully separated on a polystyrene support derivatized with an optically active amine.

Step 2: end group analysis

N-terminal residues. The development of a method for identifying N-terminal residues was introduced by Sanger in his now classical work on the elucidation of the structure of insulin. The group of choice was the 2,4-dinitrophenyl group which could be readily introduced and is stable to the condition required to hydrolyse the proteins (Fig. 3.54).

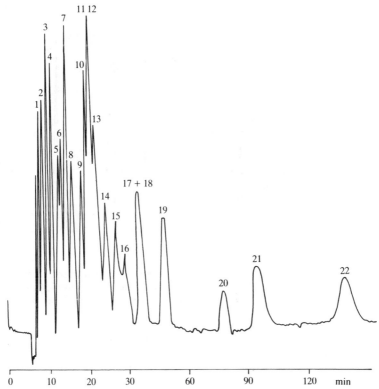

Fig. 3.50. Resolution of a mixture of 11 racemic amino acids on polystyrene bonded phase containing residues of N^1-benzyl-(R)-propanediamine-1,2. Temperature, 75°C. Elution sequence: 1 D-Pro, 2 L-Pro, 3 D-Ala, 4 L-Ala, 5 D-Ser, 6 L-Ser, 7 D-Val, 8 D-Thr, 9 L-Thr, 10 D-Leu, 11 + 12 D-Met + D-Ile, 13 L-Val, 14 D-Tyr, 15 L-Met, 16 L-Leu, 17 + 18 L-Tyr + L-Ile, 19 D-Phe, 20 L-Phe, 21 D-Trp, 22 L-Trp. Reproduced by permission of Blackie Academic & Professional (an imprint of Chapman & Hall) from W. J. Lough, *Chiral HPLC, Chiral Liquid Chromatography*, Blackie Academic & Professional, Glasgow, 1989.

Chromatography of the hydrolysate allows separation of the mixture, the N-terminal acid being readily identified by absorption spectroscopy due to the yellow colour of the 2,4-dinitrophenyl group.

Another particularly useful process is the Edman degradation, the chemistry of which is shown in Fig. 3.55. Two salient features of this process are that the N-terminal residue is obtained as an *N*-phenylthiohydantoin and that the process leaves intact the peptide chain minus the N-terminal acid. Thus by carrying out a further Edman degradation on this peptide the identity of the second amino acid can be established.

The phenylthiohydantoin can be readily identified by chromatographic procedures which include one- and two-dimensional thin-layer chromatography (TLC), and LC. In the case of LC the thiohydantoins can be detected using absorption spectroscopy. If greater sensitivity is required the thiazolines

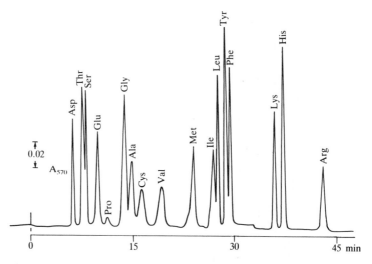

Fig. 3.51. Rapid amino acid analysis with temperature and flow programming. Reproduced by permission of International Scientific Communications Inc. from C. Bruton, *International Laboratory,* June 1986, p. 30.

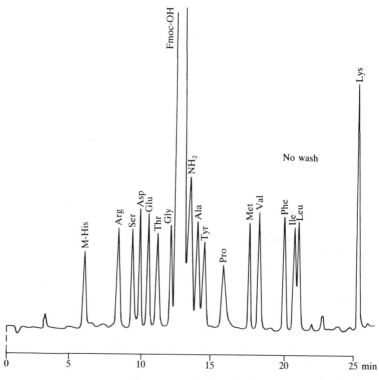

Fig. 3.52. Amino acid analysis with AminoTag pre-column derivatization. Reproduced by permission of International Scientific Communications Inc. from C. Bruton, *International Laboratory,* June 1986, p. 30.

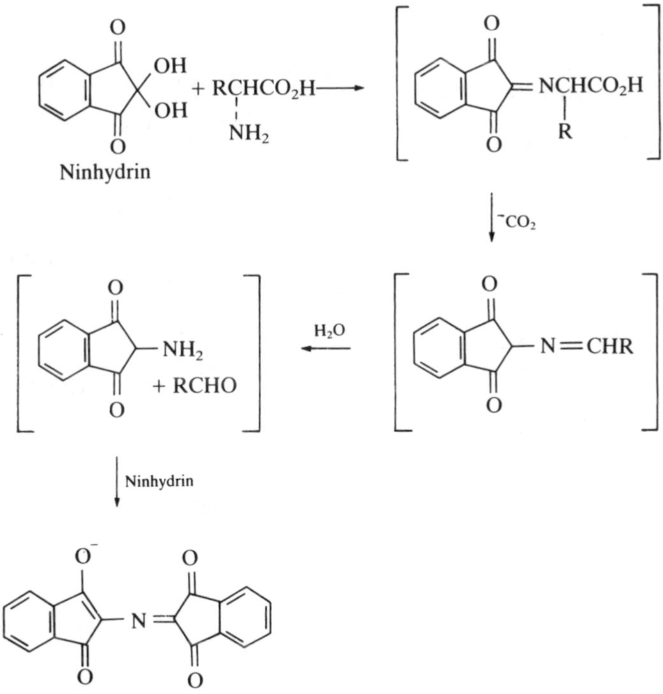

Ninhydrin
$+ RCHCO_2H$
NH_2
$=NCHCO_2H$
R
$-CO_2$
H_2O
$N{=}CHR$
NH_2
+ RCHO
Ninhydrin
O
N

Purple-coloured product

Fig. 3.53.

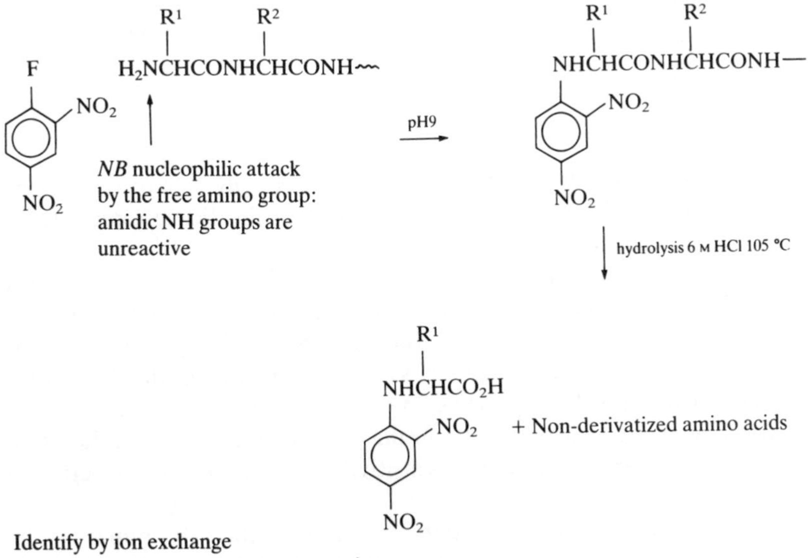

F
NO_2
R^1
R^2
$H_2NCHCONHCHCONH$
NB nucleophilic attack
by the free amino group:
amidic NH groups are
unreactive
pH9
NHCHCONHCHCONH—
hydrolysis 6 M HCl 105 °C
$NHCHCO_2H$
+ Non-derivatized amino acids

Identify by ion exchange chromatography, paper chromatography, LC, etc.

Fig. 3.54.

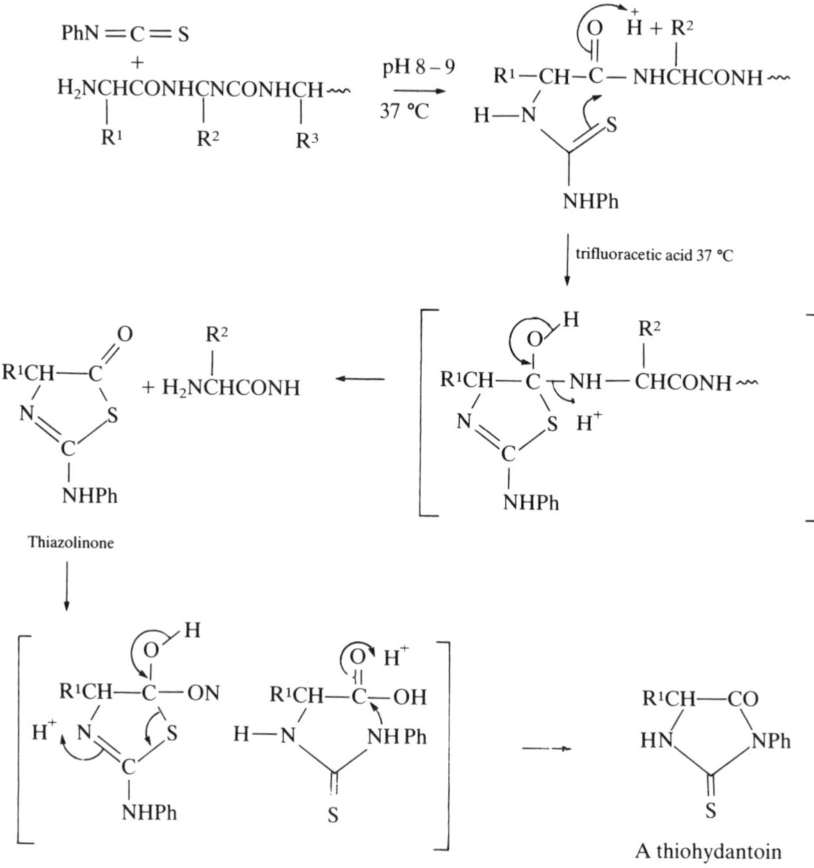

Fig. 3.55.

may be reconverted to the amino acids, derivatized with a fluorophore, and separated and identified in the usual way. Use of the Edman degradation for sequencing is dealt with later.

The use of the highly fluorescent dansyl derivatives gives a marked improvement in sensitivity of detection over the 2,4-dinitrophenylated derivatives (Fig. 3.56). A range of chromatographic techniques are available for ending identification of the dansylated acid. Using the fluorescent properties of the labelling groups allows nanomolar quantification of the derivatized acids to be detected. If, on the other hand, the dansyl derivative carries a radiolabel, even higher sensitivity (picomole concentration of derivative) can be detected.

Enzymatic degradation of the polypeptide from the N-terminal residue can also be used. Certain aminopeptidases (intracellular enzymes available from swine kidneys) cleave peptides and protein starting with release of the N-terminal residue. Clearly some of the amino acid which corresponds to

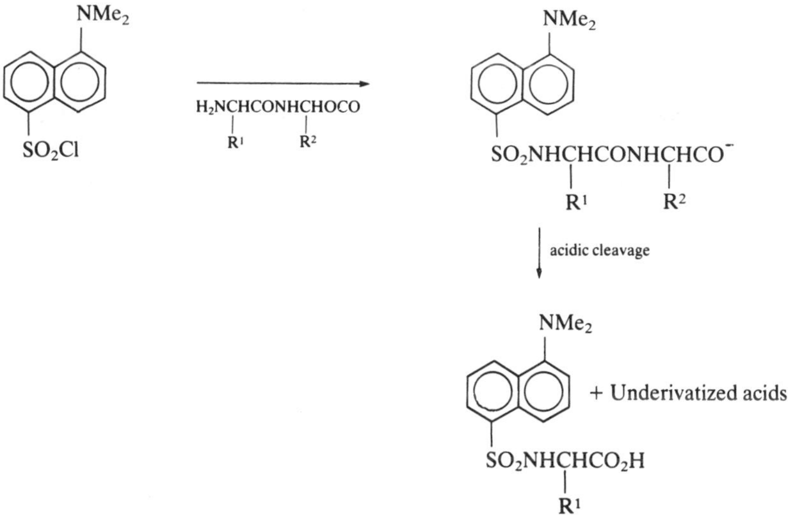

NMe$_2$

SO$_2$Cl

$\xrightarrow{}$

H$_2$NCHCONHCHCOCO
    | |
    R^1 R^2

NMe$_2$

SO$_2$NHCHCONHCHCO$^-$
    | |
    R^1 R^2

↓ acidic cleavage

NMe$_2$

+ Underivatized acids

SO$_2$NHCHCO$_2$H
    |
    R^1

Fig. 3.56.

the second amino acid residue will be released before all the N-terminal acid has been released and therefore the rate at which the various amino acids are released is measured. Detection and identification of these amino acids is aided by derivatizing with a fluorophore and sensitivity at the picomole level can thus be attained.

C-terminal residues.   When polypeptides are reacted with the strongly basic compound hydrazine, hydrazinolysis occurs (Fig. 3.57). Since the acid hydrazides are quite strongly basic the components of the reaction mixtures are amenable to separation by ion exchange chromatography and the free acid may be identified. It will be noted that the only free acid in the mixture is the C-terminal acid.

The use of carboxypeptidases to cleave amino acids sequentially from the C-terminal end is a particularly useful technique. If carboxypeptidase A is used (optimal pH 8.0) C-terminal acids are cleaved with the exception of

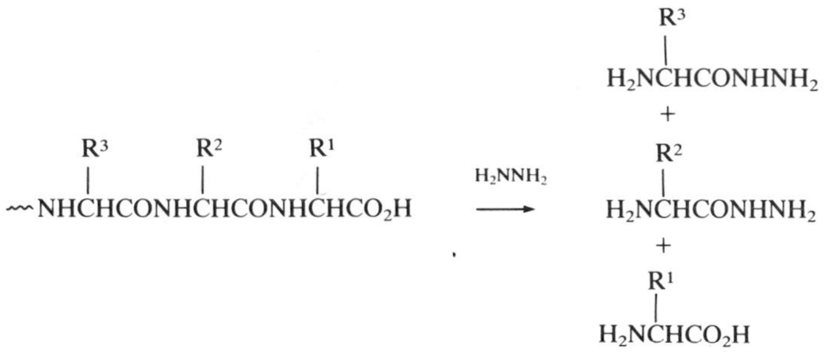

R^3 R^2 R^1
 | | |
~~NHCHCONHCHCONHCHCO$_2$H

$\xrightarrow{\text{H}_2\text{NNH}_2}$

R^3
 |
H$_2$NCHCONHNH$_2$
+
R^2
 |
H$_2$NCHCONHNH$_2$
+
R^1
 |
H$_2$NCHCO$_2$H

Fig. 3.57.

lysine, arginine, and proline. Clearly the rate at which each amino acid is liberated has to be measured if the sequence is to be obtained (Fig. 3.58).

A particularly elegant and sensitive way of identifying C-terminal residues relies on the fact that *N*-acylamino acids readily dehydrate to give azlactones which contain an acid proton. Base-catalysed tritium exchange followed by acid hydrolysis gives the C-terminal acid containing tritium (Fig. 3.59). Separation of the acids is accomplished in the usual way with the C-terminal amino acid being detected via a radiolabel.

Amount of
amino acid
liberated
(nmol mg^{-1}
of protein)

Tyr

-Asp-Phe-Tyr-CO$_2$H

Phe

Asp

Time of digestion

Fig. 3.58.

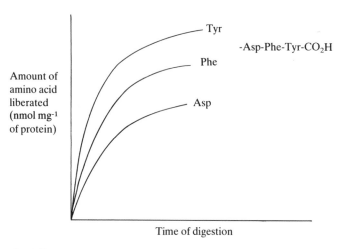

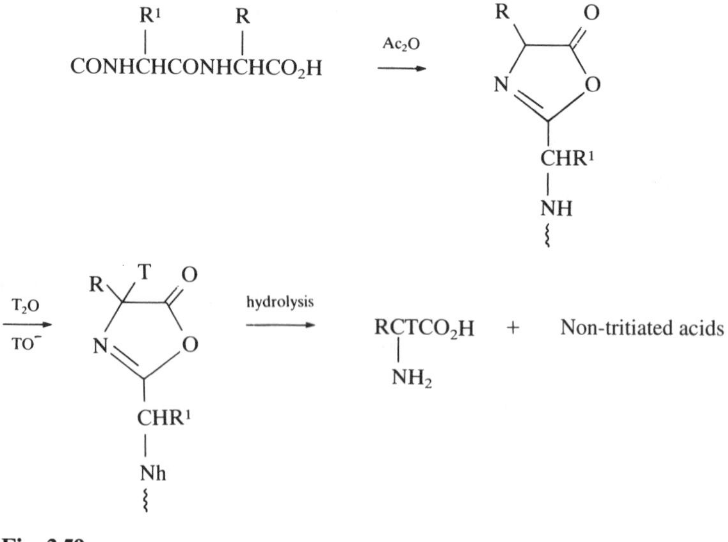

Fig. 3.59.

The determination of the identity and number of N- and C-terminal acids is of importance in determining whether the material being investigated contains more than one peptide chain. Clearly, if this is the case, the peptide chains have to be separated and the structure of each determined independently. If the chains are held together by non-covalent bonds, such as hydrogen bonds, they may be separated by treatment with either acid- or base-denaturing agents, such as urea. Usually, if chains are linked covalently it is via disulphide bridges. These can be cleaved by reduction and then oxidized or alkylated (Fig. 3.60).

End group analysis and sequencing. Some of the enzymatic N- and C-terminal determining processes also yield information concerning the sequence of the amino acids. In principle, the Edman degradation can also be used to sequence a protein from the N-terminal residue provided the thiazolone and polypeptide minus one residue can be separated. This being the case, the Edman degradation opens up a way of sequencing proteins from the N-terminal residue. There have been two important developments which have led to the widespread use of this method for sequencing. The first relates to a process for increasing the sensitivity and the second is the availability of automated machines for carrying out the sequential degradation. Increased sensitivity has been achieved by combining the Edman degradation with the process of labelling with dansyl chloride. A small portion of the polypeptide is dansylated and then hydrolysed. The dansyl amino acid is characterized. The remainder of the sample is subjected to the Edman degradation. There is now no need for the thiazolone to be characterized. The new peptide (i.e. the original peptide minus one residue) is then split into two portions. A very small portion is dansylated in order to identify the N-terminal acid. The rest of the sample is subjected to the Edman degradation to yield the peptide minus two residues. The process is then repeated. The sensitivity of this method is such that some proteins only available in nanomolar quantities have been sequenced and in some cases up to 20 residues identified.

The development of an automatic sequencer by Edman had an enormous impact since it enabled many polypeptides possessing a large number of

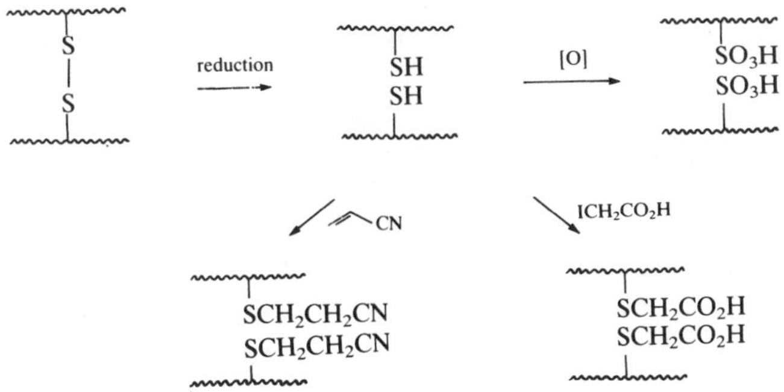

Fig. 3.60

amino acid residues to be sequenced. There are a number of examples where 50–80 residues have been sequenced. An important factor that determines the number of residues which can be sequenced is the overall yield in the degradation reaction. In order to sequence 50 residues, an overall yield of 98 per cent is required. There are now two types of sequencer available on the market. The first one developed carries out all reactions in solution, whereas in the second, the polypeptide under investigation is immobilized on a solid support: this process is called solid-phase sequencing. The cost of the equipment for solid-phase sequencing is far less than for the solution-phase sequencer. Difficulties do exist in attaching the peptides solely via their C-terminal residues to the resins but nevertheless the method allows sequencing to be carried out at the picomole level. Between 20 and 30 residues can usually be sequenced without too much difficulty. Needless to say there are examples in which a larger number of residues have been sequenced.

Steps 3 and 4: partial acid hydrolysis and enzymatic cleavage

In these steps the polypeptide is cleaved into smaller polypeptides. These smaller polypeptides have to be separated, purified, and then sequenced by identifying N- and C-terminal residues and sequencing by the methods outlined or by mass spectrometry. The separation and purification methods utilized for these smaller peptides are the same as those used to obtain pure samples of the polypeptides to be examined. Usually a combination of gel chromatography and ion exchange chromatography or reverse phase LC may be used.

Partial acid hydrolysis is carried out by reducing the stringency of the hydrolysis conditions used for total acid hydrolysis. To some extent obtaining suitable smaller peptides by this method in reasonable amounts and purity is a matter of luck.

Cleavage of polypeptides by enzymes is far more specific. For these reactions to occur with any degree of efficiency the polypeptide must be denatured, i.e. have its three-dimensional structure destroyed so that all the peptide bonds are accessible to the enzyme.

Some enzymes cleave polypeptides at specific points. Thus trypsin cleaves peptide bonds involving the carboxyl groups of the basic amino acids lysine and arginine whereas chymotrypsin hydrolyses the peptide bonds associated with the carboxyl groups of tryptophan, tyrosine, phenylalanine, and leucine. In a polypeptide containing both lysyl and arginyl residues, cleavage can be confined to the peptide bond of the arginyl residue by blocking the ε-terminal amino group of lysine by reacting it with citraconic anhydride. This blocking group is particularly useful because it can be removed, in high yield, under conditions (pH 2.0, 20 h) which do not hydrolyse peptide bonds. A protease is now available for cleaving the peptide bonds involving the carboxyl group of glutamic acid. Elastase has been used to cleave the peptide bonds involving the carboxyl group of alanine.

Several less specific enzymes are also employed. Pepsin (at pH 2) cleaves peptide bonds at both the amino and carboxyl groups of the aromatic amino acids and leucine. Thermolysin cleaves the peptide bonds involving the amino group of the hydrophobic amino acids, Ile, Leu, Val, Phe, Ala, Met, and Tyr, provided the acids are not linked to proline.

The choice as to whether a highly specific enzyme is used or not depends upon the structure of the polypeptide in question and the ability to separate and purify the enzymatically released smaller polypeptides. Use of a number of enzymes aids obtaining overlap information, which is essential if the jigsaw, of which the smaller peptides are the pieces, is to be put together. Another advantage of using specific enzymes is that the identity of some of the terminal residues will be known. The use of less specific enzymes can be very helpful if the enzymatically derived polypeptides are to be sequenced using mass spectrometry. As will be shown later, the requirement to separate and purify these polypeptides is not crucial for sequencing using this technique.

Step 5: specific cleavage of peptide bonds using chemical reagents

A particularly useful reagent for effecting a specific cleavage is cyanogen bromide which cleaves the peptide bond involving the carboxyl group of methionine. As can be seen from Fig. 3.61 the cleavage at methionine

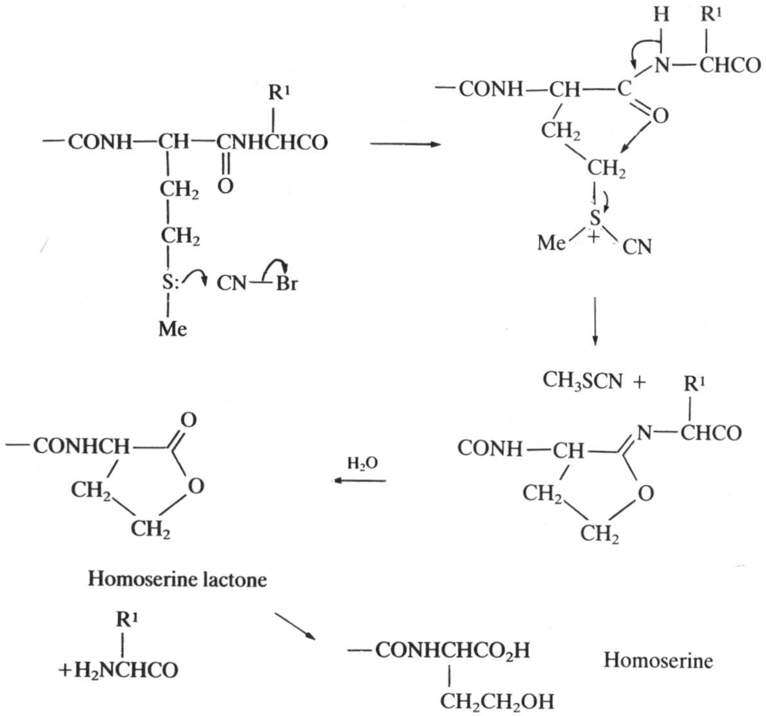

Fig. 3.61

produces two peptides for methionine residues, one of which has homoserine as the C-terminal residue. Since most polypeptides contain relatively few methionine residues, cyanogen bromide cleavage produces a few large peptides. Separation of these, followed by sequencing via the Edman procedure, usually gives information which is of value in assembling the information gained from enzymatic degradation. The production of a peptide having C-terminal homoseryl residue facilitates use of the automated solid-phase Edman sequencing procedure since it allows the ready attachment of the polypeptide to the resin:

peptide NH—⬠—O + H_2N—(R) ⟶ peptide NH—⬠—NH—(R)

where R = resin

3.7.4.2 *Sequencing using mass spectrometry*

The commonest and oldest way of generating fragments from organic compounds for analysis by mass spectrometry is by electron impact. For this process to be efficacious the material must be vaporized prior to electron bombardment. Compounds such as polypeptides, being polar and of high molecular weight, are involatile and on heating decompose. As a consequence, in their natural form they are not suitable for electron ionization (EI) mass spectrometry. However, derivatization so as to destroy all the hydrogen bonding markedly improves their volatility and processes have been developed which will cope with polypeptides containing up to ten amino acid residues. The advent of the so-called soft ionization methods, together with instruments possessing sufficient sensitivity and resolving power (high field magnet mass spectrometry), has now made it possible to record the mass spectra of compounds having molecular weights of the order of several thousands.

Sample preparation and fragmentation under EI. In order to render the polypeptides volatile they are first acetylated and then permethylated. The acetylation process acetylates all the free amino groups (N-terminal and side-chain) and the iodomethane methylates all the acidic N—H bonds, carboxyl groups, and phenolic hydroxyl groups of tyrosine (Fig. 3.62). Following this treatment the polypeptide contains no free N—H or O—H groups and consequently hydrogen bonding cannot occur.

Heating the derivatized polypeptide to 150–300°C in a mass spectrometer leads to sufficient volatilization so that electron bombardment can generate ions which fragment to give smaller ions which can be separated and analysed in the usual way (Fig. 3.63).

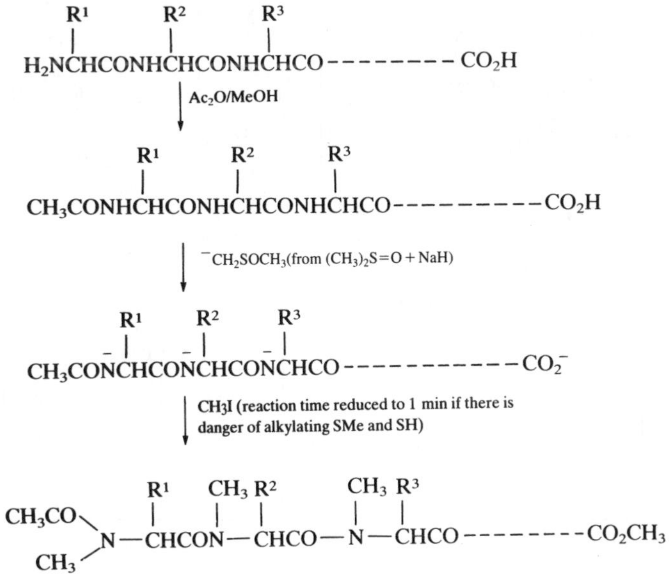

$$\overset{R^1}{\underset{|}{}}\quad\overset{R^2}{\underset{|}{}}\quad\overset{R^3}{\underset{|}{}}$$
H₂NCHCONHCHCONHCHCO – – – – – – – CO₂H

| Ac₂O/MeOH

$$\overset{R^1}{\underset{|}{}}\qquad\overset{R^2}{\underset{|}{}}\qquad\overset{R^3}{\underset{|}{}}$$
CH₃CONHCHCONHCHCONHCHCO – – – – – – – – CO₂H

| ⁻CH₂SOCH₃ (from (CH₃)₂S=O + NaH)

$$\overset{R^1}{\underset{|}{}}\qquad\overset{R^2}{\underset{|}{}}\qquad\overset{R^3}{\underset{|}{}}$$
CH₃CONCHCONCHCONCHCO – – – – – – – – – – CO₂⁻

| CH₃I (reaction time reduced to 1 min if there is
| danger of alkylating SMe and SH)

CH₃CO⏝N—CHCON—CHCO—N—CHCO – – – – – – – CO₂CH₃
CH₃⁄ (with R¹, CH₃ R², CH₃ R³ substituents)

Fig. 3.62

[CH₃CO⏝N—CHCON—CHCO—N—CHCO – – – – CO₂CH₃]
  CH₃⁄ (with R¹, CH₃ R², CH₃ R³ substituents)

↓

[CH₃CO⏝N—CH—C—N—CHCO – – – – – –]
  CH₃⁄   ‖
          O +.
  (with R¹, CH₃ R² substituents)

| Common fragmentation process

CH₃CO⏝N—CH₃—C ≡ O⁺   CH₃NCH
CH⁄   (with R¹)   (with R²)

↗ Commonly seen ion

| Common fragmentation process

CH₃CO⏝N—CH⁺ + CO
CH₃⁄   (with R¹)

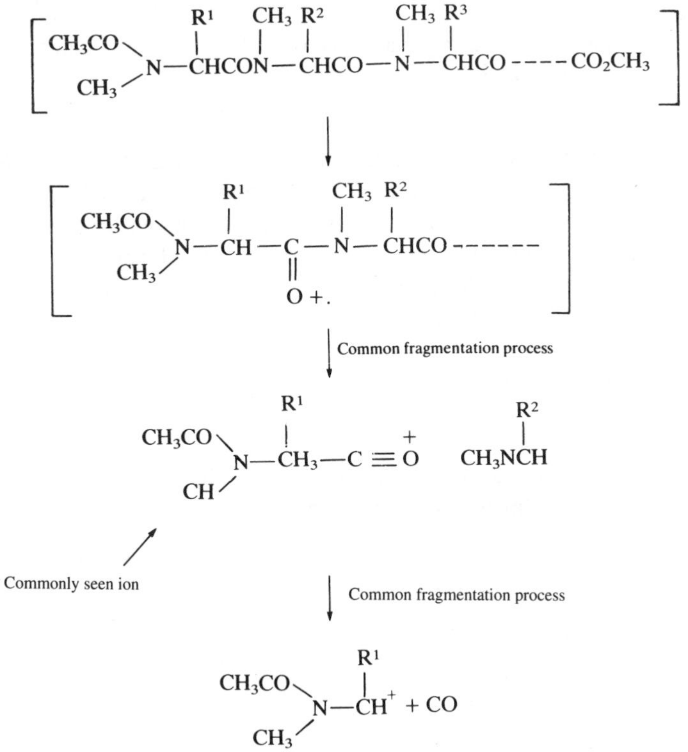

Fig. 3.63

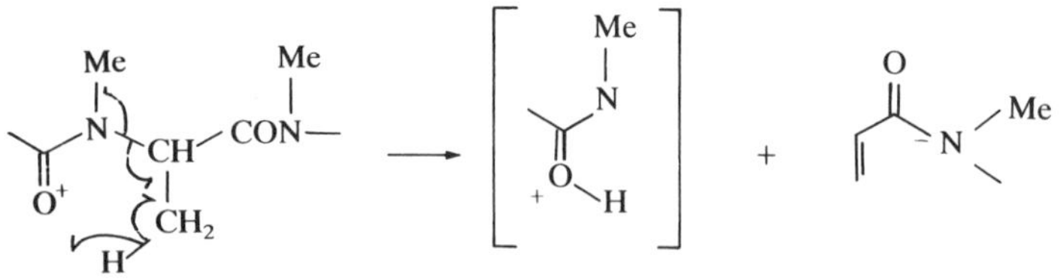

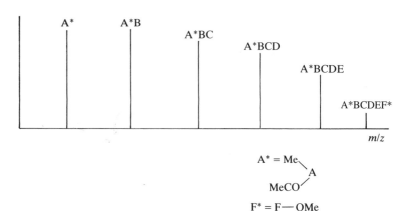

Fig. 3.64

A*
A*B
A*BC
A*BCD
A*BCDE
A*BCDEF*

m/z

A* = Me
A
MeCO

F* = F—OMe

Fig. 3.65. Idealized mass spectrum of derivatized polypeptide
A*–B–C–D–E–F*

Other fragmentation processes which occur include loss of N—Me, loss
of OMe and, of course, the McClafferty rearrangement (Fig. 3.64). Because
fragmentation of peptide bonds is a major fragmentation, the spectrum of
the fragment ions contains much structural information (Fig. 3.65). By
examining the *m/z* difference between the peaks for A and AB there is a high
probability of identifying B. Considering amino acid

$$H_2NHCHCO_2H$$
$$|$$
$$R$$

(R = alkyl) having molecular weight *M*, if it is an N-terminal amino
acid its fragmentation ion will correspond to

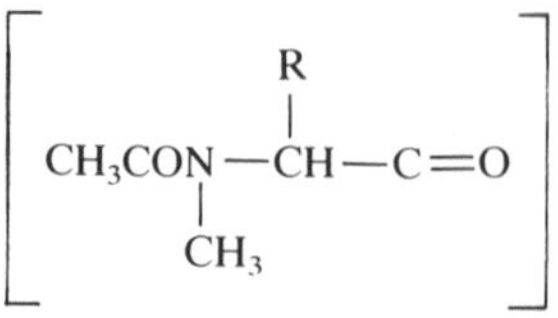

$$
\left[\begin{array}{c} R \\ | \\ CH_3CON-CH-C{=}O \\ | \\ CH_3 \end{array} \right]
$$

$= M-17 \quad +14 \quad +42$

(OH) (CH₃–H) (CH₃CO–H)

$= M + 39$

If the acid occurs in the chain, the fragment from the amino acid of molecular weight M will be:

$$
\left[
\begin{array}{c}
\underset{|}{\text{CH}_3}\ \underset{|}{\text{R}}\\[2pt]
\text{N}-\text{CH}-\text{C}{=}\text{O}
\end{array}
\right]
\qquad
\begin{array}{ccc}
= M-17 & -1 & +14\\
(\text{OH}) & (\text{H}) & (+\text{CH}_3-\text{H})
\end{array}
$$

$$= M - 4$$

If the acid is a C-terminal residue the fragment for an amino acid of molecular weight M will be

$$
\left[
\begin{array}{c}
\underset{|}{\text{CH}_3}\ \underset{|}{\text{R}}\\[2pt]
\text{N}-\text{CH}-\underset{\underset{\text{O}}{\|}}{\text{C}}-\text{OMe}
\end{array}
\right]
\qquad
\begin{array}{ccc}
= M-1 & +14 & +14\\
(-\text{H}) & (+\text{CH}_3-\text{H}) & (+\text{CH}_3-\text{H})
\end{array}
$$

$$= M + 27$$

For example, for alanine:

	N-terminal	In the chain	C-terminal
	CH₃	CH₃ CH₃	CH₃ CH₃
H₂NCHCO₂H	CH₃CONCHCO	N—CHCO	N—CHCOMe
CH₃	CH₃′		O
$M = 89$	128	85	116

If the amino acid contains functional groups which may be derivatized (e.g. phenolic hydroxyl group), introduction of the appropriate number of methyl or acetyl groups has to be taken into account.

Consider the simplified mass spectrum for a derivatized polypeptide shown in Fig. 3.66.

An N-terminal residue corresponding to m/z difference = 170 is leucine.

Fragment in the chain corresponding to m/z difference = 157 corresponds to glutamic acid.

Fragment in the chain corresponding to m/z difference = 170 corresponds to glutamine.

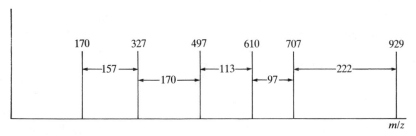

Fig. 3.66.

Fragment in the chain corresponding to m/z difference = 113 corresponds to valine.

Fragment in the chain corresponding to m/z difference = 97 corresponds to proline.

A C-terminal residue corresponding to an m/z difference of 222 is tyrosine.

Thus the sequence of the polypeptide is

Leu-Glu-Gln-Val-Pro-Tyr

Tables such as Table 3.6 have been drawn up which enable the amino acid associated with a particular mass loss to be quickly identified. The whole process has been greatly simplified by the use of computers which can, with great rapidity, match mass losses with particular amino acids.

Derivatization of the peptide by peracetylation and permethylation makes it difficult to assess whether the original peptide contained an *N*-acetyl or *N*-methyl group. Indeed, N-terminally blocked peptides are not uncommon and pose a problem for those seeking to determine peptide structure by wet methods such as N-terminal acid analysis. To overcome this problem, the peptides are derivatized using perdeuterated acetic anhydride $((C^2H_3CO)_2O)$ and perdeuterated iodomethane (C^2H_3I).

In this way the number and position of the derivatized nitrogen atoms can be easily determined; e.g. consider an N-terminally blocked peptide:

$$\overset{\overset{\textstyle R}{|}}{CH_3CONHCHCO-}$$

Treatment of the peptide with the perdeuterated reagents will in this case give a product whose mass spectrum will show the N-terminal acid having a mass 3 units higher than that observed from the peptide derivatized by non-deuterated reagents. If the *N*-acetyl group had not been present in the original peptide, use of perdeuterated reagents would have given an N-terminal residue having a mass 6 units higher.

The presentation of how to interpret the spectra has been perforce simple. However, it should be remembered that the mass spectrometers used for this type of work give m/z values with considerable accuracy and that many of the amino acids undergo complex fragmentation processes under electron impact. If accurate interpretations of spectra are to be made, it is essential that the fragmentation pathways available for each amino acid be understood as well as temperature dependence exhibited by these processes. Once again the computer is invaluable in storing such information and then using it to identify residues, not only by mass difference but by the observation of particular fragment ions.

The fact that volatility prevented the sequencing of peptides containing more than nine amino acid units until a few years ago seemed to place a restriction upon the use of the technique. Nevertheless the structures of some important small peptides were established, e.g.:

HTyr-Gly-Gly-Phe-MetOH ⎫
⎬ enkephalins
HTyr-Gly-Gly-Phe-Leu-OH ⎭

PCA-Leu-Asn-Phe-Thr-Pro-Asn-Trp-Gly-Thr-NH$_2$ locust adipokinetic hormone

(PCA = pyrrolidonecarboxylic acid)

From these examples it can be seen that the mass spectrometry offers a way of dealing with N- and C-terminally blocked peptides.

Clearly the capability of electron impact mass spectrometry for sequencing small peptides makes it ideal for dealing with peptides produced by partial acid or enzymatic hydrolysis. For sequencing via the Edman or Dansyl–Edman procedure the use of specific enzymes is preferred so that the number of peptides produced is not so large that they cannot be separated effectively, and time thus wasted by repeatedly sequencing units of known composition. The specific enzymes usually produce peptides which are too large for handling by the conventional electron impact mass spectrometry. Thus in the early days many enzymatically produced peptides could not be sequenced. The mass spectroscopists therefore turned to the use of less specific enzymes so that smaller peptides would be produced. The use of an enzyme such as cathepsin produces dipeptides which are ideal for analysis by mass spectrometry but of course all the 'overlap' information is lost. Success was obtained by using non-specific proteases such as thermolysin, elastase, and subtilisin. Usually successful sequencing requires the polypeptides to be produced by the use of more than one enzyme. Chemical degradation reagents such as cyanogen bromide are also used to increase the amount of 'overlap' information, thereby enabling a complete sequence to be obtained.

The enzymatic and chemical degradation of polypeptides to give smaller peptides, needless to say, produces a complex mixture. Remarkably, mass spectrometry can be used to sequence the peptides constituting the mixture. The usual protocol for enzymatic hydrolysis mixtures is to subject them to ion exchange chromatography. The fractions so obtained are analysed by electrophoresis, and those fractions which contain mixtures in common are pooled. A fraction containing a particular mixture is derivatized in the usual way (peracetylation and permethylation) and is then subjected to analysis by mass spectrometry. Sequencing is commenced by running the mass spectra of the mixture at as low a temperature as possible; this gives information about the most volatile peptide. The temperature is raised and the spectrum re-run. Peaks appear in this spectrum because of peptides volatile at this temperature. By gradually increasing the temperature, spectra of all the constituent peptides are accumulated. Figure 3.67 gives an indication of how temperature programming can be used to handle mixtures. Thus, in the spectrum run at 200°C, the major peaks relate to the peptide Val-Gly-Leu-Leu-Ala-Pro-Val-Ala. Other peaks are clearly present at m/z 128, 440 etc.

Table 3.6 Integral mass numbers corresponding to ions derived from derivatized amino acids. Reproduced by permission of the Biochemical Society and Portland Press from H. R. Morris, D. H. Williams and R. P. Ambler, *Biochem J.*, 1971, **125**, 189.

Amino acid	Structure of derivative	*A* N-Terminal mass	*B* Mass of residue	*C* C-Terminal mass	Associated ions[a] (commonly observed)
Glycine	CH₃ $\vert$ —N—CH₂—CO—	114	71	102	
Alanine	CH₃ $\vert$ —N—CH—CO— $\vert$ CH₃	128	85	116	Loss of CO from *A* or *B*
Valine	CH₃ $\vert$ —N—CH—CO— $\vert$ CH(CH₃)₂	156	113	144	Loss of CO from *A* or *B*
Leucine[b]	CH₃ $\vert$ —N—CH—CO— $\vert$ CH₂—CH(CH₃)₂	170	127	158	Loss of CO from *A* or *B* plus further loss of ketene from *A* (m/z 100)
Serine	CH₃ $\vert$ —N—CH—CO— $\vert$ CH₂—O—CH₃	158	115	146	Loss of CH₃OH from *A* or *B* (increases with increasing temperature)
Threonine	CH₃ $\vert$ —N—CH—CO— $\vert$ CH—CH₂—O—CH	172	129	160	Loss of CH₃OH from *A* or *B* (increases with increasing temperature); occasional loss of side chain minus H

	Structure				Notes
Aspartic acid	$\begin{array}{l} CH_3 \\ -N-CH-CO- \\ \quad\quad CH_2-CO-O-CH_3 \end{array}$	186	143	174	Loss of CO from *A* or *B*
Glutamic acid	$\begin{array}{l} CH_3 \\ -N-CH-CO- \\ \quad\quad CH_2-CH_2-CO-O-CH_3 \end{array}$	200	157	188	Loss of CO from *A* or *B*
Asparagine	$\begin{array}{l} CH_3 \\ -N-CHCO- \\ \quad\quad CH_2-CO-N(CH_3)_2 \end{array}$	199	156	187	Loss of CO from *A* or *B*
Glutamine	$\begin{array}{l} CH_3 \\ -N-CH-CO- \\ \quad\quad CH_2-CH_2CO-N(CH_3)_2 \end{array}$	213	170	201	Loss of CO from *A* or *B*
Phenylalanine	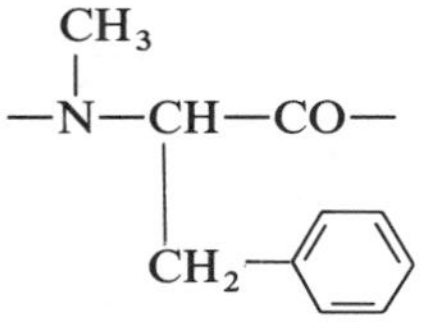 $\begin{array}{l} CH_3 \\ -N-CH-CO- \\ \quad\quad CH_2- \end{array}$ (phenyl)	204	161	192	Loss of CO from *A*; m/z 91, 162
Tyrosine	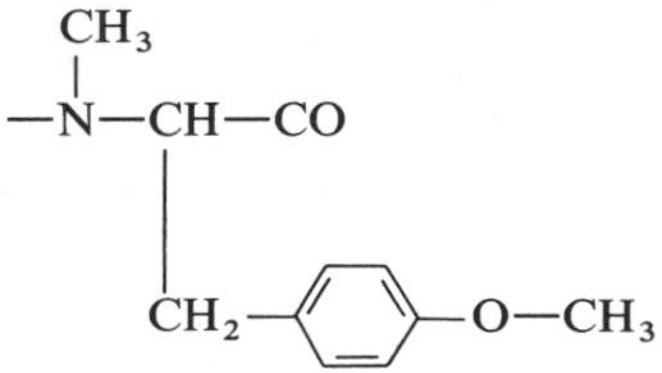 $\begin{array}{l} CH_3 \\ -N-CH-CO \\ \quad\quad CH_2- \end{array}$ (4-$O-CH_3$ phenyl)	234	191	222	Loss of CO from *A*; m/z 121, 161, 192

Table 3.6 *continued*

Amino acid	Structure of derivative	A N-Terminal mass	B Mass of residue	C C-Terminal mass	Associated ions[a] (commonly observed)
Tryptophan	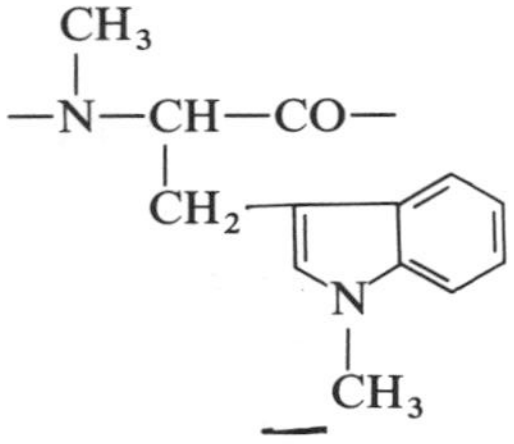 CH_3–N–CH–CO– ; CH_2 ; indole N–CH_3	257	214	245	Loss of CO from A; m/z 144, 215
Lysine	CH_3–N–CH–CO ; $[CH_2]_4$–N(CH_3)–CO–CH_3 ; CH_3	241	198	229	Loss of CO from A; m/z 171
Histidine[c]	–N–CH–CO ; CH_2–imidazole–CH_3	208	165	196	m/z 95; satellites of B at ± 14 mass units
Ornithine[d]	CH_3–N–CH–CO ; $[CH_2]_3$–N(CH_3)–CO–CH_3 ; CH_3	227	184	215	Loss of CO from A
Proline	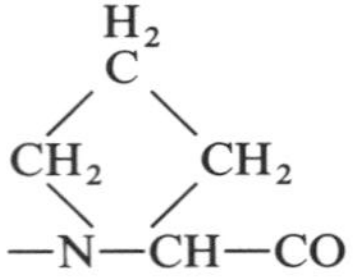 CH_2 CH_2 ; H_2C ; CH_2 –N–CH–CO	140	97	128	

[a] ... indicated by ... [b] Not differentiated from isoleucine. [c] Permethylation of this residue was not always ...

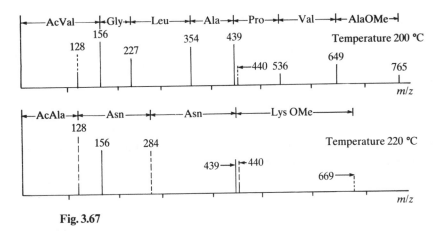

Fig. 3.67

but are relatively weak. On increasing the temperature by 20°C, these peaks increase intensely and other peaks e.g. m/z appear, whilst the major peaks associated with the spectrum run at 200°C diminish in relative intensity e.g. m/z 156 and 439. The peaks in the spectrum relating to the 220°C run which show increased intensity or are new, reveal the identity of the peptide of lower volatility i.e. Ala-Asn-Asn-Lys.

3.7.4.3 *Fast Atom Bombardment Mass Spectrometry (FABS)*

The introduction of this technique in 1981, together with the development of high performance mass spectrometers (i.e. those able through the capacity of their analysers to record ions having very high m/z values) has revolutionized the analysis of complex natural products such as peptides, glycopeptides and glycosides. The FABS technique enables underivatized peptides to be sequenced via mass spectrometry, and the extended mass range has reduced the analyst's dependence upon hydrolytic processes used to produce low molecular weight peptides.

The FABS technique employs a beam of fast neutral atoms, e.g. argon, xenon, possessing energies in the range 2–8 keV. The generation of these energetic atoms is accomplished as follows. The inert gas is ionized under electron impact, and the ionized gas accelerated so as to give ions possessing a controlled amount of energy in the range 2–10 keV. The ions are focused into a collision chamber which contains a high pressure (10^{-3}–10^{-4} Torr) of the inert gas. The ions and inert gas interact and energy is exchanged so that energized inert gas atoms are produced. The ionized species can be separated from the energized neutral atoms by an electrostatic deflector. The highly energetic atoms are then made to impinge upon the substrate which is contained in a suitable matrix, e.g. glycerol, triethanolamine, thioglycerol. In this way positive $(M + H)^+$ and negative $(M - H)^-$ ions are produced, and positive or negative ion analysis carried out. It has often proved most fruitful to examine both types of spectra to aid unequivocal identification.

The spectral intensity can be varied by varying the matrix. It is believed that the high energy atoms impinging on the matrix leads to ionization of the latter giving a radical cation and a low energy electron. It is these species which react with the substrate to generate positive and negative ions respectively. Since the ions of the peptide are generated with little kinetic energy there is little fragmentation other than via the main processes (e.g. in positive ion spectrometry the generation of acylium ions). The positive ion spectra are produced via similar fragmentation processes to those obtained with CI, i.e. one gets ions which contain the N-terminal amino acid and others which contain the C-terminal amino acid.

High molecular weight materials analysed by FABS include melittin (2845 u), bovine insulin (5730 u) and adrenocorticotropic hormone (ACTH) (4538 u).

FABS–mass spectrometry has not made the older methods of sequencing obsolete, but rather has shown that, by selecting the appropriate method for an appropriate subject ion, FABS–mass spectrometry can be used to maximum effect.

3.8   Nuclear magnetic resonance spectroscopy of peptides*

The reader will be aware of the many applications of NMR to organic chemistry and may wonder if any useful information can be gained from the NMR spectra of peptides since they are undoubtedly going to be complex with many of the signals occurring at similar chemical shifts. The advent of high field multinuclear Fourier transform spectrometers, the use of spin–spin decoupling, chemical shift reagents (some chiral), n.O.e. (nuclear Overhauser effect) difference spectroscopy, a wide variety of one-dimensional pulse sequences (DANTE, DEPT, INADEQUATE) and multidimensional NMR (commonly by two dimensions but with applications of three dimensions increasing) have facilitated the application of NMR to peptides. A particularly valuable feature of NMR is that it allows the peptide to be examined in solution and therefore conformational information and distances between particular groups often become accessible.

^{1}H and ^{13}C NMR spectroscopies are invaluable to the peptide chemist and by appropriate isotopic substitution ^{2}H NMR can also yield valuable information. Figure 3.68 shows, by way of example, how some of the above techniques have been applied to the tripeptide N-formyl-Met-Leu-Phe-OMe.

For ^{1}H NMR the chemical shifts of protons found in individual amino acids are well documented. Useful information about a peptide can be obtained from the chemical shifts, the effects of spin–spin coupling and, of course, from the integration of its ^{1}H spectrum. Figure 3.69 shows a 270 MHz ^{1}H spectrum of the tripeptide N-formyl-Met-Leu-Phe-OMe.

* Written with the help of Dr D.O. Smith, University of Kent at Canterbury.

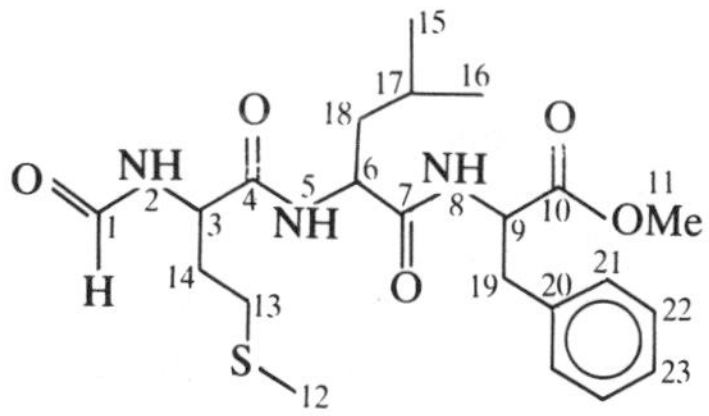

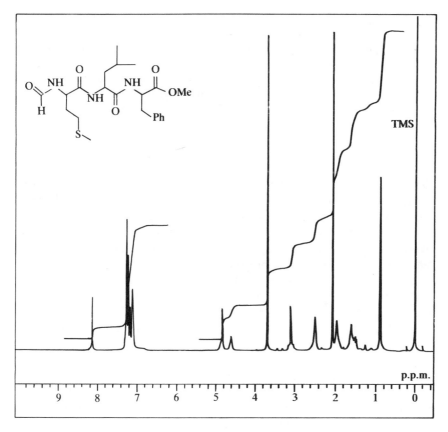

Fig. 3.68.

Fig. 3.69

Table 3.7 shows how the assignments can readily be made and Table 3.8 shows how other assignments can *tentatively* be made on the basis of chemical shift. Identifying the remaining resonances, which include the methine and amide N—H signals, is much more difficult.

However, a complete assignment of the ¹H spectrum becomes almost trivial when a ¹H/¹H COSY experiment is performed (Fig. 3.70). A brief

Table 3.7.

Shift (δ p.p.m.)	Multiplicity	Assignment		Amino acid residue
0.9	$2 \times d$	$(C\underline{H}_3)_2CH$	15, 16	Leu
2.06	s	$S\text{-}C\underline{H}_3$	12	Met
3.69	s	$O\text{-}C\underline{H}_3$	11	Phe-OMe
7.1–7.3	m	Ph	Ph	Phe
8.14	s	$C\underline{H}O$	1	N-Formyl-Met

Table 3.8.

Shift (δ p.p.m.)	Multiplicity	Assignment		Amino acid residue
2.5	m	$CH_3SC\underline{H}_2CH_2CH$	13	Met
~1.99	m	$CH_3SCH_2C\underline{H}_2CH$	14	Met
3.09	m	$PhC\underline{H}_2CH$	19	Phe

description of the two-dimensional spectrum is necessary to clarify interpretation. Along the top is seen the high resolution spectrum shown in Fig. 3.69. Beneath is the two-dimensional spectrum bounded by a square border. It forms a contour plot in which the peaks come out of the paper like mountains on a relief map. Along the diagonal, running from bottom left to top right, appears the one-dimensional spectrum. However, it is the cross-peaks, those peaks that lie off the diagonal, which are of importance. The spin system found within the N-formyl methionine residue can be used to show how the two-dimensional spectrum is interpreted. The formyl proton at $\delta 8.14$ serves as the starting point on the diagonal. Coupling between this proton and the vicinal N—H proton is indicated by the cross-peak $(1, 2)$ which is located by moving to the right from the diagonal. Moving vertically from the cross-peak $(1, 2)$ allows the N—H resonance to be identified beneath the phenyl protons on the diagonal. Moving right from this new position, on the diagonal, cross-peak $(2, 3)$ allows the methine proton (3) to be identified as the low field part of the multiplet at ~ 4.9 p.p.m. which integrated for two protons. This information will prove to be of great significance later. In similar fashion, cross-peaks $(3, 13)$ and $(13, 14)$ allow the positions of resonances 13 and 14 to be confirmed at 2.55 and 1.99 p.p.m. in the high resolution spectrum. Thus, it can be seen that neighbouring protons within the amino acid spin system can be systematically correlated. Suggested starting points for the two remaining amino acids in the tripeptide are the *iso*-propyl resonances, 0.9 p.p.m., and the benzylic methylene protons, 3.09 p.p.m.

Figure 3.71 shows the ^{13}C NMR spectrum of the tripeptide as well as the

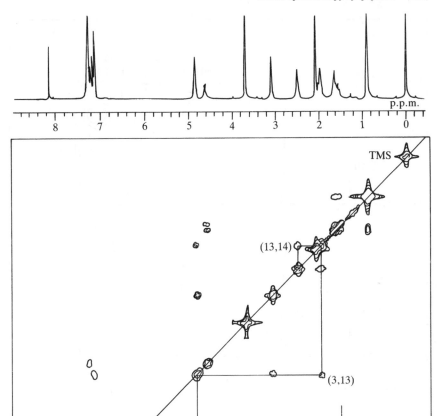

Fig. 3.70

DEPT spectrum. The latter was obtained using a pulse sequence with conditions chosen such that the signals due to CH$_3$ and CH groups are positive and those due to CH$_2$ groups are negative. This allows immediate identification of the methylene resonances. Another characteristic of the experiment is that quaternary carbon atoms can be assigned by their absence from the DEPT spectrum and presence in the standard ^{13}C spectrum.

Comparison of the two spectra in Fig. 3.71 clearly shows which resonances are associated with amide carbonyl groups (4, 7, 10), and the formyl carbonyl (1). The aromatic ring methine carbons (21, 22, 23) can all be identified

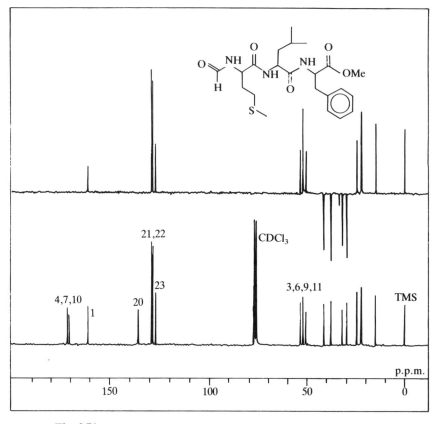

Fig. 3.71

from their chemical shifts and the quaternary ring carbon (20) is also noted for its absence in the DEPT spectrum. The signal at 77 p.p.m. is due to the deuterated solvent ($CDCl_3$) used to prepare the solution. The four resonances between 50 and 55 p.p.m. comprise the backbone methine carbons (3, 6, 9) and the methoxyl carbon (11). However, detailed assignments of these signals are very difficult since they are so close together. Theoretical chemical shift calculations may allow assignment of the remaining carbons; however, a two-dimensional $^1H/^{13}C$ COSY will allow unambiguous assignment even of the backbone methines.

The two-dimensional $^1H/^{13}C$ COSY spectrum, Fig. 3.72, shows within the rectangular boundary a typical contour plot. Along the top is the one-dimensional ^{13}C spectrum, while down the side is the ^1H spectrum. In common with the DEPT experiment quaternary carbons do not show cross-peaks in the contour plot. Interpretation is simple. For example, traversing horizontally from the formyl proton at 8.19 p.p.m. in the one-dimensional ^1H spectrum a cross-peak is encountered (1). Vertically above this, in the ^{13}C spectrum, is the formyl carbon. Since, in this system, the entire proton spectrum has been assigned from the $^1H/^1H$ COSY

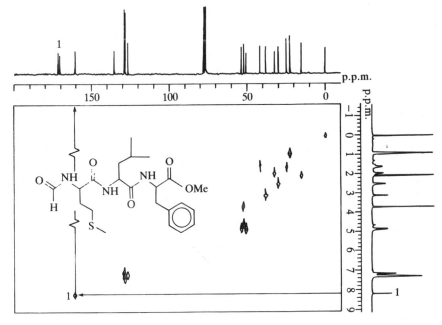

Fig. 3.72.

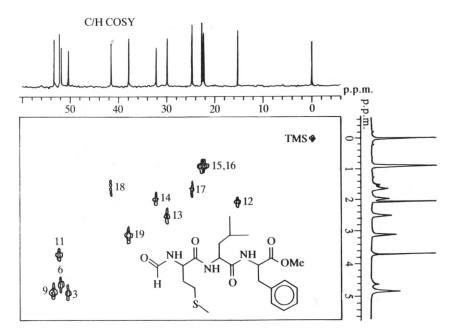

Fig. 3.73.

experiment all of the carbon signals can be identified in a similar manner. The power of the technique can readily be seen by detailed examination of the cross-peaks associated with the carbon resonances between 50 and 55 p.p.m. The chemical shifts for protons 3 and 9 were noted from the $^1H/^1H$ COSY to be different. This information, when applied to the $^1H/^1H$ COSY spectrum (Fig. 3.73), allows all four carbon resonances to be unambiguously assigned.

Thus, it can be seen that useful information, from the routine NMR experiments described above, can indeed be obtained from peptides. However, new, and more complex, NMR experiments continue to be devised and implemented almost daily. The latest HN(CO)CA three-dimensional experiments allow correlation of the amide NH to the α-carbon of the preceding amino acid via INEPT polarization transfers. Such experiments allow sequencing of the amino acids to be achieved in isotopically enriched samples. Unfortunately, such experiments cannot as yet be considered routine.

3.9   Polypeptide conformation

An analysis of the bond lengths of the peptide group using X-ray diffraction methods reveals the distances indicated in (**3.80**). If these values are compared with those of the 'normal' C–C bond lengths found between sp^3-hybridized carbon atoms in a long-chain alkane (0.154 nm) and the 'normal' C–N bond length in a primary amine (0.149 nm), it is seen that the figure of 0.132 nm for the N–C(O) bond length is much shorter than one would anticipate for the bond of order one which is formally drawn. This contraction cannot be wholly ascribed to the sp^2-hybridization of the carbonyl carbon. The value is, in fact, far closer to the value for a C=N double bond (0.127 nm). Moreover, the three atoms bonded to nitrogen lie in a plane, as (almost) do the three bonded to the carbonyl carbon, indicating that both are sp^2-hybridized, and that the resonance hybrid (**3.81**) more nearly represents the bonding state in a peptide bond. The peptide bond is thus a rigid planar unit, and is normally found in a *trans*-configuration in proteins, although the *cis*-isomer, which is ca. 8 kJ mol^{-1} less stable than the *trans*-arrangement, also occurs naturally.

Thus, every third bond in the polypeptide backbone is the central bond

(**3.80**)   (**3.81**)

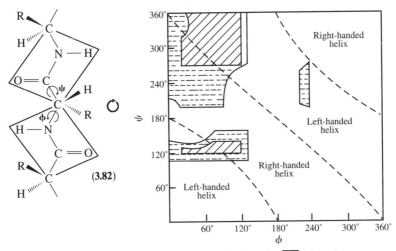

Fig. 3.74   Ramachandran plot: ▨, allowed; ⊡, higher energy; ▢, disallowed

in a rigid, planar unit, incapable of free rotation. Putting this another way, the α-carbon atoms of all the non-terminal amino acid residues are attached to two planar units (**3.82**). While there is, in principle, free rotation about the C_α–C and C_α–N bonds, this is, in practice, limited by steric hindrance. Two torsional angles, ϕ and ψ, may be defined as indicated in (**3.82**). By convention, these angles are assigned values of 0° when the main-chain atoms are *cis*, or eclipsed; both can have values varying from $-180°$ to $+180°$, and in the illustration (**3.82**) which shows the completely extended form of the chain, both ϕ and ψ have their maximum values of $\pm180°$ (but *caveat lector*: another convention, in which values of ϕ and ψ in the extended chain are assigned values of 0°, and ϕ and ψ are measured as the clockwise rotations, seen from C_α, of the two planes in (**3.82**), is also widely used). When $\psi = 180°$ and $\phi = 0$, overlap of the carbonyl carbon atoms results in severe steric hindrance, and prevents adoption of this conformation. At values of ϕ between $-40°$ and $-160°$, the carbonyl group exhibits minimal steric interaction with the side chain R (for all amino acids except glycine), and ϕ is found in this range of values in all commonly-occurring protein conformations. Computer analysis permits the sterically allowed combinations of ϕ and ψ to be calculated over the entire range of possible values, and this was performed by G.N. Ramachandran and his colleagues, resulting in plots of ϕ vs. ψ ('Ramachandran plots' or conformational maps) such as Fig. 3.74. These plots provide a valuable interpretation of the steric constraints which govern conformation in polypeptides.

3.9.1 Secondary structure in polypeptides

In a classical investigation, L. Pauling and R.B. Corey studied the possible ways in which the backbone of the polypeptide chain could be arranged

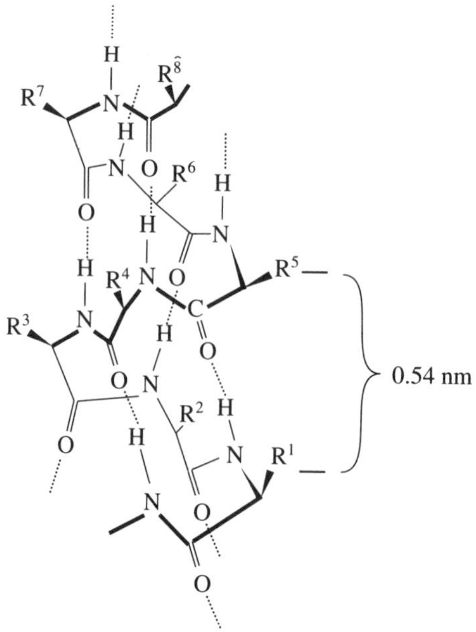

Fig. 3.75.

spatially to account for the strong reflections obtained from X-ray diffraction studies on fibres of a α-keratin, which indicates repeat units of ca. 0.55 nm, given the constraints of peptide bond planarity and the principle that hydrogen-bonding of the peptide groups should be maximized in order to stabilize the structure. The simplest arrangement envisaged was a right-handed helical arrangement, the *α-helix*, Fig. 3.75, having a helical pitch (i.e. distance corresponding to one complete turn, measured along the helical axis) of 0.54 nm, and having ca. 3.6 amino acids per turn. The NH of each amino acid residue forms a hydrogen bond with the carbonyl group of the fourth residue further along the sequence, in the direction of the N-terminus. Every peptide bond is involved in intrachain hydrogen-bonding in this arrangement, stabilizing the helical array. The values of $\phi = -57°, \psi = -47°$ in the Ramachandran diagram correspond to the most stable arrangement of the right-handed α-helix. A left-handed helix with values of ϕ and ψ of about $+60°$ is also possible for L-amino acids, but is far less stable and is not found in nature. The α-helix is the characteristic secondary structure found in the α-keratins and, indeed, one of the most important found in proteins generally. Other helical arrangements, such as the 3_{10} helix (3 amino acids per turn) and the π helix (4.4 amino acids per turn) are possible and have been observed in nature, but are less significant than the α-helix.

A different way to optimize hydrogen bonding in polypeptides is to line up fully-extended chains ($\phi = \psi = 180°$, as in (**3.82**)) side-by-side, to permit hydrogen bonding between them. In the resulting structure, the chains run antiparallel, Fig. 3.76. However, such an *antiparallel β structure* is only feasible if the R groups in Fig. 3.76 are hydrogen atoms, i.e. in polyglycine.

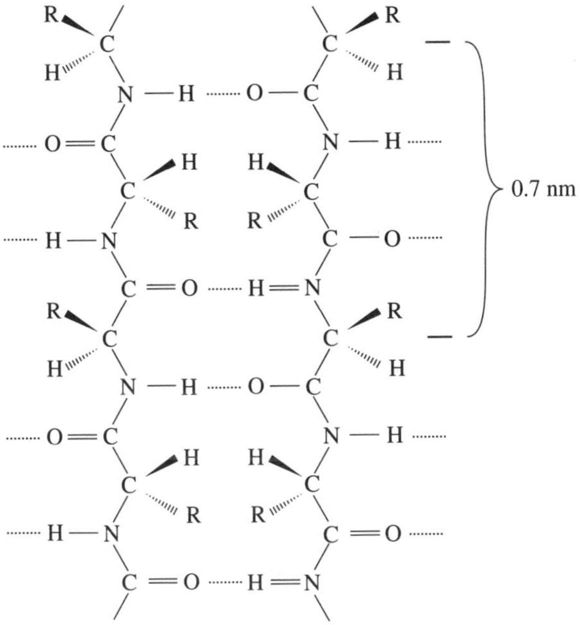

Fig. 3.76.

Otherwise, steric hindrance occurs, which can be relieved if ϕ is rotated from $-180°$ to $-140°$ while ψ is rotated to ca. $+135°$. As a result, instead of the flat fully-extended form, the antiparallel β structure forms a sheet with a regularly-folded appearance, usually described as a *'pleated sheet'*, with the R groups being 'up' on the upfolded pleat and 'down' on the downfold, Fig. 3.77. A 'parallel' arrangement of the chains is also feasible, Fig. 3.78, but the hydrogen-bonding arrangement is energetically less favoured than that in Fig. 3.76. It will be noted that in both Figs. 3.77 and 3.78 the 'R' side-chains are fully eclipsed, and this imparts a degree of steric hindrance which is energetically unfavourable and prevents adoption of these

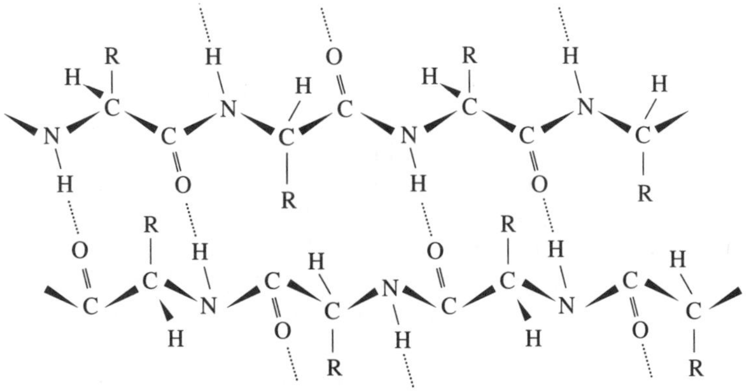

Fig. 3.77.

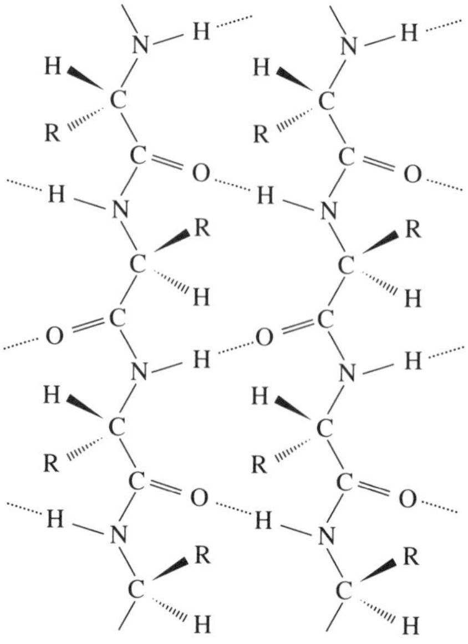

Fig. 3.78.

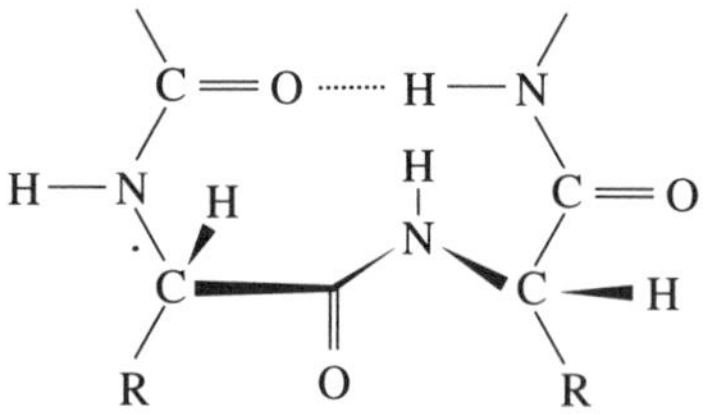

Fig. 3.79.

conformations precisely as illustrated: in practice this hindrance is relieved by the imparting of a slight twist to the sheets, and the twisted *β-pleated sheet* is a common structural arrangement. The antiparallel arrangement of Fig. 3.76 can be achieved by a single polypeptide chain folding back on itself, and the point at which this occurs is called a '*β-turn*', the plane of the peptide unit at the centre of the turn being almost perpendicular to that of the pleated sheet (Fig. 3.79). Such turns occur commonly where the spatial direction of a polypeptide chain is reversed. The antiparallel β-structure is the characteristic arrangement found in the β-keratin fibroin from the silkworm, and is a structure of major importance in proteins. Of the three types of reverse turn which have been observed, each involving four amino acids, with a hydrogen bond formed between the carbonyl group of the first residue and the amide nitrogen of the fourth, the β-turn is the most significant. The '3_{10}' helix represents one of these types of reverse turn, and indeed this conformation is rarely found elsewhere.

The secondary structures found in polypeptides are critically dependent on the nature of the amino acid side chains. For instance, cumulative runs of amino acids bearing side chains of the same charge, or branched at the β-carbon atoms, disfavour α-helix formation by electrostatic interaction and steric hindrance, respectively. At pH 7, poly-L-glutamic acid and poly-L-lysine are negatively and positively charged respectively, and have random conformation, but at pH values below the pK_a of poly-L-glutamic acid and above the pK_a of poly-L-lysine, where both polymers are neutral, each forms an α-helix.

Many attempts of varying degrees of sophistication have been made to devise a set of rules permitting the secondary structure present in a protein to be predicted from inspection of its primary structure. Chou and Fasman devised a pragmatic, but rather successful, approach by analysing the known X-ray structures of fifteen proteins to determine how frequently each of the twenty common amino acids occurred in α-helical, β, or coil locations. The ratio of the frequency with which a particular amino acid occurs in a particular location – say, the α-helix – to that of the frequency averaged over all twenty amino acids, affords a 'conformational parameter', $\underline{P_\alpha}$, which is simply a measure of the amino acid's propensity for being found in an α-helical location, but which does not take into account the effect of

Table 3.9.

	α-Helix	$\underline{P_\alpha}$	β-Structure	$\underline{P_\beta}$
Formers	Glu	1.53	Met	1.67
	Ala	1.45	Val	1.65
	Leu	1.34	Ile	1.60
	His	1.24	Cys	1.30
	Met	1.20	Tyr	1.29
	Gen	1.17	Phe	1.28
	Trp	1.14	Glu	1.23
	Val	1.14	Leu	1.22
	Phe	1.12	Thr	1.20
			Trp	1.19
Indifferent	Lys	1.07	Ala	0.97
	Ile	1.00	Arg	0.90
	Asp	0.98	Gly	0.81
	Thr	0.82	Asp	0.80
	Ser	0.79		
	Arg	0.79		
	Cys	0.77		
Breakers	Asn	0.73	Lys	0.74
	Tyr	0.61	Ser	0.72
	Pro	0.59	His	0.71
	Gly	0.53	Asn	0.65
			Pro	0.62
			Glu	0.26

neighbouring residues. It follows that an amino acid with $\underline{P_\alpha} > 1$ has a higher than average chance of being found in an α-helical region, and amino acids with $\underline{P_\alpha} < 0.75$ are 'helix breakers', the remainder being more or less 'indifferent'. Similarly, calculation of $\underline{P_\beta}$ residues permits 'β-formers', 'β-breakers' and 'indifferent' residues to be classifed (see Table 3.9). Thus, if four helix-formers occur in a consecutive run of six residues, helix nucleation is likely, or if three β-formers occur in a run of five residues, β-sheet formation is likely, so long as the remaining two residues are not helix breakers, or β-breakers, respectively. Termination of an α-helix or a β-structure occurs if a consecutve run of four residues contains two or less helix formers or β-formers respectively. Proline does not occur within α-helices or at the C-terminus, and usually breaks β-structures also. It is noteworthy that the four strongest helix-breaking residues are the four amino acids most frequently found in reverse turns.

While this account is something of a simplification of the rules, it summarizes the essentials, and rigorous application of the model permits predictions of regions of occurrence of α- and β-structures with some 80% accuracy.

When three-dimensional protein structures are depicted on the printed page without recourse to stereo images, tracts of α-helix are usually depicted as cylinders and tracts of β-structure as flattened arrows, thus permitting the reader to appreciate at once the location and role of these secondary structures as elements of the overall conformation.

3.9.2   *The prediction of protein folding*

Generally soluble proteins fold in such a way as to minimize the exposure of the hydrophobic side-chains to the aqueous medium. In 1982, Kyte and Doolittle devised a set of values ('Hydropathy indices', Table 3.10) to indicate the hydrophobic or hydrophilic nature of an amino acid residue in a protein,

Table 3.10. Hydropathy scale for amino acid side chains

Side chain	Hydropathy	Side chain	Hydropathy
Ile	4.5	Trp	−0.9
Val	4.2	Tyr	−1.3
Leu	3.8	Pro	−1.6
Phe	2.8	His	−3.2
Cys	2.5	Glu	−3.5
Met	1.9	Gln	−3.5
Ala	1.8	Asp	−3.5
Gly	−0.4	Asn	−3.5
Thr	−0.7	Lys	−3.9
Ser	−0.8	Arg	−4.5

Source: J. Kyte and R. F. Doolittle, *J. Mol. Biol.*, 1985, **157**, 105–32.

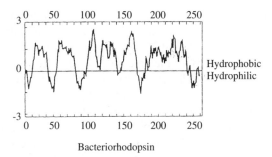

Bacteriorhodopsin

Fig. 3.80.

representative of the free energy change of its side chain in going from a non-polar to a polar environment. The values range from $+4.5$ for isoleucine, the most hydrophobic amino acid, to -4.5 for arginine, the most hydrophilic. A 'hydropathy plot' for a protein may be constructed by averaging the hydropathy index for the residues in a tract of protein sequence of chosen length (commonly six to nine residues), and plotting this value for the first residue in the tract. So, for instance, the mean hydropathy index for residues 1 to 7 (i.e. choosing a seven-residue tract) would be plotted for residue 1, the value for residues 2 to 8 plotted for residue 2 etc., along the entire length of the protein. Large positive peaks or tracts then indicate hydrophobic regions, and large negative peaks or tracts indicate hydrophilic regions. In practice, the strongly hydrophobic tracts revealed by hydropathy plots have been found to correlate well with interior regions of proteins, while the strongly hydrophilic tracts correlate with surface regions. This has practical application: most antigenic tracts in proteins are found, not surprisingly, in hydrophilic, surface-exposed regions, while membrane proteins are found to have a number (commonly seven) of lengthy (≈ 20 residue) tracts of hydrophobic amino acids which constitute their transmembrane domains. As an illustration, Fig. 3.80 shows a hydropathy plot for the photosynthetic bacterial transmembrane protein bacteriorhodopsin.

Discussion of further aspects of the higher organization of protein structure, such as 'supersecondary structure', and the use of the 'helical wheel' to identify hydrophobic or hydrophilic faces of α- and β-structures, and the sequence/structural motifs now recognized as occurring in certain proteins with special function, such as the 'zinc finger' and the 'leucine zipper' in DNA-binding proteins, are beyond the scope of this chapter, but the interested reader is encouraged to consult good modern biochemical texts for further information.

3.9.3 X-ray diffraction methods of structure analysis

The most exact methods available for the determination of protein conformation utilize the diffraction of X-rays. A crystal of the protein of interest is mounted in a quartz capillary tube in a beam of monochromatic X-rays. The lattice of the crystal acts as a three-dimensional diffraction grating, and the pattern of the scattered X-rays may be recorded either on a photographic plate or using comparable detecting devices. Rotation of the crystal in three dimensions permits a large number of diffraction photographs to be taken. The scattering of the X-rays is governed by the electron density in the crystal, and is thus greatest in regions where atoms of higher atomic number are situated, the scattering intensity being roughly proportional to the square of the atomic number. Where the beams being scattered from the atoms in the crystal reinforce each other, an intensity maximum in the scattered rays may be measured. Where the scattered rays interfere destructively, so as to cancel, no scattering is observed on the photographic plate. The phases of the scattered rays, and their amplitudes, thus determine the scattering pattern observed.

A theorem due to Fourier states that irregular continuous functions, such as the function $p(x, y, z)$ which measures the electron density at the point with coordinates (x, y, z) in the crystal, may be expressed as a sum of harmonic functions:

$$p(x, y, z) = \frac{1}{V} \sum_{h,k,l-\infty}^{h,k,l=\infty} F_{h,k,l} e^{i\alpha(hkl)} \cdot e^{-2\pi i(hx + ky + lz)}$$

where F_{hkl} are the amplitudes associated with the crystal reflection of Miller indices h, k and l, $\alpha(hkl)$ are the associated phases, and V is the volume of the unit cell. The summation is taken over all the observable reflections obtainable on the diffractometer. The problem is that while diffraction photographs may be indexed and the intensities of the scattered rays measured, the intensity I is related to F_{hkl} by the relationship:

$$I = K|F_{hkl}|^2$$

where K is a constant, and this provides no information as to the phases, $\alpha(hkl)$, of the scattered rays, thus preventing the use of the Fourier transformation.

This *phase problem* was solved in 1954 by Kendrew and Perutz using the method of *isomorphous replacements*. Heavy metal atoms, preferably one per unit cell, are introduced into the crystal structure of the protein, to bind at specific sites *without disturbing the crystal structure or packing*. The scattering caused by the introduction of this atom is superimposed on that due to the remaining atoms in the crystal, causing some scattered rays to become reinforced, and others diminished in intensity, permitting the position of the heavy atom in the unit cell to be localized. Once this has been performed, many of the phase angles of the Fourier components of $p(x, y, z)$ can be determined. Complete determination of all the phase angles usually requires the use of more than one heavy-atom isomorphous replacement: Perutz used six in solving the structure of haemoglobin, and Kendrew four for myoglobin.

The heavy atoms may be introduced either by covalent linkage of a heavy atom derivative to the protein prior to its recrystallization in a form isomorphous with that of the unmodified protein, or by crystallizing the protein from solutions containing the heavy metal ion, or by permitting heavy metal ions to diffuse from solution into the protein crystal. Haemoglobin was tagged with mercury by reaction of its sulphydryl groups with *p*-chloromercuribenzoate, for instance, and carboxypeptidase A, which normally contains zinc ions, was heavy-atom-labelled by replacement of the zinc with heavier metals.

However, the necessity of preparing heavy-atom-labelled derivatives to solve the phase problem has often proved the downfall of attempts to solve protein crystal structures by X-ray diffraction. It was realized many years ago that the required phase information is, in fact, contained in the X-ray scattering pattern, but its derivation and use to determine the electron density map of the crystal awaited the development of the mathematical and computational methods required. This has now bean realized, a feat recently recognized by the award of a Nobel Prize to Hauptmann and Karle for their development of the 'direct method', just as Kendrew and Perutz were awarded similar recognition for their development and use of the 'isomorphous replacement' method in the 1950's.

Once the electron density map of the protein molecule has been constructed, the known primary structure of the protein is fitted to the electron density contours, to reveal the conformation of the protein molecule. The degree of accuracy attained depends on the resolution of the structure. At 3.0 Å resolution the path of the polypeptide backbone is usually clear and some side chains can be resolved. At 2.5 Å resolution the side chains are well resolved, and individual atoms can be located to about ± 0.4 Å accuracy. At 1.5 Å, which is about the current limit of resolution attained, individual atoms are located to about ± 0.1 Å. As the degree of resolution is increased, the reflections to be analysed become of increasingly weaker intensity, owing to the X-rays having been scattered through greater angles, and their number also increases as the third power of the resolution. The method of isomorphous replacement cannot be used effectively at resolutions below about 2.5 Å, and further refinement below this figure has generally utilized computational methods in which electron densities predicted from model building studies based on the results at 2.5 Å are matched against the intensities at higher resolution.

Further reading

1. K.D. Kopple, *Peptides and Amino Acids*, W.A. Benjamin, New York, 1966 (an old text, but nevertheless covers many salient points in a clear, concise way); J.B. Dence, *Steroids and Peptides*, Wiley-Interscience, New York, 1980;

P.D. Bailey, *An Introduction to Peptide Chemistry*, Wiley, Chichester, 1990; E. Haslam (ed.), *Comprehensive Organic Chemistry*, vol. 5, part 23, Proteins, Amino Acids and Peptides, Pergamon Press, Oxford, 1979, pp. 177–385.

2. Synthesis and resolution of amino acids: the general methods available for the synthesis of α-amino acids are to be found in any good general text on organic chemistry, e.g. R.T. Morrison and R.N. Boyd, *Organic Chemistry*, Allyn and Bacon, Boston; M.D. Trevan, *Immobilised Enzymes*, Wiley, Chichester, 1980, chapter 3 (use of immobilized enzymes in resolution); E.J. Corey, R.J. McCaully, and H.S. Sachdev, *J. Amer. Chem. Soc.*, 1970, **92**, 2476 (an enantioselective synthesis of α-amino acids); J.E. Baldwin, R.M. Adlington, and B.J. Rawlings, *Tetrahedron Lett.*, 1985, **26**, 481 (coronamic acid); S. Hanessian and S.P. Sahoo, *Tetrahedron Lett.*, 1984, **25**, 1425 (L-vinylglycine); H. Urbach and R. Henning, *Tetrahedron Lett.*, 1984, **25**, 1143 (an example of an ACE inhibitor); A. Shanzer and J. Libman, *J.C.S. Chem. Comm.*, 1983, 846 (a synthesis of enterobactin); W.J. Lough (ed.), *Chiral hplc: Chiral Liquid Chromatography*, Blackie, London and Glasgow, 1989; E. Gross and J. Meienhofer (eds), *The Peptides: Analysis, Synthesis and Biology*, vol. 4, chapter 3, Academic Press, London, 1981. See chapter by V. Toome and M. Wiegele, p. 186 (determination of the absolute configuration of α-amino acids).

3. Peptide sequencing: S.B. Needleman (ed.), *Protein Sequence Determination*, Springer-Verlag, Berlin, 1975; R.A. Laursen, in *The Peptides: Analysis, Synthesis and Biology* (eds E. Gross and J. Meienhofer), vol. 4, chapter 6, p. 261, Academic Press, London, 1981 (solid-phase sequencing); L.R. Croft, *Introduction to Protein Sequence Analysis*, Wiley, Chichester, 1980.

4. For application of mass spectrometry see: H.R. Morris, *Nature*, 1980, **286**, 447; H.R. Morris, P.H. Williams, and R.P. Ambler, *Biochemical J.*, 1971, **125**, 189; D.H. Williams, S. Santikarn, P.B. Oelrichs, F. De Angelis, J.K. MacLeod and R.J. Smith, *JCS Chem. Comm.*, 1982, 1394.

5. For application of NMR see: H. Kessler, W. Bennel, A. Muller and K.-H. Pook, in *The Peptides: Analysis, Synthesis and Biology* (eds E.S. Udenfriend and J. Meienhofer), vol. 7, chapter 9, Academic Press, London, 1985.

6. Chemical synthesis of peptides: E. Gross and J. Meienhofer (eds), *The Peptides, Analysis, Synthesis and Biology*, Academic Press, London, 1979 (vols 1 and 2 of this series are devoted to syntheses; the use of protecting groups is elaborated in vol. 3); L.A. Carpino, *Accounts of Chem. Res.*, 1987, **20**, 401 (a review of the chemistry of the 9-fluorenylmethoxycarbonyl group); A. Bodanszky, M. Bodanszky, Z.J. Kwei, J. Martinez, and J.C. Tolle, *J. Org. Chem.*, 1980, **45**, 72 (synthesis of chicken VP); G. Barany and R.B. Merrifield, in *The Peptides: Analysis, Synthesis and Biology* (Eds E. Gross and J. Meienhofer), vol. 2, chapters 1 and 3, Academic Press, London, 1979 (solid-phase peptide synthesis using the Merrified approach); G.R. Marshall and R.B. Merrified, *Biochemistry*, 1965, **4**, 2395 (a synthesis of angiotensin); E. Atherton and R.C. Sheppard, *Solid Phase Peptide Synthesis – A Practical Approach*, IRL, Press, Oxford, 1989 (solid-phase peptide synthesis using a polyacrylamide support); H.-D. Jakubke, P. Kuhl, and A. Konnecke, *Angew. Chem. Int. Edn*, 1985, **24**, 85 (use of enzymes in peptide synthesis); G.A. Grant (ed), *Synthetic Peptides: A User's Guide*, Freeman, 1992

4 Fatty acids and their derivatives

J. Hobbs

4.1 Introduction

Fatty acids – long chain alkanoic acids – are ubiquitous in nature. They are found only in trace amounts in living cells in their free, unesterified form, but are of the greatest importance as components of lipids which, upon alkaline hydrolysis (saponification), afford the alkali metal salts of the fatty acids and other components. These include the acylglycerols, the phosphatidylglycerols and sphingolipids, the waxes, and other species. Their molecular constitution is summarized in Table 4.1.

Acylglycerols occur most commonly as triacylglycerols. Those that are solid at room temperature are called fats, and those that are liquid, oils. Clearly these, with the waxes, are of substantial commercial importance, as also are the alkali metal (usually sodium) salts of the fatty acids, familiarly called soaps. All of these species, and indeed all the lipids, are essentially water-insoluble, due in part or *in toto* to the hydrophobic fatty acid chains, and the capacity of phosphatidylglycerols, sphingolipids, and soaps to form hydrophobic aggregates, either spherical (as micelles) or as bilayers (in liposomes and cell membranes), is central to their functions.

The common structural features of these lipid species are the fatty acid components. This chapter will consider the structures and features of the fatty acids, and of certain species derived from the polyunsaturated fatty acids which are similarly hydrophobic but non-hydrolysable, and thus classified among the 'simple lipids' together with the steroids and terpenes (see Chapter 5). These derivative species include the prostaglandins, the leukotrienes, and the thromboxanes, the identification, investigation, and synthesis of which have formed some of the most fruitful and exciting areas of natural product research in recent years.

Table 4.1 Naturally occurring conjugates of fatty acids.

Species	Products of alkaline hydrolysis	Biological function
Acylglycerols (glycerides)	Glycerol and fatty acid salts	Storage form of metabolic fuel
Phosphatidylglycerols (phosphoglycerides, phosphatides)	L-Glycerol-3-phosphoric acid (*sn*-glycerol-3-phosphoric acid) and fatty acid salts (and 'head groups', if present)	Cell membrane components
Sphingolipids	Sphingosine or a related derivative, fatty acid salts, and other components	Cell membrane components
Waxes	Long-chain monohydric or steroidal alcohol and fatty acid salts	Protective coatings, on skin, fur, feathers, leaves, insects, etc.

4.2   Fatty acid structures

4.2.1   *Straight chain saturated fatty acids*

The major proportion of the naturally occurring fatty acids consists of straight-chain alkanoic acids, containing an even number of carbon atoms. A number are listed in Table 4.2, together with the trivial names by which they are often known. Butyric acid (German *Buttersäure*) is found at high levels in the milk fat of ruminants, and thus in milk-fat-derived products, such as butter. The characteristic smell of rancid butter is due to butyric acid released, in part, by the slow hydrolysis of the butyryl glycerides in butter by water. The most widespread and important members in Table 4.2 are the C10–C20 acids, with palmitic and stearic acids being of particular significance. Fatty acids of chain length greater than C20 are rare, with the longest members occurring mainly in waxes. The predominance of even-numbered straight chain acids is a consequence of their mode of biosynthesis. Odd-numbered straight chain acids are also found in nature, particularly in marine organisms and bacteria. Examples are *n*-pentanoic acid (valeric acid), *n*-heptanoic acid (oenanthic acid), and *n*-nonanoic acid (pelargonic acid).

The fatty acid composition of bacterial membranes is sufficiently characteristic of species to permit it to be used in a newly developed diagnostic identification system in which the membrane fatty acids are released by saponification, methylated to increase their volatility, and separated and

Table 4.2 Even-numbered straight chain saturated fatty acids.

$$CH_3(CH_2CH_2)_nCOOH$$

n	Total number of carbon atoms	Systematic name	Trivial name
1	4	*n*-Butanoic	Butyric
2	6	*n*-Hexanoic	Caproic
3	8	*n*-Octanoic	Caprylic
4	10	*n*-Decanoic	Capric
5	12	*n*-Dodecanoic	Lauric
6	14	*n*-Tetradecanoic	Myristic
7	16	*n*-Hexadecanoic	Palmitic
8	18	*n*-Octadecanoic	Stearic
9	20	*n*-Eicosanoic	Arachidic
10	22	*n*-Docosanoic	Behenic
11	24	*n*-Tetracosanoic	Lignoceric
12	26	*n*-Hexacosanoic	Cerotic
13	28	*n*-Octacosanoic	Montanic
14	30	*n*-Triacontanoic	Melissic

quantified by high resolution gas chromatography. Comparison by computer of the resultant 'fingerprint' with those of standards obtained for known species permits genus, species, and frequently subspecies to be assigned.

4.2.2 Unsaturated fatty acids

Unsaturated fatty acids are similarly widespread in nature, and a representative list is given in Table 4.3. For the most part, only monoenoic acids have been found in bacteria, however. Oleic, linoleic, and α-linolenic acids are of particular significance in mammalian diet, linoleic acid because it can act as a biosynthetic precursor of arachidonic acid, which in turn gives rise to the prostaglandins, the thromboxanes, and leukotrienes. Note that the common unsaturated fatty acids fall into families which are related by the distance between the double bond of the highest number and the length of the carbon chain. Thus, oleic and erucic acid both belong to the *n*-9 family, while linoleic, γ-linolenic, and arachidonic acids are members of the *n*-6 family. Note also that the typical pattern in polyunsaturated fatty acids is to have one or more methylene groups flanked by two double bonds – a 'methylene-interrupted' pattern of unsaturation. As a consequence, abstraction of hydrogen from these methylene groups can occur easily to give resonance-stabilized radicals, and the polyunsaturated fatty acids are prone to ready oxidation and peroxide formation, an initial step in the development of rancidity (discussed later).

Table 4.3 Some unsaturated fatty acids.

Total number of carbon atoms: double bonds[a]	Systematic name	Trivial name
16:1	9(Z)-Hexadecenoic	Palmitoleic
18:1	9(Z)-Octadecenoic	Oleic
18:1	11(Z)-Octadecenoic	*cis*-Vaccenic
22:1	13(Z)-Docosenoic	Erucic
18:2	9(Z),12(Z)-Octadecadienoic	Linoleic
18:3	9(Z),12(Z),15(Z)-Octadecatrienoic	α-Linolenic
18:3	6(Z),9(Z),12(Z)-Octadecatrienoic	γ-Linolenic
20:4	5(Z),8(Z),11(Z),14(Z)-Eicosatetraenoic	Arachidonic
22:5	7(Z),10(Z),13(Z),16(Z),19(Z)-Docosapentaenoic	Clupadonic

[a] Fatty acids are often designated by these numerical symbols. The number before the colon is the number of carbon atoms present, and the number following the colon is the number of double bonds present. Thus, eicosapentaenoic is 20:5, and docosahexaenoic is 22:6, etc. The position at which a double bond is located in a fatty acid may be indicated by a Greek capital delta (Δ), with the numbering position of the bond being superscripted: thus, oleic acid is Δ^9, linoleic acid is $\Delta^{9,12}$, etc. Obviously these abbreviations give less information than the systematic name!

It will be noted that all the double bonds in the examples given possess the *cis* stereochemistry. While *trans* double bonds are found in fatty acids, they occur far more rarely. The *cis* double bond has an important biological significance: by introducing a 'bend' in the alkyl chain, it prevents the hydrophobic chains of the acyl groups in fats, phosphoglycerides, and sphingolipids associating to form compact, high melting aggregates, and thus serves to depress the melting point in these species, maintaining the fluidity of depot fats and cell membranes. Experiments using the bacterium *Escherichia coli* have shown that, when grown at low temperatures (within its viable temperature range!) the membrane lipids of *E. coli* contain a higher proportion of unsaturated fatty acids than when grown at high temperatures. Evidently the bacterium can regulate its membrane composition to maintain membrane fluidity at a level sufficient to permit the transport processes essential for its metabolism to continue, and does so by the device of varying its unsaturated fatty acid content.

4.2.3 *Fatty acids of unusual structure*

Branched-chain fatty acids, the majority of which are monomethyl substituted acids, have been found in microorganisms. Two distinct classes may be distinguished, the *iso*-series possessing an isopropyl terminal group, and the *anteiso*-series with a secondary butyl terminal group. Occasionally branching takes place nearer the middle of the chain, as in tuberculostearic acid (**4.1**), and cyclopropane rings, as in lactobacillic acid (**4.2**) and dihydrosterculic acid, and cyclopropene rings as in sterculic acid (**4.3**) are also found. The cyclopropane rings, which always have *cis*-stereochemistry, are found in bacteria or protozoa, while the cyclopropene fatty acids have been found in certain seeds and plant tissues of the order *Malvales*. Species containing cyclopentene rings such as chaulmoogric acid (**4.4**) from *Chaulmoogra* seed oil have been described. Many unusual structures have been distinguished among the fatty acids from *Corynebacteria* (e.g. corynomycolic acid (**4.5**)) and *Mycobacteria* (the mycolic acids). Crepenynic acid (**4.6**), found in daisy seeds, contains a triple bond, and a number of other fatty acids containing triple bonds have been described. Oxy acids, containing keto, hydroxy, and epoxy functions, which in many cases are clearly closely related to, and derived from, unsaturated fatty acids (e.g. 12-hydroxyoleic, 10-hydroxy-stearic, *cis*-12,13-epoxyoleic, and *cis*-9,10-epoxystearic acids) have also been described in large numbers.

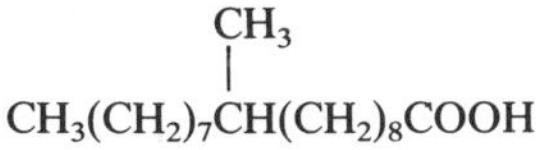

$$CH_3$$
$$|$$
$$CH_3(CH_2)_7CH(CH_2)_8COOH$$

(**4.1**)

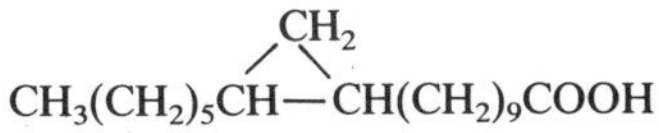

$$CH_2$$
$$CH_3(CH_2)_5CH-CH(CH_2)_9COOH$$

(**4.2**)

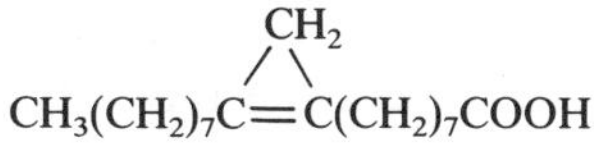

$$CH_2$$
$$CH_3(CH_2)_7C=C(CH_2)_7COOH$$

(**4.3**)

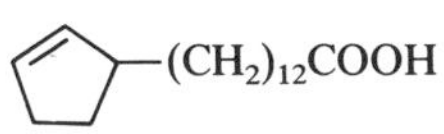

$$-(CH_2)_{12}COOH$$

(**4.4**)

$$OH$$
$$|$$
$$CH_3(CH_2)_{14}CH-CHCOOH$$
$$|$$
$$(CH_2)_{13}$$
$$|$$
$$CH_3$$

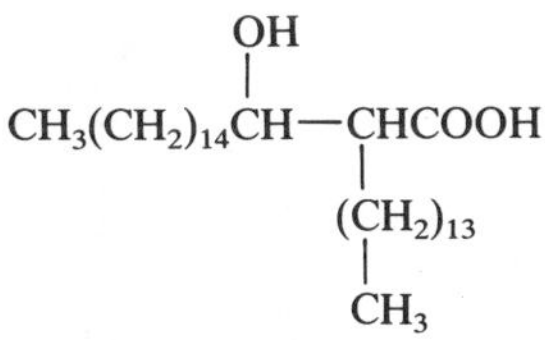

(**4.5**)

$$CH_3(CH_2)_4C\equiv C-CH_2-CH\overset{z}{=}CH(CH_2)_7COOH$$

(**4.6**)

As with *cis*-double bonds, the introduction of chain branching and modification generally depresses the melting point of a fatty acid, or of a fatty acyl species, below that of the unmodified species.

4.3   Fatty acid biosynthesis

Fatty acids are biosynthesized primarily from acetyl CoA units, and this is the reason why the commonest fatty acids contain even numbers of carbon atoms. Early hypotheses that the biosynthetic process was simply the reverse of the catabolic β-oxidation process (discussed later) could not be substantiated, and it was eventually found that the presence of bicarbonate and Mn^{2+} ions was obligatory for fatty acid synthesis from acetyl CoA to occur in avian liver systems, although addition of $^{14}CO_2$ did not result in incorporation of radioactivity. It was also recognized that the carboxylation of acetyl CoA to form malonyl CoA by a biotin-linked carboxylase represented the first step in the pathway, and that the bicarbonate was required for this purpose.

Fatty acid biosynthesis *de novo* is accomplished by multi-enzyme complexes, the components of which are separable in bacterial systems but tightly organized complexes in eukaryotes. Both acetyl and malonyl groups are transferred from the thiol function of coenzyme A to the thiol group of 4'-phosphopantetheine (see **4.7**), which is the prosthetic group of an 'acyl carrier protein' (ACP), or its functional equivalent. The acetyl group is then transferred from this CH_3CO-*S*-ACP to a thiol group on the enzyme β-ketoacyl-ACP synthetase, and acetoacetyl-*S*-ACP formed in a reaction in which malonyl-*S*-ACP is decarboxylated, with loss of the CO_2 previously introduced from bicarbonate, and with concomitant attack of the resulting carbanion on the acetyl group to displace the enzyme thiol. The overall

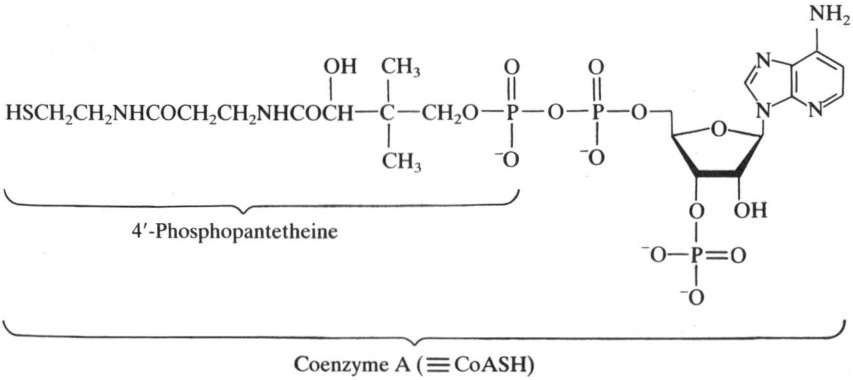

4'-Phosphopantetheine

Coenzyme A ($\equiv$ CoASH)

(4.7)

scheme of fatty acid biosynthesis is shown in Fig. 4.1. The decarboxylation in step 4 drives the reaction in the direction of synthesis of acetoacetyl-*S*-ACP. Following a reduction step (which forms specifically 3(R)-hydroxybutyryl-*S*-ACP), dehydration, and further reduction, butyryl-*S*-ACP is formed, which can then take the place of acetyl-*S*-ACP in step 3 for a further elongation cycle.

Repetition of this process leads to the formation of palmitic (*c*. 80 per cent) and stearic (*c*. 20 per cent) acids in plants and animals, and biosynthesis

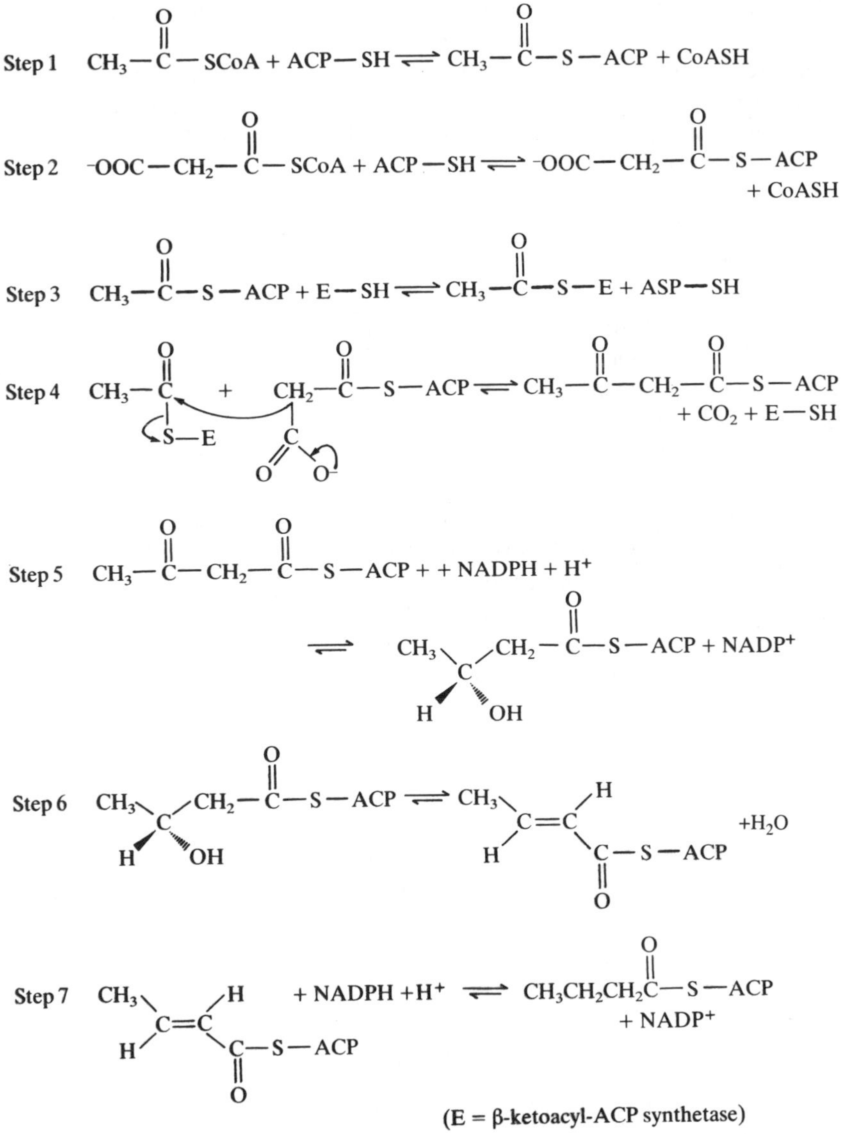

Step 1 $CH_3-\overset{\overset{\displaystyle O}{\|}}{C}-SCoA + ACP-SH \rightleftharpoons CH_3-\overset{\overset{\displaystyle O}{\|}}{C}-S-ACP + CoASH$

Step 2 $^-OOC-CH_2-\overset{\overset{\displaystyle O}{\|}}{C}-SCoA + ACP-SH \rightleftharpoons {}^-OOC-CH_2-\overset{\overset{\displaystyle O}{\|}}{C}-S-ACP + CoASH$

Step 3 $CH_3-\overset{\overset{\displaystyle O}{\|}}{C}-S-ACP + E-SH \rightleftharpoons CH_3-\overset{\overset{\displaystyle O}{\|}}{C}-S-E + ASP-SH$

Step 4 $CH_3-\overset{\overset{\displaystyle O}{\|}}{\underset{\underset{\displaystyle S-E}{|}}{C}} + CH_2-\overset{\overset{\displaystyle O}{\|}}{C}-S-ACP \rightleftharpoons CH_3-\overset{\overset{\displaystyle O}{\|}}{C}-CH_2-\overset{\overset{\displaystyle O}{\|}}{C}-S-ACP + CO_2 + E-SH$

Step 5 $CH_3-\overset{\overset{\displaystyle O}{\|}}{C}-CH_2-\overset{\overset{\displaystyle O}{\|}}{C}-S-ACP + + NADPH + H^+$

$\rightleftharpoons \underset{H \quad\quad OH}{CH_3\diagdown{}_{\overset{\textstyle C}{}}{}\diagup CH_2}-\overset{\overset{\displaystyle O}{\|}}{C}-S-ACP + NADP^+$

Step 6 $\underset{H \quad OH}{CH_3\diagdown{}_{\overset{\textstyle C}{}}{}\diagup CH_2}-\overset{\overset{\displaystyle O}{\|}}{C}-S-ACP \rightleftharpoons \underset{H}{CH_3\diagdown}{}_{C=C}{}\underset{C-S-ACP}{\diagup H} + H_2O$

Step 7 $\underset{H}{CH_3\diagdown}{}_{C=C}\underset{\overset{\displaystyle \|}{O}}{\underset{C-S-ACP}{}}\diagup H + NADPH + H^+ \rightleftharpoons CH_3CH_2CH_2\overset{\overset{\displaystyle O}{\|}}{C}-S-ACP + NADP^+$

(E = β-ketoacyl-ACP synthetase)

Fig. 4.1.

de novo then ceases. Elongation to give longer chain acids can, however, be effected by a mitochondrial system which utilizes acetyl CoA as the source of C_2-units in a process which is closely similar to the reverse of β-oxidation, or by a microsomal system which uses malonyl CoA and which appears mechanistically similar to that in Fig. 4.1.

Note that certain uncommon and branched-chain fatty acids can be generated if the acetyl CoA 'starter group' of step 1 is replaced by another starter group. Propionyl CoA, used as a starter, will give rise to the odd-numbered fatty acids; isovaleryl CoA (formed during the degradation of leucine) to the *iso*-series; and α-methylbutyryl CoA (formed during the degradation of isoleucine) to the *anteiso* series. Use of propionyl groups to elongate the chain leads to the highly branched acids found in *Mycobacteria*. An alternative route to branched chain acids such as (**4.1**) is described below. Corynomycolic acid (**4.5**) arises via 'head to tail' condensation of two molecules of palmitic acid.

4.3.1   *Mono-unsaturated fatty acids*

Two different pathways exist for the formation of mono-unsaturated fatty acids. The more widespread, and important, process occurring in plants, higher animals, yeasts, algae, protozoa, and bacteria is one of *oxidative desaturation*, in which a double bond is introduced directly into the preformed saturated fatty acid chain by an enzyme system using oxygen and NADPH as cofactors. This is apparently accomplished by the *syn*-elimination of a vicinal pair of *pro*-R hydrogen atoms, resulting in the formation of a *cis*-double bond (Fig. 4.2), exclusively between C-9 and C-10 in animals, and generally in other species, although desaturation at other positions may occur, rarely, in plants, algae, and bacteria. While the desaturase reaction has the characteristics of a mixed function oxygenation, no oxygenated intermediate has been demonstrated. The most widespread desaturase shows

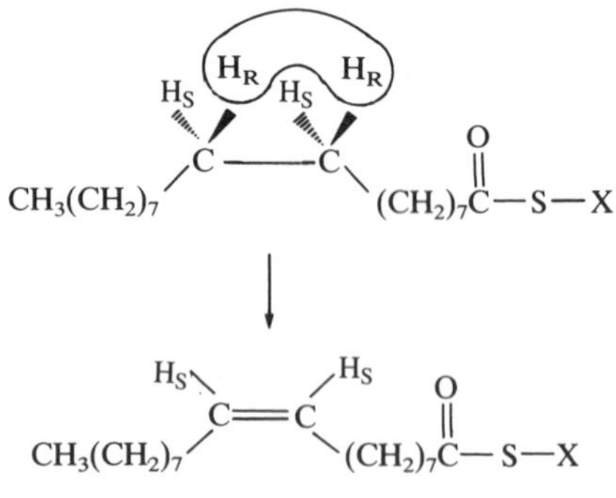

X = ACP (higher plants); CoA (yeasts, animals, bacteria)

Fig. 4.2.

chain length specificity decreasing from C_{18} to C_{14}, and thus stearoyl and palmitoyl thiol esters are the principal substrates, as esters of ACP (in higher plants) or coenzyme A (in other species), with stearoylthiol esters giving rise to oleoylthiol esters and palmitoylthiol esters to palmitoleoylthiol esters.

The second pathway, the 'anaerobic route', is found in certain aerobic and anaerobic bacteria – the *Eubacteriales* – and represents a variant of the normal biosynthetic route (Fig. 4.1). After several elongation cycles, the $3(R)$-hydroxyacyl-S-ACP species formed during elongation at the C_8 or C_{10} stage is dehydrated not to the Δ^2-*trans*-enoyl S-ACP as in step 6, but instead to the Δ^3-*cis*-enoyl-S-ACP (**4.8**). While (**4.8**) is not a substrate for the reductase catalysing step 7 of Fig. 4.1, it can be elongated in the usual manner. Thus after a further three rounds of elongation, (**4.8**, $n = 5$) gives rise to palmitoleic acid and (**4.8**, $n = 7$) to oleic acid. A single enzyme activity seems to be responsible for dehydrating 3-hydroxydecanoyl-S-ACP to a mixture of the Δ^2-*trans*- and Δ^3-*cis*-decenoates, and for their interconversion. This 'branching' of the normal biosynthetic process at the C_8 or C_{10} stage was revealed by the finding that $(1\text{-}^{14}C)$-labelled octanoic acid gave rise to labelled Δ^9-C_{16} and Δ^{11}-C_{18} acids, and $(1\text{-}^{14}C)$-labelled decanoic acid to labelled Δ^7-C_{16} and Δ^9-C_{18} acids, while $(1\text{-}^{14}C)$-labelled C_{12}- and C_{14}-acids afforded only labelled saturated fatty acids after elongation.

$$\underset{\displaystyle CH_3(CH_2)_n}{H}\diagdown \underset{\displaystyle}{C} = C \diagup \overset{\displaystyle H}{\underset{\displaystyle CH_2}{}} \overset{O}{\underset{\|}{C}} - S - ACP$$

$$n = 5 \text{ or } 7$$

(4.8)

4.3.2 *Polyunsaturated fatty acids*

Anaerobic bacteria contain only saturated and mono-unsaturated fatty acids, but all other organisms are capable of introducing more than one double bond into the chain. The resultant double bonds are usually interrupted by a methylene group, as noted previously, to give a divinylmethane system. The 'extra' double bonds are introduced by a desaturase system of the type described above, and the positional specificity of these desaturases varies importantly with species. Animal enzymes normally introduce new double bonds between an existing double bond and the carbonyl function; plant enzymes generally introduce new double bonds between an existing double bond and the methyl terminus. A few primitive organisms such as *Euglena* can desaturate in either direction. As described above, oxidative desaturation leads to a Δ^9 double bond (i.e. between C-9 and C-10), and animals *cannot directly introduce double bonds beyond Δ^9* in monoenoic fatty acids. However, by combining desaturation and elongation a number of metabolic transformations are observed, as exemplified in part for palmitoleic acid

(a)

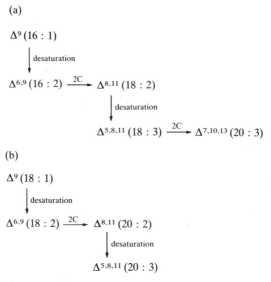

Fig. 4.3 (a) Palmitoleic acid transformations; (b) oleic acid transformations.

(16:1, Δ^9) and, more importantly, for oleic acid (18:1, Δ^9) in Fig. 4.3. Note, however, that elongation and desaturation of these acids cannot give rise to arachidonic acid (20:4, $\Delta^{5,8,11,14}$) although linoleic acid (18:2, $\Delta^{9,12}$) and γ-linolenic acid (18:3, $\Delta^{6,9,12}$) do so readily when transformed according to the same scheme. Arachidonic acid is the precursor to the prostaglandins, thromboxanes, and leukotrienes discussed later in this chapter, and the absence from mammalian diet of arachidonic acid and of fatty acids which can be transformed into it (the *essential fatty acids*) results in deficiency symptoms which, if untreated by supplement of these acids, can result in death. Consequently, these fatty acids which are essential to diet are normally provided by plant sources.

Plant (*Chlorella*) desaturases are able to convert Δ^9-fatty acids of various chain lengths to $\Delta^{9,12}$-acids, or to introduce a double bond at the (*n*-6) position (i.e. the sixth carbon atom from the methyl end of a C_n fatty acid, also designated $\omega6$) in a fatty acid containing a pre-existing double bond at the (*n*-9) (or $\omega9$) position. Thus either desaturase activity will convert oleic acid to linoleic acid. Extensive investigation of unsaturated fatty acids for 'essential fatty acid' activity has shown that all active acids possess double bonds in the (*n*-6, *n*-9) positions, and thus the products of the plant desaturases afford, or serve as precursors for, arachidonic acid.

4.3.3 'Odd' fatty acids

The 10-methyl group in tuberculostearic acid (**4.1**) and the methylene groups of the cyclopropane acids (e.g. (**4.2**)) and cyclopropene acids (e.g. (**4.3**))

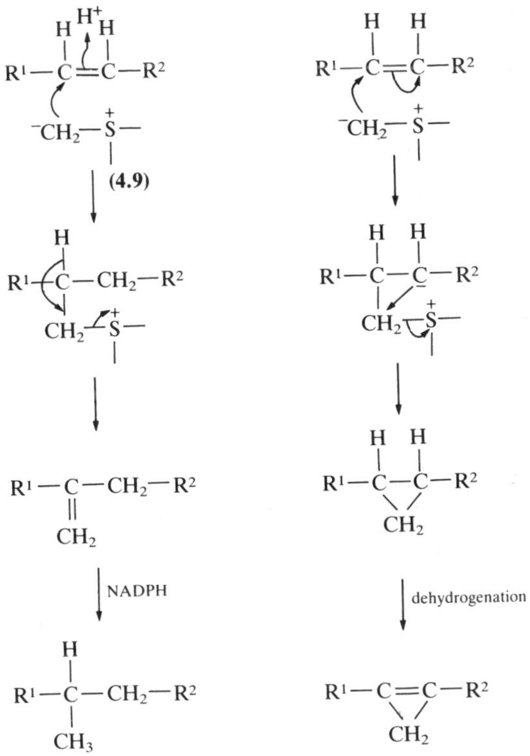

Fig. 4.4.

seem, from labelling experiments, to originate from the *S*-methyl group of *S*-adenosylmethionine. Methylation of an oleoyl moiety at C-10 probably occurs via the sulphur ylid (**4.9**) or an equivalent species: if the *S*-adenosylmethionine used contains a trideuteromethyl group, only two of the deuterium atoms are retained in tuberculostearic acid. 10-Methylene-stearic acid is formed as an intermediate and reduced to tuberculostearic acid. A minor variation on this sequence of events affords the cyclopropane fatty acids: both routes are depicted in Fig. 4.4. The cyclopropene fatty acids seem, from biosynthetic sequence studies, to arise from the cyclopropane fatty acids by desaturation.

Crepenynic acid (**4.6**) is a key intermediate in the supposed route from fatty acids to the polyacetylenes. Tracer studies have established that the polyacetylenes, which are always linear, are derived from acetyl CoA and malonyl CoA, like the fatty acids, and the proposed route involves biosynthesis of linoleate, further desaturation to crepenynate (**4.6**), and subsequent desaturation on the methyl side of the triple bonded unit followed by oxidative degradation of the octanoate unit, possibly involving an autoxidative process.

4.4   Fatty acid catabolism

Fatty acid catabolism in animals occurs at the inner mitochondrial membrane by a process of β-oxidation. The fatty acid is degraded as its thioester with coenzyme A(7), 'fatty acyl CoA'. The process is summarized in Fig. 4.5. Oxidation by a flavin-linked dehydrogenase affords the *trans*-Δ^2-enoyl CoA (**4.10**) which is stereospecifically hydrated to afford the 3(*S*)-hydroxyacyl CoA (**4.11**). Dehydrogenation by a pyridine-linked dehydrogenase which is specific for the *S*-isomer then gives the 3-ketoacyl CoA (**4.12**) which undergoes thiolytic cleavage by another molecule of coenzyme A to afford acetyl CoA and a new fatty acyl CoA species (**4.13**) which is now two carbons shorter than the original. Even-numbered straight-chain acids are thus degraded completely to acetyl CoA. Odd-numbered straight chain acids are degraded similarly, but give rise to a molecule of propionyl CoA in the final thiolase reaction. This is carboxylated in, for instance, liver by a biotin-requiring enzyme to afford (*S*)-methylmalonyl CoA (**4.14**), which is racemized enzymically to afford its (*R*) isomer (**4.15**), which is a substrate for the vitamin B$_{12}$-linked methylmalonyl CoA isomerase, affording succinyl CoA (**4.16**). Branched-chain fatty acids with a methyl group on an even-numbered carbon atom may be degraded by β-oxidation, and the branching methyl then appears as C-3 in propionyl CoA. If the branching methyl group is on an odd-numbered carbon in the *iso* series, isovaleroyl CoA is eventually formed, and dehydrogenated as usual to isopentenoyl CoA (**4.17**), which is carboxylated and then hydrated to afford β-methylglutaconyl CoA (**4.18**) and β-hydroxy-β-methyl glutaryl CoA (**4.19**) respectively. Enzymic retro-aldol cleavage of (**4.19**) then affords acetyl CoA and acetoacetate.

The degradation of unsaturated fatty acids presents an interesting problem. Consider, for example, linoleic acid (Table 4.3). After three rounds of β-oxidation, this has been degraded to 3(*Z*),6(*Z*)-dodecadienoyl CoA, and thus presents a *cis*-Δ^3 double bond rather than the *trans*-Δ^2 arrangement required for further degradation. An auxiliary isomerase enzyme exists which interconverts these species, and normal β-oxidation may then proceed. However, after a further two rounds, 2(*Z*)-octenoyl CoA is formed. This is hydrated in the normal manner, but forms 3(*R*)-hydroxyoctanoyl CoA, and a second auxiliary enzyme, an epimerase, is required to convert this to the *S* isomer, before further oxidation can proceed.

Fatty acids containing three-membered rings seem to undergo only limited degradation by β-oxidation, the catabolic process ceasing when the ring is encountered.

An alternative, less significant degradative pathway for the oxidation of fatty acids consists in α-oxidation. Labelling and other studies have led to the suggestion that the *pro*-R hydrogen atom is removed from the α-carbon of a free fatty acid (**4.20**), the resulting radical reacting with hydrogen peroxide to generate an α-hydroxyperoxyacid intermediate (**4.21**) which decomposes with loss of carbon dioxide to form aldehyde (**4.22**). Oxidation

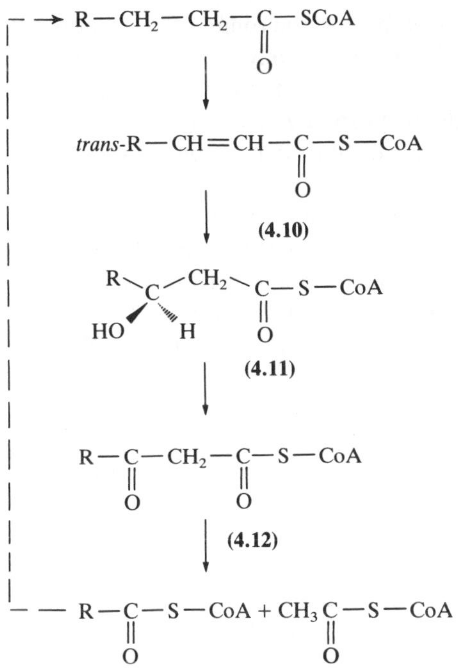

$$R-CH_2-CH_2-C-SCoA$$
$$\|$$
$$O$$

$$trans\text{-}R-CH=CH-C-S-CoA$$
$$\|$$
$$O$$

(4.10)

$$\underset{HO}{R}\diagdown\underset{H}{\overset{CH_2}{C}}\diagup C-S-CoA$$
$$\|$$
$$O$$

(4.11)

$$R-C-CH_2-C-S-CoA$$
$$\| \qquad \|$$
$$O \qquad O$$

(4.12)

$$R-C-S-CoA + CH_3C-S-CoA$$
$$\| \qquad\qquad\qquad \|$$
$$O \qquad\qquad\qquad O$$

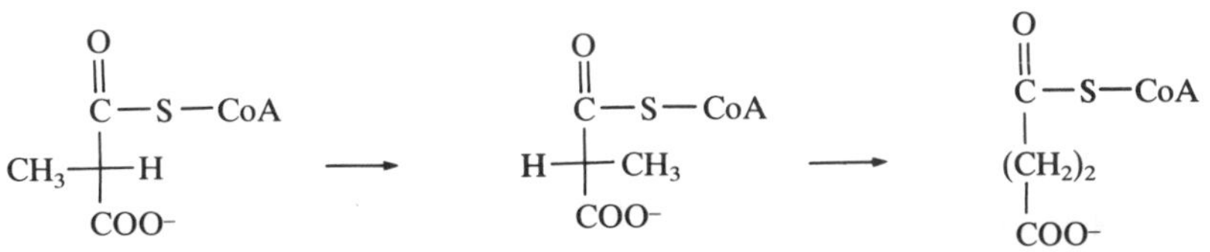

(4.13)

$$\underset{COO^-}{\overset{O}{\overset{\|}{C}-S-CoA}}\quad CH_3-\!\!\!\!\!-H$$

(4.14)

$$\underset{COO^-}{\overset{O}{\overset{\|}{C}-S-CoA}}\quad H-\!\!\!\!\!-CH_3$$

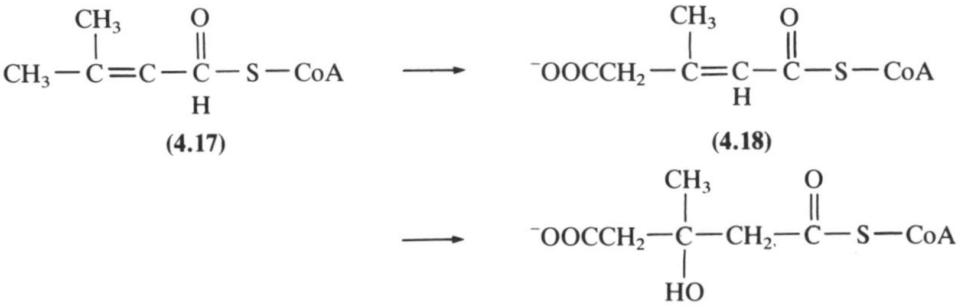

(4.15)

$$\underset{COO^-}{\overset{O}{\overset{\|}{C}-S-CoA}}\quad (CH_2)_2$$

(4.16)

$$CH_3-\underset{\underset{CH_3}{|}}{C}=C-\overset{O}{\overset{\|}{C}}-S-CoA$$
$$\qquad\qquad H$$

(4.17)

$$^-OOCCH_2-\underset{\underset{CH_3}{|}}{C}=\underset{H}{C}-\overset{O}{\overset{\|}{C}}-S-CoA$$

(4.18)

$$^-OOCCH_2-\underset{\underset{HO}{|}}{\overset{\overset{CH_3}{|}}{C}}-CH_2-\overset{O}{\overset{\|}{C}}-S-CoA$$

(4.19)

$$\underset{H_S\quad H_R}{\overset{O^-}{R}\diagup}C=O$$

(4.20)

$$\left[\ R\diagup\underset{H_S}{\overset{O}{\overset{\|}{C}}}\diagdown\underset{O-OH}{\overset{O^-}{}}\ \right]$$

(4.21)

$$R-C\diagup\underset{H}{\overset{O}{}}$$

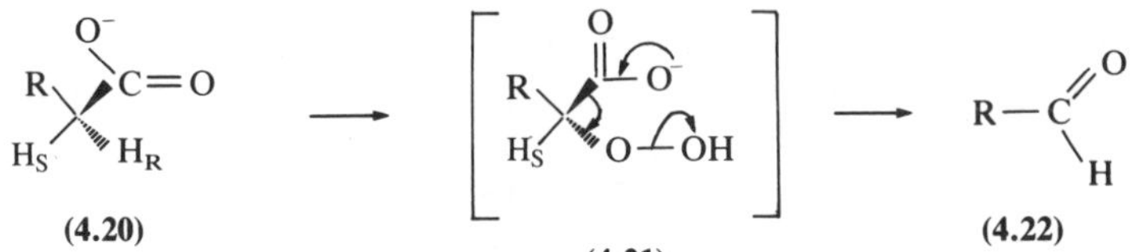

(4.22)

Fig. 4.5.

by a pyridine dehydrogenase then completes the round of α-oxidation. While the significance of the process is not completely certain, it may occur in conjunction with β-oxidation in cases where, for instance, a methyl group is present at C-3 in a branched chain. Normally this would block β-oxidation, but one round of α-oxidation would bring the methyl group to C-2, permitting β-oxidation to proceed.

ω-Oxidation, involving a mixed function oxygenase, occurs in the microsomal fraction of cells, to introduce a hydroxy group at the methyl (or penultimate methylene) group remote from the carboxyl group. Further oxidation yields, ultimately, a dicarboxylic acid. Oxidation by mixed function oxygenases, followed by β-oxidation, is an important method of degrading alkanes and alkyl chains.

Peroxidation occurs in unsaturated fatty acids, either by autoxidation, or via chemical catalysis by iron-tetrapyrrole species such as haematin, or by enzymic catalysis by lipoxygenases. *In vivo* this leads to deterioration of the tissues, and it is retarded by the presence of natural antioxidants such as vitamin E. In foods the autoxidation of fats leads to rancidity, and the free radicals generated in the process additionally damage non-fat molecules by addition reactions and by hydrogen abstraction. In order to exert its activity, lipoxygenase requires that a methylene group flanked by two *cis* double bonds must be present, and molecules of this type are also typically susceptible to autoxidation. The sequence of events occurring in linolenic acid via either agency is depicted in Fig. 4.6.

4.5   Fatty acid synthesis

In general, methods used to synthesize carboxylic acids are applicable to the synthesis of fatty acids. These may be grouped in three broad classes. In the first, chemical modification of a long chain compound of similar skeletal structure occurs without change in carbon chain length. Thus, for instance, a primary alcohol may be oxidized to the corresponding acid. This is best performed using ruthenium dioxide–potassium periodate in aqueous acetone, or potassium permanganate–crown ether in benzene, reagents which oxidize the aldehyde intermediate rapidly to the acid. Otherwise, condensation of the aldehyde with unreacted alcohol may afford the hemiacetal which is oxidized to the ester, forming an unwanted product and resulting in diminution of yield. Hydrolysis of a nitrile in acid or alkali also belongs in this class. The oxidation of aldehydes to the corresponding acids is best performed using permanganate in aqueous media, or fuming nitric acid, or silver oxide in aqueous alkali, a mild and selective procedure. Methods used specifically for the synthesis of unsaturated fatty acids without change in chain length include the dehydration of α- or β-hydroxylated acids to give α,β-unsaturated acids, or the dehydrohalogenation of an α- or β-halogenated acid to give the same products. Acetylenic acids may be obtained by

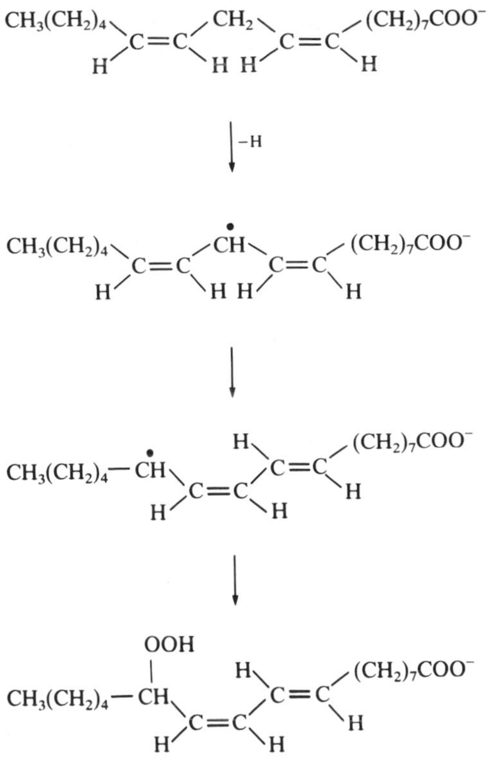

Fig. 4.6.

halogenation of ethylenic acids followed by dehydrogenation of the resultant vicinal dihaloacid. Reduction using Lindlar catalyst or a dialkylborane may be used to effect *cis*-reduction of an isolated acetylenic bond to give the ethylenic acid, and reduction with sodium and liquid ammonia, similarly, to give the *trans*-ethylenic acid. Unsaturated acids may be hydrogenated, or hydroxy-acids reduced with hydriodic acid, to afford the saturated fatty acids. The chemistry involved in all these processes is well known.

The second, larger class of methods involves carbon chain extension, which in its simplest form is the carboxylation of a lithium alkyl, a Grignard reagent, a Wittig reagent, or an acetylide using carbon dioxide. Another one-carbon-addition process is carbonylation. This often affords a secondary carboxylic acid, but certain additions utilizing transition metals in low valency states may be used to afford primary acids or their esters. For instance, the action of sodium tetracarbonylferrate on a primary halide or tosylate, followed by insertion of carbon monoxide into the iron–carbon bond and oxidation by a hypohalite, yields a primary acid. Another method affords (Z)-α,β-unsaturated fatty acid esters starting from (Z)-iodovinyl species. Homologation by two carbon atoms is easily achieved using the Arndt–Eistert synthesis, in which diazomethane attacks an acyl chloride to give a diazoketone, which undergoes silver(I)-catalysed Wolff rearrangement with hydrolysis of the resultant ketene to give the acid. Malonic and acetoacetic esters are also useful for 2-carbon homologation, and related reactions. Condensation of malonic acid with an aldehyde in the presence of pyridine (Doebner reaction) affords an α,β-unsaturated acid. Condensation of acetoacetic ester with an acyl halide, followed by retro-Claisen reaction using sodium methoxide, generates β-ketoesters of type (**4.23**), which may either be reduced using Clemmensen or Wolff–Kishner procedures to give the saturated fatty acid ester, or condensed with hydrazine to give pyrazolone (**4.24**) which can be brominated to (**4.25**), which on treatment with base gives the α,β-acetylenic acid (**4.26**). Aldehydes condense with malonic acid in the presence of triethanolamine to give β,γ-unsaturated acids, and allyl halides with malonic ester to give, after hydrolysis, γ,δ-unsaturated acids. Several of the procedures described above are exemplified in Fig. 4.7.

Carbon chain extension is, in fact, not clearly divisible from the third major means of fatty acid synthesis, namely condensation reactions. Some of the examples cited above illustrate this clearly. Again, condensation of the potassium salt of cyclohexane-1,3-dione with an alkyl iodide, followed by a retro-Claisen reaction using alkali and Wolff–Kishner reduction, results in homologation by six carbon atoms (Fig. 4.8). Attack on cyclohexanone by a Grignard reagent, followed by oxidation of the substituted cyclohexane thus formed and Clemmensen reduction, results in a similar homologation. Attack on furfural by the same reagent may be used for 5-C homologation. Another heterocycle which lends itself to such homologation is thiophene. Friedel–Crafts acylation of thiophene followed by Wolff–Kishner reduction generates a 2-alkylthiophene, which may be acetylated again under Friedel–Crafts conditions, oxidized to the corresponding acid using a hypohalite, and reduced using Raney nickel to give 5-C homologation. Use of succinic anhydride in place of acetyl chloride, with a second Wolff–Kishner reduction, permits 8-C homologation.

Reaction of acyl chlorides with enamines affords another useful route to homologation. Thus reaction of *n*-hexanoyl chloride with the enamine formed from 4-methylcyclohexanone and piperidine affords (**4.27**), which on hydrolysis with acid and Wolff–Kishner reduction gives (**4.28**). Conversion of (**4.28**) to its acid chloride and repetition of the sequence using an enamine derived from cyclohexanone affords tuberculostearic acid (**4.1**).

The useful facility of sodio-alkynes for attacking α,ω-dihalogeno compounds has been used in the synthesis of certain unsaturated fatty acids, such as oleic and linoleic acids. The use of other organometallics, such as the cadmium alkyls, to attack alkoxycarbonylacyl halides can be used to form saturated straight-chain or branched chain compounds. Both procedures are illustrated in Fig. 4.8.

Organoboron species may also be utilized for fatty acid synthesis, particularly of unsaturated fatty acids. Thus, diazoacetic ester will react with alkyl chloroboranes or alkynylboranes to give esters of saturated or β,γ-ynoic

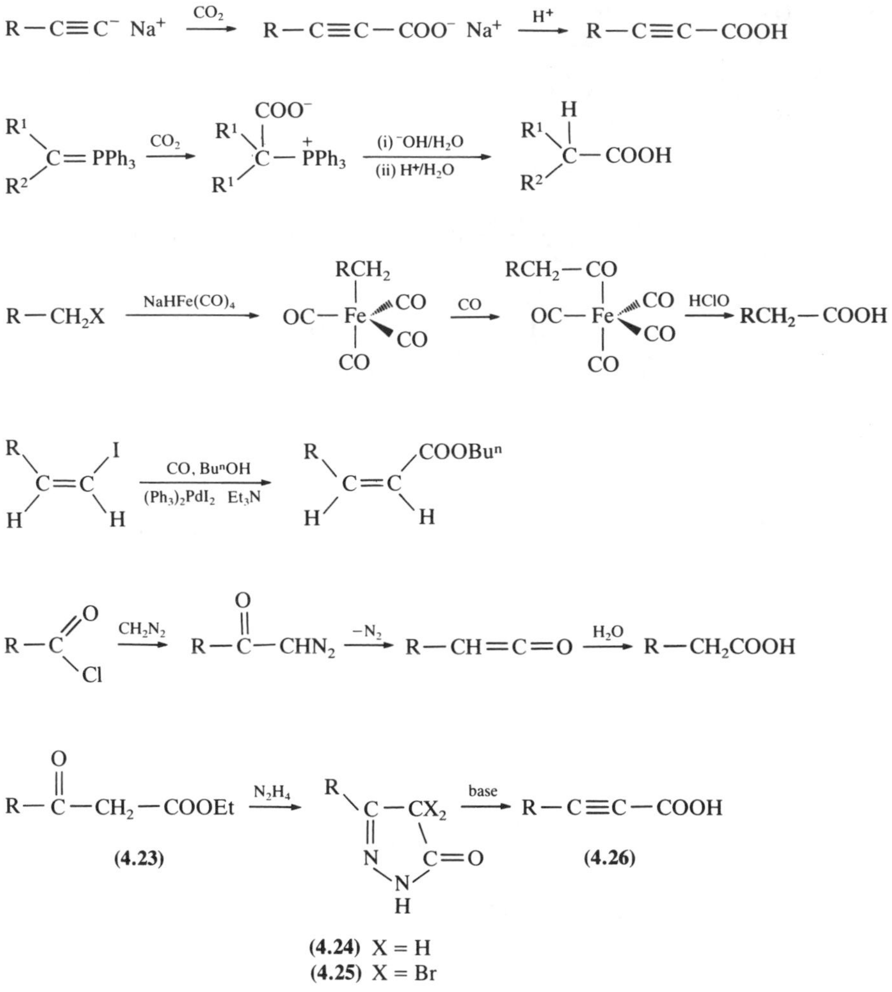

$$R-C\equiv C^-\ Na^+ \xrightarrow{CO_2} R-C\equiv C-COO^-\ Na^+ \xrightarrow{H^+} R-C\equiv C-COOH$$

$$\underset{R^2}{\overset{R^1}{\diagdown}}C=PPh_3 \xrightarrow{CO_2} \underset{R^1}{\overset{R^1}{\diagdown}}\overset{COO^-}{\underset{|}{C}}-\overset{+}{PPh_3} \xrightarrow[\text{(ii) } H^+/H_2O]{\text{(i) } ^-OH/H_2O} \underset{R^2}{\overset{R^1}{\diagdown}}\overset{H}{\underset{|}{C}}-COOH$$

$$R-CH_2X \xrightarrow{NaHFe(CO)_4} \text{OC}-\underset{CO}{\overset{RCH_2}{Fe}}\overset{CO}{\diagdown}{}_{CO} \xrightarrow{CO} \text{OC}-\underset{CO}{\overset{RCH_2-CO}{Fe}}\overset{CO}{\diagdown}{}_{CO} \xrightarrow{HClO} RCH_2-COOH$$

$$\underset{H}{\overset{R}{\diagdown}}C=C\underset{H}{\overset{I}{\diagup}} \xrightarrow[(Ph_3)_2PdI_2\ Et_3N]{CO,\ Bu^nOH} \underset{H}{\overset{R}{\diagdown}}C=C\underset{H}{\overset{COOBu^n}{\diagup}}$$

$$R-\overset{O}{\overset{\diagup\!\!\diagup}{C}}\overset{}{\diagdown}_{Cl} \xrightarrow{CH_2N_2} R-\overset{O}{\overset{||}{C}}-CHN_2 \xrightarrow{-N_2} R-CH=C=O \xrightarrow{H_2O} R-CH_2COOH$$

$$R-\overset{O}{\overset{||}{C}}-CH_2-COOEt \xrightarrow{N_2H_4} \underset{\underset{H}{\diagdown N\diagup}}{\overset{R}{\diagdown}}\overset{}{\underset{N}{\overset{||}{C}}}-CX_2\overset{}{\diagdown}_{C=O} \xrightarrow{base} R-C\equiv C-COOH$$

(4.23)   (4.26)

(4.24) X = H
(4.25) X = Br

Fig. 4.7.

acids, respectively (Fig. 4.9). Trialkyl boranes react with ethyl 4-bromocrotonate to give a β,γ-unsaturated ester in a useful 4-C homologation.

The exploitation of the Wittig reaction, mentioned above, often in conjunction with reactions involving alkyne species, has proved a consistently valuable and important route to the synthesis of unsaturated fatty acids and compounds derived from them. In an early example, alkylation of triphenylphosphine with ethyl 11-iodoundecenoate (Fig. 4.10) afforded the phosphonium species (4.29) which on treatment with sodium ethoxide in dry DMF afforded the phosphorus ylid ('Wittig reagent') (4.30). Upon condensation with long chain alkanals of general formula $Me(CH_2)_mCHO$,

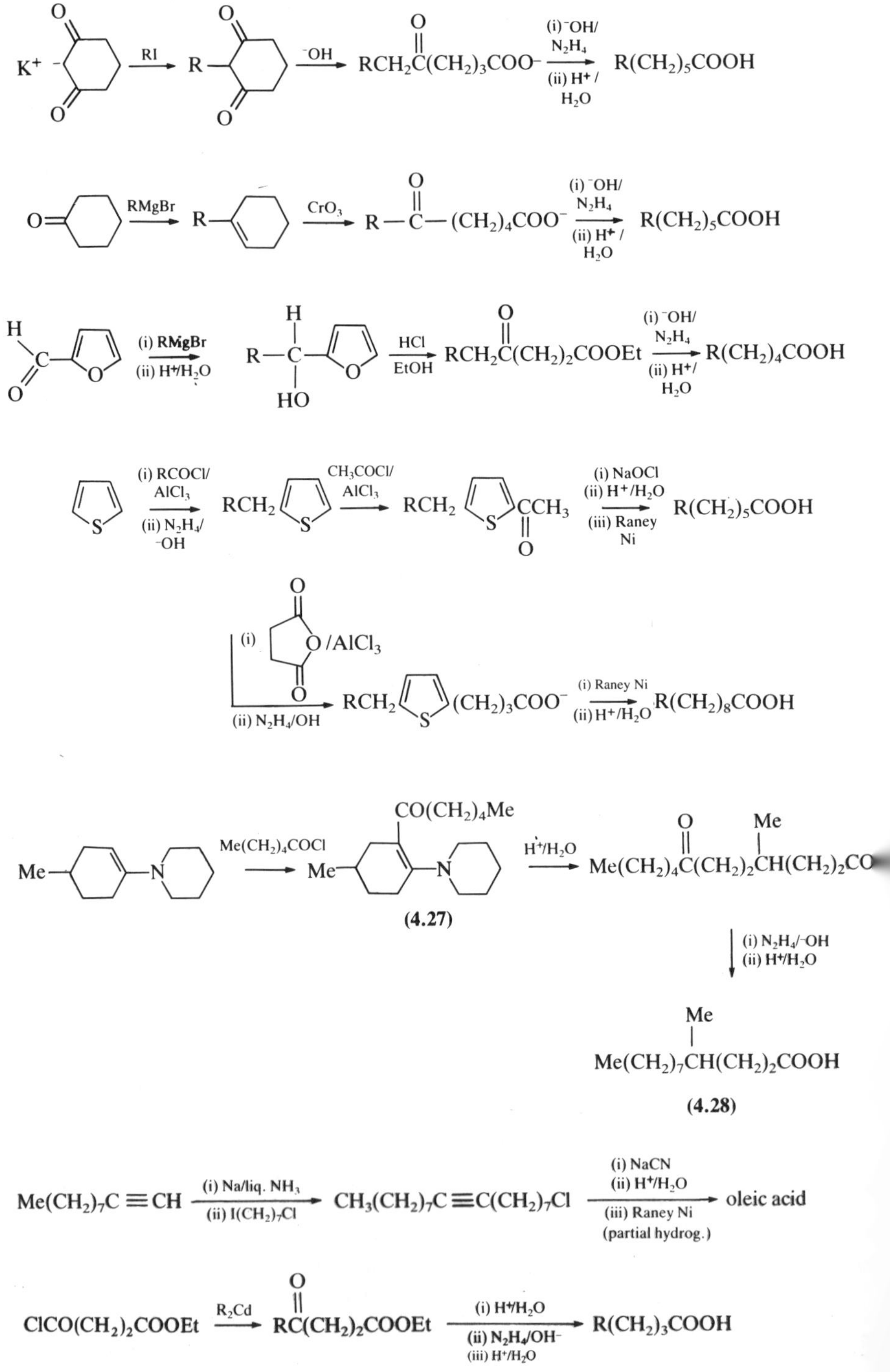

$$K^+ \xrightarrow{RI} R \xrightarrow{^-OH} RCH_2\overset{O}{\overset{\|}{C}}(CH_2)_3COO^- \xrightarrow[\text{(ii) H}^+/\text{H}_2\text{O}]{\text{(i) }^-\text{OH}/\ N_2H_4} R(CH_2)_5COOH$$

$$O \xrightarrow{RMgBr} R \xrightarrow{CrO_3} R-\overset{O}{\overset{\|}{C}}-(CH_2)_4COO^- \xrightarrow[\text{(ii) H}^+/\text{H}_2\text{O}]{\text{(i) }^-\text{OH}/\ N_2H_4} R(CH_2)_5COOH$$

$$\overset{H}{\underset{O}{\overset{\|}{C}}} \text{-furan} \xrightarrow[\text{(ii) H}^+/\text{H}_2\text{O}]{\text{(i) RMgBr}} R-\underset{HO}{\overset{H}{\overset{|}{C}}}\text{-furan} \xrightarrow[\text{EtOH}]{HCl} RCH_2\overset{O}{\overset{\|}{C}}(CH_2)_2COOEt \xrightarrow[\text{(ii) H}^+/\text{H}_2\text{O}]{\text{(i) }^-\text{OH}/\ N_2H_4} R(CH_2)_4COOH$$

$$\text{thiophene} \xrightarrow[\text{(ii) N}_2\text{H}_4/^-\text{OH}]{\text{(i) RCOCl/AlCl}_3} RCH_2\text{-thiophene} \xrightarrow{\text{CH}_3\text{COCl/AlCl}_3} RCH_2\text{-thiophene-}\overset{O}{\overset{\|}{C}}CH_3 \xrightarrow[\text{(iii) Raney Ni}]{\substack{\text{(i) NaOCl}\\ \text{(ii) H}^+/\text{H}_2\text{O}}} R(CH_2)_5COOH$$

$$\xrightarrow[\text{(ii) N}_2\text{H}_4/\text{OH}]{\text{(i) succinic anhydride/AlCl}_3} RCH_2\text{-thiophene-}(CH_2)_3COO^- \xrightarrow[\text{(ii) H}^+/\text{H}_2\text{O}]{\text{(i) Raney Ni}} R(CH_2)_8COOH$$

$$\text{Me-cyclohexene-N-piperidine} \xrightarrow{\text{Me(CH}_2)_4\text{COCl}} \text{Me-cyclohexenyl-}CO(CH_2)_4Me\text{-N-piperidine}$$

(4.27)

$$\xrightarrow{\text{H}^+/\text{H}_2\text{O}} Me(CH_2)_4\overset{O}{\overset{\|}{C}}(CH_2)_2\underset{Me}{\overset{|}{C}}H(CH_2)_2CO$$

$$\xrightarrow[\text{(ii) H}^+/\text{H}_2\text{O}]{\text{(i) N}_2\text{H}_4/^-\text{OH}} Me(CH_2)_7\underset{Me}{\overset{|}{C}}H(CH_2)_2COOH$$

(4.28)

$$Me(CH_2)_7C\equiv CH \xrightarrow[\text{(ii) I(CH}_2)_7\text{Cl}]{\text{(i) Na/liq. NH}_3} CH_3(CH_2)_7C\equiv C(CH_2)_7Cl \xrightarrow[\substack{\text{(iii) Raney Ni}\\ \text{(partial hydrog.)}}]{\substack{\text{(i) NaCN}\\ \text{(ii) H}^+/\text{H}_2\text{O}}} \text{oleic acid}$$

$$ClCO(CH_2)_2COOEt \xrightarrow{R_2Cd} R\overset{O}{\overset{\|}{C}}(CH_2)_2COOEt \xrightarrow[\substack{\text{(ii) N}_2\text{H}_4/\text{OH}^-\\ \text{(iii) H}^+/\text{H}_2\text{O}}]{\text{(i) H}^+/\text{H}_2\text{O}} R(CH_2)_3COOH$$

Fig. 4.8.

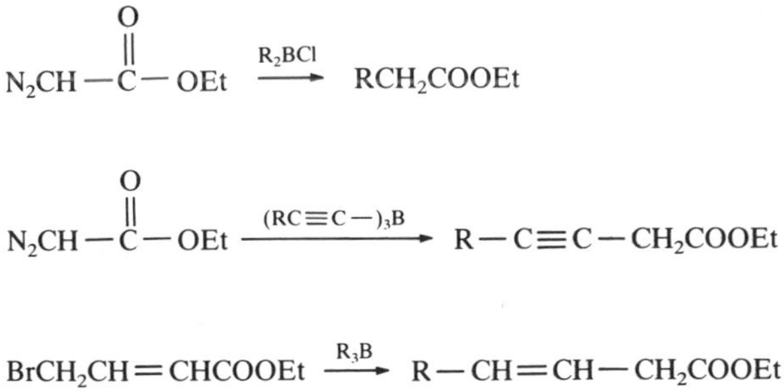

$$N_2CH-\overset{\overset{\displaystyle O}{\|}}{C}-OEt \xrightarrow{R_2BCl} RCH_2COOEt$$

$$N_2CH-\overset{\overset{\displaystyle O}{\|}}{C}-OEt \xrightarrow{(RC\equiv C-)_3B} R-C\equiv C-CH_2COOEt$$

$$BrCH_2CH=CHCOOEt \xrightarrow{R_3B} R-CH=CH-CH_2COOEt$$

Fig. 4.9.

followed by alkaline saponification, long chain fatty acids of general formula (**4.31**), with predominantly (>90 per cent) *cis* stereochemistry, were generated. Substitution of methyl alkyl ketones for the long chain alkanals afforded instead branched chain unsaturated fatty acids after saponification, with general formula (**4.32**). The method is, of course, generally applicable: the length of the iodoalkanoate ester used initially can be varied widely. Reduction of the double bonds introduced in this way permits isotopic labelling at the appropriate sites in the resultant saturated fatty acids, affording compounds useful in metabolic studies.

The stereochemical distribution in the products of the Wittig reaction is markedly dependent on the solvent employed in the reaction, on the reactivity of the ylid intermediate, and on the presence or absence of counterions (e.g. alkali metal salts or halide ions) to stabilize the transition states involved. Generally the use of a polar solvent, such as DMF, with a highly reactive ylid gives a good yield of predominantly *cis* product, but conditions can be manipulated to advantage if *trans* stereochemistry is required. The synthesis of a pheromone from the sex gland of the female grapevine leaf roller *Lobesia botrana* provides a nice example: methylation of cycloheptanone to its enol ether, followed by ozonolysis, gives (**4.33**), which condenses with formyl-methyltriphenylphosphorane to afford stereospecifically the *trans* isomer of (**4.34**). When (**4.34**) is treated with propyltriphenylphosphonium bromide and the sodium salt of hexamethyldisilazane in THF, a second Wittig reaction ensues, this time generating the *cis* product (**4.35**), which, after reduction with lithium aluminium hydride and acetylation of the product alcohol, yields the pheromone (*E*)-7,(*Z*)-9-dodecadienyl acetate (**4.36**) as 92 per cent of the isomeric mixture of products formed. Pheromones are commonly metabolites of fatty acids, and many unsaturated pheromones have been prepared by routes involving the Wittig reaction, as above. The facility to direct the stereochemistry of the reaction product by judicious choice of conditions has also been much exploited in syntheses of the leukotrienes, **discussed** later.

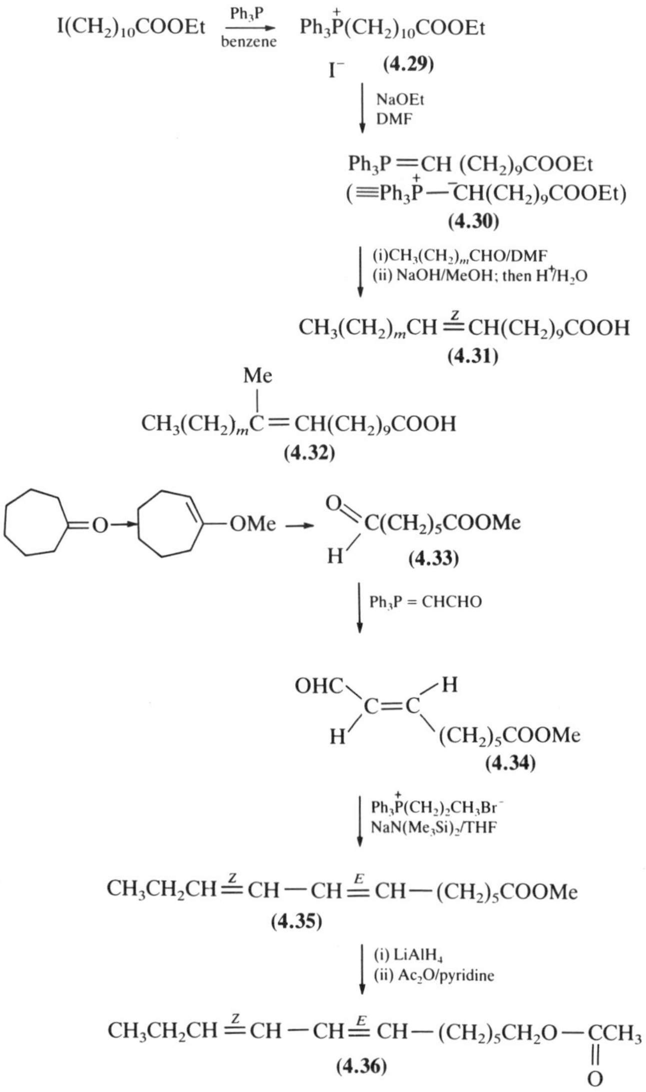

$$I(CH_2)_{10}COOEt \xrightarrow[\text{benzene}]{Ph_3P} Ph_3\overset{+}{P}(CH_2)_{10}COOEt$$

$$I^- \quad \textbf{(4.29)}$$

$$\downarrow \begin{array}{l} \text{NaOEt} \\ \text{DMF} \end{array}$$

$$Ph_3P = CH\,(CH_2)_9COOEt$$
$$(\equiv Ph_3\overset{+}{P} - \overset{-}{C}H(CH_2)_9COOEt)$$
$$\textbf{(4.30)}$$

$$\downarrow \begin{array}{l} \text{(i)}\,CH_3(CH_2)_m CHO/DMF \\ \text{(ii)}\,NaOH/MeOH;\ \text{then}\ H^+/H_2O \end{array}$$

$$CH_3(CH_2)_m CH \overset{Z}{=\!\!=} CH(CH_2)_9COOH$$
$$\textbf{(4.31)}$$

$$\overset{\displaystyle Me}{\underset{\displaystyle }{\,|\,}}$$
$$CH_3(CH_2)_m \overset{|}{C} = CH(CH_2)_9COOH$$
$$\textbf{(4.32)}$$

$$\overset{O}{\underset{H}{\diagdown}}C(CH_2)_5COOMe$$
$$\textbf{(4.33)}$$

$$\downarrow \quad Ph_3P = CHCHO$$

$$\underset{H}{\overset{OHC}{\diagdown}}C = C\underset{(CH_2)_5COOMe}{\overset{H}{\diagup}}$$
$$\textbf{(4.34)}$$

$$\downarrow \begin{array}{l} \overset{+}{Ph_3P}(CH_2)_2CH_3Br^- \\ NaN(Me_3Si)_2/THF \end{array}$$

$$CH_3CH_2CH \overset{Z}{=\!\!=} CH - CH \overset{E}{=\!\!=} CH - (CH_2)_5COOMe$$
$$\textbf{(4.35)}$$

$$\downarrow \begin{array}{l} \text{(i) LiAlH}_4 \\ \text{(ii) Ac}_2O/\text{pyridine} \end{array}$$

$$CH_3CH_2CH \overset{Z}{=\!\!=} CH - CH \overset{E}{=\!\!=} CH - (CH_2)_5CH_2O - \overset{O}{\underset{\parallel}{C}}CH_3$$
$$\textbf{(4.36)}$$

Fig. 4.10.

Most of these methods are applicable to the synthesis of acids in general, and were not specifically developed, but rather may be utilized, for the synthesis of 'fatty' acids. The above is far from being an exhaustive list! A further useful method, anodic synthesis, essentially consists of Kolbe electrolysis of a mixture of a monocarboxylic acid and the half-ester of a dicarboxylic acid, followed by hydrolysis of the product of crossed coupling (**4.37**; Fig. 4.11). In theory this permits the carboxylic acid chain to be

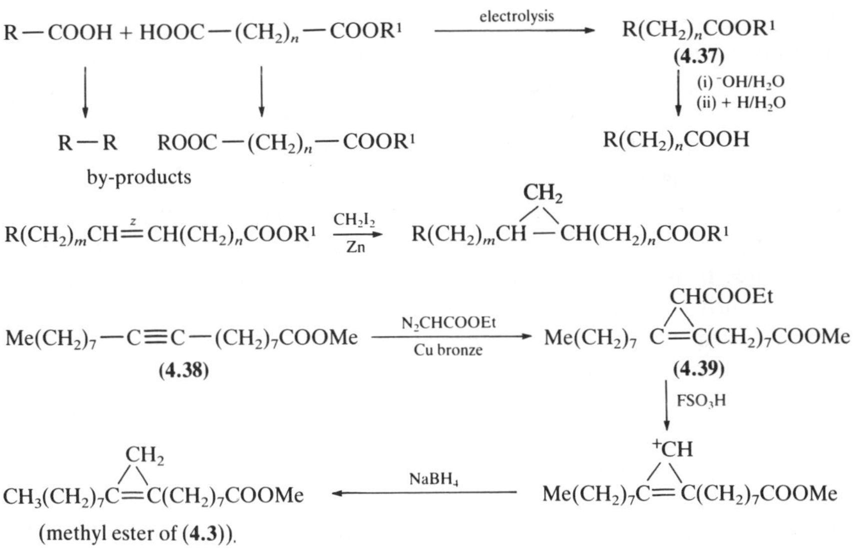

$$R-COOH + HOOC-(CH_2)_n-COOR^1 \xrightarrow{\text{electrolysis}} R(CH_2)_nCOOR^1$$

$$\textbf{(4.37)}$$

$$\downarrow \qquad\qquad \downarrow \qquad\qquad\qquad\qquad \begin{array}{l}\text{(i) }^-OH/H_2O\\ \text{(ii) } + H/H_2O\end{array} \downarrow$$

$$R-R \qquad ROOC-(CH_2)_n-COOR^1 \qquad\qquad R(CH_2)_nCOOH$$

by-products

$$R(CH_2)_mCH \overset{z}{=} CH(CH_2)_nCOOR^1 \xrightarrow[Zn]{CH_2I_2} R(CH_2)_mCH\overset{\overset{\displaystyle CH_2}{\diagup\backslash}}{} CH(CH_2)_nCOOR^1$$

$$Me(CH_2)_7-C\equiv C-(CH_2)_7COOMe \xrightarrow[\text{Cu bronze}]{N_2CHCOOEt} Me(CH_2)_7 \overset{\overset{\displaystyle CHCOOEt}{\diagup\backslash}}{C=C}(CH_2)_7COOMe$$

$$\textbf{(4.38)} \qquad\qquad\qquad\qquad\qquad\qquad \textbf{(4.39)}$$

$$\downarrow \text{FSO}_3\text{H}$$

$$CH_3(CH_2)_7\overset{\overset{\displaystyle CH_2}{\diagup\backslash}}{C=C}(CH_2)_7COOMe \xleftarrow{NaBH_4} Me(CH_2)_7\overset{\overset{\displaystyle {}^+CH}{\diagup\backslash}}{C=C}(CH_2)_7COOMe$$

(methyl ester of (**4.3**))

Fig. 4.11.

extended by any number of carbon atoms in a single step. In practice, homologation by two to eight carbons is usually performed.

Clearly, some of the more unusual fatty acid structures demand rather specific synthetic approaches. Cycloalkanoic acids such as lactobacillic acid (**4.2**) are normally prepared by a Simmons–Smith reaction which utilizes the ester of a *cis*-alkenoate, diiodomethane, and zinc. Methyl sterculate (cf. (**4.3**)) has been prepared from methyl stearolate (**4.38**) by reaction with ethyl diazoacetate in the presence of copper bronze, to give (**4.39**), which is decarbonylated by means of fluorosulphonic acid to give the cyclopropenium cation ester which on reduction with sodium borohydride gives the methyl ester of (**4.3**).

4.6 Prostaglandins

In 1930, it was reported by two American gynaecologists that fresh human semen contained a substance which provoked human uterine muscle – and indeed smooth muscle (intestinal tract, spleen, uterus, and that of the circulatory system) generally – to strong contraction or relaxation. In subsequent classical studies, von Euler in Sweden and Goldblatt in England identified this activity in seminal plasma and showed that it caused lowering of blood pressure in animals, and the former demonstrated that it was associated with the fatty acid fraction of lipid extracts of vesicular glands.

The factor responsible, which was active in minute amounts, was named 'prostaglandin' although it is now well known that it is found widely in animal tissue, and not confined to the prostate gland. In 1947 Bergström, in Sweden, started work to isolate the active species, and found it to be associated with a fraction containing unsaturated hydroxy-acids. In 1956, he was able to isolate two prostaglandins, PG-E_1 and PG-$F_{1\alpha}$, in crystalline form, and after years of painstaking effort using chemical degradation, mass spectrometry, NMR and X-ray crystallographic techniques on milligram quantitites of material, the structures of the two species were established. Subsequently a number of other prostaglandins were isolated and characterized, permitting a number of distinct classes to be recognized, as described below.

4.6.1   Prostaglandin structures

The nomenclature of prostaglandins is based on prostanoic acid, the fully saturated fatty acid in which a bond joins C-8 and C-12, forming a 5-membered ring (**4.40**; Fig. 4.12). Consequently, derivatives of this system are termed prostanoids. Classification is based on the substitution pattern in the 5-membered ring, and the structures of the chains comprising C-1–C-7 and C-13–C-20, as indicated in Fig. 4.12. Thus, prostaglandins A–C are cyclopentenone derivatives, D and E are hydroxycyclopentanones, and so on, with the ring substitution patterns indicated. The numerical subscript refers to the total number of alkene double bonds in the two side chains. Thus the '2' series illustrated contains one R^1 chain at C-8, containing one alkene double bond, and one R^2 (or R^3) chain at C-12, containing one alkene double bond, totalling two. The stereochemistry of the hydroxy group at C-15 in R^2 (and R^3 and R^5) is normally *S*. In PG-E_1, for instance, the R^1 chain is replaced by R^4 and thus one alkene unit fewer is present. In PG-E_3, the R^2 chain is replaced by R^5. In PG-$F_{2\alpha}$ the α subscript refers to the stereochemistry at C-9. So, PG-$F_{3\alpha}$ has the same structure as PG-$F_{2\alpha}$ but with R^5 replacing R^2, and so on. The chain comprising C-1–C-7 is known as the α-chain, and that comprising C-13–C-20 as the β-chain. It will be noted that in PG-I_2, the hydroxy-function at C-9 of PG-$F_{2\alpha}$ has become cyclized to the R^1 chain, to form a second five-membered ring. This compound, more commonly known as prostacyclin, has attracted considerable attention owing to its ability to lower blood pressure and inhibit the aggregation of blood platelets, and is possibly the most medicinally significant of the prostanoids.

4.6.2   Occurrence and significance of the prostaglandins

As noted above, PG-E_1 and PG-$F_{1\alpha}$ occur in seminal plasma, and PG-D_2, PG-E_2, and PG-$F_{2\alpha}$ are also widely distributed in mammalian tissue but at

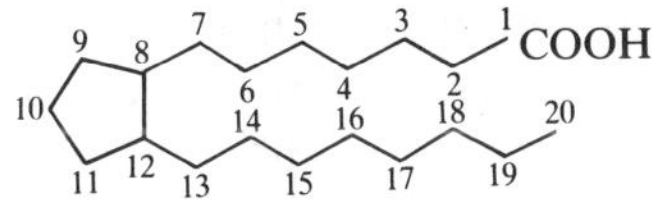
COOH

(4.40)

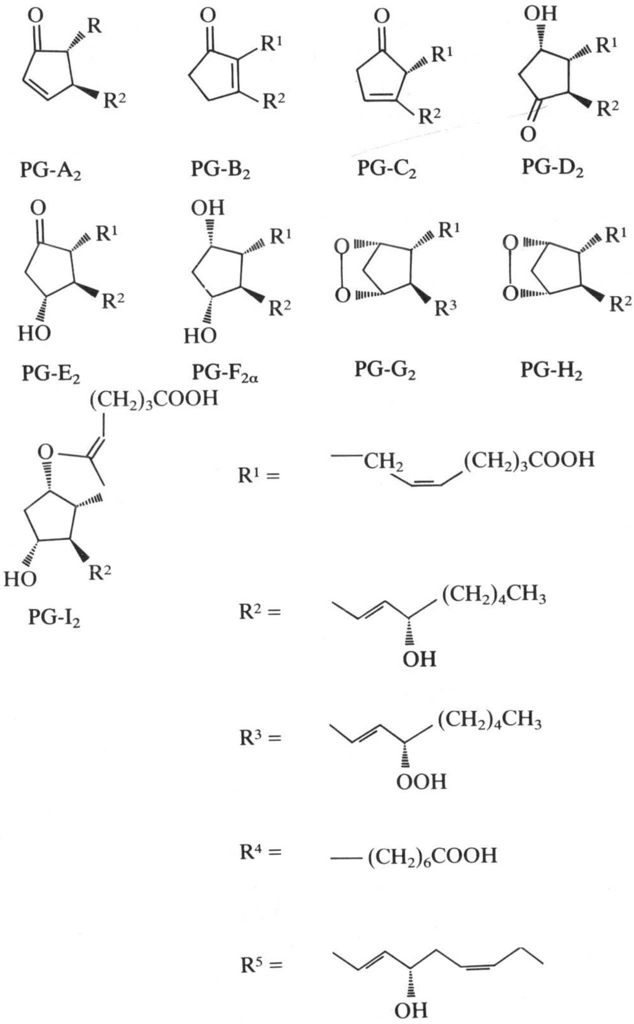
PG-A2
PG-B2
PG-C2
PG-D2
PG-E2
PG-F2α
PG-G2
PG-H2
(CH2)3COOH
PG-I2
R1 =
CH2
(CH2)3COOH
R2 =
(CH2)4CH3
OH
R3 =
(CH2)4CH3
OOH
R4 =
(CH2)6COOH
R5 =
OH

Fig. 4.12.

low concentration. The alga *Gracilaria lichenoides* has been found to contain significant amounts of PG-$F_{2\alpha}$ and PG-E_2. Onions contain PG-A_1 and a Caribbean coral, *Plexaura homomalla*, contains large quantities of the C-15 epimer of PG-A_2 and its 15-acetoxy, methyl ester, and represents a rich, if limited, natural source of these compounds.

The prostaglandins are effective at hormone-like concentrations, of the order of micrograms per kilogram of tissue (or nanomolar concentration), and frequently affect several important physiological processes. Indeed, much research into prostaglandin analogues has been performed with a view to improving or maximizing one particular activity displayed by these species while suppressing unwanted associated activities. The PG-A species seem to have a critical role in controlling blood pressure, thus offering the possibility of therapeutic use as anti-hypertensive agents. They also suppress gastric secretion. The PG-E series relax airway smooth muscle (thus offering potential use in treating bronchial spasms in asthma patients), lower blood pressure (vasodilation), stimulate smooth muscle contraction in gut and uterus, antagonize hormone-stimulated free fatty acid release from adipose tissue, inhibit acid secretion, and sensitize pain receptors. Different members of the series display these properties to varying extents in different animals, however. $PG-E_2$ has been used as an abortifacient, or to induce labour, in both humans and domestic animals. $PG-F_{2\alpha}$ is a potent luteolytic, causing regression of the corpus luteum in experimental animals and thus, by lowering progesterone levels, causing the termination of pregnancy. As noted above, $PG-I_2$, prostacyclin, not only lowers blood pressure but also suppresses platelet aggregation (a Ca^{2+}-ion-dependent process) by acting at a specific receptor on the platelets, causing cytoplasmic cyclic AMP levels to rise and Ca^{2+} levels to drop in consequence, and the potential use of such properties in treating patients at risk from vascular disease has caused particularly diligent pursuit of syntheses of prostacyclin and its analogues.

4.6.3 Prostaglandin biosynthesis

The biosynthesis of the prostaglandins has been unravelled largely by two research teams: that of van Dorp in Holland, and, particularly, that of Bergström and Samuelsson in Sweden. It is now established that the '1' series is derived from $8(Z),11(Z),14(Z)$-eicosatrienoic acid, the '2' series from $5(Z),8(Z),11(Z),14(Z)$-eicosatetraenoic acid (arachidonic acid), and the '3' series from $5(Z),8(Z),11(Z),14(Z),17(Z)$-eicosapentaenoic acid. Essentially the pathway by which the prostaglandins are derived is the same for each series. Consider, for example, the '2' series. Upon incubation of arachidonic acid with whole homogenates of sheep seminal vesicular glands, the formation of the '2' series by 'prostaglandin synthetase' was demonstrable, the first recognizable product being the hydroperoxide $PG-G_2$. When the process was performed under $^{18}O_2$, all the three oxygen atoms at C-9, C-11, and C-15 became labelled, indicating that their source was molecular oxygen, and in addition Samuelsson showed that if a mixture of $^{18}O_2$ and $^{16}O_2$ was used, the oxygen isotopes incorporated at C-9 and C-11 were always the same, demonstrating that the two ring oxygens are derived from the same oxygen molecule. The supposed pathway, catalysed by the 'cyclo-oxygenase' enzyme, is depicted in Fig. 4.13. It is seen that the

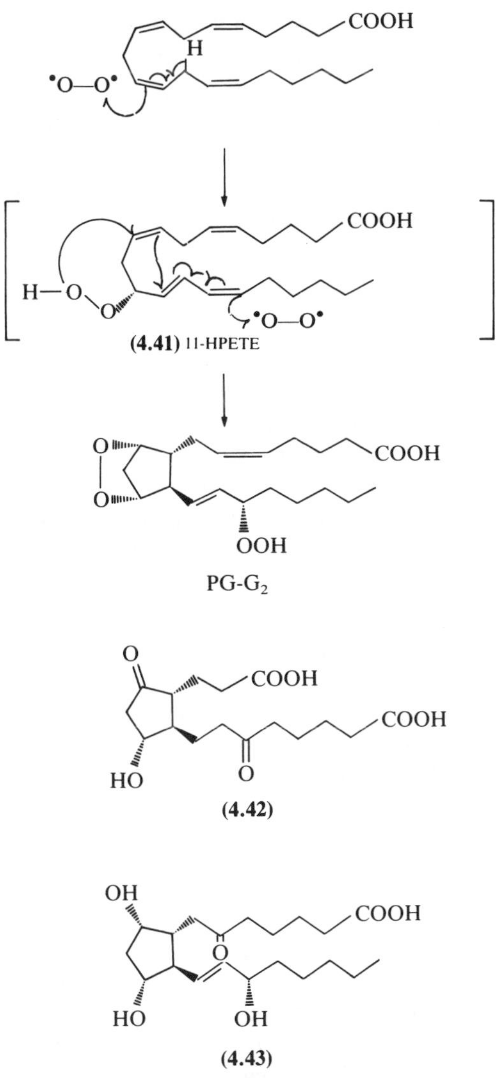

Fig. 4.13.

arachidonate structure may be arranged in a pseudo-prostanoic shape, and the initial peroxidation of the divinylmethane system is entirely analogous to that depicted in Fig. 4.6 to afford the putative 11-hydroperoxyeicosatetra-enoic acid (11-HPETE) (4.41). Further peroxidation, this time accompanied by cyclization, affords PG-G$_2$ which is then reduced by a peroxidase enzyme to afford PG-H$_2$, and it is from this that the other '2' series prostaglandins are thought to be derived by enzyme-catalysed rearrangements (PG-E$_2$, PG-D$_2$), reduction (PG-F$_{2\alpha}$, from PG-E$_2$ or PG-H$_2$), or cyclization (PG-I$_2$, via prostacyclin synthase in the blood vessels). $PG\text{-}G_2$ and $PG\text{-}H_2$ also give rise to the thromboxanes (see Section 4.7), and the effect of certain non-steroidal anti-inflammatory drugs, such as aspirin and indomethacin, is to inhibit the cyclo-oxygenase enzyme and thus the formation of the prostaglandins and thromboxanes. Anti-inflammatory steroids, in contrast, function by preventing the release of arachidonic acid from storage in cell membrane phospholipids, possibly by stimulating synthesis of an inhibitor of phospholipase A_2.

4.6.4   Prostaglandin catabolism

The prostaglandins are degraded rapidly and efficiently *in vivo* by processes which are easily understood in relation to those we have seen previously with the fatty acids. Firstly the C-15 hydroxy group is oxidized by a dehydrogenase enzyme to a 15-ketoprostanoid which is then reduced at the C-13–C-14 double bond by a hydrogenase enzyme. Typically, two cycles of β-oxidation of the α-chain and ω-oxidation of the ω-chain then ensue, the resulting metabolite being excreted in the urine. Thus, for instance, degradation of $PG\text{-}E_2$ results in the formation of (**4.42**), and the degradation of the other prostaglandins affords products of similar side-chain structure. Prostacyclin ($PG\text{-}I_2$) is unstable under physiological conditions, when hydration across the C-5–C-6 double bond followed by ring-opening affords 6-ketoprostaglandin $F_{1\alpha}$ (**4.43**) which is subsequently degraded by the same processes as described above.

4.6.5   Prostaglandin synthesis

The brilliant work of Corey and his colleagues must be cited before all else, for synthetic work not only on the prostaglandins but also on the other products of arachidonate metabolism, the thromboxanes and leukotrienes (as described in Sections 4.7 and 4.8). Clearly any total synthesis of the prostaglandins must accommodate the stereochemistry found in the natural compounds, notably at C-5, C-8, C-9, C-11, C-12, C-13, and C-15. Of these, the most serious problems are, arguably, concerned with the introduction of the required chirality at C-8, C-9, C-11, C-12, and C-15. In the earliest synthesis of prostaglandins reported during 1968–69, racemates were generated which required laborious resolution before the 'natural' stereo-isomer could be isolated. Of course this approach also yielded enantiomers of the prostaglandins, but these proved to be of very limited interest: the enantiomer of $PG\text{-}E_1$, for instance, was found to have at most 0.1 per cent of the ability of the natural isomer in stimulating smooth muscle contraction. In 1969, however, Corey's group described a stereocontrolled synthesis of DL-$PG\text{-}E_2$ and DL-$PG\text{-}F_{2\alpha}$, and the principles and approach described in

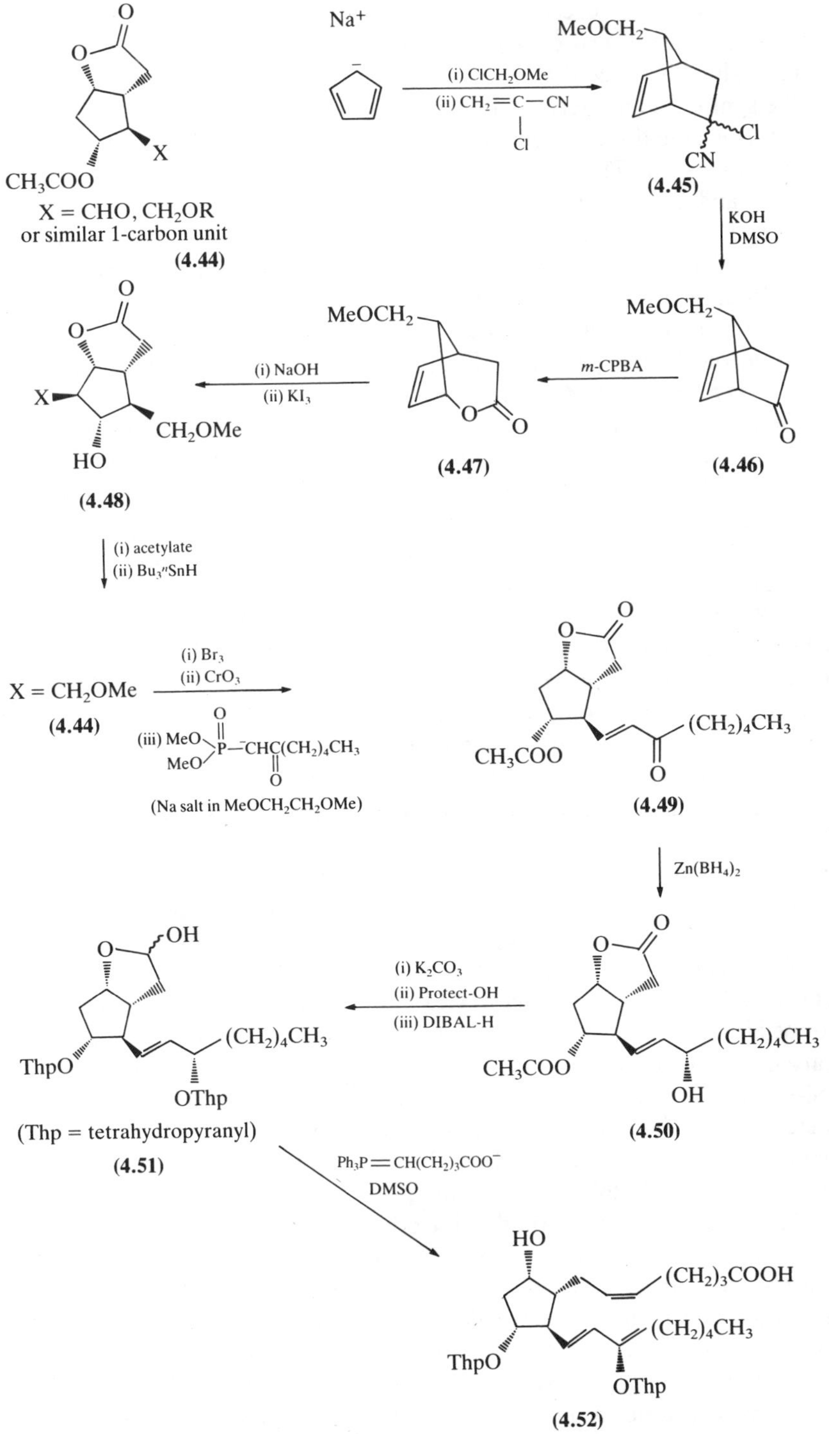
X = CHO, CH2OR
or similar 1-carbon unit
(4.44)
Na+
(i) ClCH2OMe
(ii) CH2=C—CN
Cl
(4.45)
KOH
DMSO
m-CPBA
(4.46)
(4.47)
(i) NaOH
(ii) KI3
(4.48)
(i) acetylate
(ii) Bu3ⁿSnH
X = CH2OMe
(4.44)
(i) Br3
(ii) CrO3
(Na salt in MeOCH2CH2OMe)
(4.49)
Zn(BH4)2
(4.50)
(i) K2CO3
(ii) Protect-OH
(iii) DIBAL-H
(Thp = tetrahydropyranyl)
(4.51)
Ph3P=CH(CH2)3COO⁻
DMSO
(4.52)

Fig. 4.14.

the paper have influenced much of the subsequent work in this area, and stimulated a great deal of elegant chemistry in the intervening years.

Basically, the method (Fig. 4.14) consists in generating, at an early stage in the synthesis, an intermediate species (**4.44**) in which the stereochemistry which is eventually to be required at C-8, C-9, C-11 and C-12 has already been established. The species (**4.44**) can readily be elaborated to the full prostaglandin structure, and serves as a common intermediate for generating the PG-E$_n$ and PG-F$_{n\alpha}$ ($n = 1-3$) series. It has become known as Corey's lactone. Originally, it was generated as follows: treatment of cyclopentadienylsodium with chloromethyl methyl ether afforded 5-methoxymethyl-1,3-cyclopentadiene, which underwent Diels–Alder reaction with 2-chloroacrylonitrile in the presence of copper(II) fluoroborate catalyst to yield (**4.45**). Hydrolysis of (**4.45**) with potassium hydroxide in DMSO afforded the *anti*-bicyclic ketone (**4.46**) which underwent Baeyer–Villiger oxidation using *m*-chloroperbenzoic acid (*m*-CPBA) to give lactone (**4.47**). This was then hydrolysed with dilute sodium hydroxide, and the solution neutralized and treated with potassium triiodide. Oxidative attack took place on the less hindered side of the cyclopentene ring, with concomitant cyclization, generating the iodolactone (**4.48**; X = I), which, upon acetylation of the hydroxy-group and reductive deiodination using tributyltin hydride, afforded Corey's lactone (**4.44**; X = CH$_2$OMe). Demethylation of this species using boron tribromide, and oxidation of the resulting alcohol using chromium trioxide pyridine ('Collins' reagent') gave (**4.44**; X = CHO) which was condensed with the sodium salt of dimethyl 2-oxoheptylphosphonate in dimethoxyethane (Emmons–Horner variant of the Wittig reaction) to give stereospecifically the *trans*-enone lactone (**4.49**). On reduction with zinc borohydride this afforded a 1:1 mixture of the 15-α-hydroxy-species (**4.50**) and its β-epimer, separable by thin layer chromatography. Then, deacetylation with potassium carbonate, protection of the 11- and 15-α-hydroxy groups as their tetrahydropyranyl derivatives using dihydropyran and *p*-toluenesulphonic acid, and reduction with diisobutylaluminium hydride (DIBAL-H) generated the lactol (**4.51**) which reacted with the Wittig reagent formed from 5-triphenylphosphoniopentanoic acid and sodiomethylsulphinyl carbamide in DMSO to give the bis-tetrahydropyranyl derivative of DL-PG-F$_2$ (**4.52**). This latter step is also a valuable procedure for stereoselective formation of the *cis*-olefin (cf. Fig. 4.10). Deblocking (**4.52**) with aqueous acetic acid afforded DL-PG-F$_{2\alpha}$, while chromic reagent oxidation of (**4.52**) followed by similar deblocking afforded DL-PG-E$_2$.

The extension of this method to introduce the side-chains found in PG-E$_1$, PG-E$_3$, PG-F$_1$, and PG-F$_{3\alpha}$ is clear. While the initial reaction to form (**4.45**) must of necessity generate enantiomers, note how steric effects have been used elegantly to generate the desired stereochemistry in the steps which form (**4.45**) and, particularly, (**4.48**), and hence (**4.44**). An immediate result of the description of this pathway was to stimulate a search for alternative paths to Corey's lactone. A number of elegant routes to this, and similar key compounds, have involved (2 + 2) cycloadditions of ketenes to cyclo-

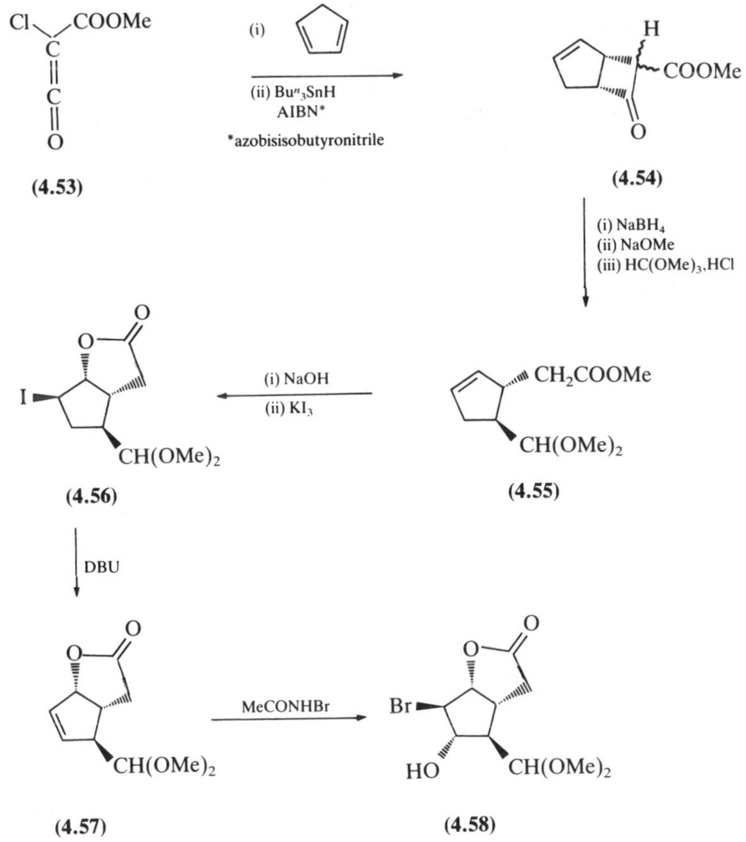

Fig. 4.15.

pentadiene. For instance (Fig. 4.15), addition of (carbomethoxy)chloro-ketene (**4.53**) to cyclopentadiene leads, after reduction, to the bicyclo-heptenone (**4.54**). Reduction with borohydride, followed by methoxide-catalysed retro-aldol cleavage of the cyclobutane ring with simultaneous conferment of *trans* stereochemistry on the ring substituents, and acetal formation, generate (**4.55**), which upon alkaline hydrolysis and iodo-lactonization (cf. (**4.47**) → (**4.48**) above) yields (**4.56**). Elimination of hydrogen iodide using 1,8-diazabicyclo[5.4.0]undec-7-ene (DBU) then gives (**4.57**), which on treatment with *N*-bromoacetamide affords regio- and stereo-selectively the bromohydrin (**4.58**). Protection of the hydroxy group followed by debromination using tri-*n*-butyl tin hydride effectively affords (**4.44**). In addition, (**4.57**) affords an intermediate for the synthesis of the PG-A series. Again, note how steric influences are used throughout to generate the desired stereochemistry.

Other unlikely-looking starting materials which have been used to prepare

Corey's lactone include (Z,Z)-1,5-cyclooctadiene (formed by cyclodimerization of butadiene) and norbornadiene. For details of these and other routes, the reader is recommended to reviews cited at the end of this chapter.

Another elegant and versatile route to prostaglandins (Fig. 4.16) also begins with $(2 + 2)$ cycloaddition, this time of dichloroketene to cyclopentadiene, to afford the dichlorobicycloheptenone (4.59) which is reduced

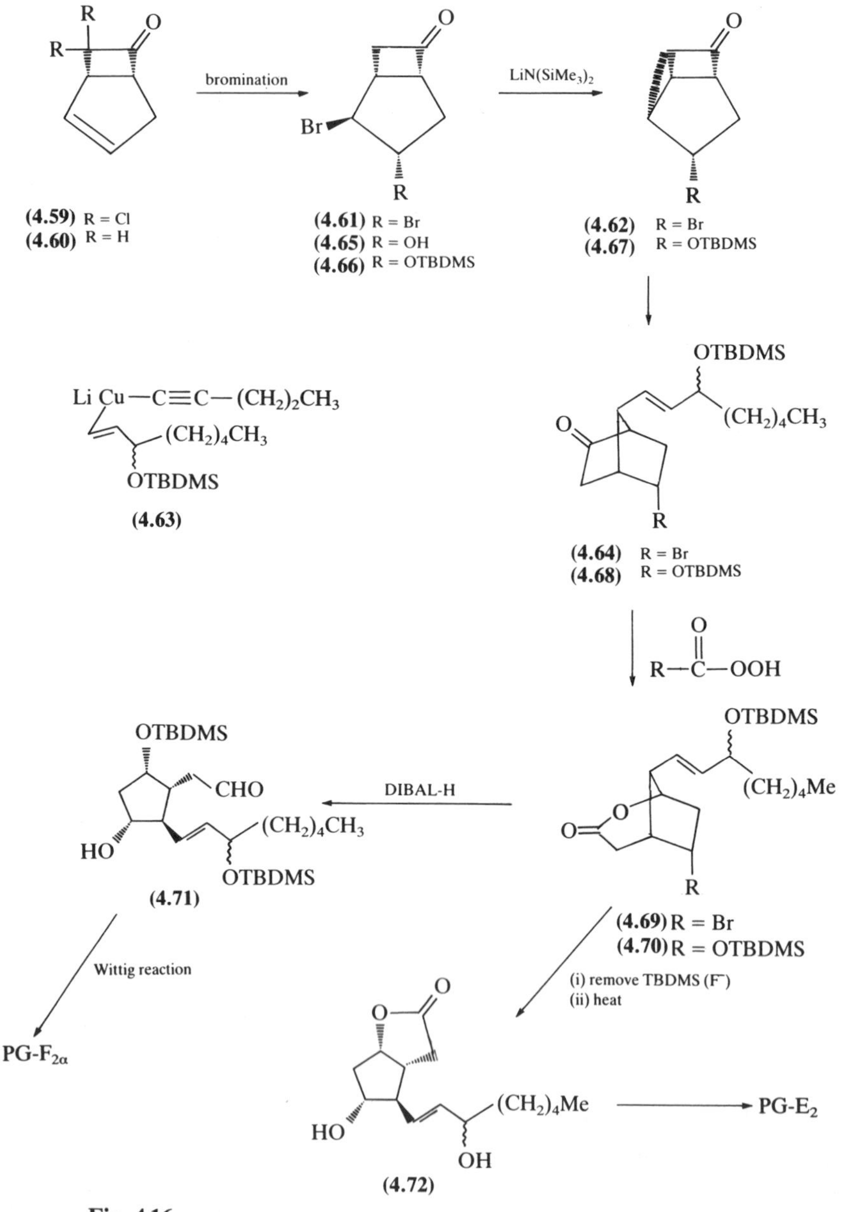

Fig. 4.16.

to bicyclo[3.2.0]hept-2-en-6-one (**4.60**). Steric hindrance by the cyclo-butanone ring dictates the course of bromination of (**4.60**), which proceeds via the *exo*-bromonium ion and nucleophilic attack at C-3 to give principally (**4.61**), in which the stereochemistry is ideal for proton abstraction using lithium bis(trimethylsilyl)amide and ring closure with expulsion of bromide to afford 3-bromotricyclo[3.2.0.0^{2,7}]heptanone (**4.62**). Such systems are cleaved stereospecifically by nucleophilic attack at C-1 with cleavage of the C-1–C-7 bond, and this is the key to this route to the prostaglandins: treatment with the lithium vinyl cuprate reagent (**4.63**) affords ketone (**4.64**), permitting stereospecific introduction of the β-chain in one step. If bromination of (**4.60**) is performed in water, the bromohydrin (**4.65**) is formed, and may be conveniently protected as its *t*-butyldimethylsilyl ether (**4.66**), and converted, in its turn, to (**4.67**) and (**4.68**). Both (**4.64**) and (**4.68**) may then be lactonized using per-acids to (**4.69**) and (**4.70**) respectively. Elimination of hydrogen bromide from (**4.69**) with DBU gives the cyclopentene lactone which isomerizes on heating in DMF to afford an intermediate which is convertible to PG-A$_2$ using the same sort of procedure as that for the conversion (**4.50**)–(**4.52**). Reduction of (**4.70**) with DIBAL-H affords the hydroxyaldehyde (**4.71**) which is convertible to PG-F$_{2\alpha}$ and its 15-epimer by a Wittig reaction followed by removal of the silyl protecting groups. Oxidation of the product of Wittig reaction on (**4.71**) affords the corresponding hydroxycyclopentanone and thus a route to the PG-D$_2$ series, while heating desilylated (**4.70**) under acidic or basic conditions results in rearrangement to (**4.72**), and hence a route to PG-E$_2$.

Clearly the processes which form (**4.60**) also form its enantiomer (**4.73**), and for an efficient synthetic route it is desirable that (**4.73**) also be utilized to yield the compounds desired. This has been demonstrated (Fig. 4.17): the bromohydrin (**4.74**), the epimer of (**4.65**), is formed from (**4.73**) by the same method as for (**4.60**)–(**4.65**), converted to the ketal, and treated with base to afford the epoxy ketal (**4.75**). The four-membered ring in this system apparently confers regioselectivity during nucleophilic attack, and on attack by (**4.63**), followed by deblocking of the ketal, (**4.75**) gives predominantly (**4.76**), which upon Baeyer–Villiger oxidation followed by desilylation affords (**4.72**). Also, irradiation of desilylated (**4.76**) in aqueous acetonitrile leads, via a supposed oxacarbene intermediate, to the lactol (**4.77**), which is one Wittig reaction removed from PG-F$_{2\alpha}$.

The enantiomeric bicycloheptenones (**4.60**) and (**4.73**) are thus used in enantiocomplementary routes which give rise to the same array of prostaglandins. Clearly an efficient resolution of a racemic mixture of (**4.60**) and (**4.73**) is required, and may be realized by utilizing enzymic stereospecificity. Reduction of the racemate using actively fermenting bakers' yeast affords the diastereoisomeric alcohols (**4.78**) and (**4.79**), which are separable by distillation and re-converted to the parent ketones, (**4.73**) and (**4.60**) respectively, by Jones's oxidation. The utilization of enzymic stereoselectivity as a tool in synthetic organic chemistry is an increasing trend.

Another synthetic route to prostaglandins which has enjoyed conspicuous

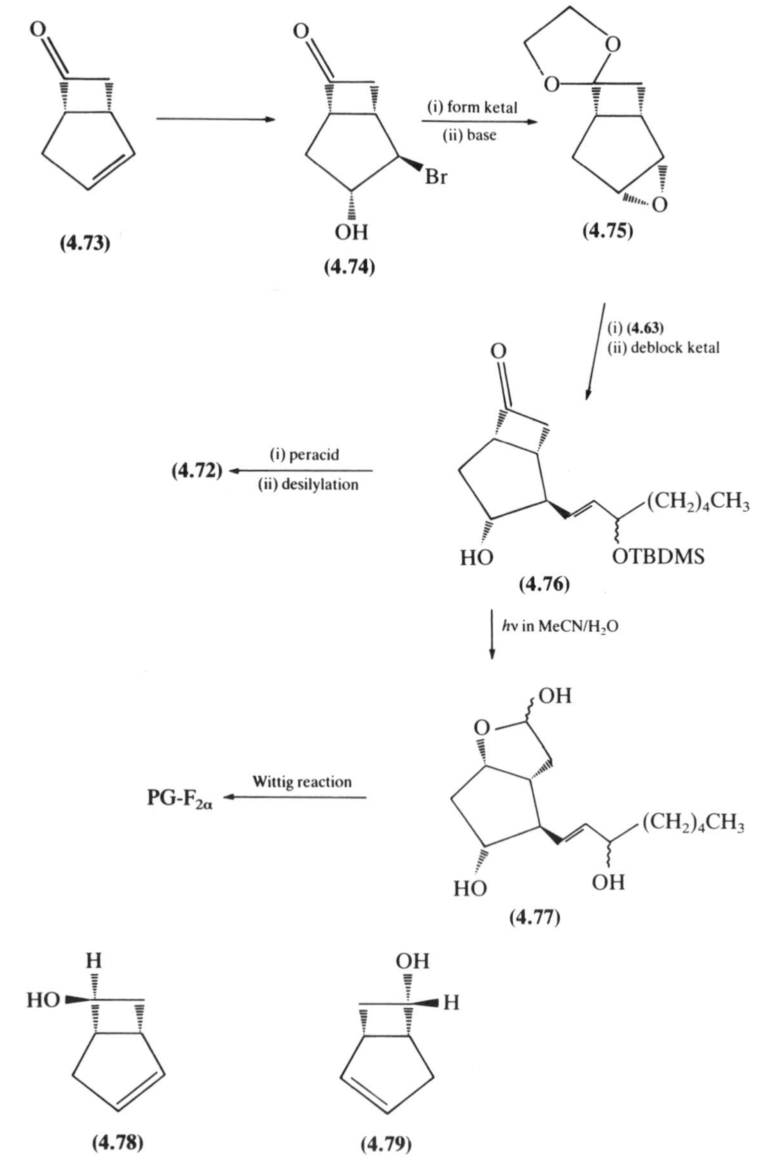

Fig. 4.17.

success involves conjugate addition to 2-alkyl-4-alkoxycyclopent-2-enones (Fig. 4.18). These are readily prepared from cyclopentadiene, which is first converted to the epoxide (**4.80**), treated with a suitable lithium alkyl, then with *m*-chloroperbenzoic acid to give the epoxide of the 2-alkylcyclopent-3-enone (**4.81**), and subsequently cleaved with triethylamine to afford predominantly (**4.82**), the kinetic product. Now, (**4.82**) may be isomerized

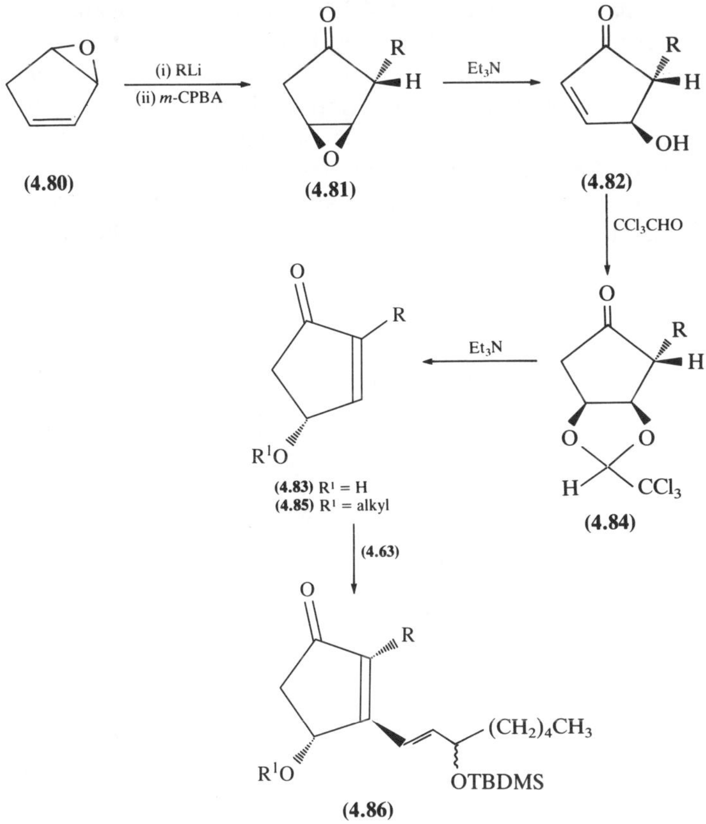

Fig. 4.18.

to form the desired (**4.83**), the thermodynamic product, in dilute sodium hydroxide, but this process only proceeds reliably when the alkyl and hydroxy groups in (**4.82**) are in a *cis* relationship, affording an unhindered face for attack by hydroxide. However, on treatment with chloral, (**4.82**) forms a hemiacetal which ring-closes to form the acetal (**4.84**), and treatment with triethylamine then affords the desired (**4.83**) in high yield. Then, attack by a lithium vinyl cuprate reagent, such as (**4.63**), permits direct introduction of the β-chain. The reagent attacks from the least hindered side of the cyclopentane ring, i.e. *trans* to the alkoxy grouping in the 4-position. Consequently the stereochemistry at C-4 of (**4.83**) dictates the stereochemistry of the adduct at C-3, and protonation of the enolate ion generated by attack of (**4.63**) on (**4.85**) then affords preferentially the thermodynamically more stable product (**4.86**) in which the groups at C-2 and C-3 are also *trans*. Thus, starting with the 4-α-alkoxy species (**4.85**) predetermines the stereochemistry at C-2 and C-3 of the cyclopentanone formed.

An alternative approach (Fig. 4.19) consists in conjugate addition to a 4-alkoxycyclopentenone, followed by trapping of the enolate ion formed with a species which is then convertible into the α-chain. The starting material is rapidly accessible: treatment of cyclopentadiene with cumyl hydroperoxide and copper(II) acetate, with addition of aqueous ferrous sulphate, affords the acetate (**4.87**) in a single step, and separation of the stereoisomers, followed by hydrolysis of the 4α-anomer with methanolic potassium hydroxide and Jones's oxidation, then yields (**4.88**). When this is treated

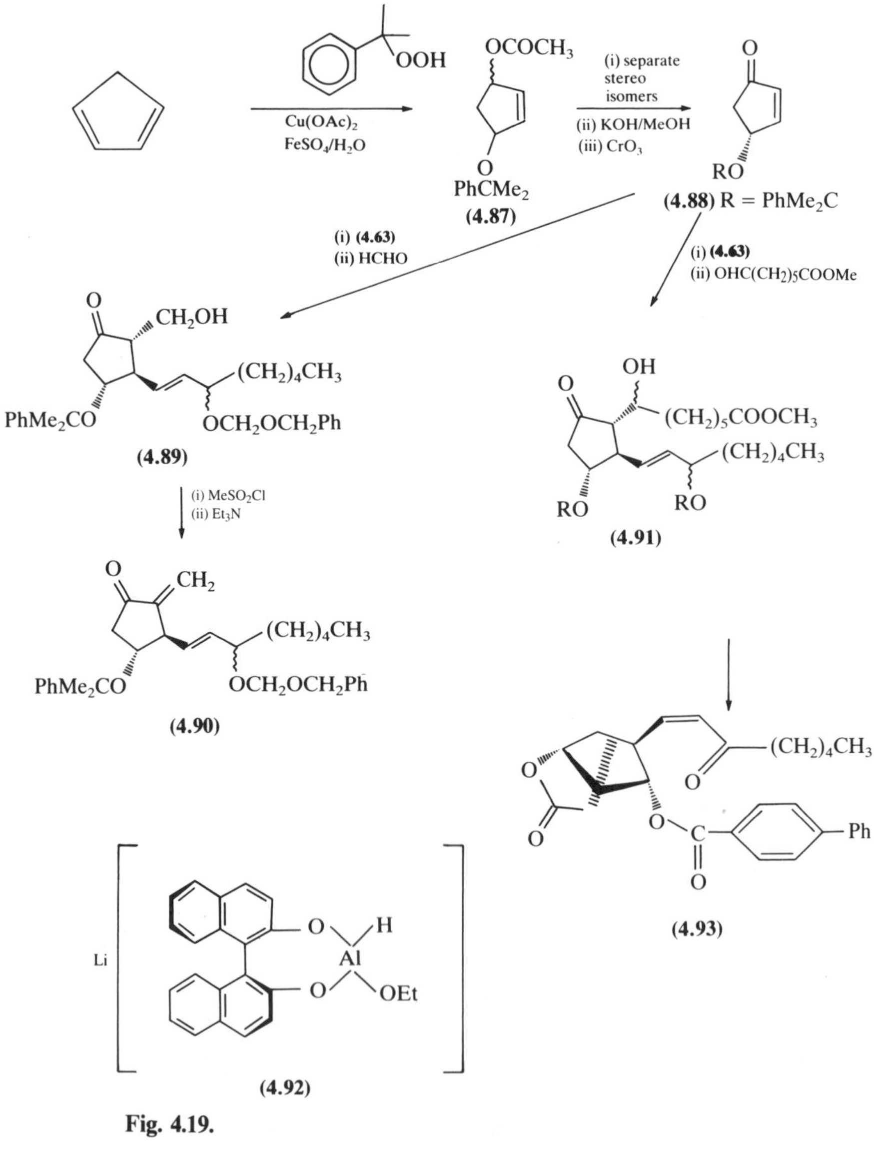

Fig. 4.19.

with a lithium vinyl cuprate reagent of type (**4.63**), and the enolate ion thus generated trapped by the addition of formaldehyde, the 4-alkoxy-3-alkyl-2-hydroxymethylcyclopentanone (**4.89**) is formed. Treatment with mesyl chloride, followed by base-catalysed elimination using triethylamine, then gives the exocyclic enone (**4.90**), to which the α-chain is added by Michael addition of a suitable organometallic reagent, when steric influences again dictate that the substituents at C-2 and C-3 of the cyclopentanone ring become *trans* to each other. The enolate ion may alternatively be trapped with an aldehyde, such as 6-methoxycarbonylhexanal, completing a 'three-component coupling' process to afford a species of type (**4.91**), which is then mesylated, the mesyl group eliminated with 4-dimethylaminopyridine to form an enone (cf. (**4.90**)), and the double bond of the enone reduced with zinc dust in isopropanol-acetic acid to afford, after deblocking, the methyl ester of PG-E_1. Again, steric influences favour the stereochemistry desired. The enolate may alternatively be used for Michael addition, in order to attach the α-chain in a variant of the above three-component condensation.

The reader cannot have failed to notice that, while steric influences may be utilized in determining stereoselectivity in a particular reaction in which one component is already chiral, some reactions which generate chiral centres must necessarily form racemates. In the one case which we have examined, both components of a racemate, (**4.60**) and (**4.73**), can be utilized to the same end. In others (as, for instance, in the case of (**4.87**)) a racemic mixture must be resolved, or else a DL-mixture of the final product is the result. The formation of a racemate, particularly in the latter stages of a multi-stage reaction, is thus to be avoided, and a major stride forward has been the recent development of reagents which exhibit stereoselectivity. These have been particularly useful in reduction processes in prostaglandin synthesis. For example, the (*S*) binaphthol ethoxyaluminium hydride (**4.92**), by virtue of the chirality imposed by the naphthalene rings, can reduce cyclopent-4-en-1,3-dione to form the (*R*) product (i.e. (**4.88**; R = H)) in 94 per cent yield. Clearly this is of huge advantage if, as in the examples examined above, the (*R*)-enantiomer is desired. Moreover, the same reagent will reduce (**4.49**) (with tetrahydropyranyl replacing the acetyl group) to (**4.50**) (ditto) in 95 per cent yield, giving a product with 99.5 per cent stereoisomeric purity. If the acetyl group in (**4.49**) is replaced by a bulky protecting group such as *p*-phenylbenzoyl or *p*-phenylphenyl carbamoyl, the proximity of this bulky group to the enone blocks approach of a reducing agent from one face of the molecule. The side-chain preferentially adopts the *S-cis*-configuration (**4.93**) and attack from the unhindered side (see arrow) with a bulky borohydride reagent permits stereocontrol, with a preponderance of the 15*S* alcohol formed. A useful reagent of this type which is highly regioselective (attacking the keto group rather than its conjugated double bond) is L-selectride (lithium tri-*sec*-butyl-borohydride). Thus manipulation of steric constraints about a keto group or enone to be reduced permits a degree of stereoselectivity.

We have now seen a number of approaches to prostaglandins A, E, and

F, and prostaglandins B and C are simply obtainable from prostaglandins E by standard transformations. Prostaglandins G and H demand the formation of a peroxide bridge, while prostacyclin, PG-I_2, also affords a special problem by virtue of being an unstable enol ether. By one ingenious method or another, these problems have been solved, but discussion of the synthesis is more properly left to specialist reviews.

This section has been necessarily selective, concentrating on routes to prostaglandins which have involved unusual precursors, or intermediates, and novel ways of solving the stereochemical problems involved. Many of the routes to prostaglandin analogues have involved the principles exemplified here. It could be argued that the major value in prostaglandin synthesis lies in the novel synthetic chemistry which it has generated, rather than in the intrinsic value of the compounds discovered, and the many hundreds of research papers dealing with prostaglandin synthesis bear witness to this. Certainly a chapter such as this can only scratch the surface of the subject, and the reader is referred to the references and reviews cited at the end of this chapter.

4.7   Thromboxanes

In the mid-1970s, Samuelsson and his colleagues identified a new group of biologically active compounds formed in blood platelets from arachidonic acid via the cyclo-oxygenase pathway. Following painstaking work using UV spectrometry, GC–MS data obtained on the silylated methyl esters, and supportive data obtained by ozonolysis, two structures, based on tetrahydro-pyran rings, were deduced, and were christened thromboxane (TX) A_2 (**4.94**) and B_2 (**4.95**) (Fig. 4.20). Labelling experiments indicated that TX-A_2 arises from PG-G_2 (by the action of 'thromboxane synthetase') and TX-B_2 in turn from TX-A_2 on nucleophilic opening of the oxetane ring. The subscripts have the same force, and R_1 and R_2 (in (**4.94**) and (**4.95**)) the same meaning, as in the prostaglandin nomenclature (Fig. 4.12). The strained oxetane ring renders TX-A_2 highly unstable, and it is easily attacked by nucleophiles such as azide, ethanol, and methanol with the introduction of azido, ethoxy, and methoxy groups in place of the anomeric hydroxy group in (**4.95**). It should be noted that the B series thromboxanes are hemiacetals. Just as the '3' series prostaglandins are formed from eicosapentaenoic acid, so, too, are TX-A_3 and TX-B_3 in blood platelets. Upon formation within blood platelets, TX-A_2 acts to increase the cytoplasmic calcium concentration which causes the cells to deform, releasing TX-A_2 into the blood plasma. The deformed platelets aggregate to form a thrombus or clot, hence the name 'thromboxanes'. Also, TX-A_2 is a powerful vasoconstrictor, and it is thought that its release may contribute to bronchial spasms in allergic conditions. Thus the effects of PG-I_2 and TX-A_2 are directly antithetic. In healthy individuals the effects are balanced, but an imbalance resulting in diminished PG-I_2 production or TX-A_2 overproduction will favour the development of

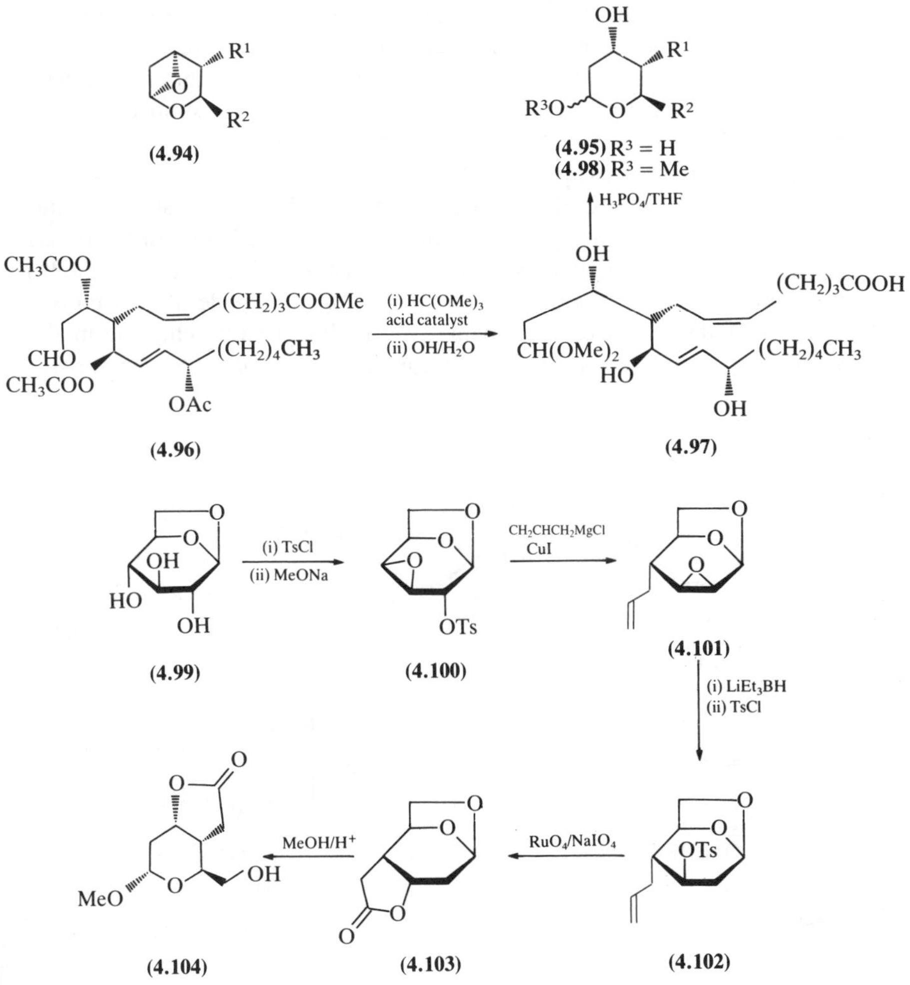

Fig. 4.20.

thrombotic conditions. By contrast, TX-A$_3$ is not strongly aggregatory, and it has been suggested that dietary eicosapentaenoic acid, by forming TX-A$_3$, may influence the prostacyclin/thromboxane balance towards a less thrombogenic state. Hopes of developing thromboxane analogues capable of antagonizing the effects elicited by TX-A$_2$ by binding to its receptors, and thereby reducing platelet aggregation, have stimulated synthetic work on the thromboxanes and their analogues.

4.7.1 Thromboxane synthesis

The strained oxetane ring in TX-A$_2$ has rendered it very difficult of synthetic access, and greater success has been attained in synthesizing TX-B$_2$. For instance (Fig. 4.20), oxidation of the methyl ester of PG-F$_{2\alpha}$-9,15-diacetate using lead tetra-acetate opens the cyclopentane ring to give the aldehyde (**4.96**) which, on treatment with trimethyl orthoformate and an acid catalyst, followed by alkaline hydrolysis, gives acetal (**4.97**). Phosphoric acid in aqueous THF converts (**4.97**) to a mixture of TX-B$_2$ (**4.95**) and the corresponding methyl pyranoside (**4.98**).

The fact that the thromboxanes are pyranose derivatives suggests that sugars may form promising starting materials for synthesis, and this has been amply demonstrated (Fig. 4.20). For instance, laevoglucosan (**4.99**) upon tosylation and subsequent treatment with methoxide affords epoxy-tosylate (**4.100**), which on treatment with allylmagnesium chloride in the presence of cuprous iodide is transformed to (**4.101**). Stereoselective

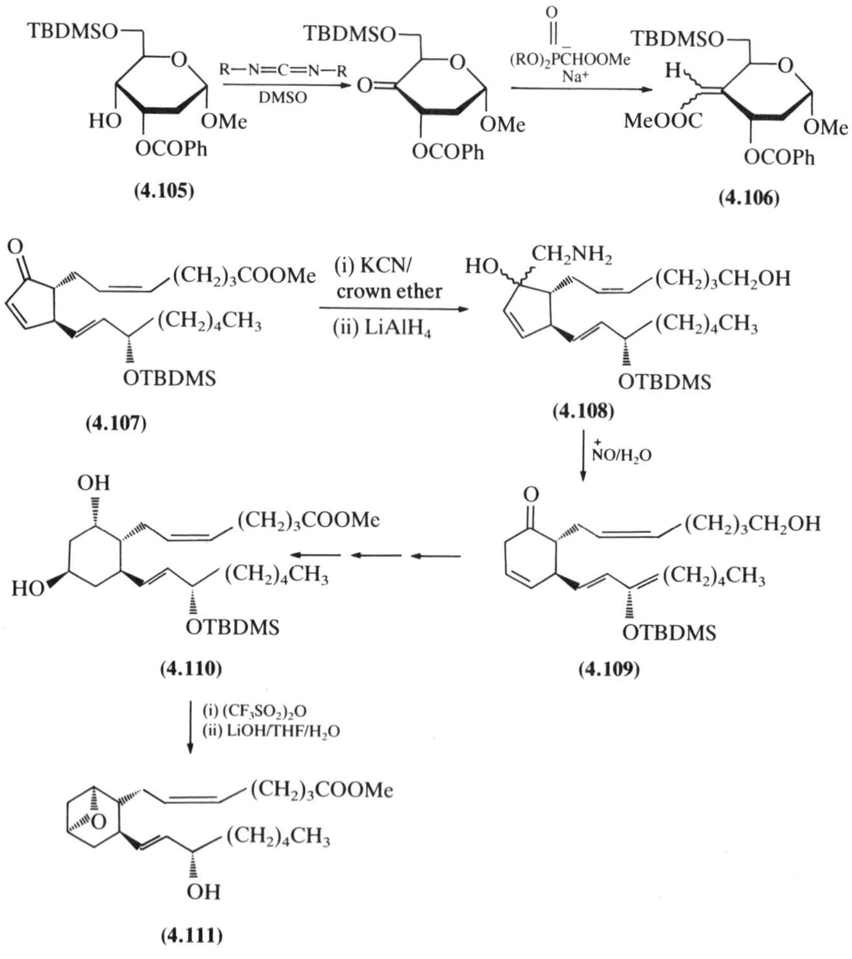

Fig. 4.21.

reduction with lithium triethyl borohydride, followed by tosylation, then affords (**4.102**). Oxidation of the vinyl group with ruthenium tetroxide and sodium periodate gives the carboxylic acid which then ring-closes with expulsion of the tosylate to afford lactone (**4.103**), and acidic methanolysis then gives the methyl pyranoside (**4.104**), convertible to TX-B$_2$ using methods described previously. Formulae (**4.99**)–(**4.103**) have been drawn as Haworth projections to emphasize the carbohydrate origins of the target species. Another noteworthy route (Fig. 4.21) utilizes the methyl β-D-2-deoxy-allopyranoside (**4.105**), which is oxidized using a carbodiimide in DMSO (Pfitzner–Moffatt oxidation) to the 4-ketosugar derivative, and then condensed with an appropriate phosphonate in an Emmons–Horner–Wittig reaction to give (**4.106**). Reduction of (**4.106**) using palladium and hydrogen, followed by solvolysis of the benzoate and ring-closure, gives the silylated derivative of (**4.104**).

While TX-A$_2$ proved relatively difficult to synthesize, a number of close structural analogues were prepared, mostly by lengthy routes. One preparation, of 11,12-methylene-TX-A$_2$, may serve as an illustration (Fig. 4.21). Addition of cyanide to the PG-A$_2$ derivative (**4.107**) using 18-crown-6, followed by lithium aluminium hydride reduction, generated (**4.108**), which on treatment with nitrous acid undergoes ring-expansion to the cyclohexenone (**4.109**). Oxidation and re-esterification of the primary alcoholic side-chain of (**4.109**), with stereospecific reduction of the keto-group and hydration of the double bond, afford (**4.110**), which is converted to the β-trifluoromethanesulphonate (sulphonation of the less hindered hydroxy group) on treatment with trifluoromethanesulphonic ('triflic') anhydride. Lithium hydroxide in aqueous THF then both hydrolyses the ester and effects ring closure to the oxetane and desired analogue (**4.111**).

4.8 Leukotrienes

As noted above, non-steroidal anti-inflammatory drugs such as aspirin and indomethacin exert their activity by inhibition of the cyclo-oxygenase enzyme responsible for the formation of prostaglandins and thromboxanes from polyunsaturated fatty acids, while steroidal anti-inflammatory drugs inhibit the release of arachidonate from its storage phospholipids. Since these two classes of drugs displayed significantly different anti-inflammatory effects, it was speculated that metabolism of arachidonate might also lead to the formation of materials, by a pathway not involving cyclo-oxygenase, which were likewise mediators of inflammation. Accordingly, Samuelsson and his colleagues studied the metabolism of arachidonic acid in polymorphonuclear leukocytes (PMNL) from the peritoneal cavity of rabbits, and in 1976 they reported that the major metabolite was a product of a lipoxygenase pathway (again, cf. Fig. 4.6), namely 5(S)-5-hydroxy(6E,8Z,11Z,14Z)-eicosatetra-enoic acid (5-HETE) (**4.112**) (Fig. 4.22). In addition, more polar products

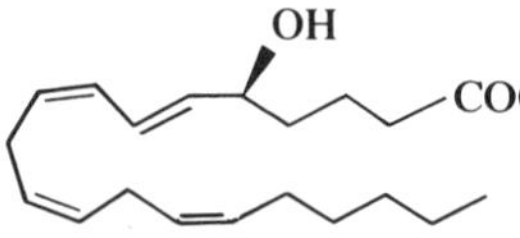

(4.112)

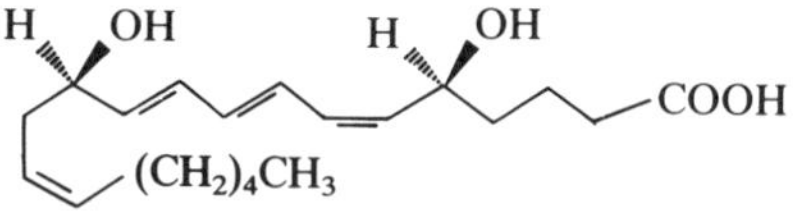

(4.113)

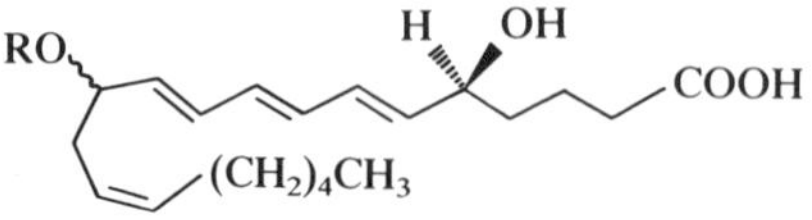

(4.114) R = H
(4.116) R = Me or Et

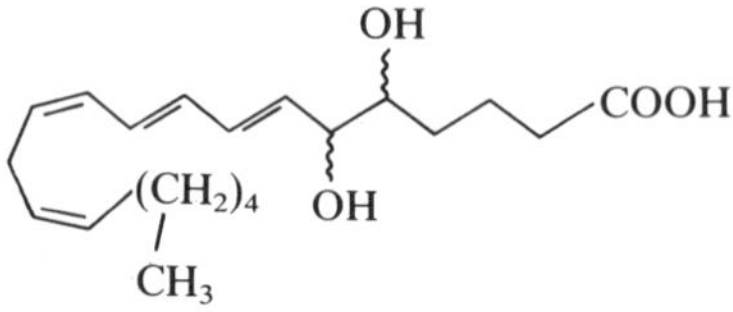

(4.115)

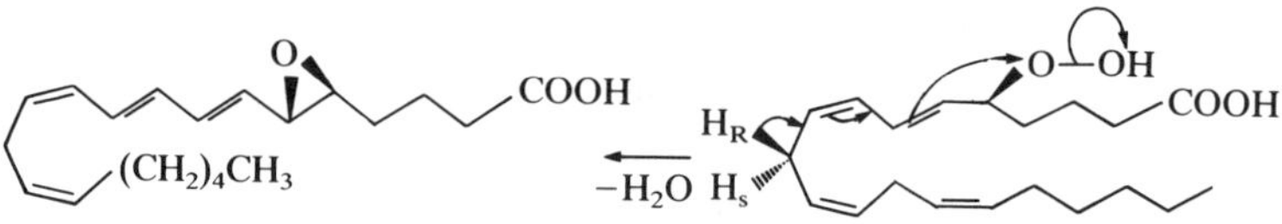

(4.117) **(4.118)**

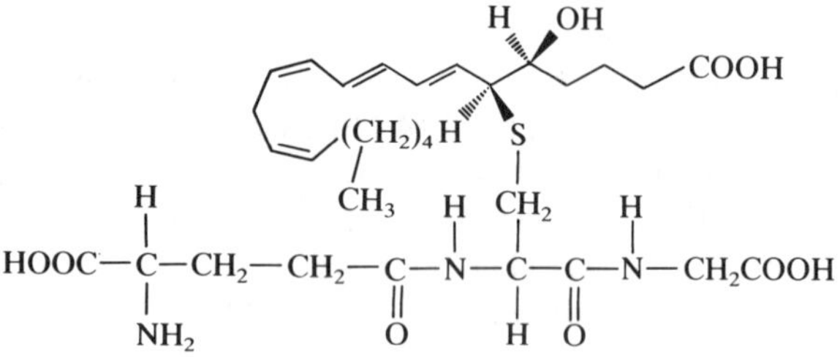

(4.119)

Fig. 4.22.

were present, which were subsequently identified as (5*S*,12*R*)-5,12-dihydroxy-(6*Z*,8*E*,10*E*,14*Z*)-eicosatetraenoic acid (now called leukotriene B_4, abbreviated LT-B_4) (**4.113**), two (5*S*)-5,12-dihydroxy-(6*E*,8*E*,10*E*,14*Z*)-eicosatetraenoic acids which were epimeric at C-12 (**4.114**), and two isomeric 5,6-dihydroxy-(7*E*,9*E*,11*Z*,14*Z*)-eicosatetraenoic acids (**4.115**). Labelling experiments using oxygen-18 established that the oxygen atom introduced at C-5 of (**4.113**) was derived from atmospheric oxygen, while that at C-12 was derived from water. It was hypothesized that this array of products arose via formation of an unstable intermediate which underwent rapid nucleophilic attack. On incubation of rabbit PMNL with arachidonic acid for 30 seconds, followed by addition of excess methanol or ethanol, or hydrochloric acid, the 15-methoxy- or ethoxy-derivatives (**4.116**) were formed in addition to (**4.112**)–(**4.115**). Thus the unstable intermediate could undergo rapid nucleophilic attack by alcohols. Analysis of the mixture of products obtained after various time intervals by reverse phase HPLC suggested that, since the concentrations of (**4.114**)–(**4.116**) increased with time, while those of (**4.112**) and (**4.113**) remained constant, the former arose by solvolytic attack on the intermediate, while (**4.113**) was formed by enzymic hydrolysis. The intermediate was therefore proposed as being $(5S)$-5,6-epoxy-$(7E,9E,11Z,14Z)$-eicosatetraenoic acid (leukotriene A_4, or LT-A_4) (**4.117**), an allylic epoxide. This arises as follows: the action of 5-lipoxygenase on arachidonic acid generates $(5S)$-5-hydroperoxy-$(6E,8Z,11Z,14Z)$-eicosatetraenoic acid (5-HPETE) (**4.118**) which undergoes loss of water, under the action of a dehydrase, to form (**4.117**). Labelling experiments have shown that in human leukocytes the *pro-R* hydrogen atom at C-10 is abstracted stereospecifically during this process. Stereospecific hydrolysis of (**4.117**) by a hydrolase then generates (**4.113**) (LT-B_4; $5S,12R$-DHETE), while non-enzymatic hydrolysis of (**4.117**) leads, possibly via an allylic carbonium ion intermediate, to (**4.114**), (**4.115**), or (**4.116**), depending on the nucleophiles present.

Leukotriene A_4 is, in turn, the progenitor of other biologically active substances. Following treatment with cobra venom, a smooth muscle contracting factor appears in the perfusate of guinea pig lung, for which the name 'slow reacting substance' (SRS) was coined as long ago as 1938. When released immunologically, as for instance following interaction between receptor-bound immunoglobulin E (IgE, a class of antibodies associated with allergic reactions) and an antigen such as pollen, SRS is referred to as SRS-A (slow reacting substance of anaphylaxis). The substances making up SRS are mediators of immediate hypersensitivity reactions such as asthma. Structural studies on SRS were initially limited by the very tiny quantities of the purified substances available, but labelling studies showed that arachidonic acid and cysteine were incorporated into the structure, and the UV spectrum closely resembled that of (**4.115**) but with the maximum shifted 10 nm to longer wavelength, consistent with a sulphur atom situated α- to a conjugated triene. Desulphurization on Raney nickel afforded 5-hydroxyeicosaenoic acid, suggesting linking of cysteine to the arachidonate-derived species via a thioether bond. However, the other product was not alanine: in fact, degradation of SRS by hydrochloric acid released the tripeptide glutathione (γ-glutamylcysteinylglycine). Treatment of SRS with a 15-lipoxygenase from soybeans led to formation of a 15-hydroperoxide in which the chromophore of a tetraene was not present, and since the transformation of *cis*-divinyl methane to *cis,trans*-diene hydroperoxide (as in Fig. 4.6) was well defined for this enzyme, the position of the triene chromophore was established, and the structure of SRS proposed, and subsequently confirmed by total synthesis, as $(5S,6R)$-6-S-glutathionyl-5-hydroxy-$(7E,9E,11Z,14Z)$-eicosatetraenoic acid (**4.119**), called leukotriene C_4 (LT-C_4). Leukotriene D_4, formed by incubation of LT-C_4 with γ-glutamyl transpeptidase, has cysteinylglycine attached at C-6 in thioether linkage, the glutamyl residue having been lost, and leukotriene E_4, formed from LT-D_4

by the actions of a dipeptidase, contains only cysteine attached similarly at C-6. However, LT-E_4 can act as an acceptor of a γ-glutamyl residue to form the γ-glutamylcysteine derivative, LT-F_4. The substance SRS-A is, in fact, a mixture of LT-C_4, LT-D_4, and LT-E_4. Attack of the sulphydryl group of glutathione at C-6 of LT-A_4 catalysed by glutathione S-transferase generates LT-C_4 *in vivo*.

The name 'leukotrienes' was coined because the substances were first recognized in leukocytes and contained a conjugated triene. However, these species can arise not only from arachidonic acid, but in general from C_{20} fatty acids possessing the $5Z,8Z,11Z$-triene system with no, or more than one, double bonds at or beyond C-14. Hence LT-A_3, LT-A_4, and LT-A_5 are known, all based on $5(S)$-5,6-epoxy-$(7E,9E,11Z)$-eicosaenoic acid (LT-A_3). As we have seen, LT-A_4 possesses, in addition, a $14(Z)$-double bond, and LT-A_5, formed from eicosapentaenoic acid, has $14(Z)$ and $17(Z)$ double bonds. Extrapolation to the structures of the LT-B_n, LT-C_n, LT-D_n, and LT-E_n ($n = 3$–5) series is obvious.

The pathway described above is, however, by no means the only route by which leukotrienes are formed. Human leukocytes contain 12-lipoxygenase and 15-lipoxygenase, which use arachidonic acid as substrate to form 12-HPETE and 15-HPETE, respectively, each of which possesses S chirality. Similar investigations to those described above indicate that these form 11,12-oxido and 14,15-oxido species (11,12-LT-A_4 and 14,15-LT-A_4) respectively, in analogy to formation of the 5,6-oxido species (**4.117**), and that these in turn give the dihydroxy species 5,12-LT-B_4, and 8,15-LT-B_4 and 14,15-LT-B_4 respectively, again in analogy to the formation of (**4.113**) and (**4.115**) from (**4.117**). Nor is this the end of the story: other lipoxygenases can introduce the hydroperoxy group at positions 8, 9, and 11 of arachidonic acid. New metabolites of this species are still being found, as shown by the discovery of the trihydroxytetraenes, also formed in human leukocytes, and the lipoxins, and it seems likely to be some while before the full range of products of arachidonate metabolism has been described and thoroughly characterized.

4.8.1   *Biological effects of the leukotrienes*

As indicated above, the leukotrienes are mediators of allergic responses and inflammation. The cysteine-containing leukotrienes LT-C_4, LT-D_4, and LT-E_4 are potent bronchoconstrictors and vasoconstrictors, although, for instance, the bronchoconstrictor activity of LT-C_4 may, in fact, be due in some cases to stimulation of the release of TX-A_2. LT-C_4, LT-D_4, and LT-E_4 cause the walls of microcapillary blood vessels to become permeable, causing plasma leakage, at much lower concentrations than histamine. LT-B_4 appears to have powerful chemotactic stimulant properties, causing leukocyte adhesion to endothelium, and may mediate the migration of leukocytes from the blood areas of inflammation.

The biological effects of the leukotrienes, and other metabolites of arachidonic acid, continue to be researched, and increasing knowledge of the cellular responses elicited by these agents offers the possibility of developing therapeutic agents which act either as enzyme inhibitors, to control arachidonate metabolism, or as antagonists to control or counter the response of the organism to its products.

4.8.2 Leukotriene synthesis

The first synthesis of ($\pm$) LT-A$_4$ (**4.117**), by Corey's group in 1978, used a Wittig reaction to construct the TBDMS derivative of (2E,4E,6Z,9Z)-pentadecatetraen-1-ol, which was converted to the sulphonium ylid (**4.120**) (Fig. 4.23), which in turn condensed with methyl-5-oxopentanoate to afford

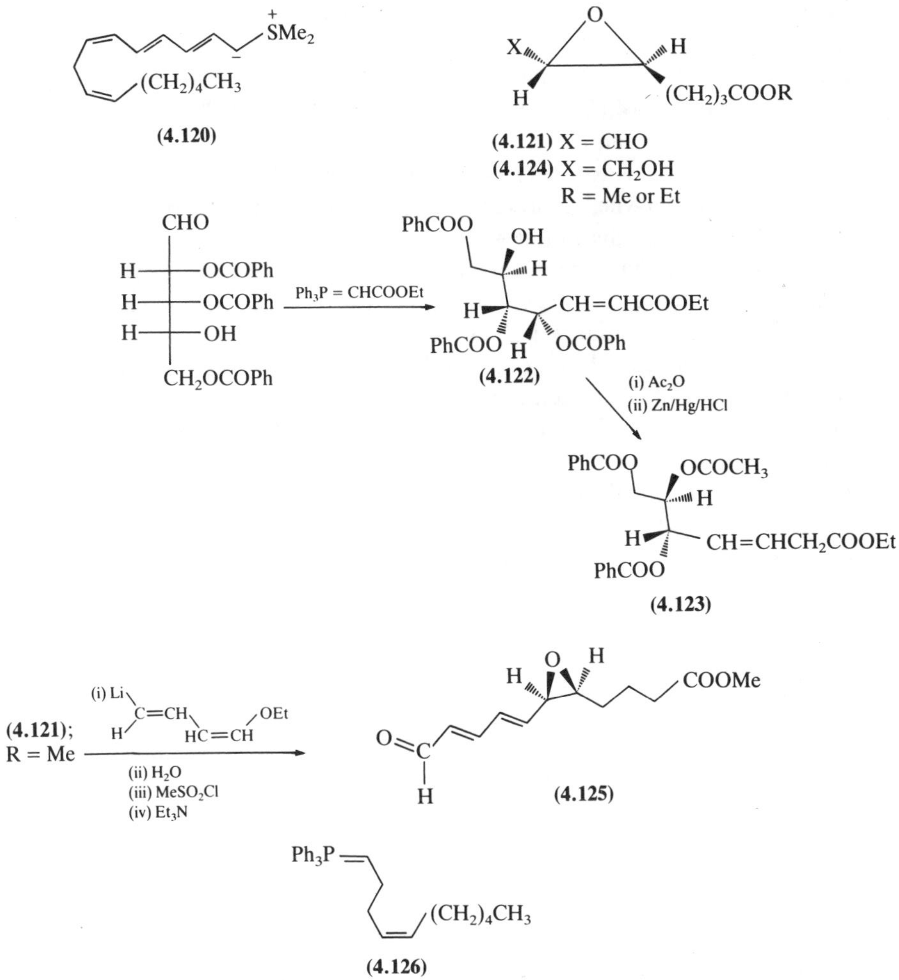

Fig. 4.23.

the methyl ester of the racemic epoxide. Since such a method, involving ylid addition to an aldehyde unit, does not afford the natural enantiomer selectively, subsequent syntheses have tended to involve preparation of the epoxy-ester (**4.121**) from a suitable chiral starting material, and to use this as the key synthon. These chiral precursors have usually been carbohydrate in nature. Thus, 2,3,5-tri-*O*-benzoyl-D-(−)-ribose upon reaction with ethoxycarbonylmethylenetriphenylphosphorane affords a mixture of the *E* and *Z* isomers of (**4.122**), which upon acetylation and reduction with zinc amalgam in ether saturated with HCl gives the β,γ-unsaturated ester (**4.123**), again as an *E/Z* mixture. Catalytic reduction of (**4.123**), followed by deacetylation, tosylation, and debenzoylation with potassium carbonate with subsequent ring-closure with expulsion of the tosyl group, gives the *trans*-epoxide (**4.124**: R = Et), which is oxidized to (**4.121**) with Collins' reagent. The stereochemistry at C-5–C-6 of LT-A$_4$ is thus established. Then, treatment of (**4.121**) with 1-lithio-4-ethoxybutadiene, followed by aqueous quenching of the reaction, mesylation of the secondary alcohol formed and elimination with triethylamine, gives the epoxydienal (**4.125**) which, upon condensation with (**4.126**), gives the methyl ester of LT-A$_4$. This is easily hydrolysed by cold aqueous base to give LT-A$_4$. An efficient alternative route from (**4.121**) to (**4.125**) involves condensation with 1-lithio-4-methoxybut-3-en-1-yne, followed by reduction with Lindlar catalyst, mesylation, and elimination in weak aqueous base.

Glyceraldehyde has also been used, as its acetonide, to prepare (**4.121**), although the route is rather complex. 2-Deoxy-D-ribose, however, is a versatile starting material (Fig. 4.24), and on treatment with ethoxycarbonyl-methylidene triphenylphosphorane followed by catalytic reduction affords the triol ester (**4.127**), which is not only useful as a source of (**4.121**), but may also be converted to the 5*R*,6*S*-, 5*S*,6*S*-, and 5*R*,6*R*-stereoisomers of (**4.121**) and thus, eventually, to the corresponding stereoisomers of LT-A$_4$. If, however, treatment with the same triphenylphosphorane is followed by tosylation, (**4.128**) is obtained, and subsequent treatment with lithium diisopropylamide (LDA) at low temperature removes a proton α- to the ethoxycarbonyl group to open the furanose ring, the 6-alkoxide thus generated displacing tosylate to give epoxide (**4.129**). After catalytic reduction of the double bond, treatment with methoxide results in rearrangement to (**4.124**). Note that the 3-*O*-tosylated isomer of (**4.128**), treated first with LDA and then reduced catalytically, affords the enantiomer of (**4.124**).

The recent development of reagents which perform enantioselective epoxidation has also facilitated the synthesis of compounds such as (**4.124**). For instance, oxidation of the alcohols (**4.130**) by Sharpless's method, using *t*-butyl hydroperoxide, L(+)-diethyl tartrate, and titanium tetraisopropoxide, affords the epoxides (**4.131**), which are easily converted to (**4.124**). While many syntheses of the leukotrienes have utilized Wittig reagents, other organometallic reagents such as the lithium alkenides and lithium vinyl cuprates have recently been used to good effect. An example is lithiation of the vinyl iodide (**4.132**) to afford (**4.133**) which condenses with (**4.121**) to

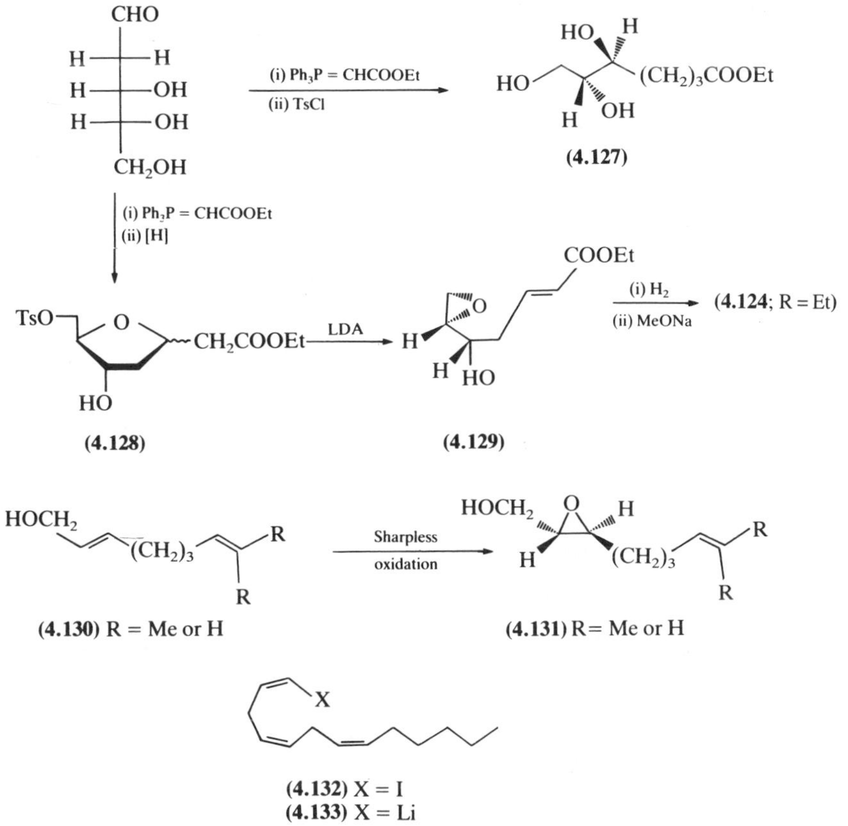

Fig. 4.24.

give a diastereoisomeric mixture of secondary alcohols, which, upon mesylation and treatment with DBU, undergo stereospecific elimination to give the methyl ester of LT-A_4.

The original preparation of LT-B_4 (Fig. 4.25) was of particular importance since it established unequivocally the configuration of the double bonds, a feature which previous structural studies had been unable to resolve. 2-Deoxy-D-ribose was converted to its acetonide and then to (**4.134**) by a Wittig reaction, followed by reduction. Tosylation of (**4.134**), followed by removal of the protecting group, closure with base to form the epoxide, and benzoylation, then gave (**4.135**), which upon hydrolytic ring-opening of the epoxide and oxidation with lead tetra-acetate gave (**4.136**; R = Me). Upon Wittig reaction with (**4.137**; R = H), hydrolysis of the product afforded LT-B_4 (**4.113**) and 6E-LT-B_4. The original synthesis of (**4.137**; R = H) as described by Corey is somewhat tortuous, but an elegant route was subsequently developed by Rokach, another major contributor to this field. Silylation of (**4.128**), followed by displacement of tosylate with iodide, affords the corresponding iodomethyl derivative, which reacts with the appropriate

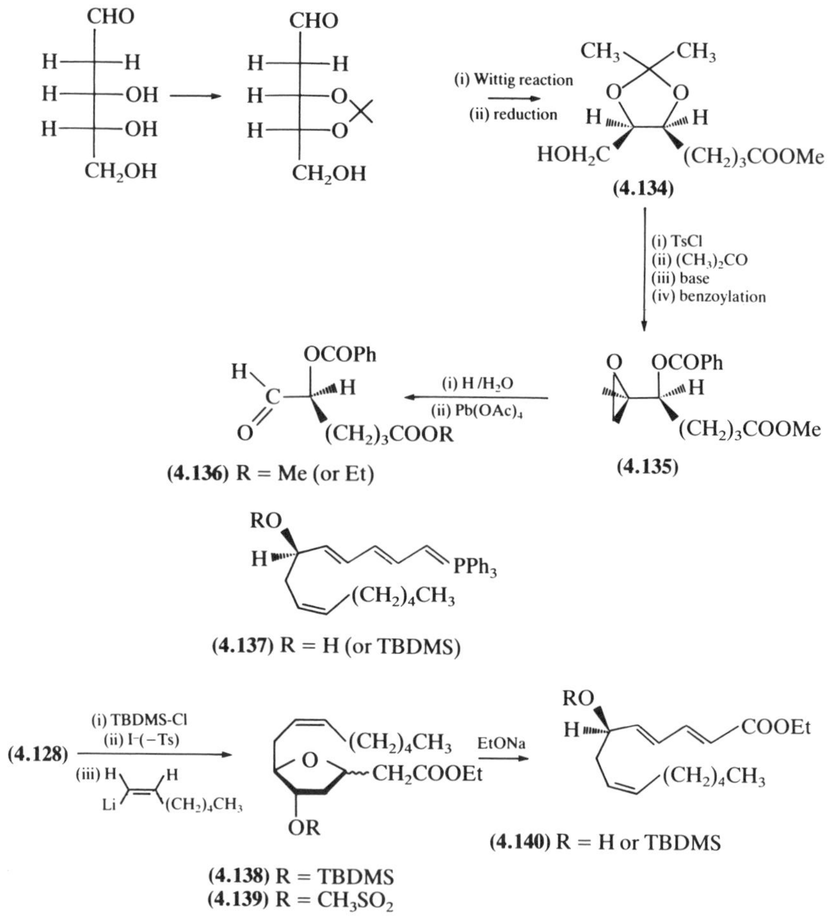

Fig. 4.25.

lithium vinyl cuprate to give (**4.138**). Desilylation of (**4.138**) using tetra-*n*-butylammonium fluoride, followed by mesylation, gives (**4.139**). This species is, in effect, a masked diene: treatment with ethoxide affords (**4.140**; R = H) which is easily converted to (**4.137**; R = H).

Reference was made earlier in this chapter to the convenient way in which Wittig reaction conditions can be exploited to achieve the stereochemistry desired. A synthesis of LT-B$_4$ provides a nice example (Fig. 4.26). The aldehyde (**4.141**), prepared from D-mannitol, was condensed with formyl-methylenetriphenylphosphorane in benzene, giving the *trans* product (**4.142**), which in turn underwent Wittig–Emmons–Horner reaction with diethyl ethoxycarbonylmethylenephosphonate in acetonitrile in the presence of DBU to afford the correspondingly silylated (**4.140**): again, the *trans* product. This was converted to the Wittig reagent (**4.137**; R = TBDMS) by standard methods, after which condensation with (**4.136**; R = Et) (also prepared from D-mannitol) in hexamethylphosphorous triamide (HMPT) afforded (**4.143**),

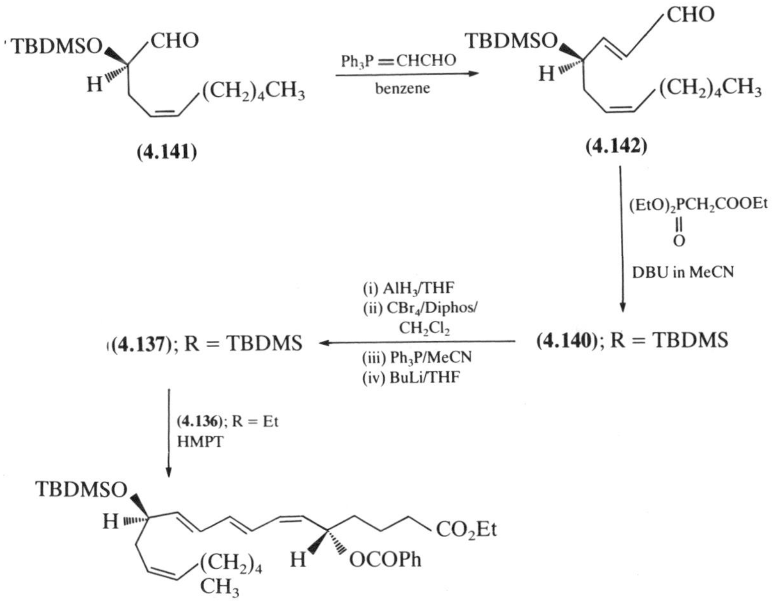

(4.141) (4.142) (4.137); R = TBDMS (4.140); R = TBDMS (4.143)

Fig. 4.26.

the 14-*cis* product predominating over *trans* in 7:3 ratio. Standard desilylation and deprotection procedures then afforded LT-B_4 (**4.113**).

A number of other syntheses have used Lindlar reduction of polyynes to achieve the desired double bond stereochemistry in leukotriene syntheses. For instance (Fig. 4.27), Lindlar reduction of undeca-2,4-di-yn-1-ol (prepared from propargyl alcohol in three steps), followed by Sharpless epoxidation, gives (**4.144**), which after Collins' oxidation and a Wittig reaction gives (**4.145**). Treatment of (**4.145**) with hydrogen bromide gives (**4.146**), which on reaction with triphenylphosphine and proton abstraction gives Wittig reagent (**4.137**; R = H).

Leukotrienes of the C, D, and E series have generally been obtained by treating the corresponding leukotriene A methyl ester with glutathione and triethylamine in methanol (for the C series) or with *N*-trifluoroacetyl glutathione dimethyl ester or the correspondingly protected L-cysteinyl glycine or L-cysteine, again with triethylamine in methanol, for the C, D, and E series respectively.

The precursors to the leukotrienes have also been synthesized: iodolactonization of arachidonic acid itself affords (**4.147**), which loses hydrogen iodide on treatment with DBU to give the conjugated diene (**4.148**). Treatment of (**4.148**) with methanolic triethylamine gives the methyl ester of racemic (**4.112**), (±)5-HETE, and subsequent mesylation of this ester, followed by treatment with hydrogen peroxide, then affords the methyl ester of racemic (**4.118**), (±)5-HPETE. The methyl esters may be hydrolysed

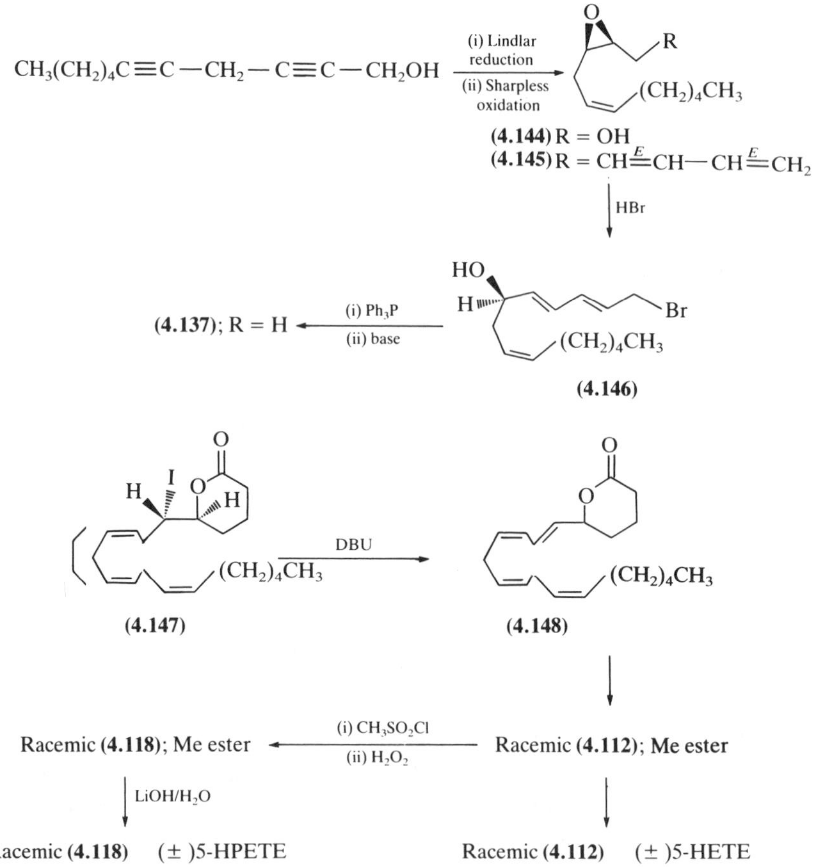

Fig. 4.27.

with lithium hydroxide at room temperature to give (±)5-HETE and (±)5-HPETE, respectively. In a biomimetic synthesis, the methyl ester of (**4.118**) has been treated with 1,2,2,6,6-pentamethylpiperidine in the presence of triflic anhydride at low temperature to give the methyl ester of (**4.117**) in moderate yield, together with other products.

The few examples of leukotriene syntheses instanced here have of necessity been highly selective and many others could have been described, but the key reactions, such as Wittig and Emmons–Horner condensations, Sharpless epoxidation, stereoselective reduction, and the use of Lindlar catalyst on alkyne intermediates, which recur constantly have all been exemplified above. This area of research has been very active for a considerable time and new eicosanoid natural products continue to be described. The finding that certain arachidonate analogues, such as the acetylenes eicosatetra-5,8,11,14-ynoic acid and 5,6-dehydroarachidonic acid, and the allene 4,5-dehydroarachidonic acid, are powerful inhibitors of the leukotriene pathway offers hope for the development of agents which can control some allergic and inflammatory reactions and is spurring much research to this end.

Further reading

1. Fatty acids, etc.: W.W. Christie in *Topics in Lipid Chemistry* (ed. F.D. Gunstone), vol. 1, Logos Press, London, 1970; W.W. Christie, *Lipid Analysis*, Pergamon Press, Oxford, 1973; D.G. Bishop and P.K. Stumpf in *Biochemistry and Methodology of Lipids* (eds A.R. Johnson and J.B. Davenport), Wiley-Interscience, New York, 1971; N.A. Porter, 'Mechanisms for the autoxidation of unsaturated lipids', *Acc. Chem. Res.*, 1986, **19**, 262.
2. Prostaglandins and thromboxanes: R. F. Newton and S. M. Roberts (eds), *Prostaglandins and Thromboxanes: An Introductory Text*, Butterworth Press, Sevenoaks, 1982 ; J.S. Bindra and R. Bindra, *Prostaglandin Synthesis*, Academic Press, London, 1977; A. Mitra, *The Synthesis of Prostaglandin Derivatives*, Wiley, Chichester, 1978; S. M. Roberts and F. Scheinmann, *Recent Synthetic Routes to Prostaglandins and Thromboxanes*, Academic Press, London, 1982; M.P.L. Caton, *Tetrahedron*, 1979, **35**, 2705; R.F. Newton and S.M. Roberts, *Tetrahedron*, 1980, **36**, 2163; R.F. Newton and S.M. Roberts, *Annual Reports of the Royal Society of Chemistry*, 1981, **78B**, 347; W. Bartmann and G. Beck, *Angew. Chem. Int. Ed. Engl.*, 1982, **21**, 751; R. Noyori and M. Suzuki, *Angew. Chem. Int. Ed. Engl.*, 1984, **23**, 847; J.E. Pike and D.R. Morton, *Chemistry of Prostaglandins and Leukotrienes*, Raven Press, New York, 1985.
3. For some recent elegant and short routes to prostaglandins see: M. Suzuki, A. Yanagisawa and R. Noyori, *J. Amer. Chem. Soc.*, 1985, **107**, 3348; E.J. Corey *et al.*, *Tetrahedron Lett.*, 1986, **27**, 2199; S.S. Bhagwat, P.R. Hamann, and W.C. Still, The long elusive thromboxane A_2 has also been prepared, *J. Amer. Chem. Soc.*, 1985, **107**, 6372.
4. Leukotrienes and related compounds: B. Samuelsson, *Angew. Chem. Int. Ed. Engl.*, 1982, **21**, 902; B. Samuelsson, *Science*, 1983, **220**, 568; R.H. Green and P.F. Lambeth, *Tetrahedron*, 1983, **39**, 1687; J. Ackroyd and F. Scheinmann, *Chem. Soc. Rev.*, 1982, **321**; E.J. Corey, *Experientia*, 1982, **38**, 1259; J. Rokach and J. Adams, *Acc. Chem. Res.*, 1985, **18**, 87.
5. Among more recently identified classes of eicosanoids, the lipoxins have been described by Samuelsson and his group. See, for instance: *Proc. Natl. Acad. Sci. USA*, 1984, **81**, 5335; 1986, **83**, 1983, and *J. Biol. Chem.*, 1986, **261**, 16340. Further references relating to these, to the hepoxilins and trioxilins (formed when 12-(S)-hydroxyperoxy-5(Z),8(Z),10(E),14(Z),17(Z)-eicosapentaenoic acid (12-(S)-HPEP) is incubated with rat pancreatic extracts), and to the clavulones, punaglandins, and chlorovulones (prostaglandin-related species found in marine organisms) may be found in P. Barraclough, *Annual Reports of the Royal Society of Chemistry*, 1986, **83B**, 331.
6. For a recent survey of the medicinal properties and importance of arachidonate-derived metabolites, see G.A. Higgs, E.A. Higgs, and S. Moncada, in *Comprehensive Medicinal Chemistry*, vol. 2: 'Enzymes and Other Molecular Targets' (eds P.G. Sammes and J.B. Taylor; ser. ed. C. Hansch), Pergamon Press, Oxford, 1990, chapter 6, section 6.2.

5 Terpenoids

D. V. Banthorpe

5.1 Introduction

The diverse, widespread, and exceedingly numerous family of natural products constructed from five carbon building-units and so comprising compounds with $C_5, C_{10}, C_{15}, C_{20}, \ldots, C_{40}$ skeletons (together with a few higher members) are synonymously termed terpenoids, terpenes, or isoprenoids, with the important subgroup of steroids sometimes singled out as a class in its own right. There is no agreement on the basic nomenclature and the various subgroups are often given the -oid or -ene suffixes interchangeably, e.g. monoterpenoids ≡ monoterpenes. Here we shall use the -oid suffix for both general class and subgroups: this is the logical system, cf. alkaloids, flavonoids. The -ene suffix should be restricted to the alkenes of the family.

These compounds are typically found in all parts (i.e. seed, flowers, foliage, roots, wood) of higher plants and also occur in mosses, liverworts, algae, and lichens, although some are of insect or microbial origin. Steroids are widespread in both animal and plant kingdoms and also in many microorganisms. Members of the class, as components of oils or in extracts, have been used since antiquity as ingredients of flavours, preservatives, perfumes, medicines, narcotics, soaps, and pigments. Camphor (**5.1**),* easily

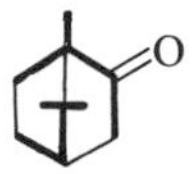

(5.1)

obtainable virtually pure as a crystalline and very aromatic solid, was introduced into Europe from the Orient in the 11th century AD, and by the late Middle Ages numerous essential oils (i.e. essences) from common herbs such as lavender, rosemary, thyme, and wormwood were used as perfumes

---

* Many terpenoids occur naturally in both enantiomeric forms. In this chapter, the absolute configuration is sometimes not specified, but when it is it corresponds to the more common natural isomer. The sign of the specific optical rotation is only quoted when one isomer has some particular biological property.

and in folk medicine. Consequently, the lower terpenoids (C_{10} and C_{15} compounds) especially have been the subject of study since the dawn of modern chemistry. Many mono- and a few sesqui- and diterpenoids (C_{10}, C_{15}, C_{20} compounds respectively) were isolated and studied in the last century and by 1887 Wallach could propose an 'isoprene rule' that the monoterpenoids were hypothetically constructed by the linkage of isoprene (**5.2**; 2-methylbuta-1,3-diene) units. By *c.* 1894, the structures of camphor and α-pinene (**5.3**; obtained from turpentine, hence the derivation of the name 'terpene') had been elucidated. In these, the isoprene units were found to be linked 'head-to-tail' (bars in **5.1** and **5.3**) and this orientation was later shown to be general. Citral (**5.4**) was synthesized in 1896 and camphor in 1904. The sesquiterpenoid α-santonin (**5.5**), although previously studied for nearly a century, did not yield its structure until 1929, and details of its complex photochemistry were not elucidated until *c.* 1957. The steroids (degraded C_{30} compounds) had also been long known although their suspected relationship with the other terpenoids was not demonstrated until the late 1950s. The carotenoids (C_{40} compounds; strictly tetraterpenoids, but historically named as the former) which occur as plant pigments were also studied in the last century, although again their kinship with the lower terpenoids was not appreciated until much later.

CHO

(5.2)   (5.3)   (5.4)   (5.5)

Slow progress was made on the characterization and correlation of these classes until the period following 1950 when the general application of chromatographic techniques of separation (LC, GC, TLC, HPLC) and of spectroscopic methods of structure determination (IR, UV, NMR, MS), and the advent of commercially-available radioisotopes of carbon and hydrogen led to an ever-accelerating increase in knowledge of structural types, biosynthesis, and ultimately of biological significance that has continued to the present. Over the past 60 years, terpenoids have also provided substrates for numerous studies on the fundamentals of mechanistic organic chemistry; e.g. on the nature of carbocations and their rearrangements, on the relationships between structure and colour, on conformational analysis of the reactivity of cyclic systems, and on the Woodward–Hoffmann rules for correlation of orbital symmetries during organic reactions. At a more homely level they have become articles of commerce in the perfumery and food industries as being the basis of cosmetics, soaps, flavours, and pigments as well as being used in disinfectants, detergents, and in many medical preparations and vitamin supplements. Consequently, although some terpenoids of commerce are still isolated from plant sources, many such compounds are now synthesized industrially on a large scale. Such ubiquitous compounds would be expected to play significant roles within the organisms producing them, and indeed, over the past 25 years in particular terpenoids have been implicated in such diverse functions as being mammalian sex hormones, insect pheromones, plant-growth substances, natural insecticides, defensive secretions of fish, ancillary pigments in photosynthesis, and as receptors in visual processes in animals. The occurrence of such compounds has been the basis of taxonomic classifications of plants and microorganisms; and their implication in plant–insect and plant–plant interactions and the like has become one of the foundations of the burgeoning science of chemical ecology.

A great number of naturally occurring terpenoids (over 15 000) have been isolated and characterized with varying degrees of rigour and dozens more are reported weekly: many of these compounds have been functionalized or derivatized in the laboratory in the hope of preparing compounds with enhanced or desirable biological properties. The prolificity of nature is illustrated by the occurrence of at least 38 distinct skeletal types of monoterpenoids increasing to over 200 for the sesquiterpenoids, and by the discovery of over 500 individual members of the iridoids – a mere subgroup of the monoterpenoids! It will be appreciated that this chapter can present but a brief summary of the chemistry and significance of few (but nevertheless important and representative) members of the family.

Like all chemical families, the terpenoids have a systematic nomenclature that is used in *Chemical Abstracts* and in retrieval and data banks. Thus α-pinene (**5.3**) is a derivative of pinane (numbered as in **5.6**) or more rigorously of the bicyclo-[3.1.1]-heptane skeleton. However, trivial or at best semi-systematic names are invariably used colloquially and in research publications and these are often based on the plant genus or species used as source. Thus α-pinene is isolated from *Pinus* (≡ pine) species and isothujone (**5.7**) from the *Thuja* genus of conifers.

(**5.6**)   (**5.7**)

5.2 General routes of biogenesis

The subgroups of terpenoids are listed in Table 5.1. As the number of known individual terpenoids increased, Wallach's modified isoprene rule (in the head-to-tail form) was found not always to be obeyed and the structural interrelationships posed baffling problems. The various subclasses and the

Table 5.1 Classes of terpenoids.

C_n	Name	Parent(s)	Occurrence in plants
C_5	Hemiterpenoid	IPP (**5.9**); DMAPP (**5.10**)	(Few) emissions; oils
C_{10}	Monoterpenoid	GPP (**5.11**)	Oils; petals
C_{15}	Sesquiterpenoid	FPP (**5.12**)	Oils; resin; petals
C_{20}	Diterpenoid	GGPP (**5.13**)	Oils; resin; heart-wood
C_{25}	Sesterterpenoid	GFPP (**5.14**)	Oils; resin; heart-wood (rare)
C_{30}	Triterpenoid	Squalene (**5.17**)	Resin; heart-wood; leaf wax
C_{40}	Carotenoid	Phytoene (**5.18**)	All green tissue; roots; petals
C_n ($n = 45$ to 10^5)	Polyisoprenoid	GGPP (**5.13**)	Latex; leaf wax

ever-increasing structural types within each class were finally rationalized by Ruzicka in 1953 in his 'biogenetic isoprene rule', which although based on the fragmentary knowledge of biosynthesis available at the time proved to be brilliantly correct in conception. This unifying principle stressed that each member of a terpenoid subgroup was derived from a single parent compound that was unique to that group, and that the various parents were related in a simple homologous fashion. For example, the plethora of sesquiterpenoids were all derived from 2*E*,6*E*-farnesyl pyrophosphate (**5.12**; FPP) by a sequence of straightforward cyclizations, functionalizations, and sometimes rearrangements of types that were well known (in simpler examples) from mechanistic organic chemistry. All these steps were correctly presumed to be enzymatically mediated and Ruzicka identified the parents and defined many of the branching routes from them to known products by mechanistically reasonable pathways. This seminal work emphasized that the understanding of terpenoid (and indeed of natural product) chemistry would only be achieved on biogenetic principles and this approach has continued to the present day and forms the basis of subsequent discussion in this chapter.

In present-day terms, Ruzicka's ideas can conveniently be summarized and extended thus:

(i) Mevalonic acid (MVA; $3R(+)$-isomer; **5.8**; Fig. 5.1), a C_6-acyclic compound, is the precursor of all terpenoids.

(ii) The parents of the various subclasses (see Table 5.1; Fig. 5.1) are: hemiterpenoids, isopentenyl pyrophosphate (**5.9**; IPP) and 3,3-dimethylallyl pyrophosphate (**5.10**; DMAPP); monoterpenoids, geranyl pyrophosphate (**5.11**; GPP); sesquiterpenoids, 2*E*,6*E*-farnesyl pyrophosphate (**5.12**; FPP); diterpenoids, 2*E*,6*E*,10*E*-geranylgeranyl

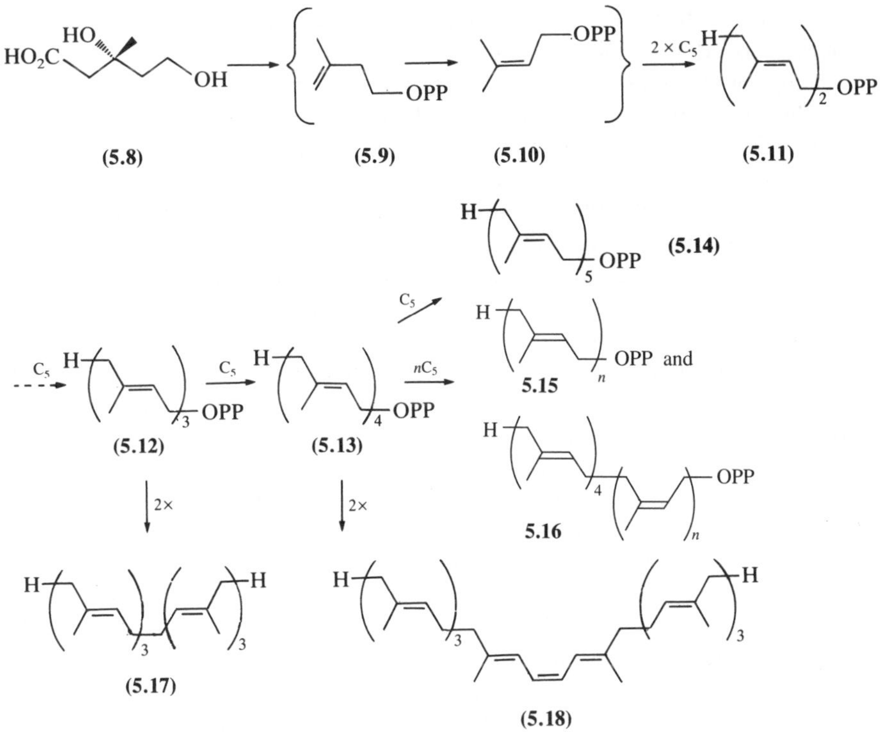

Fig. 5.1. OPP ≡ pyrophosphate.

pyrophosphate (**5.13**; GGPP); sesterterpenoids, 2E,6E,10E,14E-geranylfarnesyl pyrophosphate (**5.14**; GFPP); triterpenoids, squalene (**5.17**); and carotenoids, phytoene (**5.18**).

This implies:

(iii) That the central or 'core' pathway up to C_{25} compounds is formed by sequential addition of C_5-moieties derived from IPP (**5.9**) to a starter unit derived from DMAPP (**5.10**) (for development of this idea, see Section 5.5: note that this subsumes the Wallach isoprene rule).

(iv) That the parents of the C_{30} and C_{40} compounds are formed by reductive coupling of two FPP (i.e. C_{15}-residues) or GGPP (i.e. C_{20}-moieties) respectively. This means that the condensing enzymes have evolved to couple two equivalent units only and that generation of C_{25} or C_{35} compounds by condensation of two sizeable unequal units is not possible.

The rubbers and other polyisoprenoids (e.g. **5.15**; **5.16**), which were not specifically considered by Ruzicka, are constructed by repeated additions of C_5 units to a starter unit, which has recently been characterized as GGPP. The resulting pathway (Fig. 5.1) is present in animals and many microorganisms as a route to the physiologically essential steroids (see Section 5.9) but it is generally only in the plant kingdom that side branches occur from GPP, FPP, GGPP, squalene, and phytoene to yield accumulations of the derived terpenoids, although not all of these latter branch systems are usually significantly displayed in any one plant species.

The core pathway has been amply demonstrated in many living systems, although there is a paucity of direct experimental evidence for the subsequent branches that elaborate the large numbers of individual terpenoids. But there is very little doubt that the overall picture represents, very closely, the situation *in vivo*. Not only would nature have missed a very elegant opportunity if these routes had not evolved, but there is no realistic alternative proposal as to how these compounds could be formed. *In toto*, the Ruzicka proposals and their developments that are outlined in the following sections provide one of the most elegant rationalizations and generalizations of data in the whole of science. As for other classes of natural products, the tools for elucidation of the routes include use of ^{14}C and ^{13}C-labelled precursors (such as MVA) and the location of tracer by degradative or spectroscopic (non-degradative) methods respectively. Use of 1,2-[^{13}C$_2$]-acetate to define the position of the individual acetate units in terpenoids is also applicable, although the general low uptake in studies involving plants has limited this approach. Considerable work has also been carried out on the use of cell-free extracts and purified enzyme systems from plants, microorganisms and tissue cultures in attempts to elucidate the individual biochemical steps. However, these studies have been restricted by the universal presence in plant material of phenolics that denature proteins when cells are disrupted.

The lower terpenoids are widely and evenly distributed across the 94 orders of flowering plants that are generally acknowledged by taxonomists. However, not only are compilations out of date but only a small proportion of the plant kingdom has been screened for secondary metabolites, let alone terpenoids: perhaps only 15 per cent of the estimated half million of higher plants have been examined even to unexacting standards and there is immense scope for detailed studies using modern techniques such as GC–MS, and GC–FT-IR (Fourier transform infrared spectroscopy). Specific terpenoids that are growth-regulators, e.g. abscisic acid (Section 5.6) and the gibberellins (Section 5.7) together with the plant steroids (Section 5.9) and carotenoids (Section 5.11) are probably ubiquitous in higher plants in view of their essential physiological functions.

The subgroups listed in Table 5.1 do not exhaust the possibilities. Numerous degraded terpenoids are known. Thus C$_9$ compounds such as cryptone (**5.19**) can occur in plant oils along with its obvious parent β-phellandrene (**5.20**); the C$_{13}$ compound β-ionone (**5.21**) – a component of violet oil – is similarly a degraded sesquiterpenoid; and carotenoids can be broken down *in vivo* to yield C$_{18}$ compounds based on trisporic acid (**5.22**) that are fungal sex markers. Some gibberellins (Section 5.7) and the vitamins A (Section 5.11) are other examples of important degraded terpenoids. Terpenoids can also be incorporated into molecules of mixed biosynthetic origin to yield so-called meroterpenoids. These compounds will

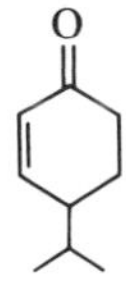

(5.19) 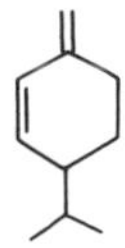   **(5.20)** 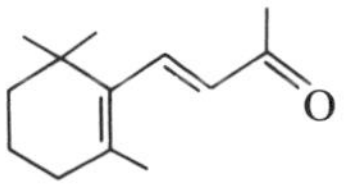   **(5.21)** 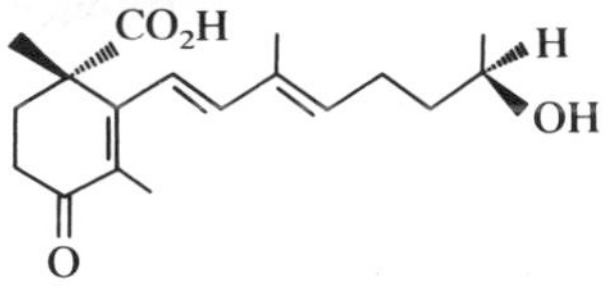   **(5.22)**

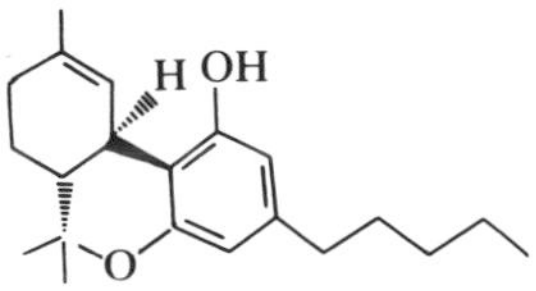

(5.23) 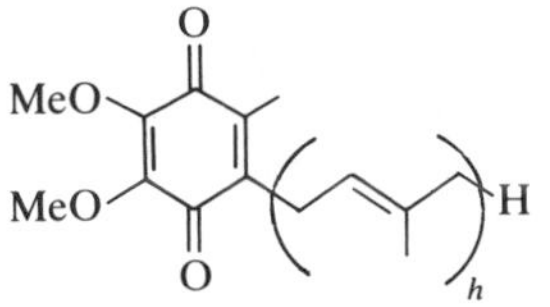   **(5.24)**

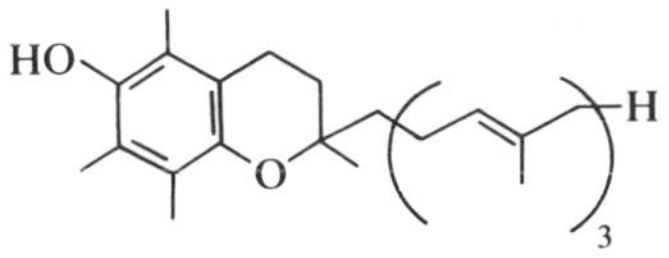

(5.25) 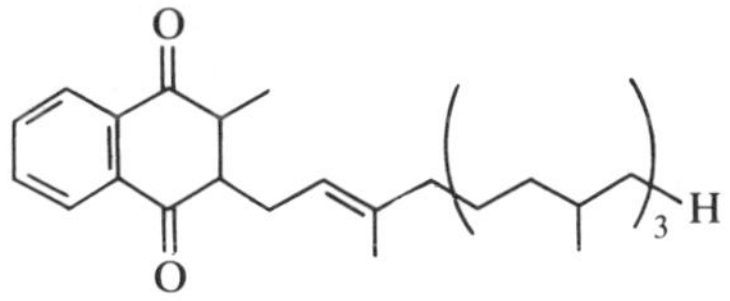   **(5.26)**

not be discussed in this chapter but examples are the incorporation of a C_5 unit (from DMAPP) or a C_{10} unit (from GPP) into terpene alkaloids (see Chapter 7) and the incorporation of the latter unit into cannabinoids (e.g. (−)-Δ'-tetrahydrocannabinol; **5.23**) and related narcotics. Other extremely important meroterpenoids are the ubiquinones (**5.24**; $n = 8$–10) which are vital components of phosphorylation and electron-transfer associated with the respiratory chain; α-tocopherol (**5.25**; vitamin E) which is a free-radical scavenger present in mammalian blood that removes hydroxyl radicals derived from hydrogen peroxide; and vitamin K_1 (**5.26**) which is implicated in blood clotting.

Many normal terpenoids occur *in vivo* as glycosides, usually β-D-glucosides, and the appendage confers a degree of water solubility, and hence the possibility of intra- and inter-cellular transport, as well as acting as a protecting group, and providing a possible 'key' to penetrate membranes to achieve access to metabolic sites. Cleavage of the aglycone may occur during extraction due to acidic reaction conditions or to the presence of endogenous glycosidases, but often the sugar moiety survives extraction and has to be intentionally removed to generate the terpenoid fragment.

5.3   Structure determination

There are no set recipes for the isolation and purification of terpenoids as the members span such a range of molecular masses and chemical types. The classical procedure has been to grind the source tissues after freezing (liquid N_2) or to macerate them in a blender followed by extraction with suitable solvents (Soxhlet), or (for lower terpenoids) by use of steam distillation. With micro-apparatus the latter technique can be performed such that the oil from a small sample of plant material (*c.* 0.5 g) may be concentrated in little (*c.* 2 ml) solvent. A modern, very effective, procedure is to extract with supercritical CO_2 (i.e. above the critical point; T_c 31°C, P_c 73.8 × 10^5 Pa): thus low temperature extraction can be efficiently performed with an inert, easily removable solvent. Particular compounds can then be removed by specific chemical methods; e.g. ketones with Girard reagents. The initial breakthrough in monoterpenoid chemistry followed the demonstration (by Tilden in 1877) that these volatile, difficult-to-separate liquids could often be converted into crystallizable nitrosochlorides. Care must be taken that artefacts are not produced during extraction: in the absence of buffer, the acidity generated during steam distillation may increase to pH 2, and this, for example, can induce the formation of sylvestrene (**5.28**) from car-3-ene (**5.27**). Carotenoids, too, are often light-labile and can dehydrogenate and isomerize, especially if subject to acid conditions in work-up.

(5.27)   (5.28)

Use of standard chromatographic techniques guided by experience and the literature should then lead to isolation of a sample (usually 50–100 mg) of a pure compound suitable for spectral examination. In particular, since the late 1960s the impact of ^1H and then ^{13}C NMR spectroscopy at high fields (200–600 MHz for ^1H) has revolutionized the elucidation of structures. Some 20 years ago, the use of ^{13}C FT–NMR spectroscopy at natural abundance allowed the assignment of the signal for each carbon in a steroid skeleton; and such correlations allow the positions of substitution or unsaturation in the skeleton to be determined by inspection. The sensitivity of the technique is well illustrated by the total analysis of the ^1H FT-NMR spectrum at 300 MHz for vitamin D_3 (*c.* 5 µg) achievable with standard NMR tubes fitted with glass inserts and using a solvent that was highly isotopically enriched (*c.* 99.96 per cent $CDCl_3$). Such achievements and the applications of COSY, INADEQUATE, and n.O.e. techniques and the like are nowadays almost routine and the exercises in ingenuity may be found in any book on NMR spectroscopy and its applications. Such methods are immensely powerful, although it is worth noting that often – especially for the sesqui- and diterpenoids – one skeleton (of the several possible) is implicitly assumed on the basis of the source and the structure of known co-occurring compounds, and the NMR and other spectral data are then hung on this in order to yield a solution. And even today, spectral analysis does not always yield unambiguous results, and recourse has to be made to the ultimate arbiter – X-ray crystallography. It is also worth noting that proposed structures should be consistent with current biogenetic theory. Here we shall outline several classical approaches to structure determinations that made little use of spectroscopic techniques. These were the procedures whereby most of the common compounds and skeletal types were elucidated, and their perusal is not only of historical interest but provides insights into terpenoid chemistry.

Monoterpenoids were initially opened-up at double bonds by ozonolysis or other oxidative procedures followed by reduction and elimination to yield new unsaturated centres for recycling ultimately to generate identifiable products. In a variation, α-terpineol (**5.29**), a component of citrus oils, was converted into lactonic acids (Fig. 5.2) which were characterized by unambiguous synthesis.

The number of rings in a terpenoid (or an organic compound generally) may easily be obtained from the molecular formula (which nowadays would routinely be determined by mass spectrometry). An acyclic alkane has the formula C_nH_{2n+2}: if a terpenoid analyses as C_nH_m, then $(2n + 2 - m)/2$ gives the number of rings plus the number of double bonds as each alkenic group or ring fusion reduces the number of hydrogens by 2. Thus α-pinene (**5.3**) is $C_{10}H_{16}$; and as catalytic hydrogenation reveals one double bond, the molecule must be bicyclic.

A third general approach, much used by Ruzicka in his pioneering studies on sesquiterpenoids in the 1920 period, involved destructive degradation by heating with sulphur or selenium (Vesterberg method). Aromatization occurred and, with due regard for loss or migration of methyl groups at bridgeheads, examination of the product could reveal the skeleton of the

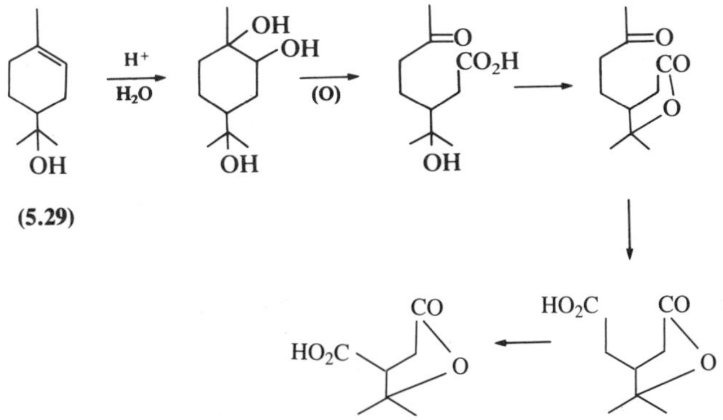

Fig. 5.2.

parent. Thus the sesquiterpenoids (**5.30**) and (**5.31**) yielded tractable aromatics (Fig. 5.3). The location of a ketone group in the original molecule could be 'marked' by methylation (via MeMgBr), when subsequent dehydrogenation would yield a product carrying an extra methyl at the original oxygen site. The method was also often cleverly adapted: a crucial part of the structural proof of guaiol (**5.32**) required conversion into a diketone, followed by intramolecular cyclization and aromatization (Fig. 5.3). Some substrates underwent dehydrogenation without the need for the full treatment. Thus matricin (**5.33**) readily yielded the intensely blue azulene (**5.34**; Fig. 5.4), but sometimes things went wrong: pyrethrosin (**5.35**)

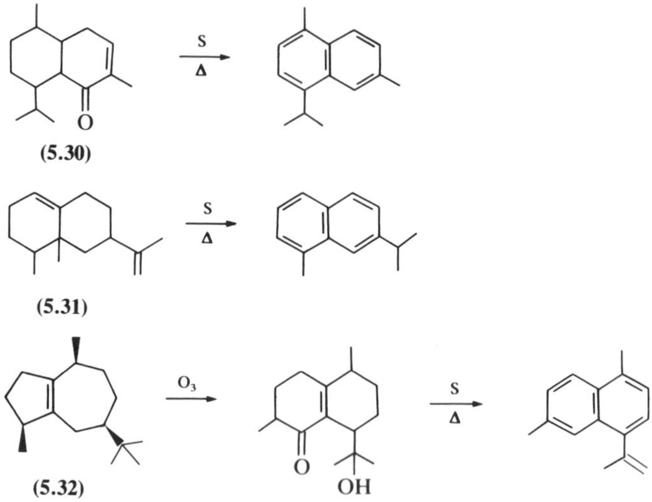

(5.30)

(5.31)

(5.32)

Fig. 5.3.

(5.33)

(5.34)

(5.35)

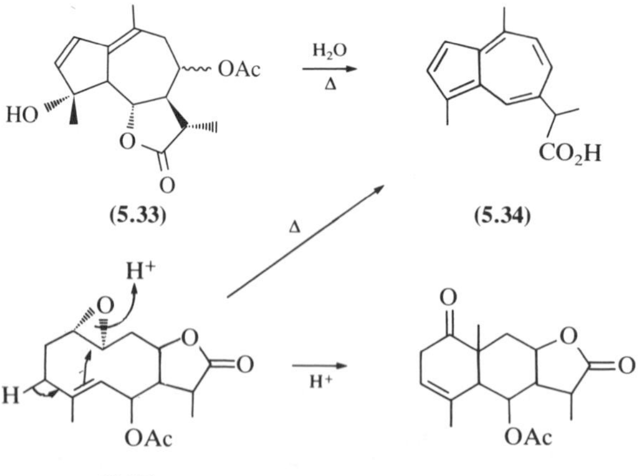

Fig. 5.4.

similarly formed (**5.34**) – a finding which led, not unnaturally, to a completely incorrect structural assignment. Matters were complicated by another rearrangement of (**5.35**) that occurred on treatment with acid (Fig. 5.4).

An exemplary structure determination is that of the sesquiterpene α-cedrene (**5.36**) which co-occurs with its β-isomer (**5.37**) and related alcohol cedrol (**5.38**) in cedarwood oil. The molecule is tricyclic and the only point of attack (in ring A) was exploited as outlined in Fig. 5.5; the nature of the oxygenated degradation products was determined; and the latter were further converted into anhydrides and other derivatives. The UV spectrum of the ketone (**5.39**) exhibited λ_{max} at 240 nm and application of the Feiser–Woodward rules proved ring A to be six-membered. Bromination of certain degradation products followed by alkene-formation via dehydrobromination and subsequent oxidation then allowed attack on rings B and eventually C. The whole elegant procedure takes many pages to discuss in a standard text but the result is the structure (**5.36**) which was confirmed by total synthesis. Cedrol is a tertiary alcohol (viz., its failure to oxidize to a carbonyl compound under mild conditions), and the interrelationships in Fig. 5.6 prove the

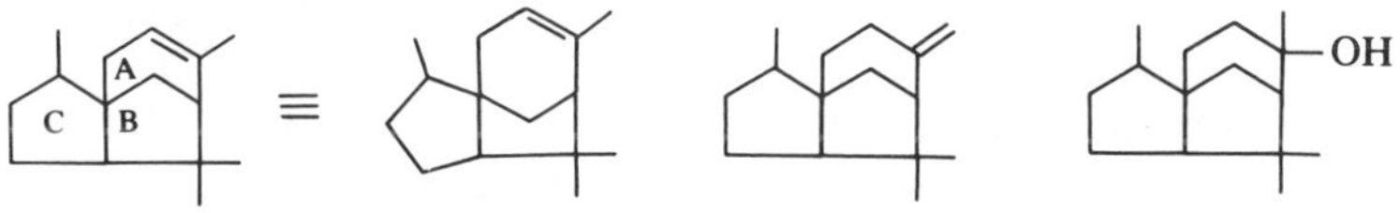

(**5.36**)   (**5.37**)   (**5.38**)

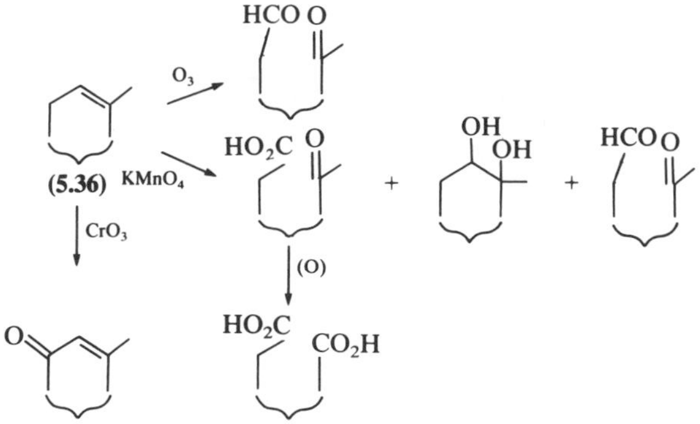

(**5.39**)

Fig. 5.5.

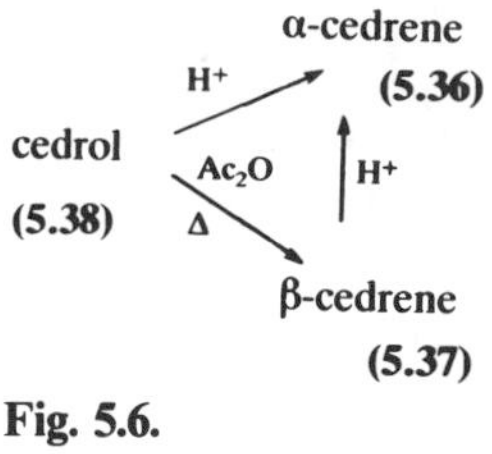

α-cedrene

cedrol   (**5.36**)

(**5.38**)

β-cedrene

(**5.37**)

Fig. 5.6.

structures (**5.37**) and (**5.38**). However, these studies tell us nothing about the relative or absolute configurations of the naturally occurring (+)-α-cedrene. The latter has four asymmetric carbons and so 2^4 ($\equiv 16$) stereoisomers, i.e. eight pairs of enantiomers, are theoretically possible. However, diaxial-bridging requires that rings A and B be *cis*-linked and this reduces the situation to four pairs: these can be represented by the four *relative* configurations in Fig. 5.7, each of which represents a single member of a pair of mirror images which differ in *absolute* configurations. In the event, a detailed set of arguments based on conformational analysis led to the deduction of the relative configuration as in (**5.40**). The absolute configuration was shown also to be (**5.40**) by chiro-optical studies (optical rotary dispersion, ORD, and circular dichroism, CD) involving use of the octant rule which correlates the sign of the Cotton effect with the absolute stereochemistry. This conclusion was verified by X-ray studies, and the configurations of α-cedrene and cedrol followed.

The use of ORD curves to assign absolute configurations is routine: thus the similarity of the ORD curves of (+)-carissone (**5.44**; Fig. 5.8) and 1-(+)-α-cyperone (**5.45**) indicates that the molecules share the same absolute configuration. Absolute configurations may also be assigned by chemical methods. (+)-Citronellal (**5.46**) can be correlated stereochemically with many other terpenoids and can also be converted into (+)-methylsuccinic acid (**5.47**) without configurational change: the latter, in turn, has been correlated into (+)-glyceraldehyde of known absolute configuration (the configuration of the last was elucidated by X-ray crystallography).

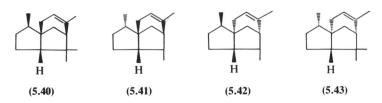

H	H	H	H
(**5.40**)	(**5.41**)	(**5.42**)	(**5.43**)

Fig. 5.7.

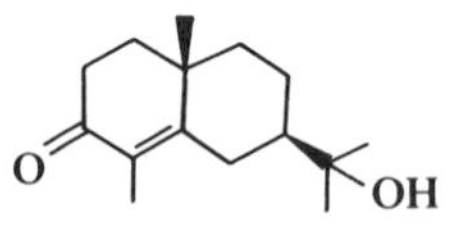

(**5.44**)	(**5.45**)

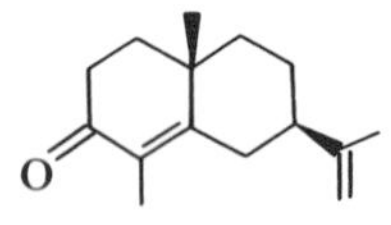

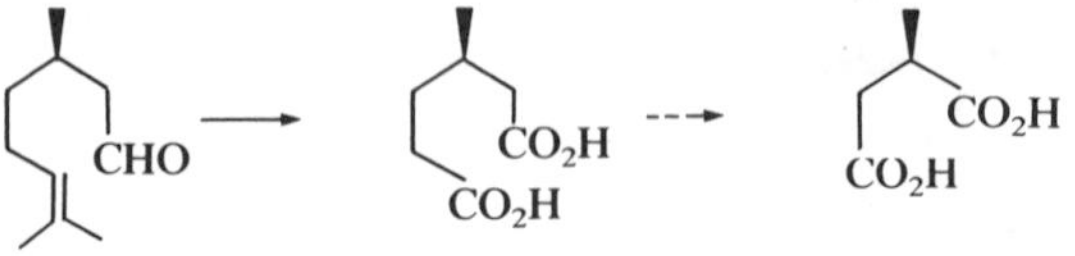

(**5.46**)		(**5.47**)

Fig. 5.8.

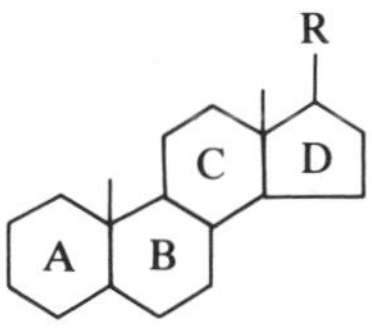

(5.48)

The resolution of the structures of the steroids which possess the skeleton **5.48** (R various; see Section 5.9) proved difficult but was completed using classical methods by the early 1930s. The onslaught on these tetracyclic compounds was initiated by Windaus, Wieland, and Diels from about 1903 and much use was made of oxidative cleavage of rings as described for α-cedrene. Most studies were made on cholesterol, the most readily accessible member which also happens to be the parent of the whole class (see Section 5.9). Use was made also of the (now notorious) Blanc rule which purported to distinguish between the cyclization reactions of dicarboxylic acids of different chain length: this was employed to assign the ring size in cyclic ketones that were obtained during the degradation procedure (see Fig. 5.9). Thus the sequence **(5.49)** → **(5.50)** was taken to prove that the ring A in cholesterol was six-membered. The rule was also applied in conjunction with the elegant Barbier–Wieland degradative procedure to assign the size of ring D (which carries the side-chain). The latter technique is exemplified for an acidic side-chain (in a bile acid) in Fig. 5.10 **(5.51** → **5.52)** and allows the length of the side-chain to be determined by clipping-off one carbon at a time. Extension of the sequence to **(5.51)** and formation of an anhydride on pyrolysis allowed ring D to be assigned as five membered, and so on. The net result was the structure **(5.54)** for cholesterol. Hardly had Wieland and Windaus collected Nobel prizes for their immense labours when it was shown, in the first crystallographic study

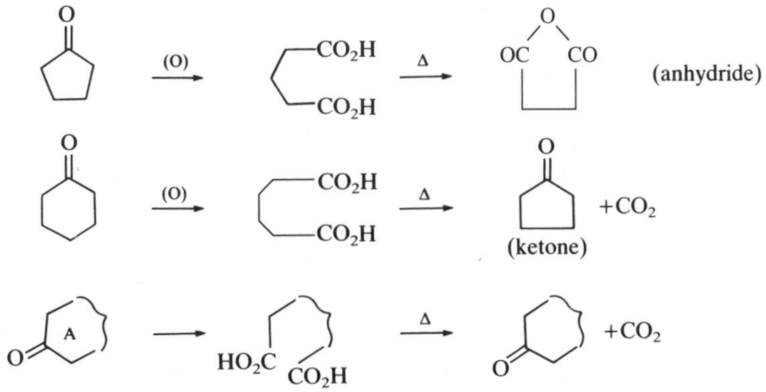

(5.49)   **(5.50)**

Fig. 5.9.

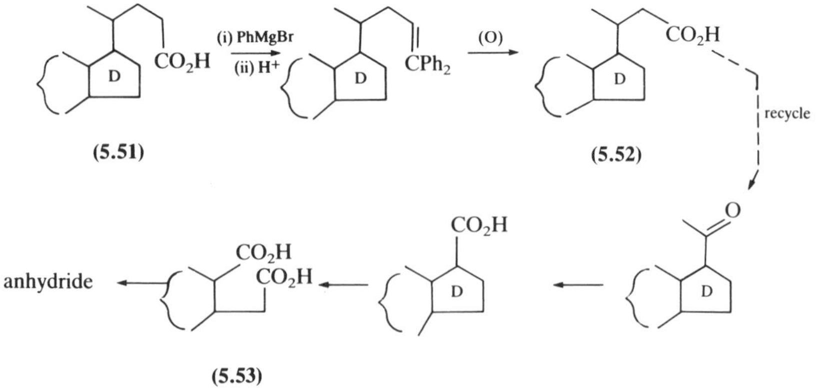

(5.51)   (5.52)

(5.53)

Fig. 5.10.

to provide novel structural information, that (**5.54**) was unlikely to be correct. Thus ergosterol, a steroid from yeast that presumably shares the same skeleton as cholesterol, could not have the structure shown, as the quaternary centre would have resulted in a portly molecule that could not be accommodated within the calculated molecular dimensions (*c.* $0.4 \times 0.7 \times 2$ nm). The reason for the erroneous structure was that the Blanc rule, although applicable to rings A and D, failed for ring B. The rule is now known to be completely unreliable.

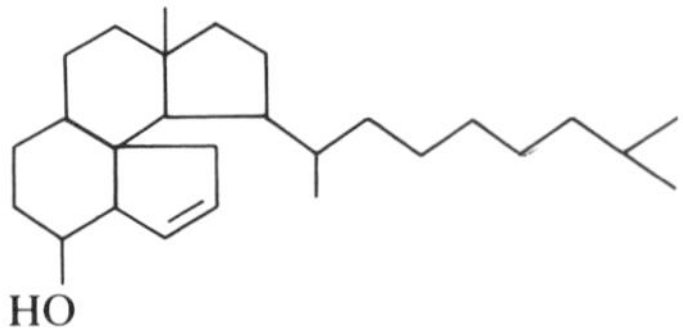

(5.54)

However, the solution was near to hand. Selenium dehydrogenation of cholesterol and other steroids had long before yielded the aromatics (**5.55**) and (**5.56**) and the elucidation of the especially significant structure of the latter (Diels's hydrocarbon) by synthesis and a reappraisal of the previous results led to the proposal of the correct structure for cholesterol (**5.57**). The *trans*-ring junctions in the molecule could be inferred from the molecular dimensions and also from the isolation of fragments such as (**5.58**) from multi-step degradation. Additionally, the *trans* B,C-junction followed from the observation that the 7,12-diketo derivative (**5.59**) was not isomerized by hot alkali and so must contain the more stable orientation at the above junction which is adjacent to the C-7 keto group. These studies revealed the relative configuration: the absolute configuration was shown to be as in (**5.57**) by correlation of cholesterol as in Fig. 5.11 with (+)-citronellal (**5.46**), the configuration of which was known (see above). Similar studies on other

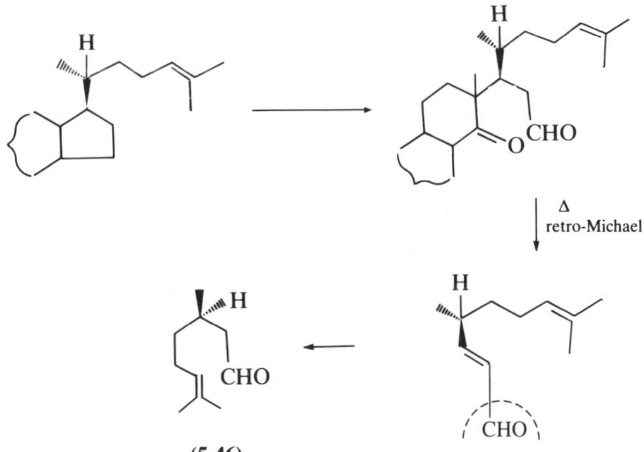

Fig. 5.11.

(5.46)

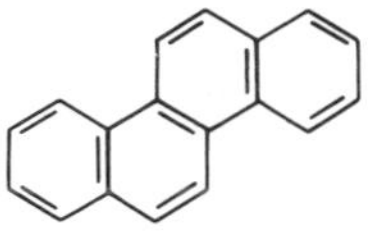

(5.55)

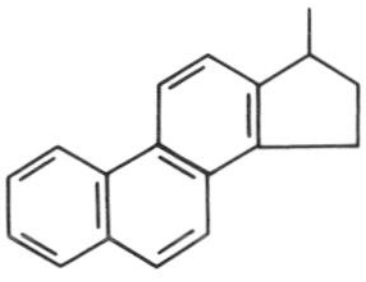

(5.56)

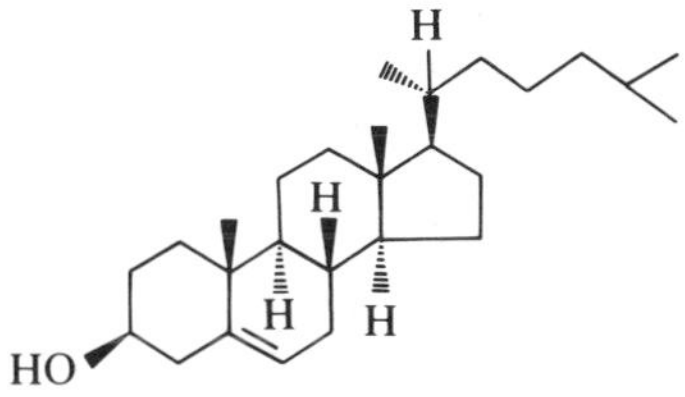

(5.57)

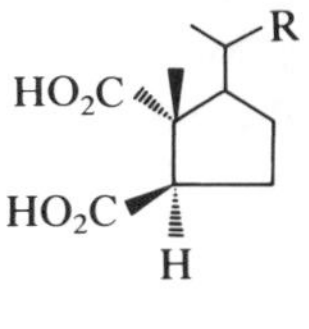

(5.58)

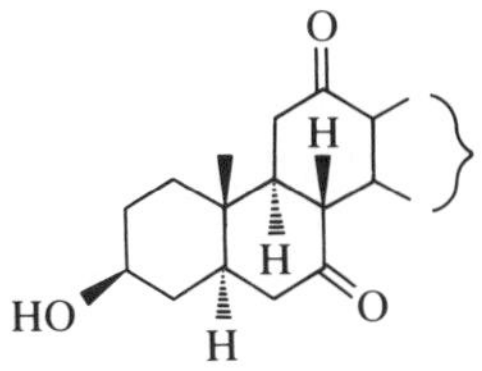

(5.59)

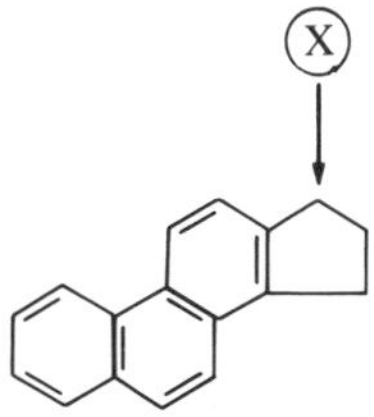

(5.60)

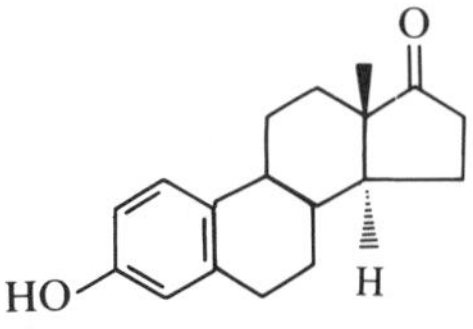

(5.61)

steroids led to consistent conclusions. For example, the sex hormone oestrone ($C_{18}H_{22}O_2$) contains one keto group and one phenolic hydroxyl and no other unsaturation except one aromatic ring (detected by its UV absorption). Wolff–Kishner reduction followed by selenium dehydrogenation led to (5.60) and the position of the original keto group was shown by the marker method to be at position X. Thus the molecule was characterized as (5.61).

The above classical methods, without use of any beyond the most rudimentary spectroscopic techniques, allowed the correct deduction of the structure of numerous terpenoids but were not always successful. Thus the tricyclic sesquiterpenoid longifolene (see Section 5.6), which has only one position susceptible to chemical attack and is prone to rearrangement, was studied most exhaustively with albeit inconclusive results and X-ray studies were necessary to elucidate its structure.

Many terpenoids, especially mono- and sesqui-compounds, occur naturally in an optically impure form, i.e. as a mixture of enantiomers with one isomer predominating. Such mixtures can often be resolved by chromatography (LC, GC, or HPLC) on columns containing chiral phases or resolution may be effected by conventional derivatization and recrystallization. In the absence of these frequently onerous procedures, the enantiomeric excess in a mixture can often be rapidly and accurately measured using chiral lanthanide-shift reagents in NMR spectroscopy. These reagents bind to polar sites and may split the resonance signals due to certain ^1H or ^{13}C atoms, each component of the resulting doublet corresponding to a particular enantiomer: integration of the signals then yields the optical purity.

5.4 Hemiterpenoids – C_5 compounds

The parent of the terpenoids is the C_6-compound $3R$-(+)-mevalonic acid (**5.8**) which was isolated in 1956 as a metabolite of a *Lactobacterium* species and was found to be a potent growth factor for yeast. It is usual to write this in the anionic form (**5.64**; shown in Fig. 5.12 with pro-chiral hydrogens) since the un-ionized acid rapidly lactonizes and is biosynthetically inert. The unnatural $3S$-(−)-acid or its lactone is not similarly utilized. Before the discovery of the role of MVA, the identity of the biogenetic equivalent of the C_5 unit implicit in the isoprene rules had been much disputed. C_5 compounds such as 3,3-dimethylacrylic, tiglic, and angelic acids had been proposed and amino acids such as valine and leucine had been suggested as precursors of the building block, but studies on incorporation of labelled precursors lent no support. Isoprene itself, although a known product of pyrolysis of monoterpenes, was never seriously considered as no remotely likely biochemical routes utilizing this compound could be suggested.

Detailed enzymatic studies using cell-free extracts from yeast and liver (the results of which are generally applicable) were carried out from the 1950s by Lynen, Bloch, and Cornforth and their coworkers to define the metabolism of MVA, cf. Fig. 5.12. Acetyl-coenzyme A (CH_3CO-S-CoA; **5.62**) undergoes Claisen condensations to yield 3-hydroxy-3-methyl-glutaryl-coenzyme A (HMG-CoA; **5.63**) which is reduced to MVA. The last steps – mediated by HMG-coenzyme A reductase – are especially important. They are essentially irreversible and rate-limiting for the sequence. Also they commit C_2 units to MVA, which has no known anabolic role other than the

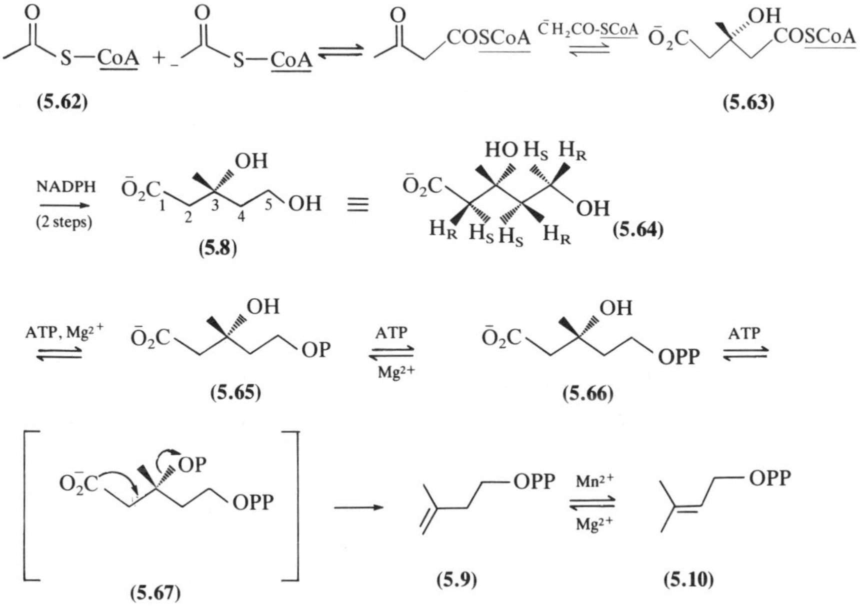

(OP = phosphate, OPP = pyrophosphate, CoA−SH = coenzyme A)

Fig. 5.12.

synthesis of terpenoids (MVA cannot be degraded by reversal of steps
(5.64) → **(5.63)** – although it can be broken down in some organisms in an
independent 'shunt', the significance of which is not clear). Thus, the activity
of HMG-CoA reductase which exists in interconvertible pools of active and
inactive forms under allosteric control at different intracellular sites
determines the throughput to terpenoids.

MVA is then phosphorylated in two steps to form the isolable mono and
pyrophosphates, MVAP **(5.65)** and MVAPP **(5.66)** but triphosphorylation
leads to no stable intermediate: **(5.67)** is immediately broken down by a
decarboxylase to yield isopentenyl pyrophosphate **(5.9**; IPP) which is
isomerized into 3,3-dimethylallyl pyrophosphate **(5.10**; DMAPP). **(5.9)** and
(5.10) together represent the equivalent of the isoprene unit, and the latter
ester especially (favoured *c.* 13:1 at equilibrium) yields hemiterpenoids.

Replacement of each of the six pro-chiral (*R* or *S*; see **5.64**) methylene
hydrogens in MVA with ³H or ²H has been achieved synthetically and the
fate of the six possible labelled MVAs followed into terpenoids by co-feeding
with ¹⁴C-MVA and measuring isotope ratios (³H:¹⁴C) in products. Not
surprisingly, all the enzymic steps that involve these pro-chiral hydrogens
are stereospecific. Thus the isomerase that interconnects IPP and DMAPP
abstracts the pro-(*R*) hydrogen from the C-2 of IPP (originally the
4-pro (*S*) hydrogen of MVA) and the stereochemistry is as in Fig. 5.13. Note
that the CH₃ in **(5.10)** (which originates from C-2 of MVA) finishes *trans* to

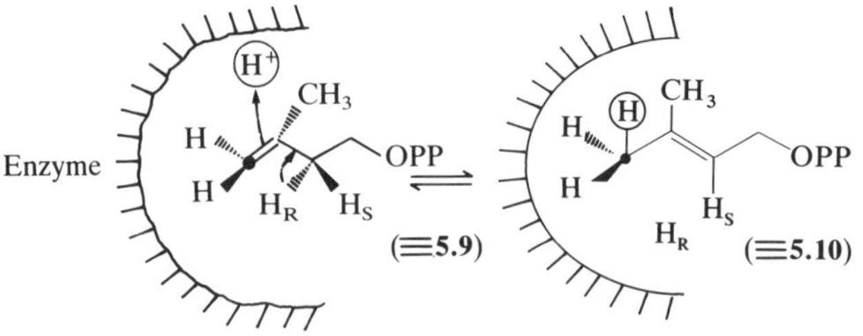

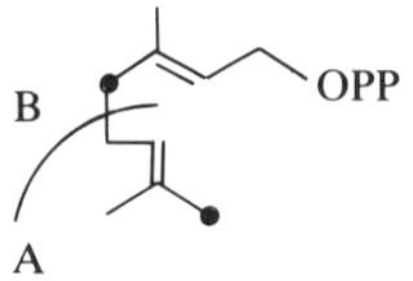

Fig. 5.13.

the ester group. This was demonstrated by feeding [2-^{14}C]-MVA to *Rosa* spp., when coupling of DMAPP and IPP (moieties A and B; see Section 5.5) led to the labelling of geraniol as in (**5.68**), whereupon chemical or enzymatic excision of the marked carbon eliminated tracer from A.

Not surprisingly, few true hemiterpenoids are known. Isoprene is emitted from leaves of many plant species and may play a role as a plant hormone similar to ethylene. Isopentenol and 3,3-dimethylallyl alcohol (from cleavage of **5.9** and **5.10**) occur in some oils either free or as esters and may be more widespread in view of their ready loss during work-up due to volatility. The acetate of the latter C_5-alcohol is a bee alarm pheromone where its volatility (rapid dispersal and fading of signal) could be advantageous. DMAPP (a potential source of $Me_2C{=}CHCH_2^+$) is an excellent electrophile and readily forms meroterpenoids (cf. Section 5.2): thus humulone (**5.69**) is a bitter principle from hops. The C_5 unit is more concealed in oroselone (**5.70**). Other well-known examples occur in other polyketides and in the ergot alkaloids (see Chapter 7).

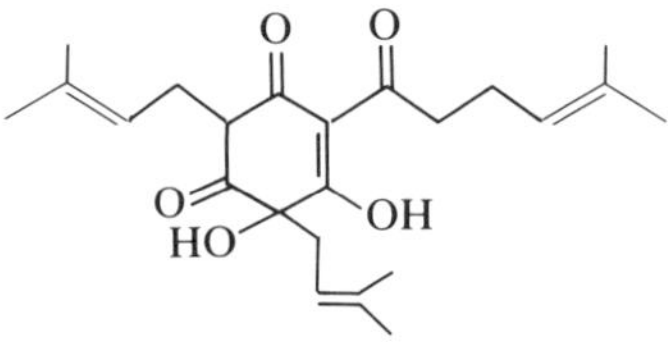

(**5.69**) 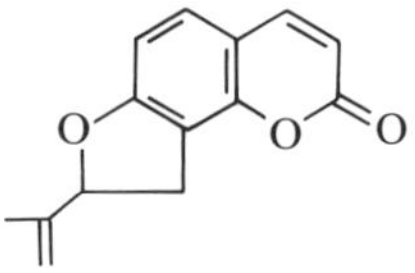   (**5.70**)

5.5   Monoterpenoids – C_{10} compounds

The discovery that unrearranged crystalline derivatives could often be recovered by treatment of fractions from plant oils with nitrosyl chloride (HCl gave rearranged products) gave impetus to studies of this class from 1880 onwards (by Wallach, Semmler, and Tilden), but pure specimens of the hydrocarbons of the family were rarely obtained before the advent of GC (in 1953): the alcohols, ketones, etc., had been isolated much earlier as crystalline solids or by derivatization.

Often both enantiomers of a monoterpenoid occur naturally and sometimes they differ in odour and taste: thus (+) and (−)-carvone are characteristic of caraway and spearmint. During the last decade it has become apparent that monoterpenoids play an important role in chemical ecology. α-Pinene, myrcene, geraniol, citral, and others (see **5.71, 5.74, 5.75, 5.78**; Figs 5.14 and 5.17) can act variously as aggregation, trail, alarm and sex pheromones for bees, weevils, silkworms, and the like and occur as defensive secretions in ants and beetles. Male cotton boll weevils are attracted to their host plants by the secretion of (+)-α-pinene (**5.71**) which they ingest and metabolize to form unusual monoterpenoids (Fig. 5.14) that in turn attract the female. Other monoterpenoids, e.g. 1,8-cineol and camphor (**5.81** and **5.83**; Fig. 5.17), are emitted by assorted (predominantly arid zone) plants like eucalyptus and sagebrush and inhibit germination and development of seeds of competing species. Yet others act as attractants to insect pollinators or repellants of insect or larger predators. Patterns of co-evolution of plant and insect species are slowly emerging.

Enzymic studies of reaction rates with chemically-modified substrates have shown that GPP, the parent of the class, is generated by electrophilic addition of the C$_5$-moiety of DMAPP to IPP and subsequent elimination of the pro-2R hydrogen of the latter (Fig. 5.15; this latter atom corresponds to the pro-4S hydrogen of MVA). The parent can yield acyclic monoterpenoids directly, but can also be converted *in vivo* into over 30 skeletal types – although only 6 are common. The formation of all but one of the latter can be rationalized by a series of additions, eliminations, and rearrangements of hypothetical carbocations as in Fig. 5.16 which is largely due to Ruzicka (1959). Free carbocations are too reactive to be free intermediates in these reactions and it is likely that enzyme-bonded equivalents or (especially, see later) ion-pairs are the actual species.

Some important monoterpenoids are shown in Fig. 5.17. Limonene (**5.79**) and α-terpineol (**5.80**) are components of citrus oils, whereas 1,8-cineol (**5.81**) occurs in wormwood and eucalyptus. Limonene is one of 14 structural isomers of *p*-menthadiene, 6 of which occur naturally. Borneol (**5.82**),

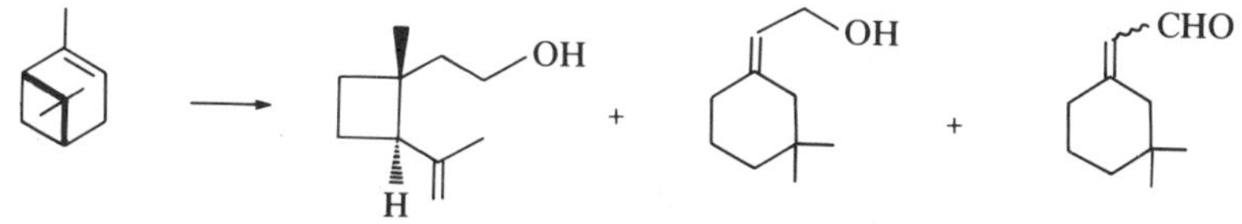

(5.71)

Fig. 5.14.

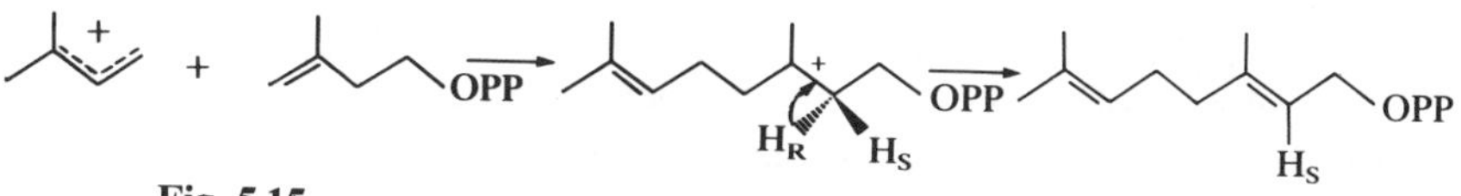

Fig. 5.15.

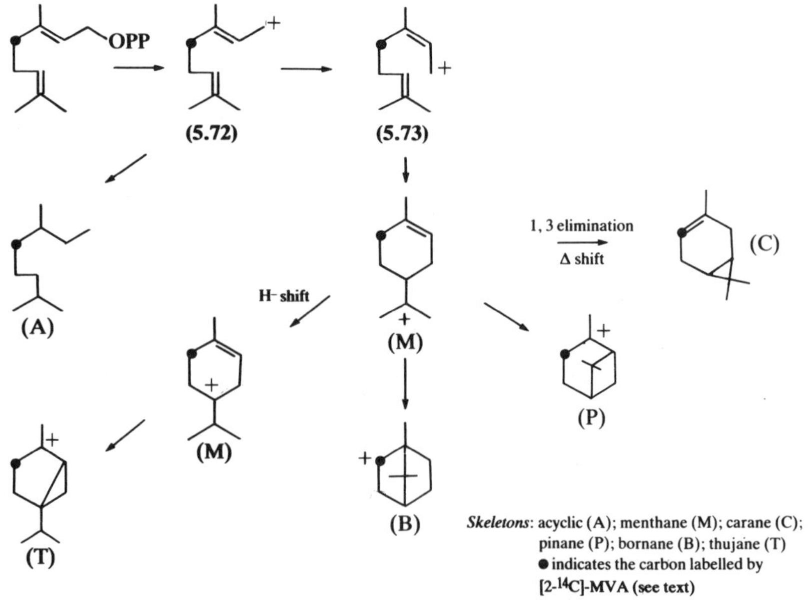

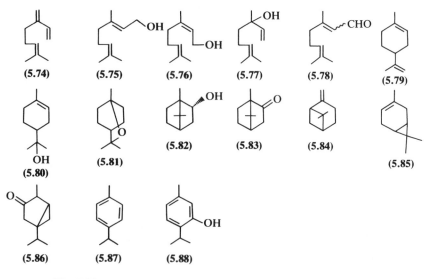

Fig. 5.16.

Fig. 5.17.

camphor (**5.83**) and β-pinene (**5.84**) are widespread, whereas car-3-ene (**5.85**) and thujone (**5.86**) predominate in only a few oils from few genera. *p*-Cymene (**5.87**) and thymol (**5.88**) are aromatic menthane derivatives.

The final commonly occurring class possesses the iridane skeleton (**5.89**), which is not derivable from GPP by any simple route involving carbocations.

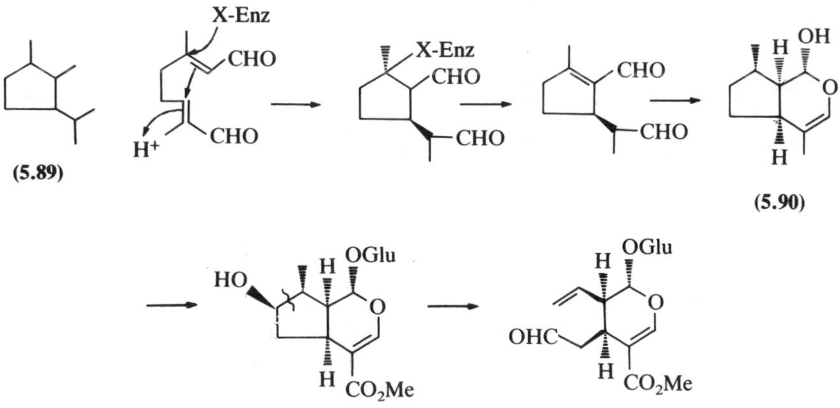

Fig. 5.18.

Its biogenesis is outlined in Fig. 5.18, wherein the dialdehyde derived from geraniol undergoes a double Michael addition ($-X \equiv$ enzyme-linked nucleophile, such as $-\bar{S}$) to yield iridodial (a defensive secretion of the devil's coach horse beetle), loganin (**5.91**) and, by cleavage, secologanin (**5.92**) which is the building brick for terpene alkaloids (see Chapter 7). Nepetalactone (the ketone of **5.90**) is an insect repellant and a notorious stimulator of cats! These iridoids (the first examples of which were isolated from the *Iridomyrnex* genus of ants) are rare except in plants that accumulate terpene alkaloids where they occur as water-soluble β-glucosides.

All seven of these primary skeletons have been shown, in specific compounds, to be constructed from MVA although *in vivo* the incorporation is often only *c*. 0.1 per cent, perhaps owing to compartmentation restricting access of exogenous precursor to the biosynthetic sites. Nevertheless, feeding of flowerheads of *Rosa* species has resulted in uptakes of *c*. 22 per cent 3R-MVA into geraniol (**5.75**) and nerol (**5.76**). Although generally low, the incorporation of tracer from [2-¹⁴C]-MVA was position-specific with the surprising restriction that only the moiety derived from IPP was usually labelled: thus the skeletal types in Fig. 5.16 were labelled as shown. This asymmetry probably arises from condensation of IPP (from MVA) with DMAPP from an endogenous pool. Recently the details of Fig. 5.16 have been partly filled in. It is stereochemically impossible for GPP to cyclize to form cyclohexyl rings, and the roles of the pyrophosphates of nerol (**5.76**) and linalool (**5.77**) – for which there is no such restriction – have been long debated. One possibility is that geraniol is converted into nerol by a redox sequence (Fig. 5.19): the crucial interconversion of the aldehydes, step (i), is known to be spontaneous, possibly involving Michael addition of water and its retro-elimination. Tracer studies with R and S-[5-³H₁]-MVA have shown that this route occurs in *Rosa* and other species *in vivo*, but in other

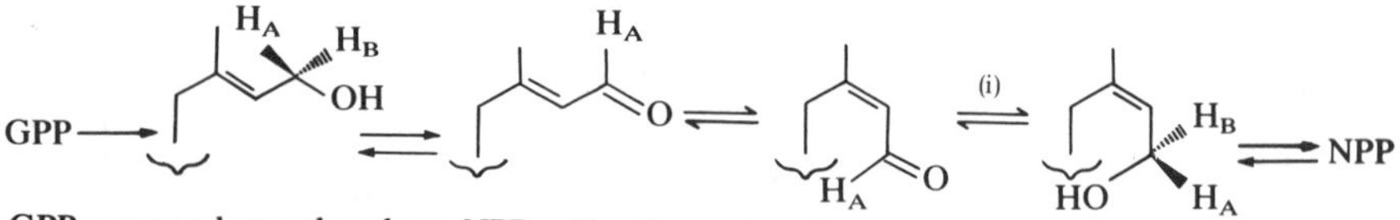

GPP = geranyl pyrophosphate, NPP = Neryl pyrophosphate

Fig. 5.19.

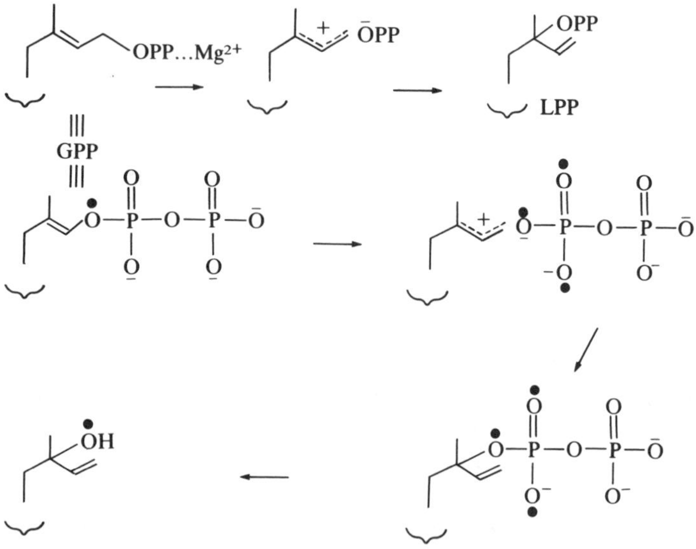

Fig. 5.20.

plants the use of crude cell-free extracts has shown an isomerization–cyclization route (involving a linked enzyme system) whereby, in effect, GPP is directly converted into linoloyl pyrophosphate (LPP).

The latter route envisages formation of a tightly bonded ion-pair generated by Mg^{2+}-assisted cleavage of GPP (Fig. 5.20). This was demonstrated by studies on GPP and the analogous C_{15} compound labelled with ^{18}O in the ester group as shown. The product alcohol contained 33 per cent of the initial tracer and this showed that an ion-pair, rather than a direct 1,3- or 1,6-sigmatropic shift, occurred, in which there was sufficient time for free rotation about the P—O—P bond but not for tumbling of the anion component before the ions collapsed to yield the isomeric product. The cyclization of LPP (Fig. 5.21) or its equivalent ion pair then yielded (+)-bornyl pyrophosphate (**5.93**) which was demonstrated to be a precursor of the free alcohol. The same cyclase enzyme gave (+)-α-pinene and here it was supposed that the alternative cyclization of the ion pair was not completed by ion-collapse to a pyrophosphate because of steric hindrance to formation at the tertiary ester: now proton-loss directly yielded the alkene (**5.94**). Enzyme systems of opposite stereospecificity were proposed to be

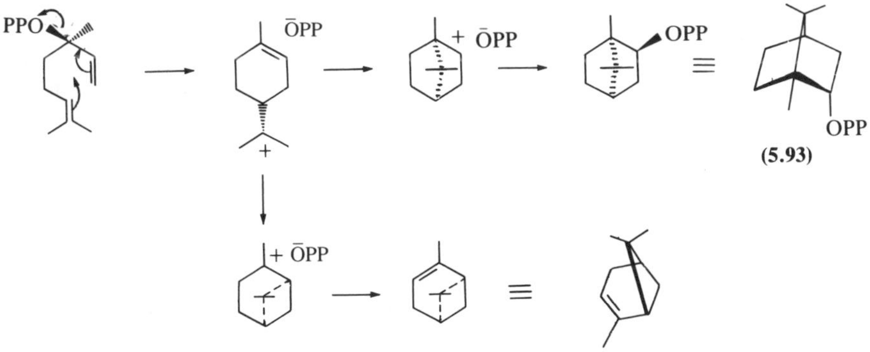

Fig. 5.21.

present in other, or sometimes the same, plant species that converted the optically inactive GPP to the other enantiomer of LPP: then the same route led to the enantiomers of (**5.93**) and (**5.94**). These proposals were consistent with previously reported labelling patterns whereby tracer from [2-^{14}C]-MVA specifically resided at C-3 for both ($+$) and ($-$)-α-pinene and at C-6 for ($+$) and ($-$)-camphor.

Similar routes via LPP may occur for the construction of other skeletons. The menthyl carbocation, perhaps better represented as the biogenetically equivalent protein-bonded species, was shown to undergo *anti*-specific shifts to yield α-thujene (**5.95**).

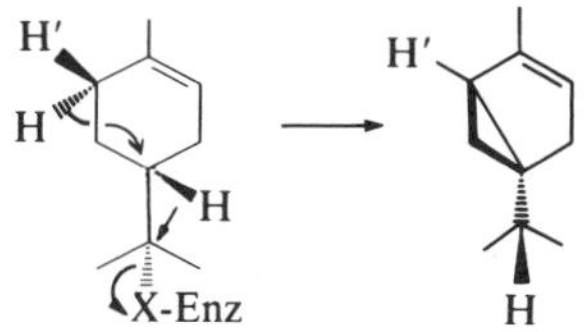

(**5.95**)

Cyclizations to give the main skeletons are often accompanied *in vivo* by secondary transformations (also enzyme-catalysed and so genetically controlled) that yield other skeletons or interconvert known ones *via* Wagner–Meerwein rearrangements. Some examples applied to the pinane skeleton are shown in Fig. 5.22: fenchols, fenchanes, and santanes – compounds based on such secondary skeletons – are found in oils from fennel and sandalwood respectively.

Compounds with these primary and secondary skeletons are then elaborated and functionalized (perhaps by relatively unspecific enzyme systems constituting a metabolic grid) to give the whole complex array of the monoterpenoids. Mostly the parents of each skeletal type are alkenes which are initially oxidized with cytochrome P-450 dependent oxygenases.

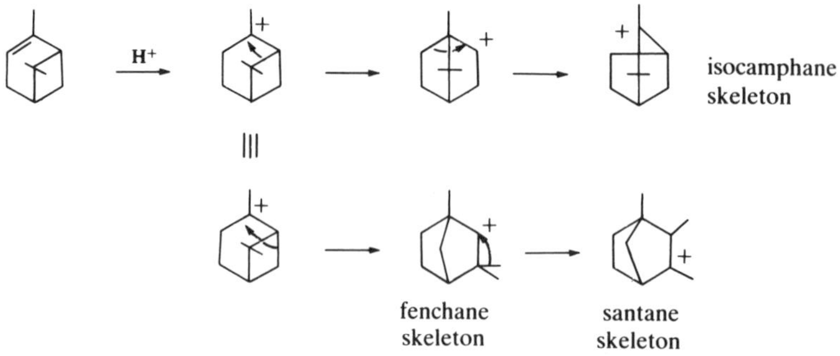

isocamphane
skeleton

fenchane
skeleton

santane
skeleton

Fig. 5.22.

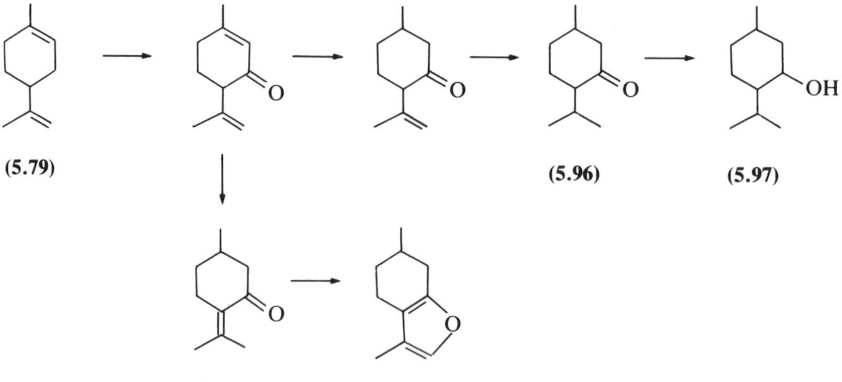

(5.79)   (5.96)   (5.97)

(5.98)   (5.99)

Fig. 5.23.

Limonene (**5.79**) yields the sequence (Fig. 5.23) to the isomeric menthones (**5.96**), menthols (**5.97**), pulegones (**5.98**), and menthofurans (**5.99**) amongst others. Different species of the *Mentha* (mint) genus possess different enzymatic complements and hence different members of the sequence are accumulated in, for example, spearmint, peppermint, pennyroyal, and water mint, by naturally occurring genetic blocks. Oxidation in spearmint can also occur at C-2 of limonene to form carvones (**5.100**), whilst oxidation of the isomeric γ-terpinene leads to the peroxide ascaridole (**5.101**) in *Chenopodium* species. α-Pinene can likewise be oxidized at A, B (with double bond rearrangement) or C (**5.102**) to form myrtenyl, pinocarvyl, and verbenyl compounds respectively and many other examples occur in nature.

The essential oils of commerce often also contain artefacts due to collection, storage, and photodecomposition (aside from straightforward adulterants!) and this results in even more complex mixtures than nature provides. Rose oils may contain up to 400 components of which over 80 per cent may be monoterpenoids, and the minor compounds – even artefacts – may play a

(5.100)　　　(5.101)　　　(5.102)　　　(5.105)

(5.103)　　　　　(5.104)　　　(a)　(b)　(c)

subtle role in odour quality. An example is the presence of the thio-analogue of α-terpineol (**5.80**) in grapefruit oil; although present at levels below GC detection, this imparts the characteristic flavour and is detectable by the human nose and palate at dilutions of parts per billion.

All the above monoterpenoids are regular in the sense of obeying the Ruzicka isoprene rule. A few dozen irregular compounds are nevertheless known that are formed not from GPP (i.e. IPP + DMAPP) but from condensation of DMAPP (i.e. DMAPP + DMAPP). Very few analogous irregular compounds are known for the other classes of terpenoids. All the irregular monoterpenoids are believed to be formed from *trans*-chrysanthemol which comprises the terpenoid part of the natural insecticides pyrethrin I and II (**5.103**; R = H; CH=CH₂) which occur in East African plants. *trans*-Chrysanthemol – the biosynthesis of which, for reasons which will become clear, is discussed in Section 5.9 – is believed to be enzymatically cleaved to give three irregular skeletons (**5.104**). The most important of the derived compounds is lavandulol (**5.105**) which is a component of French lavender oil.

So far, we have ignored stereochemical details of the various mono-terpenoids, but those are well exemplified by the menthones (**5.96**) and menthols (**5.97**). Menthone has two chiral centres and hence four stereoisomers (two pairs of enantiomers). If we consider one set of enantiomers, the eponymous menthone which is the more stable (methyl and iso-propyl groups both equatorial) must have the relative configuration (**5.106**), whereas isomenthone will be (**5.107**; Fig. 5.24). On reduction, enantiomeric pairs of menthols are formed. These have been assigned the relative configurations menthol (**5.108**), neomenthol (**5.109**), isomenthol (**5.110**), and neoisomenthol (**5.111**) on the basis of chemical transformations. Menthyl derivatives were resistant to E2 elimination (*anti*-periplanar leaving groups) and when this occurred only formed menth-2-ene, whereas

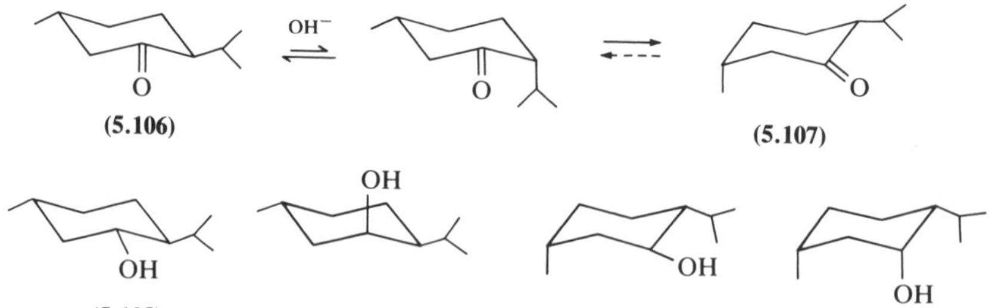

Fig. 5.24.

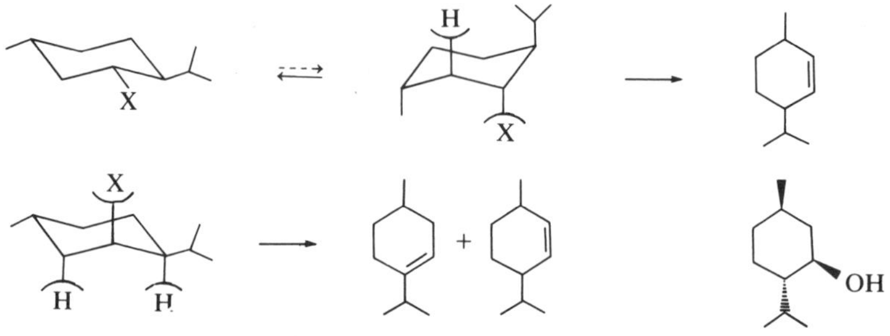

Fig. 5.25.

neomenthyl compounds readily reacted thus to give both 2-ene and 3-ene isomers (Fig. 5.25), and so on. The absolute configuration of (−)-menthol was shown to be (**5.108**) (≡**5.112**) by correlation with (+)-glyceraldehyde.

The detailed chemistry of the monoterpenoids was developed as a consequence of their commercial importance. The discovery of the Wagner–Meerwein rearrangement in the laboratory predated by 40 years its application in biogenetic theory, and similarly the acid-catalysed transformation of α-pinene into borneol, camphene, α-terpineol and α-fenchyl alcohol, and many similar sequences were long-known. Wagner–Meerwein shifts involving methyl groups were originally called Nametkin rearrangements, and an early example provided the explanation of why (+)-α-pinene in the above reaction yielded (±)-camphene. (+)-Camphene (**5.113**) may be written in a projection (Fig. 5.26) in which it is obvious that protonation–deprotonation, coupled with Nametkin shifts, all at equilibrium, will lead to racemization.

Another feature of terpenoid chemistry is the occurrence of non-classical carbocations as reaction intermediates. The first example found in 1939 resulted from the observation that isobornyl chloride (**5.114**) underwent S_N1 up to 10^3-fold faster than its epimer bornyl chloride (**5.115**) under identical conditions (Fig. 5.27), and, unlike the latter, largely yielded rearranged

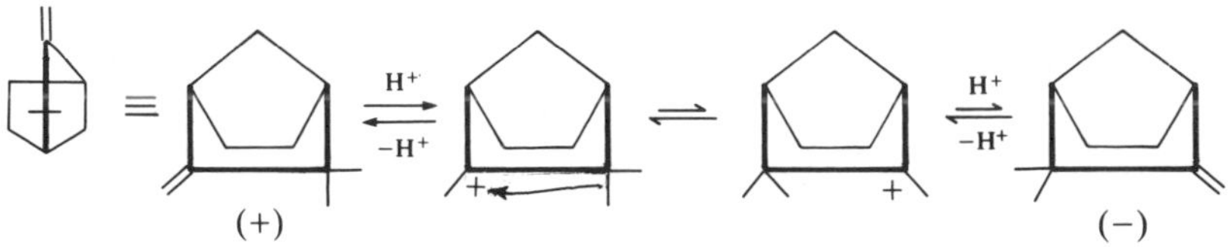

(5.113)

Fig. 5.26.

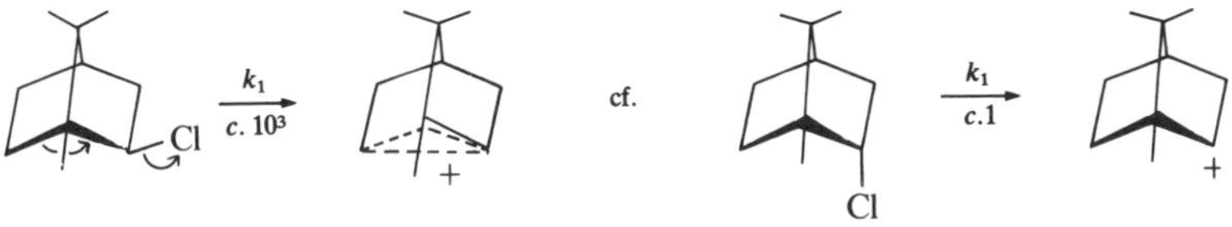

(5.114)   (5.115)

Fig. 5.27.

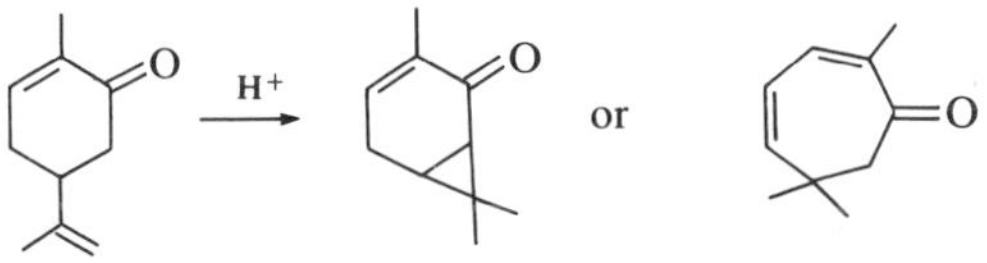

(5.116)

Fig. 5.28.

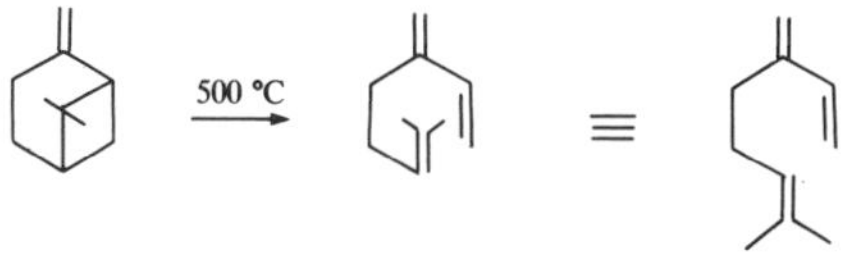

(5.84)   (5.74)

Fig. 5.29.

products (e.g. camphene). The rate enhancement was correlated with the perfect orientation in (5.114) for σ-bond assisted ionization and this led to the development of the whole tendentious concept. Although few mechanistic studies have been performed, it is likely that non-classical ions commonly occur in the reactions of mono- and higher terpenoids as the rigid skeletons and orientations necessary for the phenomenon so frequently occur.

A further selection from monoterpenoid chemistry could include the dispute over the structure of eucarvone (Fig. 5.28) which could only be settled in favour of (5.116) by ¹H-NMR (in one of the first applications of the technique to structural problems) as chemical evidence (ozonolysis; hydrogen-isotope exchange, etc.) had yielded conflicting results. The pyrolytic cleavage of β-pinene (5.84) to myrcene (5.74) by a retro-cycloaddition (Fig. 5.29), and the differing routes of pyrolysis of the thujenes

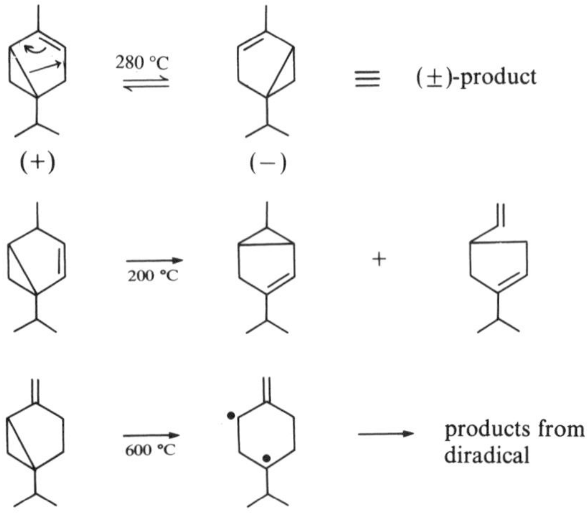

Fig. 5.30.

(5.74)

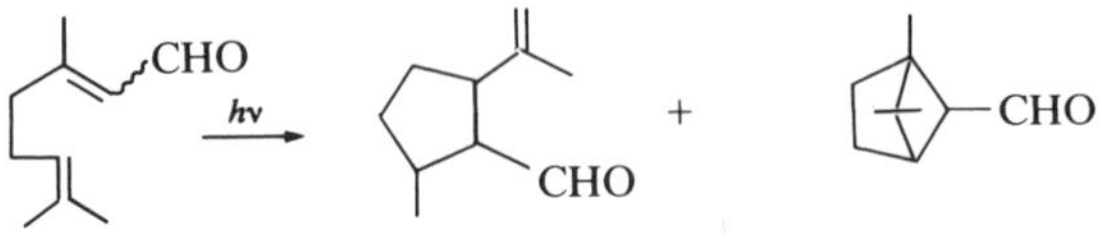

(5.117)

(Fig. 5.30) are also noteworthy. Monoterpenoids also frequently exhibit a fascinating photochemistry, *viz.*, the reactions of myrcene (**5.74**) and of citral (**5.117**) to form photocitrals A and B (Fig. 5.31), which provide entries into less usual skeletal types. Most of these reactions will have analogies for the higher terpenoids although as the simplest class (excluding the C_5 type) monoterpenoids have been the most extensively studied as prototypes.

5.6   Sesquiterpenoids – C_{15} compounds

Addition of IPP to GPP as in Fig. 5.15 yields 2E,6E-farnesyl pyrophosphate (FPP; **5.118**), the parent of the sesquiterpenoids. Only a few sesquiterpenoids were characterized by 1920, but application of the Vesterberg dehydro-genation method (Section 5.3) led to a gradual acceleration of information and now over 200 skeletal types are known – the most for any class of terpenoids. Sesquiterpenoids are widespread but usually minor (< 5 per cent w/w) components of plant oils but often endow crucial flavour characteristics (e.g. in oils of ginger, cloves, citronella and hops). Tracer studies have indicated that they are synthesized at intracellular sites denied to monoterpenoids.

The bewildering complexity of the class was rationalized elegantly by Ruzicka and, as for monoterpenoids, their formation may best be pictured in terms of the reactions of hypothetical carbocations as models for the enzymatic processes. Additions to accessible double bonds to yield cyclic intermediates, followed by hydride shifts to generate new cationic centres, recyclization, elimination, and hydrolysis can then result in the observed compounds; all the reactions being governed by the stereoelectronic and steric factors well known from mechanistic organic chemistry. Although relatively few routes have been investigated by tracer or enzymic studies, there is no doubt as to the general validity of the interpretation. The whole pattern illustrates how enzymes exploit the innate reactivities of their substrates.

Our description will be in terms of the ions derived from (**5.118**), but as for monoterpenoids the allylic isomer, i.e. nerolidyl pyrophosphate (**5.119**), is probably often the enzyme-bonded species involved in an isomerase–cyclase system. In contrast to the situation for the lower class, hydrocarbons do not seem to be the typical initial stable products; instead the carbocations are usually captured by water to yield alcohols rather than undergoing elimination to alkenes. Alternatively (cf. monoterpenoids), collapse to the pyrophosphate ester may be followed by hydrolysis.

Farnesol (derived from **5.118**) is fairly rare in higher plants, and other acyclics of the class include *trans-β*-farnesene (**5.120**; R = CH₃), a potent aphid-repellant present in hops and sweet potatoes, and *β*-sinesal (**5.120**; R = CHO), a flavour component of oranges.

(**5.118**) (**5.119**) (**5.120**)

The major cyclic sesquiterpenoids can conveniently be classified into seven groups on biogenetic grounds. In the first the oxygen of the pyrophosphate moiety of the parent (2*E*,6*E*-FPP) is retained, but in the others a carbocation may be formally generated which cyclizes in different ways for the (putative, see above) 2*E*,6*E*- and 2*Z*,6*E*-ions (Fig. 5.32, routes 2–7). These types are itemized below. It must be appreciated that many of the schemes, although very reasonable, are hypothetical and other possibilities can be (and have

ex 2*E*, 6*E*-FPP   ex 2*Z*, 6*E*-FPP   ex 2*E*, 6*E*-FPP

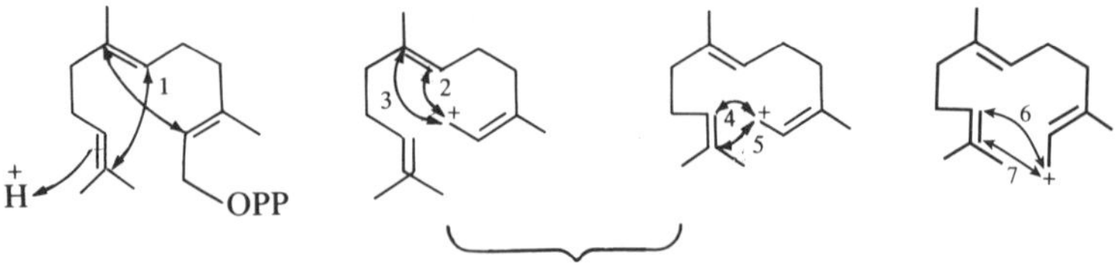

(5.121)

Fig. 5.32. Double headed arrows indicate bond linkages, not electron flow.

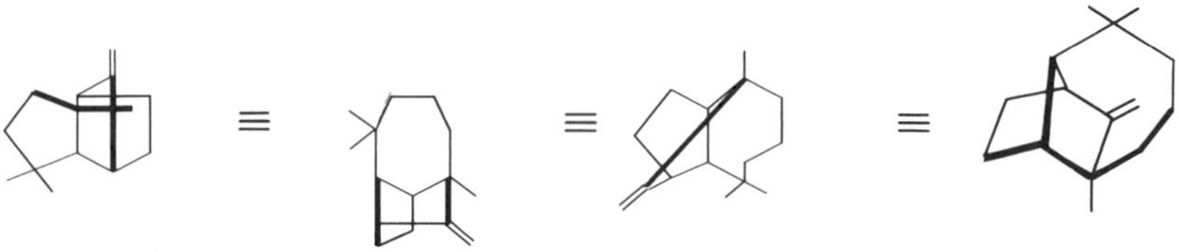

Fig. 5.33.

been) suggested. Also, as sesquiterpenoids frequently occur in enantiomeric forms in different plant genera or orders, the relative configurations given in the following discussion are more significant than the absolute configurations shown. Another point is that a molecule of particular configuration may be represented by many different two-dimensional projections. Four such visualizations of (+)-longifolene are shown in Fig. 5.33: that chosen in a particular discussion will depend upon one's viewpoint of the reaction pathway and its relationship to other structures.

Let us take routes 1 to 7 in turn. (i) The first group of cyclic sesquiterpenoids comprise a small number of compounds and no formal carbocationic intermediate need be postulated. It is convenient to represent 2*E*,6*E*-FPP in a different conformation (Fig. 5.34) from that shown in (**5.118**), and typical products of cyclization are drimenol (**5.122**) and the aphid-inhibitor polygodial (**5.123**). The cyclization yields a *trans*-ring junction reminiscent of that found in higher terpenoids (Sections 5.7 and 5.9). Partial cyclization followed by functionalization could lead to abscisic acid (ABA; **5.124**), an important regulatory compound ubiquitous in higher plants that antagonizes gibberellins (see Section 5.7) and acts to initiate dormancy in autumn and to slow down transpiration in drought. During the latter, and in wilting foliage, the level of free ABA enormously increases (it being released from the stored β-glucoside) and the effect is to close the stomata. Although some studies have suggested that ABA is a sesquiterpenoid, as here described, others have suggested that it is a degraded carotenoid (cf. **5.18**) and the situation is unresolved. Relatives are the degraded sesquiterpenoids (or

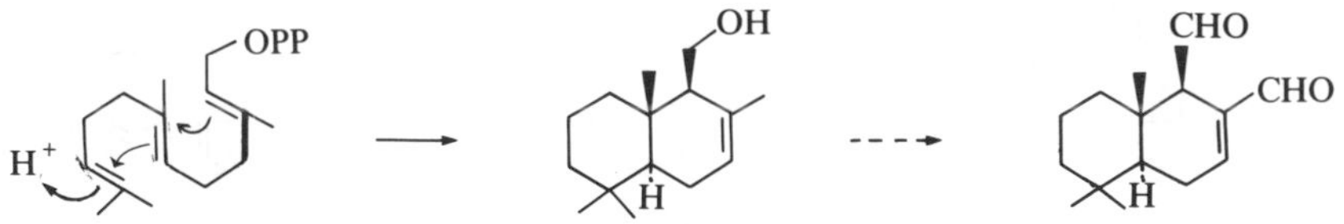

OPP

(5.118)

OH

(5.122)

CHO
CHO

(5.123)

CO$_2$H

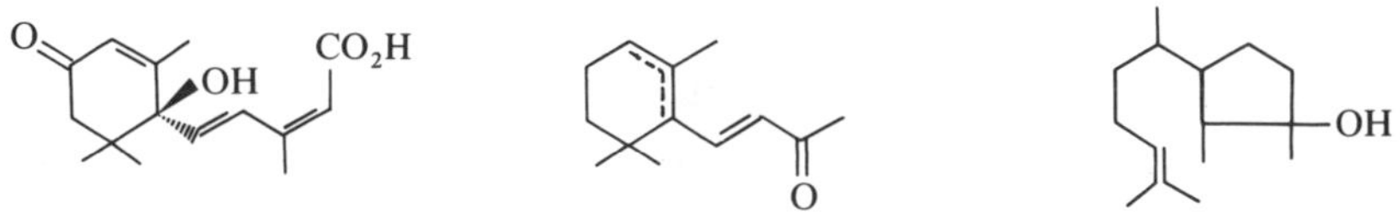

(5.124)

(5.125)

OH

(5.126)

Fig. 5.34.

carotenoids!) α- and β-ionones (**5.125**) which contribute importantly to the odour of violets. The fungal metabolite cyclonerodiol (**5.126**) is formed via cyclization at the distal end of (**5.118**).

(ii) Compounds of the second type are formally derived from 2Z,6E-FPP (**5.121**) (or its allylic isomer – see before) via cyclization to the bisabolene skeleton (**5.127**; Fig. 5.35). The ion or the derived γ-bisabolene (**5.128**) can then undergo a sequence of rearrangements to the bicyclic compounds cuparene (**5.129**), widdrol (**5.130**), and the tricyclic thujopsene (**5.131**) which accumulate in different plant species; and further functionalization of (**5.129**) results in the mycotoxin trichothecin (**5.132**; R = C$_3$H$_5$CO). The latter route has been confirmed not only by detailed double-label tracer studies but also by feeding radioactive compounds that were considered likely precursors. Additionally, α-cedrene and cedrol (**5.133**, **5.134**; from cedar wood oil) can be generated by tricyclization from a suitable (enzyme-induced) conformation of (**5.128**; Fig. 5.36).

Other pathways of reaction may be clarified if the ion (**5.127**) is thought of as a sesquimenthane derivative (Fig. 5.37). Then folding as for the monoterpene analogues (Section 5.5) yields the sesquipinane α-bergamotene (**5.135**) and sesquibornane camphenilol (**5.136**) amongst other options. Wagner–Meerwein rearrangement of (**5.136**) yields the sesquiisocamphane α-santalene (**5.137**) which is doubly renowned as being the first sesquiterpenoid to be assigned a correct structure in 1910 and being instrumental in the development of Ruzicka's rule in 1953. These reaction schemes have been confirmed in several examples by tracer studies. The bisabolane skeleton is labelled as in (**5.138**) by different feeding experiments (carbons derived from C-2 of MVA show up as singlets in the ^{13}C NMR spectrum of the metabolite formed by incorporation of [^{13}C]-acetate). This pattern follows through as expected into all the products that were examined. Similar studies showed that the fungal product ovalicin (**5.139**; Fig. 5.38) was formed by route (a), rather than the vague pathway (b).

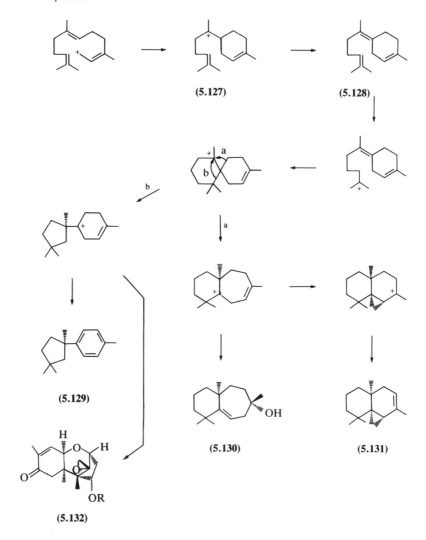

(5.127)   (5.128)

(5.129)

(5.130)   (5.131)

(5.132)

Fig. 5.35.

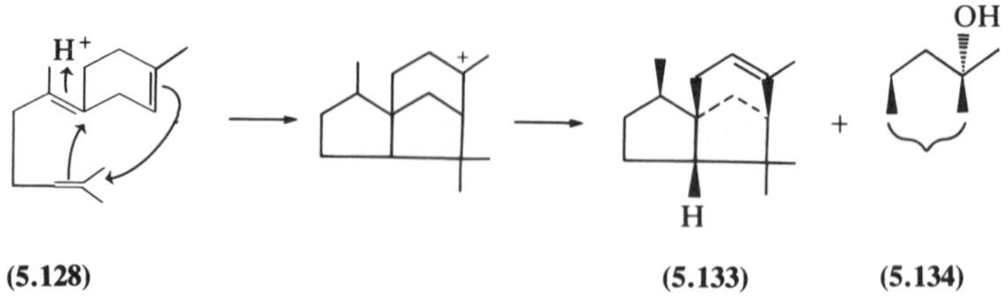

(5.128)   (5.133)   (5.134)

Fig. 5.36.

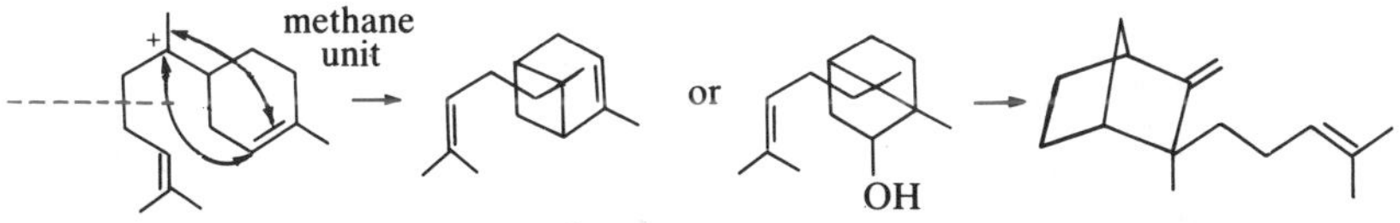

(5.135) (5.136) (5.137)

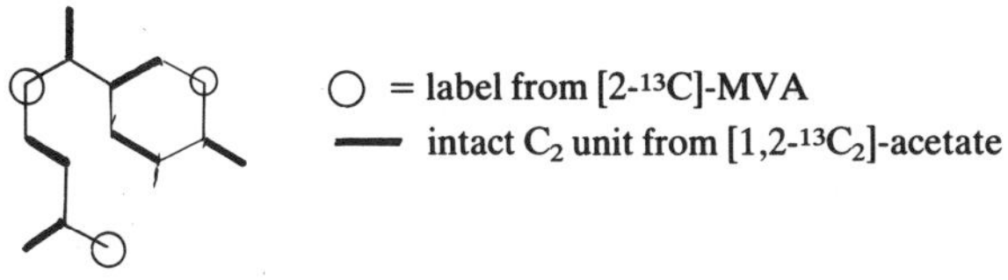

○ = label from [2-^{13}C]-MVA
— intact C$_2$ unit from [1,2-^{13}C$_2$]-acetate

(5.138)

Fig. 5.37.

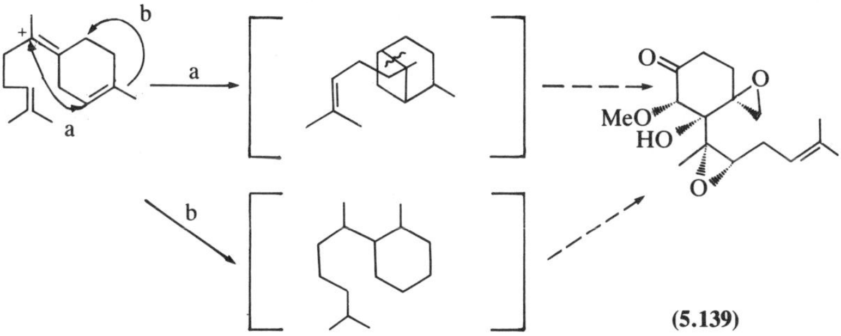

(5.139)

Fig. 5.38.

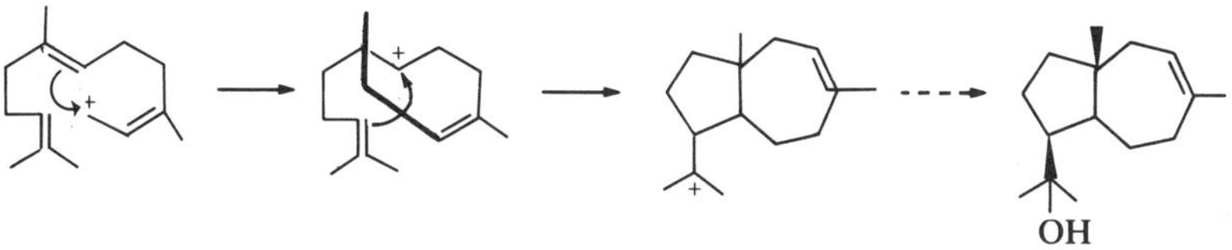

OH

(5.140)

Fig. 5.39.

(iii) Compounds of the third of our defined classes comprise a small group derived from (**5.121**) by the alternative mode of cyclization (Fig. 5.39). Typical is caratol (**5.140**).

(iv) The fourth cyclic class of sesquiterpenoids arises by addition at the pendant double bond of the ion derived from (**5.121**) (Fig. 5.40). Following 1,3-hydride shifts, compounds with a variety of relative stereochemistries based on *cis* or *trans*-decalins (e.g. **5.141**) may be formed and further

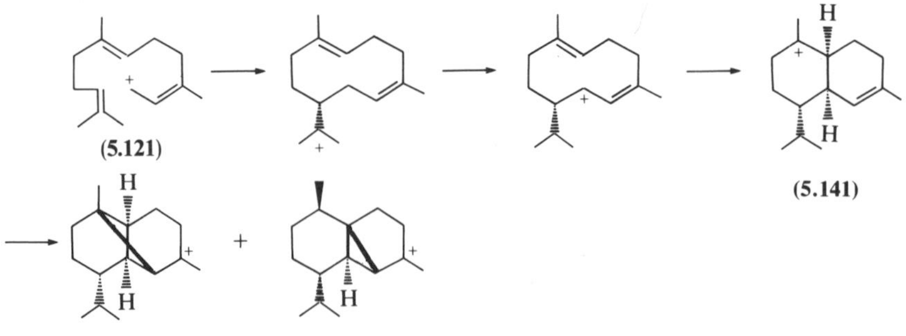

(5.121)

(5.141)

(5.142)   (5.143)

Fig. 5.40.

elaborated to such as the copaane and cubebane skeletons (**5.142, 5.143**). An alternative secondary cyclization followed by ring-cleavage yields helminthosporal (**5.144**), a microbial toxin responsible for damage to cereal crops.

(v) The fifth primary skeletal type arises from the alternative mode of cyclization to the alkene group (Fig. 5.41). One pathway of secondary cyclization then yields caryophyllene (**5.145**), a major component of oil of

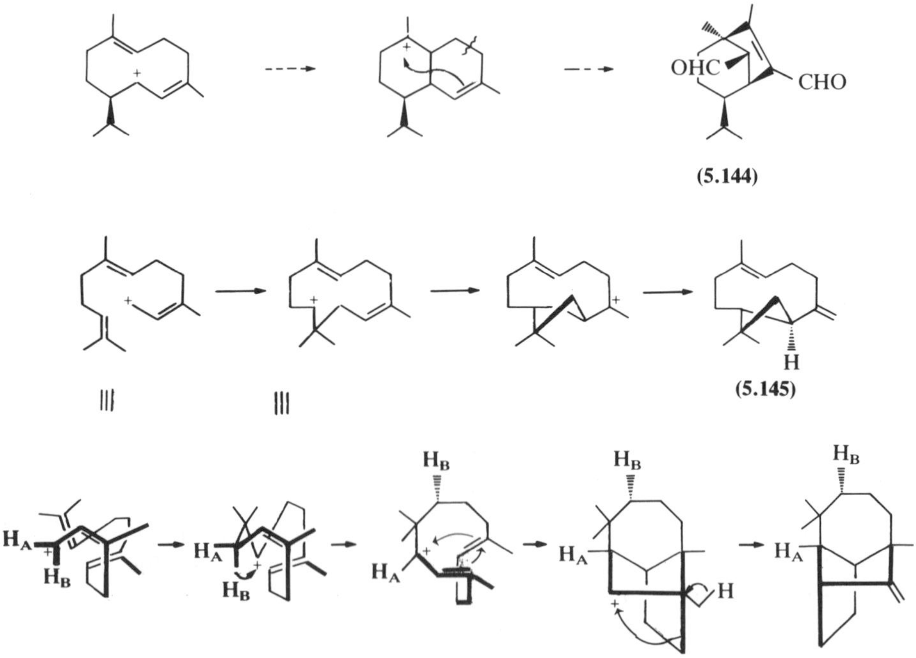

(5.144)

(5.145)

(5.146)   (5.147)   (5.148)

Fig. 5.41.

cloves. *epi*-Caryophyllene which possesses a *cis* ring-junction does occur naturally, but is rare. A more complex pathway is to (+)-longifolene (**5.148**, where an attempt is made to provide stereochemical perspective). The transannular hydride migration (**5.146**) → (**5.147**) has been demonstrated by tracer studies.

The remaining two primary skeletons of the sesquiterpenoids result from the ion derived from 2E,6E-FPP (Fig. 5.42; **5.149**).

(vi) One mode of cyclization gives the germacrene skeleton (**5.150**) which can cyclize to the eudesmane skeleton (**5.151**): the latter can then rearrange to the eremophilane skeleton (**5.152**). It was a compound of the last class that, in 1939, was the first sesquiterpenoid found to abrogate the simpler version of the isoprene rule and this led to the biogenetic reasoning codified by Ruzicka. (**5.150**) can alternatively cyclize to the epimeric valencanes (skeleton **5.153**) that occur in grapefruit or to *spiro*-compounds, e.g. β-vetivone (**5.154**), that are important in perfumery. A Cope rearrangement (Fig. 5.43) can give the elemanes (skeleton **5.155**). The last products have been reported in several oils but it is not clear whether they are natural

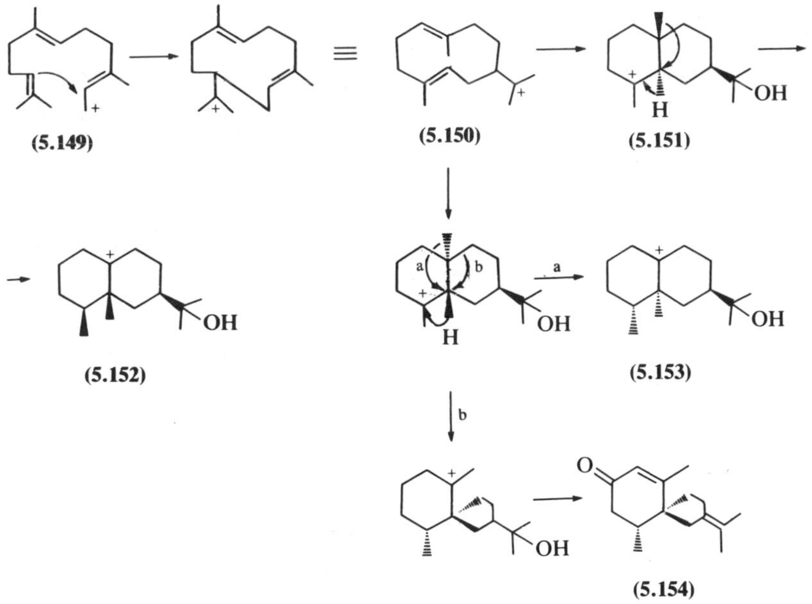

Fig. 5.42.

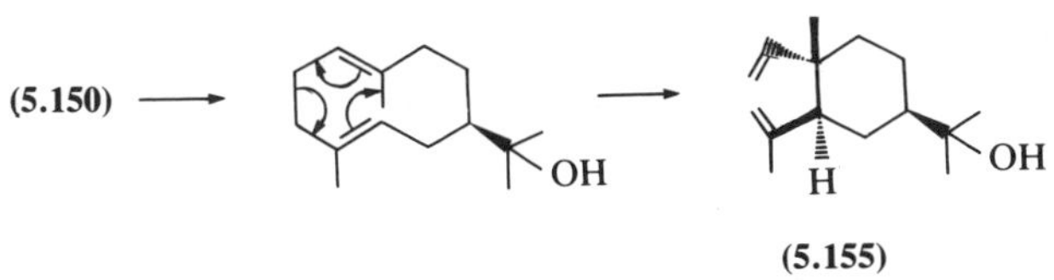

Fig. 5.43.

products or artefacts of isolation. A large number of important related compounds result from a different conformation of the initially cyclized ion (Fig. 5.44). Bulnesol (**5.156**) possesses a seven-membered ring and β-patchoulene (**5.157**) and patchouli alcohol (**5.158**) are tricyclic. The accepted structure of the last compound was for long incorrect: the assignment was based on the formation of (**5.157**) and isomeric alkenes (the structures of which were secure) on pyrolysis of the acetate and so the alcohol was presumed to have the same skeleton. Later X-ray studies conclusively demonstrated the structure (**5.158**) and so an unprecedented rearrangement must have occurred during the elimination.

(vii) The seventh and last class of cyclic sesquiterpenoids is derived from the alternative cyclization of 2*E*,6*E*-FPP. The commonest example is humulene (**5.159**) from oil of hops (Fig. 5.45). The all *trans*-configuration was proved by X-ray studies of the silver salt; and the last two examples illustrate the impact such studies have had on modern structural chemistry. Another member of the class is illudin S (**5.160**) which occurs in luminescent mushrooms. The proposed route (Fig. 5.46) was supported by detailed tracer studies: in particular, incorporation of 1,2-[$^{13}C_2$]-acetate led to identification of the intact C_2 units located as in (**5.161**) – which is consistent with incorporation of three C_5 units as indicated.

Sesquiterpenoids exhibit a rich variety of biological properties. Over 500 lactones have been isolated and many are cytotoxic and play a role in

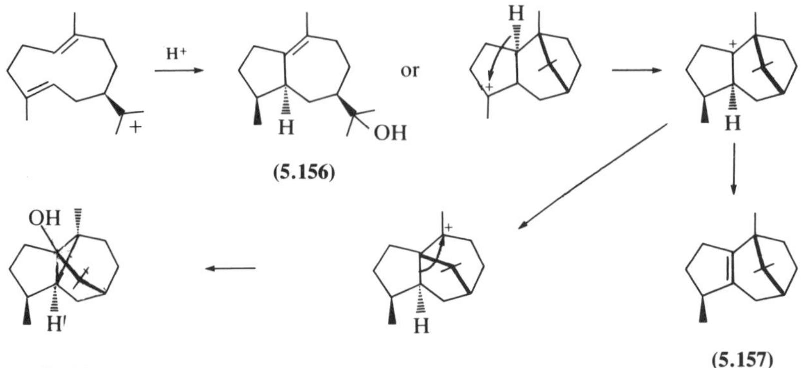

Fig. 5.44.

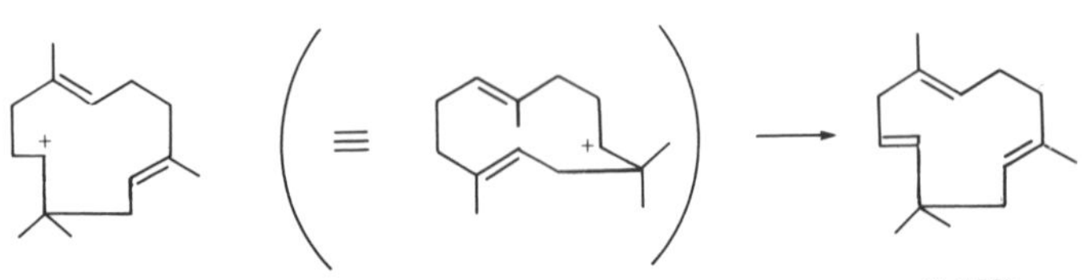

Fig. 5.45.

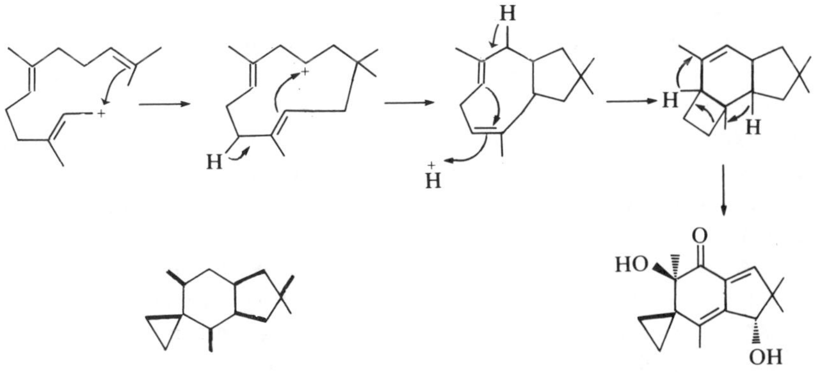

(5.161)   (5.160)

Fig. 5.46.

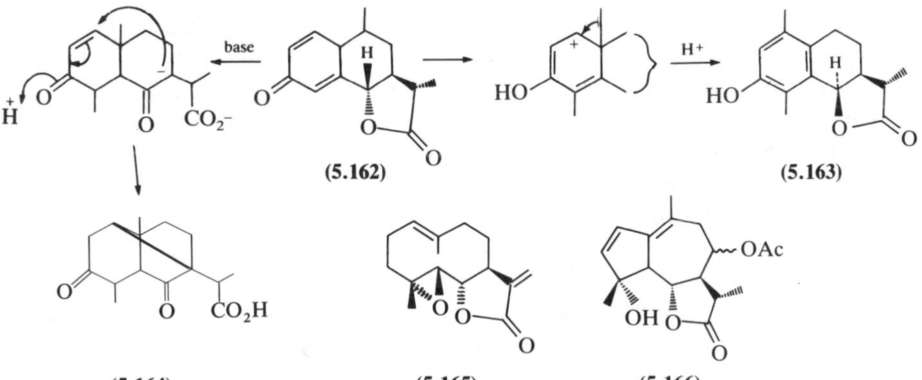

(5.162)   (5.163)

(5.164)   (5.165)   (5.166)

Fig. 5.47.

deterring herbivores (deer, locusts, etc.). Their pharmacological properties have also ensured their frequent use, in crude preparations, in folk medicine. The most famous is α-santonin (5.162; Fig. 5.47) which is widely used in the Far East to combat intestinal worms. α-Santonin rearranges in the presence of acid to (5.163) in what was the first-recorded example of the dienone–phenol rearrangement, which was followed by epimerization of the benzylic centre to form the more stable *cis*-fusion of the γ-lactone ring. In the presence of base, the lactone ring of (5.162) opens and an internal Michael addition leads to (5.164). Artemisinin, the Chinese drug Qinghaosa, which is a 1,2,4-trioxane derived from the skeleton of (5.141) is a valuable anti-malarial drug; other important C₁₅-lactones are parthenolide (5.165), an anti-migraine compound from the herb feverfew, and matricin (5.166) from chamomile. The latter lactone readily oxidizes during extraction from the plant to form a deep blue azulene derivative.

Many phytoalexins, i.e. stress metabolites produced by higher plants after fungal infection, are sesquiterpenoids. Capsidiol (**5.167**; actually a *nor-*sesquiterpenoid), rishitin (**5.168**), and ipomeamarone (**5.169**) are such examples produced by red peppers, tobacco, and sweet potatoes. Another important group of sesquiterpenoids are the juvenile hormones which control the metamorphoses of insects. The first such hormone isolated in 1967 (JHI; **5.170**) was purified (*c.* 300 μg) from some 10^8 specimens of the giant silkworm moth and was a homologue of the sesquiterpenoids. The two 'extra' methyls in the main skeleton were derived from homomevalonate (the C_7-homologue of MVA): thus the skeleton is constructed from two C_6 units plus one C_5 unit rather than the conventional three C_5 units. Five juvenile hormones are now known with different homologies, C_{15} to C_{18}, and they probably act in the majority of insects. JHIII (a C_{15} compound) is the most common and has also been detected in certain plants. A bizarre observation led to the discovery of the phyto-JHs (as opposed to the above zoo-JHs). An insect line established in the USA did not achieve maturity but rather formed giant larvae which subsequently perished. This was traced down to the insects' ingestion of paper from the cage linings and this paper contained a JH-type compound named juvabione (**5.171**) derived from balsam fir that was used in its manufacture. The phenomenon had not been observed in European laboratories as different species of conifer – deficient in juvabione – are there used in the pulp industry. Some 20 such phyto-JHs are now known and in some plants (e.g. sedge) the levels are at least 150-fold greater than those in insect tissues. Presumably the evolution of phyto-JHs is part of the plant's defence mechanisms to deal with insect predation. Insects ingesting such compounds suffer the same hormonal imbalances leading to death that befell the unfortunate insects exposed to the 'paper factor'. An interesting point is that the phyto-JHs (and other synthetic JHs that have been evaluated) may not act as primary hormones but may rather block the enzyme systems that degrade the natural insect JHs at their full term: the result is that the natural hormone levels are maintained beyond the normal periods for their

(5.167)   (5.168)   (5.169)

(5.170)   (5.171)   (5.172)   (5.173)

presence and action. Attempts to use these natural insecticides in the field have been disappointing, probably owing to the difficulty of targeting exogenously supplied JHs (or their analogues) to the active sites of the insect prey.

Numerous other biological effects have been attributed to individual sesquiterpenoids. Two examples are the action of the germacrene diepoxide (**5.172**) as a sex pheromone of the cockroach and the observation that hernanduluin (**5.173**) from a Mexican plant is some 10^3-fold sweeter than sucrose: the sweetness of such plant extracts had been known since antiquity.

The chemistry of the sesquiterpenoids is highly developed. Typical of its sophistication are the patterns of acid-catalysed rearrangements of caryophyllene (**5.154**; Fig. 5.48) to yield caryolan-1-ol (**5.174**), clovene (**5.175**), and neoclovene (**5.176**), and the generation of new ring systems such as in α- and β-bourbonenes (**5.177**) on photolysis of germacrene D (**5.178**; Fig. 5.49).

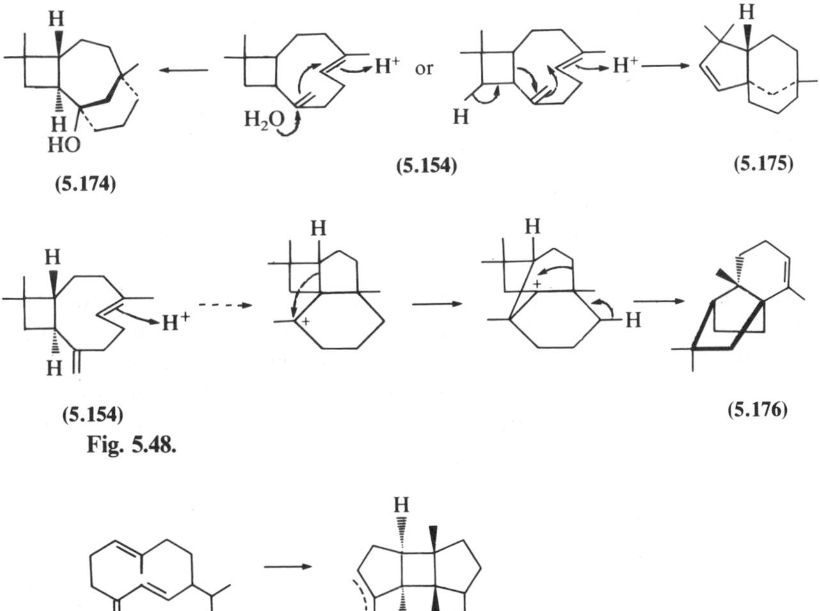

(5.174)   (5.154)   (5.175)

(5.154)   (5.176)

Fig. 5.48.

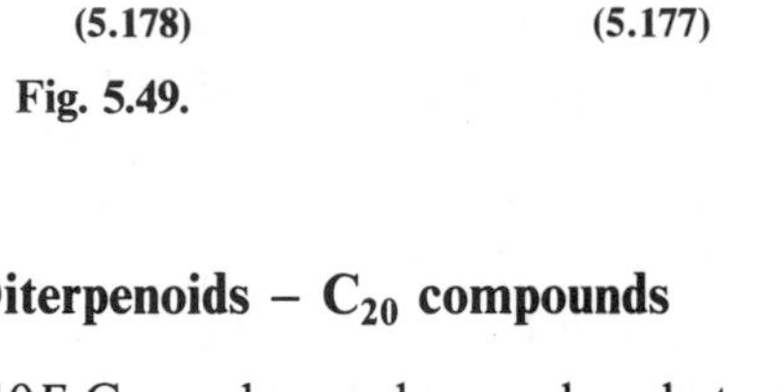

(5.178)   (5.177)

Fig. 5.49.

5.7   Diterpenoids – C₂₀ compounds

2E,6E,10E-Geranylgeranyl pyrophosphate (GGPP; **5.179**), or its allylic geranyl linaloyl isomer, is the parent of this class and is formed by condensation of IPP with 2E,6E-FPP. A few members, e.g. phytol (**5.180**)

(5.179)   (5.180)   (5.181)   (5.182)

(5.183)   (5.184)   (5.185)

which constitutes the lipophilic side-chain of chlorophyll, are acyclic, but most involve cyclization either to the normal or to the less common *ent*-series (**5.181**, **5.182**). Such reactions to form decalins are enzymic but must be highly favoured stereoelectronically as they can often be carried out with great success under purely chemical conditions of acid catalysis. Simple cyclized diterpenoids are sclareol (**5.183**), an anti-fungal compound from sage, and abietic acid (**5.184**), a major component of the resin acids from the residues of turpentine distillations. An elegant demonstration of the α-orientation of the acid group in the latter was the finding that oxidation yielded a tribasic acid (**5.185**) that was optically inactive (and so had a plane of symmetry).

Ruzicka (from 1920 on) and Arigoni and Barton (1950 on) and their coworkers, amongst many others, have explored the more esoteric diterpenoids but the structural complexity of the class is much less than that of the sesquiterpenoids. Important tricyclic members (e.g. phyllocladene, **5.186**) are formed by a route involving an unexpected ring rearrangement rather than the more obvious methyl shift (Fig. 5.50), and the skeleton produced can be hypothetically cleaved and recycled to account for other known types with skeletons (**5.187**) to (**5.189**). ^{14}C-studies have shown that the pleuromutilins (e.g. **5.190**) are formed as outlined (Fig. 5.51), and the bitter principle columbin (**5.191**) is believed to result from a 'backbone' rearrangement which interconverts the normal and the *ent*-series (Fig. 5.52).

The 'decalin'-type of cyclization of GGPP is not a unique occurrence. Another conformation of the latter yields a hypothetical ion (**5.192**; Fig. 5.53) that can give the C_{14}-ring compound cembrene (**5.193**) that occurs in sources as diverse as pine exudate, tobacco, and corals, and a derived lactone which has aroused interest as a tumour inhibitor. Further ring formation yields the skeletons (**5.194**) and (**5.195**) – the latter being that of taxinine, the main toxic component of yew. Cembrene is also the hypothetical parent of the tetracyclic rippertane derivative (**5.196**) which acts as a defensive compound for certain termites. The elaboration of a biogenetic route from (**5.193**) to (**5.196**) is left as an exercise for the reader.

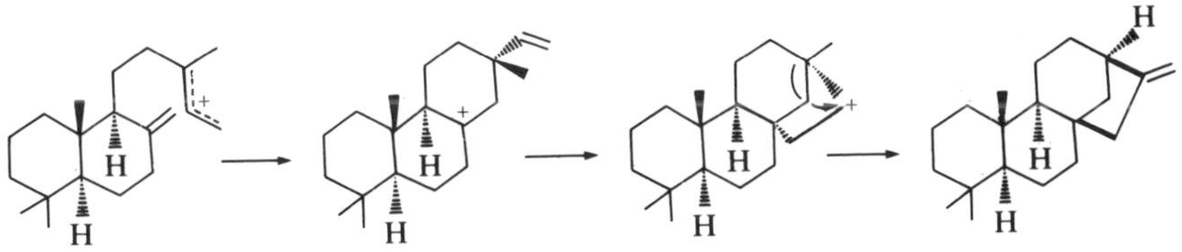

(5.186)

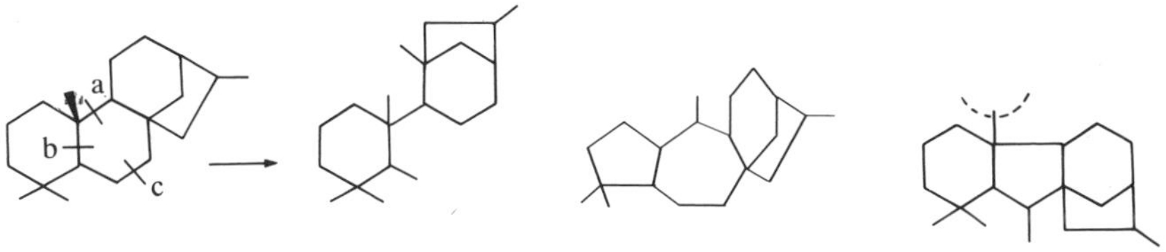

(5.187)   (5.188)   (5.189)

Fig. 5.50.

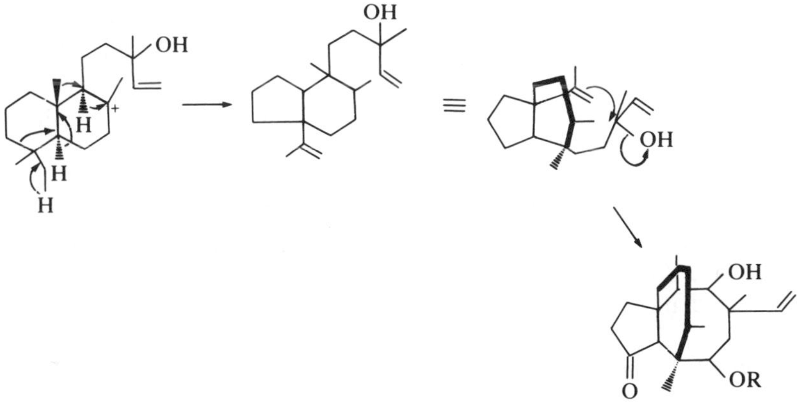

(R = · H
  = · COCH₂OH)   (5.190)

Fig. 5.51.

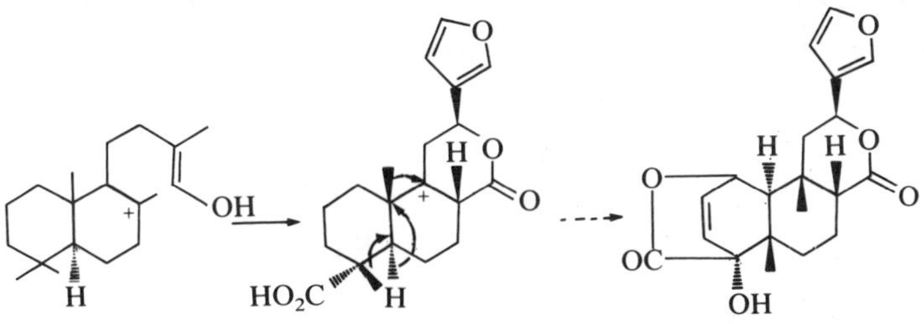

(5.191)

Fig. 5.52.

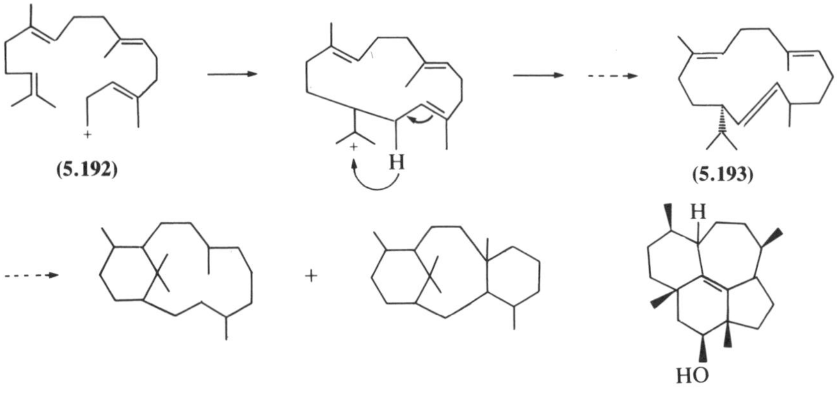

(5.192)

(5.193)

(5.194)   +   (5.195)   (5.196)

Fig. 5.53.

The most biologically significant class of diterpenoids is the gibberellins, the prototype of which is gibberellic acid (**5.197**), nowadays classified as GA$_3$. This was isolated from the culture medium of a fungus of the *Gibberella* species that infects rice seedlings and causes rapid spindly growth followed by collapse and death. GA$_3$ was later shown to be ubiquitous in higher plants and to control cell division, growth and the induction of flowering, and also to be antagonistic towards abscisic acid (Section 5.6). GA$_3$ is a C$_{19}$ compound, but (by 1991) 80 related gibberellins (GA$_1$ to GA$_{80}$) had been characterized including both C$_{19}$ and C$_{20}$ compounds. All possess the gibberellane skeleton (**5.189**) that is generated by cleavage c, Fig. 5.50; and their immediate parent is *ent*-kaurene (**5.198**) which has the unusual geometry of linkage of its A and B rings that is enantiomeric with that in most diterpenoids found in higher plants and fungi. Not all naturally occurring GAs have an OH-group at C$_3$ (as does GA$_3$) and this indicated that the initial cyclization is direct and does not involve an epoxide (for relevance, see Section 5.9); but members devoid of the OH-group did not show plant-regulatory activity and were presumably hydroxylated *in vivo* to active compounds. Low intrinsic levels of GAs are responsible for dwarf varieties of some plants. It has not been found possible usefully to enhance crop yield by application of GAs but the compounds, or their synthetic analogues, have

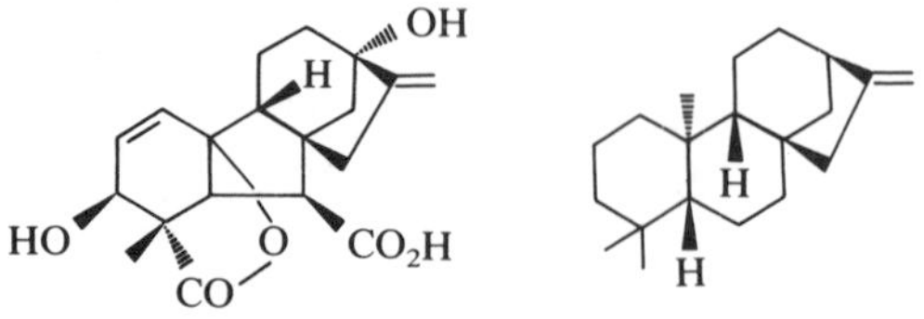

(5.197)   (5.198)

found use as weedkillers in grass or cereal crops. The latter are relatively unresponsive to applied GA (in carefully monitored levels) under conditions where broadleaved plants (cornfield weeds, etc.) elongate and die.

5.8   Sesterterpenoids – C_{25} compounds

For a long time it was thought that, with the exception of the polyisoprenoids (Section 5.12), terpenoids with more than 20 carbons were formed by dimerization of equal-sized moieties. Thus, the sequence was C_5, C_{10}, C_{15}, C_{20}, C_{30}, and C_{40}. Since 1965, the existence of C_{25} (and a few C_{35}) compounds has been recognized, although relatively few structural types are known and the sources seem often rather inaccessible (e.g. pathogenic fungi, marine organisms). Few tracer studies have been carried out, but the compounds appear to be derived from a C_{25} homologue of GGPP – geranylfarnesyl pyrophosphate (GFPP) – all the C_5 units being linearly *trans*-linked. Of some hundred sesterterpenoids reported, many possess the skeleton exemplified by the fungal metabolite ophiobolin A (**5.199**).

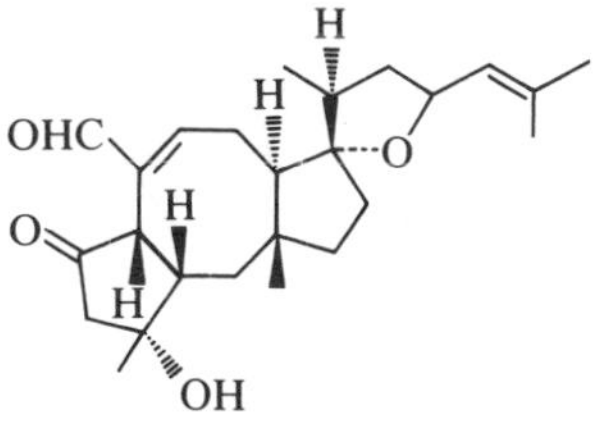

(**5.199**)

5.9   Steroids – C_{18} to C_{29} compounds

Steroids are a subclass of the triterpenoids (C_{30} compounds) based on the skeleton (**5.200**) (R = C_0 to C_{10}). Many are degraded forms and a few are homologous. Consequently they usually fall within the range of carbon content as above. They are considered separately because of their outstanding chemical, biological, and medical importance: indeed for a long time they

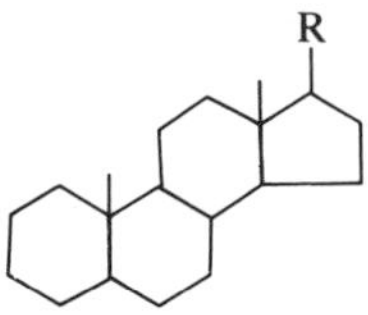

(**5.200**)

were considered a separate class of natural products and it was only the deciphering of the biogenetic pathways that made the relationship apparent. They are here discussed before the residual (non-steroidal) triterpenoids because the known intimate details of their biosynthesis illuminate the relatively unexplored latter class. Lengthy studies on the structure of the steroid skeleton were carried out using the readily available cholesterol (from gallstones) and cholic acid (from bile) and success was achieved in 1932 (see Section 5.3). Cholesterol was shown to be (**5.201**) and the skeleton was numbered as in (**5.202**). Reduction of cholesterol yields 5α- or 5β-cholestanols (**5.203, 5.204**): the former in small amounts accompanies cholesterol in most tissues, whereas the 5β-alcohol (coprostanol) is formed from cholesterol in the digestive tract and occurs in the faeces of most animals. The *trans*-ring junctions in (**5.201**) ensure a flat, rigid molecule (**5.205**) and these properties

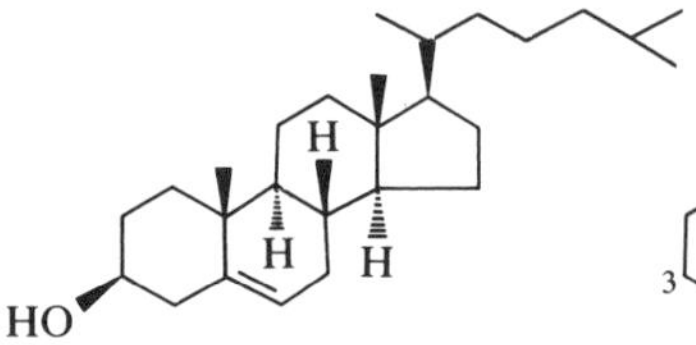

(**5.201**) 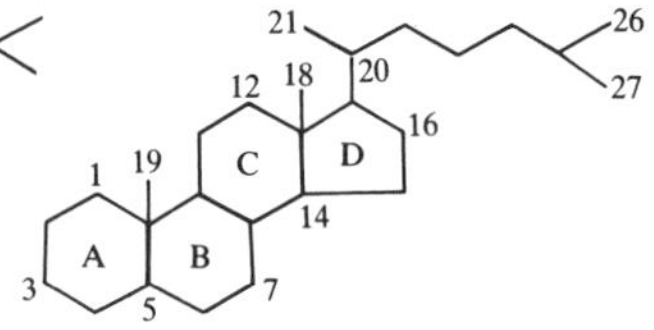   (**5.202**)

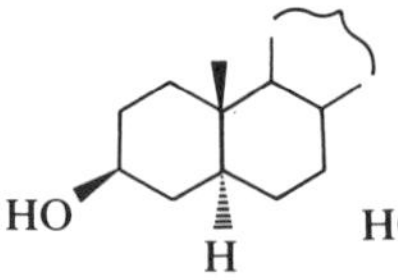

(**5.203**) 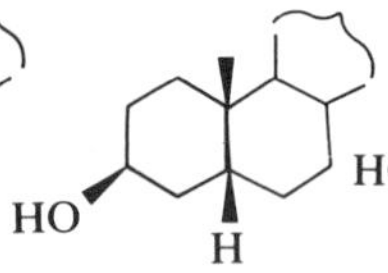   (**5.204**) 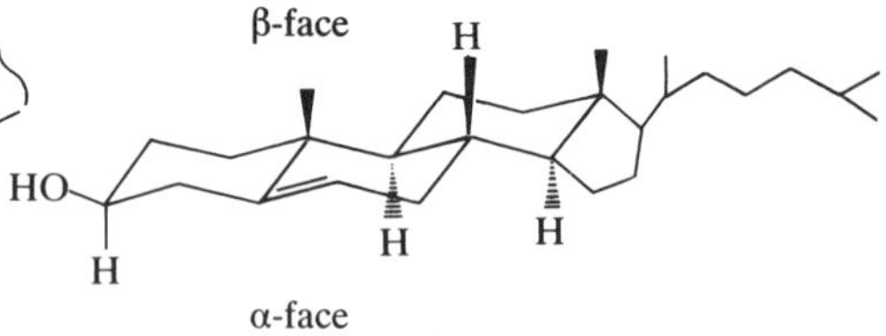   (**5.205**)

may reflect the fact that cholesterol occurs in membranes of animals and plants where it plays an important structural role. Its chemistry is dominated by steric hindrance at the β-face caused by the angular methyls: thus, for example, hydride reduction of oxo-groups at C-3 or C-11 occurs with addition

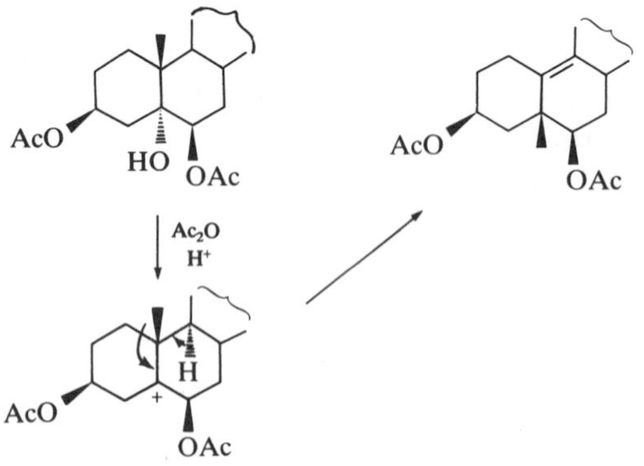

Fig. 5.54.

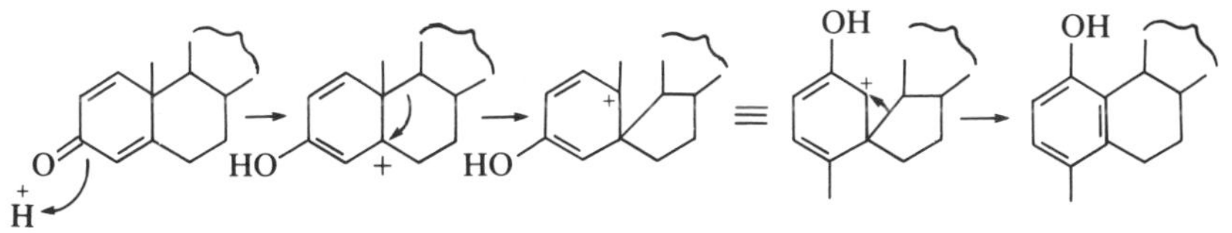

Fig. 5.55.

from the α-face to yield predominantly the β-alcohols. Rearrangements of the steroid skeleton are uncommon but two important types are the Westphalen and the dienone–phenol rearrangements (Figs 5.54 and 5.55).

The biosynthesis of cholesterol has been studied in exceptional detail (by Bloch, Popjak, Cornforth, Lynen *et al.*, 1950 onwards) as a consequence of its implication in arterio-cardiac disease and the realization that it is the parent of all steroids. The all-*trans* polyene squalene (**5.206**) had been discovered in shark liver oil in the 1920s and it was later boldly speculated that this C$_{30}$ compound was a precursor of cholesterol, and indeed an almost correct biogenetic scheme had been proposed linking the two compounds (by Robinson in 1935; Ruzicka in 1953). Later tracer and enzymic studies confirmed that squalene is indeed an obligatory intermediate and the reductive condensation of two molecules of 2E,6E-FPP to this compound is believed to take place through the mediation of a nucleophilic centre at the active site of the enzyme (X-group mechanism; Fig. 5.56). 1,3-Elimination of hydrogen (originally the 5-*pro-S* atom of MVA) then yields presqualene alcohol pyrophosphate (**5.207**) and thence squalene in which one of the central hydrogens is derived from NADPH. Presqualene alcohol had first been found in NADPH-deficient mutants of certain fungi and its role *en route* to cholesterol was unexpected: this should not have been so (with

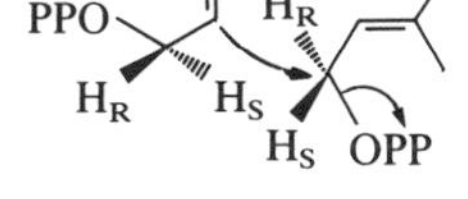

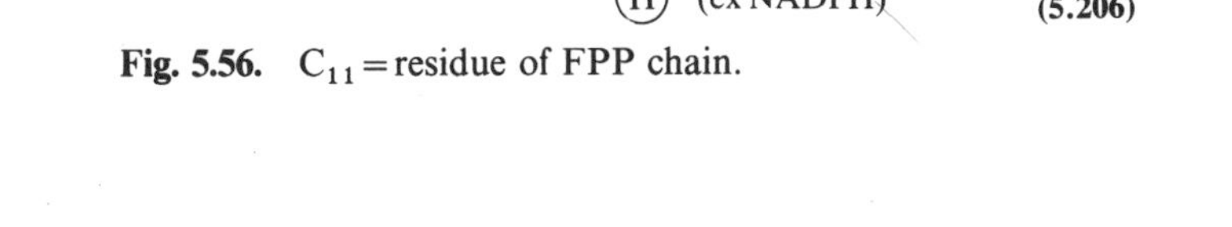

(5.206)

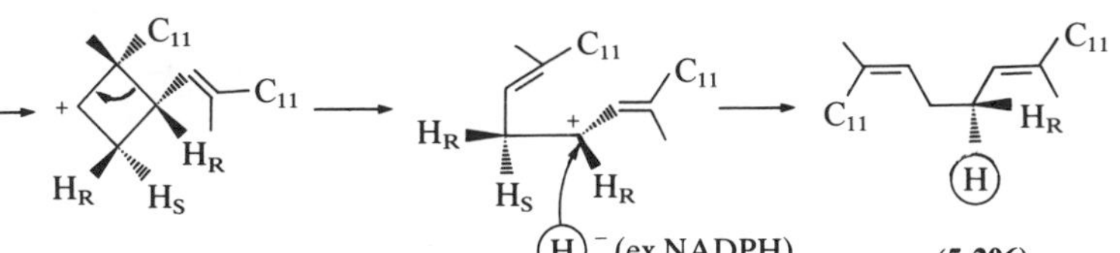

Fig. 5.56. C$_{11}$ = residue of FPP chain.

hindsight!) as the condensation to yield (**5.207**) was the exact analogue of the linking of two DMAPP units in a tail-to-tail fashion to form the known irregular monoterpenoid *trans*-chrysanthemol (Section 5.5). Squalene is then converted into the 3S-oxide (using O_2 from the air as shown by ^{18}O studies) and this is held in a (quasi) chair–boat–chair–boat (c–b–c–b) conformation (**5.208**) on an enzyme surface and cyclized to the hypothetical protosterol (**5.210**; Fig. 5.57) which then synchronously undergoes a series of *anti*-migrations to yield (**5.211**): note that if the β-H at C-9 had migrated, the angular (marked) methyl at C-10 could not have followed suit in a concerted manner as it is *cis* to the former atom: hence the hydrogen does not migrate but is eliminated to place the double bond in (**5.211**). Thus squalene oxide zips-up without detectable intermediates and the product (**5.211**) is a known triterpenoid – lanosterol – a component of lanolin from oil of sheep's wool. Lanosterol has seven chiral carbons and hence there are 128 possible stereoisomers (64 enantiomeric pairs); but the stereoelectronic control in the coupled cyclization rearrangement is such that only one occurs in nature – that with the absolute configuration shown which is also the configuration of cholesterol.

Lanosterol must lose three methyl groups to achieve the steroid skeleton. That at C-14 is lost with cleavage of formic acid (Fig. 5.58), followed by the 4α-methyl (Fig. 5.59) with cleavage of CO_2 and then the 4β-group likewise. The double bond in the side-chain is reduced by uptake of H^+ from water and *anti*-addition of H^- from NADPH. These modifications may not always take place in a defined sequence but may involve a metabolic grid of relatively unspecific enzymes that trace different pathways to cholesterol in different organisms or tissues and so can yield a wide range of possible intermediates.

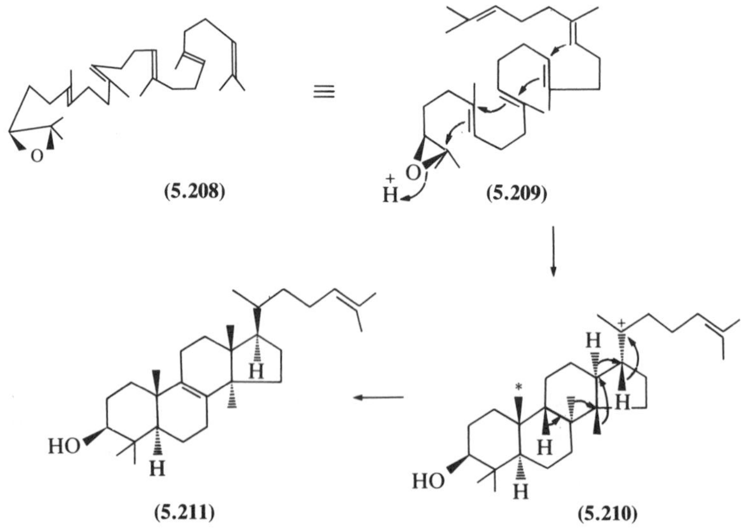

Fig. 5.57.

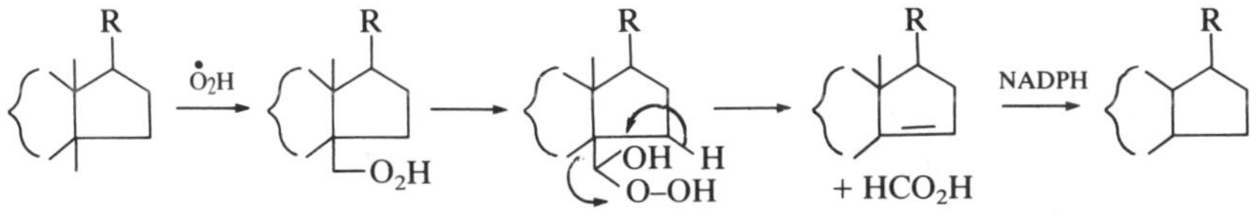

Fig. 5.58.

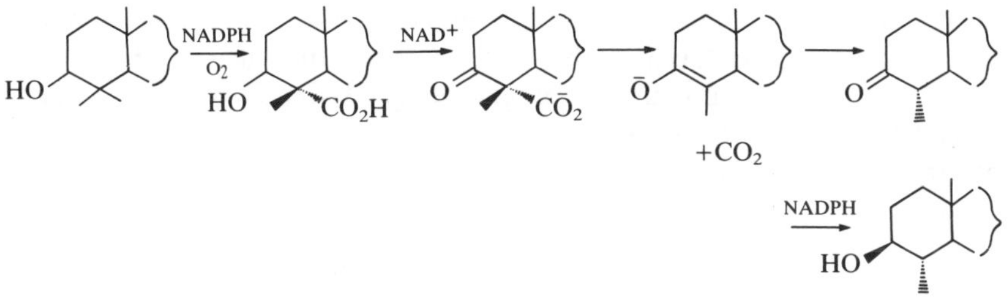

Fig. 5.59.

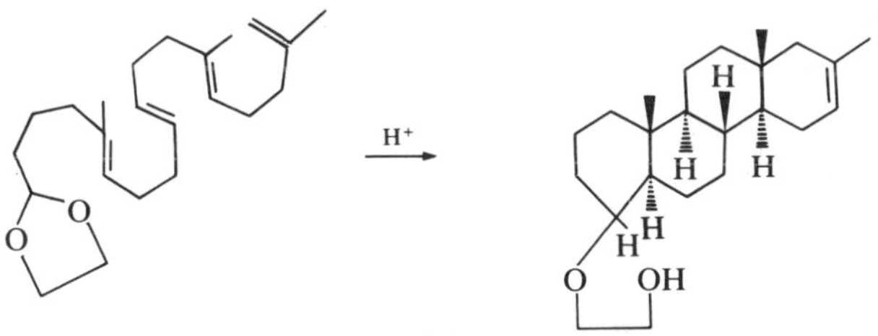

Fig. 5.60.

The evidence for the pathway from FPP is conclusive: most of the individual enzymes have been well studied and the migrations shown in Fig. 5.57 have been demonstrated by studies using ^{14}C and ^{13}C-labelled acetate MVA as precursors. As for the diterpenoids, the formation of *trans*-fused six-membered rings seems favoured even in non-enzymic studies: thus the model system in Fig. 5.60 yielded *c.* 30 per cent of such products on acid treatment.

A great deal of work has been carried out on the factors that control steroid synthesis, as the subject has obvious medical implications. The activity of HMG-CoA reductase (see Section 5.4) is crucial and this enzyme exists *in vivo* in both active and inactive forms – the levels being modulated by phosphorylation regulated by MVA or by low density lipoprotein, and by feedback control by products. Minor metabolites such as 25-hydroxy-cholesterol also inhibit the rate of cholesterolgenesis by increasing the rate of degradation or decreasing the rate of synthesis of HMG-CoA reductase. The whole picture is extraordinarily complex.

5.9.1 Bile acids

Cholesterol can be converted into water-soluble conjugates by linkage to sulphate, sugar, or a specific sterol carrier protein and thus can be transported intracellularly to other metabolic sites. One class of compound derived from cholesterol is the bile acids, most of which are hydroxylated derivatives of cholanic acid (**5.212**; NB A:B junction is *cis*): thus cholic acid has α-OH groups at C-3, C-7, and C-11. The side-chains vary from C_5 to C_8 atoms depending on the animal, and very rarely are 3β-OH compounds found.

The acids are produced in the liver and form the bile salts by peptide linkages to the amino groups of glycine (NH_2CH_2COOH) or taurine ($NH_2CH_2SO_3H$). The function of these salts is to emulsify fats and promote absorption of lipids in the gut.

(5.212)

5.9.2 Sex hormones

An immense amount of work has centred on the mammalian sex hormones derived from cholesterol. These fall within three classes: the oestrogens and the androgens which govern the primary female and male sex characteristics respectively and the progestogens which regulate various functions of the female reproductive cycle. All are remarkably similar in chemical structure, seem to be universally occurring in mammals, and are produced in balanced proportions under the control of gonadotrophic hormones which are peptides secreted by the pituitary gland. The oestrogens are formed in the mammalian ovary and their levels in pregnancy urine (free or as glycouronides) may be enhanced 500-fold over that normally found. The first to be isolated from such urine (by Butenandt, in 1929) was oestrone (**5.213**; $R^1, R^2 = O$; $R^3 = H$) followed by oestriol (**5.213**; $R^1, R^2 = OH, H$; $R^3 = α$-OH) but these are metabolites of the more active primary hormone oestradiol (**5.213**; $R^1 = OH, R^2 = H$; $R^3 = H$), *c*. 15 mg of which was isolated from ovarian tissue of 50 000 sows. Late-pregnancy urine from mares contains oestrogens not found in humans such as equilin and equilenin (**5.214, 5.215**): these are less active than oestrone and may be secondary products.

The first androgen isolated (in 1931) was androsterone (**5.216**): again prepared in small yield (15 mg) from urine (15 000 l) in another epic of separation and characterization. This is also a secondary product, now of

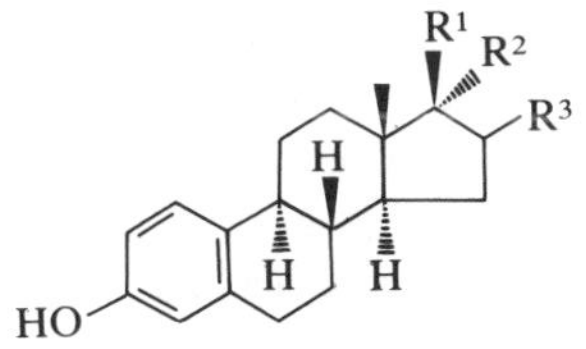

(5.213)

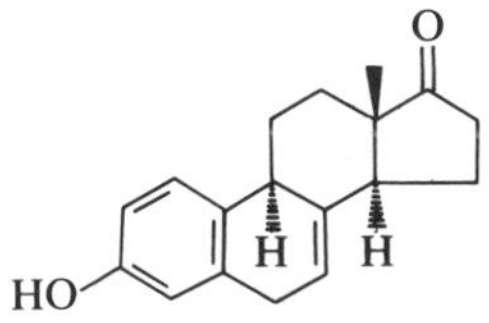

(5.214)

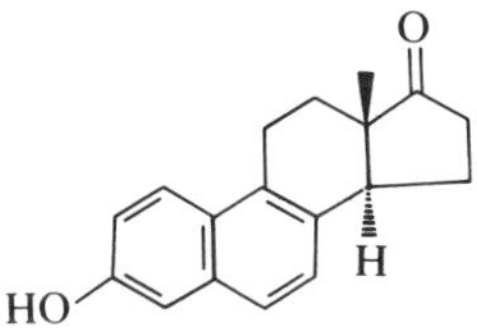

(5.215)

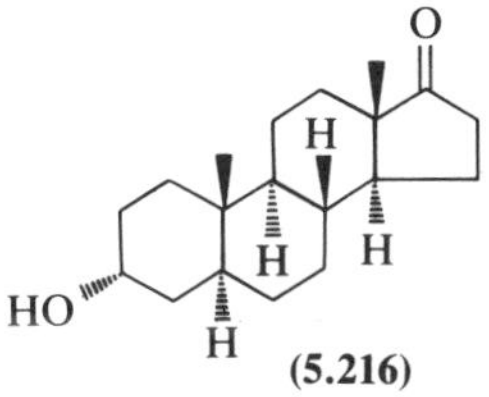

(5.216)

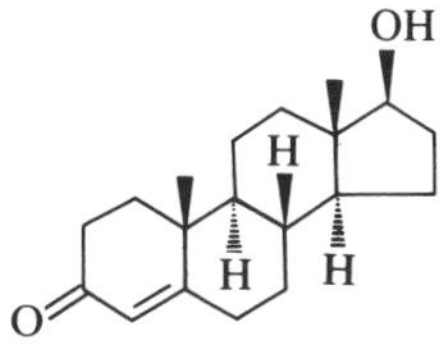

(5.217)

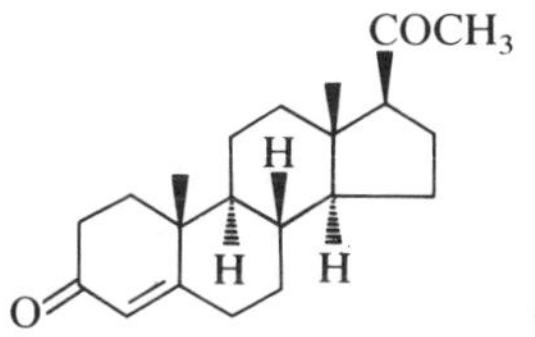

(5.218)

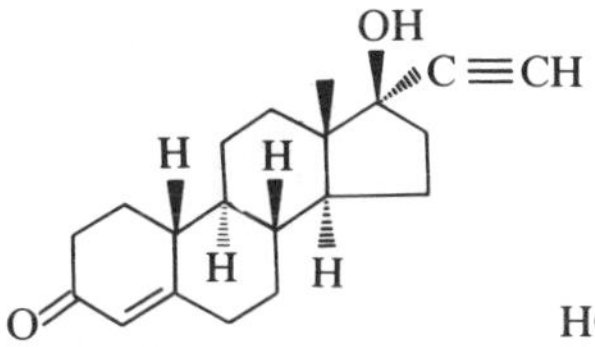

(5.219)

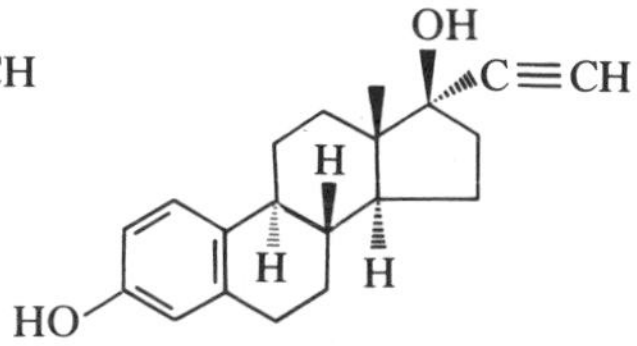

(5.220)

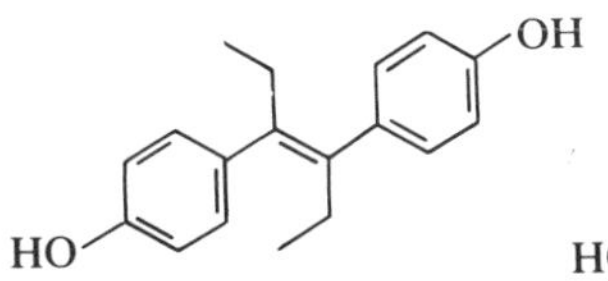

(5.221)

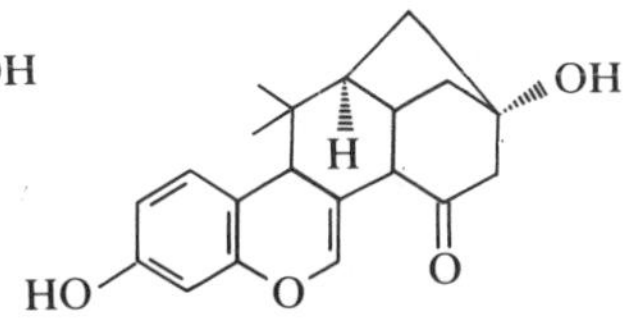

(5.222)

the primary hormone testosterone (**5.217**) that was isolated in 1935 (10 mg from 100 kg of steer testes). The main progestogen is progesterone (**5.218**) which is secreted by the ovary and prepares the uterus for pregnancy, and in the event suppresses ovulation and controls lactation. This compound was isolated in 1934 (in typically minute yield) from sows' ovaries. The prevention of ovulation caused by increased levels of this compound led to the idea that it or synthetic analogues could be used to control pregnancy; but as protesterone itself cannot be administered orally, the advent of such contraceptive techniques had to await the development of safe, synthetic analogues. Such compounds are norethindrone (**5.219**) and novestrol (**5.220**): the former – a 19-norsteroid – preserves the α,β-enone grouping and is a progestogen that is often used in conjunction with the latter, a synthetic oestrogen, in oral contraceptives to maintain the desired balance of hormonal types. The synthetic compound diethylstilboestrol (**5.221**) has oestrone-like activity and is used therapeutically and also for the fattening of poultry and livestock. Miroestrol (**5.222**) is a plant mimic of oestrone that is used in folk medicine in South East Asia to procure abortion: in fact, both male and female mammalian sex hormones occur in minute quantities in plants (pomegranate seed; willow flowers; apple seed) where their function is unclear. Perhaps the genes responsible for these compounds have been cross-transmitted to plants by viruses? A main component of boar odour – which sexually arouses sows – is the truncated steroid (**5.223**).

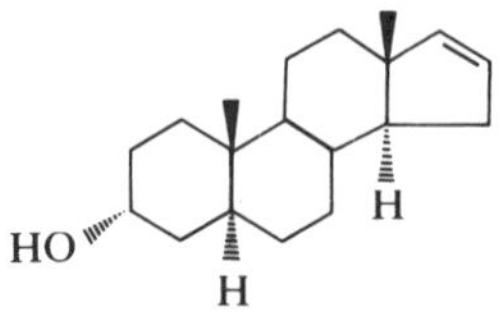

(5.223)

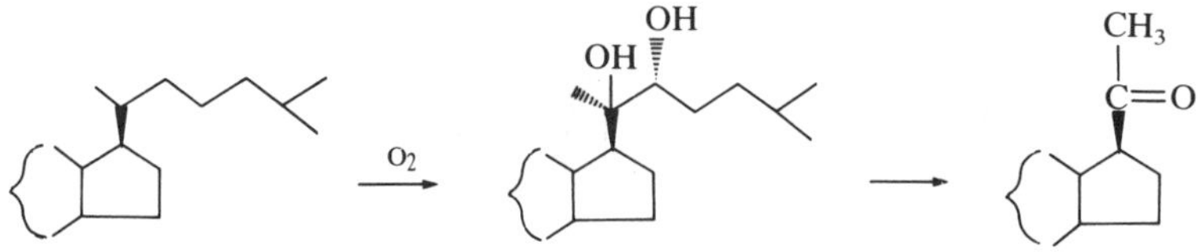

Fig. 5.61.

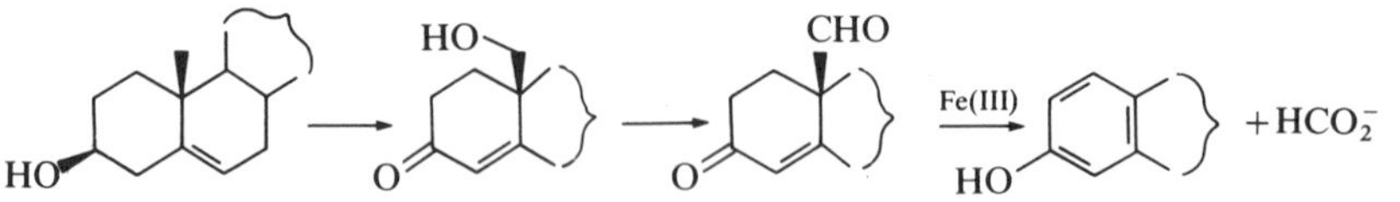

Fig. 5.62.

The side-chain of cholesterol is shortened initially by a two-step oxidation mediated by cytochrome P-450 followed by cleavage of the diol; further oxidation then leads to the C_{17}-oxo compound (Fig. 5.61). Modification at ring A is catalysed by an enzyme complex (aromatase) which utilizes NADPH, O_2, and Fe(III) (Fig. 5.62).

5.9.3 Adrenocortical hormones

The adrenal gland is essential to life and its cortex contains numerous steroids including at least seven active corticosteroids that play a role in carbohydrate metabolism, salt and water balance, and in controlling allergic and inflammation responses. The levels of these hormones are controlled by ACTH (adrenocorticotrophic hormone), a peptide also secreted by the adrenal cortex. The main corticosteroids are cortisol (**5.224**), cortisone (**5.225**), and aldosterone – which has the structure of (**5.224**) minus the

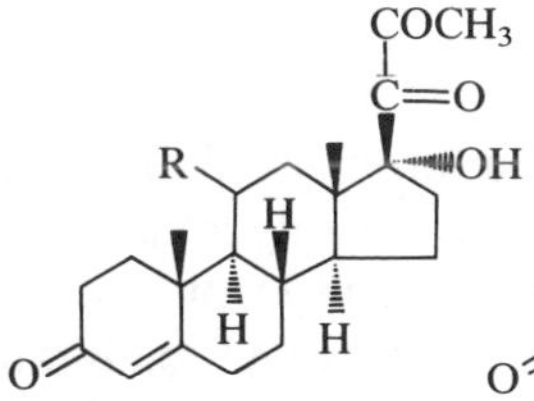

(5.224) R = β-OH
(5.225) R = O:

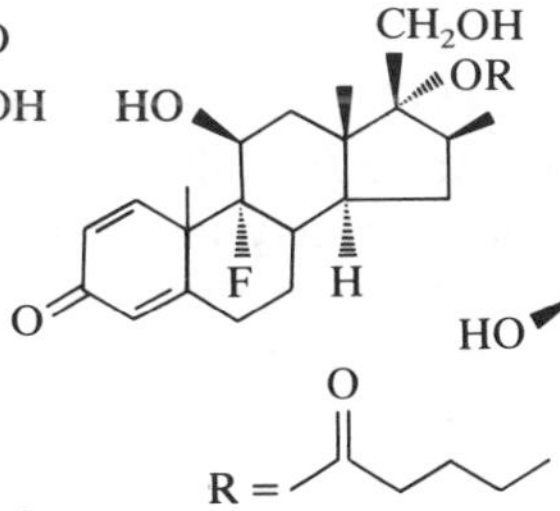

R =

(5.226)

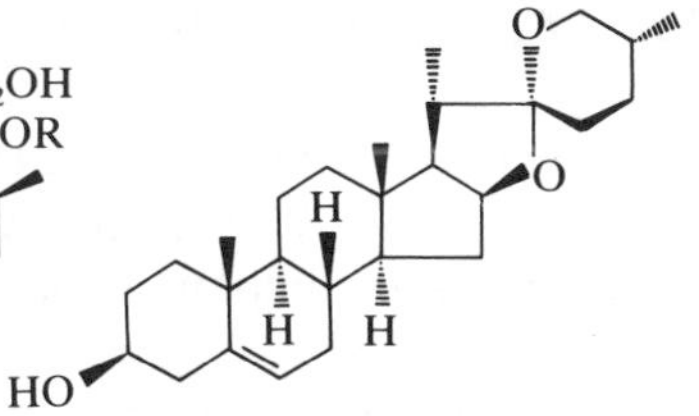

(5.227)

OH-group at C-17 and with the methyl at C-13 converted into a CHO-group (Reichstein, 1940 on). Cortisol and cortisone were early used for the treatment of rheumatoid arthritis and asthma but had undesirable side-effects. The search for synthetic substitutes led to the great upsurge in steroid chemistry in the 1950s and later, in which chemical and microbial techniques were devised for functionalization of virtually every carbon of the skeleton. Numerous synthetic corticosteroids were unearthed and Betnovate (**5.226**) has been one of the most successful.

5.9.4 *Saponins*

These surface-active compounds are distributed in several plant families, and leaves containing them have been used as natural soaps. They are composed of sugar residues attached to the β-OH group at C-3 of the C_{27}-aglycone: the latter is termed a sapogenin. An important sapogenin is diosgenin (**5.227**) which comprises up to 6 per cent w/w of the roots of *Dioscorea* (yam) species and has been commercially exploited in the synthesis of medically important steroids (Section 5.13). The modified side-chain which is characteristic of sapogenins is hypothetically constructed as in Fig. 5.63. Aqueous solutions of saponins, when injected, cause the haemolysis of red blood cells, and so their presence in plant extracts can be easily detected by spot tests on gels containing blood, although ingestion of the compounds is relatively harmless. Most molluscicides of plant origin are saponins, and trees of the *Swartzia* genus which contain saponins have been used to control the bilharzia snail in West Africa. These compounds are also very toxic to fish.

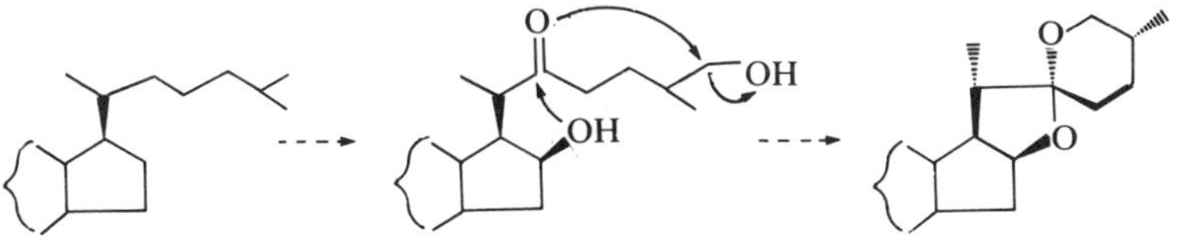

Fig. 5.63.

5.9.5   *Cardiac glycosides*

These comprise a family of C_{21} to C_{24} steroids linked through C-3 to a chain of sugars: the latter ensure water-solubility and are easily cleaved on extraction. The compounds occur widely in temperate and tropical plants and also in the skin secretions of (especially tropical) frogs and toads, and being highly toxic with a powerful effect on heart muscle are very effective deterrents to predators. Appropriate plant and animal extracts have been used as arrow poisons by native hunters, but in low concentrations the cardiac glycosides act as a heart stimulant and consequently have found use in folk and ultimately conventional medicine. A familiar example is the digitalis extract from foxglove, in which the active principle is the glycoside digitonin containing the aglycone digitoxigenin (**5.228**) with *cis, trans, cis* ring junctions. This and related compounds also occur in buttercup, wallflower, and many other familiar herbs. Such aglycones with the five-membered hetero-ring attached at C-17 are classified as cardenolides. Bufadienolides or bufotoxins, in contrast, are fewer in number and have a six-membered appendage and occur in frogs, toads, and the squill family of plants. An example is bufotalin (**5.229**) from toads. Such compounds are no doubt responsible for the inclusion of toads' organs in esoteric recipes in folk medicine and black magic. Cardenolides and bufadienolides are biosynthesized by condensation of acetate and propionate respectively with progesterone (**5.218**), cyclization to form the heterocyclic ring, and then skeletal functionalization.

Cardenolides are implicated in a classic defence system of certain butterflies. Monarch and other danaids lay eggs on plants containing these compounds and the larvae, and subsequently the adults, have evolved the capacity for detoxification and storage of the ingested cardenolides which nevertheless renders the insects unpalatable and emetic to birds. Thus a secure niche for the deposition of eggs and also defence in after-life is achieved. Also noteworthy is the ability of many cardiac glycosides, notably digitonin, to

$R^1 = \beta\text{-}$; $R^2 = H$ (**5.228**)

$R^1 = \beta\text{-}$; $R^2 = OAc$ (**5.229**)

complex with steroids containing 3β-OH groups to form insoluble crystalline products. Great use has been made of this in studies on isolation and biosynthesis.

5.9.6 *Vitamin D*

Fat-soluble products from cod and other fish liver oils were found to ameliorate the symptoms of rickets, a deformation of the bones of children due to faulty calcium metabolism, and the corresponding adult disease of osteomalacia. Exposure of the victims to sunlight or UV-irradiation of the food of experimental animals also effected a cure and these afflictions were traced to the dietary deficiency of minute amounts of the so-called vitamin D. Detailed investigations of the products of photolysis of ergosterol (**5.230**) from yeast revealed the sequence of the photochemical and thermal reactions in Fig. 5.64. Vitamin D_2 was characterized as calciferol (**5.233**) and vitamin D_1 as the 1:1 complex of (**5.233**) with lumisterol (**5.232**). These two vitamins are obtained by humans from ingestion of plants. Vitamins D_3 and D_4 are present in fish oils and also are formed in the skin on exposure to sunlight: they are analogous compounds to (**5.233**) but are derived from 7-dehydrocholesterol (**5.234**) and 7,22-dehydrocholesterol, respectively. Almost alone of the reactions discussed in this chapter, these are not enzymic processes, and as deep sea fish must experience very little sunlight it is

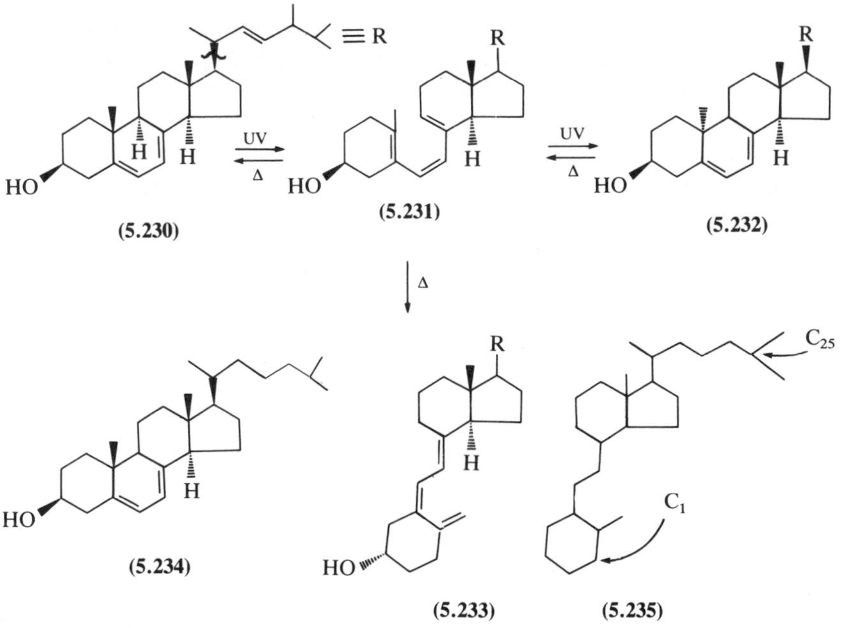

Fig. 5.64.

presumed that the vitamins in the liver oils are accumulated from a diet of plankton. The vitamins D should be classified as hormones as their active principles have recently been shown to be the 1,25-dihydroxy or 25-dihydroxy compounds (cf. **5.235**) produced by *in situ* modification followed by secretion.

Rickets is a condition of malnutrition, coupled with a life largely spent under artificial light or in conditions of atmospheric pollution. Consequently, it is seldom encountered nowadays in developed countries. Nevertheless large quantities of vitamin D_3 are produced for food supplements by bromination of cholesterol at C-7 followed by elimination to the 7-dehydro compound and photolysis. The sequence of specific photochemical and thermal cyclizations and isomerizations (only partly shown in Fig. 5.64) was instrumental in the development of the Woodward–Hoffmann rules for the conservation of orbital symmetry whereby organic processes involving ground or excited states can follow different but predictable stereochemical routes.

5.9.7 *Phytosterols*

Ergosterol (**5.230**) is a member of the phytosterols which as a class can play a role in the structure and permeability of membranes in higher plants, algae, and fungi corresponding to that played by cholesterol and other sterols in animals. Fungi and animals synthesize lanosterol, as previously discussed, *en route* to cholesterol and other sterols. In contrast, higher plants and algae utilize a route whereby the hypothetical protosterol (**5.210**) rearranges and cyclizes to cycloartenol (**5.236**) which is then metabolized to the phytosterols or cholesterol. Construction of the C_3 ring that was concerted with the other migrations would entail the *syn*-loss of a hydrogen and it seems likely that formation of cycloartenol involves the intrusion of an 'X-group' mechanism whereby enzyme-linked intermediate is formed that can be broken-down in an *anti*-elimination (Fig. 5.65): thus in plants the angular hydrogen at C-8 migrates to C-7 rather than being lost (with double bond formation) to form lanosterol. The reason for this complication is not understood.

Phytosterols also differ from sterols in that the side-chain is alkylated at C-24. Ergosterol illustrates this, and sitosterol, stigmasterol, and fucosterol possess the cholesterol nucleus with the side-chain modified as in (**5.237**)–(**5.239**). The additional carbons are introduced stepwise using the common biological methylating agent, *S*-adenosylmethionine (Fig. 5.66), and tracer studies have revealed the hydride shift shown. The production of alkylated steroids may well be part of the plants' defence system. Insects,

(**5.237**)   (**5.238**)   (**5.239**)

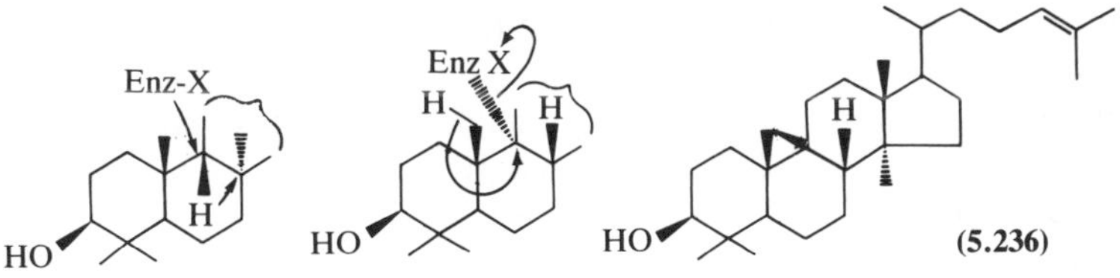

Fig. 5.65.

(5.236)

$$CH_3-\overset{+}{S}(Me)-R$$

recycle
reduce

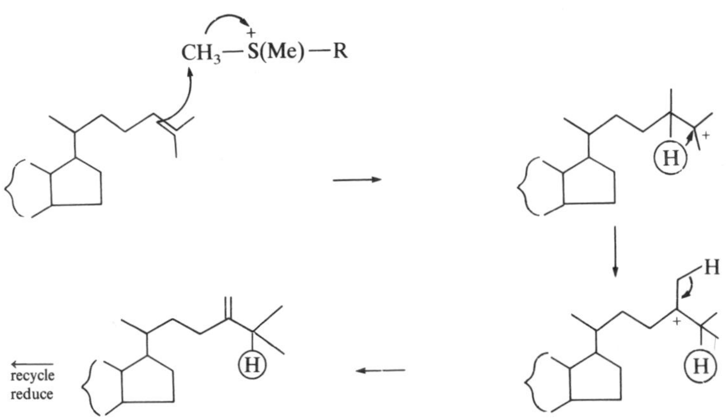

Fig. 5.66.

protozoa, and bacteria cannot biosynthesize sterols *de novo* and need to accumulate them from their diet (i.e. as vitamins!). Thus all such existing phytophagous organisms must possess enzyme systems capable of dealkylating (via epoxidation) these side-chains to produce cholesterol or its close relatives. During evolutionary history, the appearance of modified sterols, i.e. phytosterols, in plants must have endowed great survival value in the constant struggle with insect predators, and the relicts of these defensive mechanisms still exist.

5.9.8 *Miscellaneous*

As shown in the above discussion, steroids possess numerous physiological activities, doubtless due to the nature of the skeleton which allows substituent groups (which interact with receptors) to be held in rigidly stereochemically defined orientations and, in addition, endows a favoured conformation for the whole molecule to allow, for example, insertion into membranes. Of several other subgroups, the most explored is possibly the ecdysterols or insect moulting hormones. These are modified *in vivo* from ingested cholesterol, and act on epidermal cells to induce skin-shedding such that, in conjunction with the juvenile hormones, they control the life-cycle from larva through pupa to adult. In 1954, small quantities (*c.* 25 mg) of α-ecdysone (**5.240**) were isolated from a tonne of silkworm moths, but more recently the same mass has been recovered from only 10 g of yew root or of the rhizomes of ferns! Some 100 such phytoecdysones have been characterized, compared with 6 zooecdysones, and the former are undoubtedly part of the chemical defence systems involved by plants, cf. the discussion of juvenile hormones. The active hormonal principle seems to result from hydroxylation at C-20 *in situ* of (**5.240**) and its relatives.

Other examples of biologically active steroids are pavoninin (**5.241**), a shark-deterrent released (at the initiation of attack) by sole; and the potent salamander toxin (**5.242**). The latter is perhaps best classified as a terpene alkaloid (see Chapter 7).

(**5.240**)   (**5.241**)   (**5.242**)

5.10   Non-steroidal triterpenoids – mainly C_{30} compounds

These, the rump of the metabolites derived from squalene, are numerous and widespread in the plant kingdom but most do not appear to possess any particular biological properties. Before the advent of NMR and MS, the structures of many were intractable as there are usually few positions for chemical attack to open up the ring systems. The characteristic 3β-OH; 4,4-dimethyl grouping in ring A was often detected by rearrangement and oxidation to liberate acetone.

Three structural types may be discerned. The first has a homosteroidal skeleton. Thus lanosterol is formed by folding 2,3-oxosqualene into the (c–b–c–b) conformation whereas a (c–c–c–b)-folding leads to a different protosterol ion and results in euphol which is epimeric with lanosterol at C-17 and at the C–D ring junction. Many nortriterpenoids are formed by loss of the methyls at C-4 and C-14, and truncation of the side-chain by 4-carbons followed by furan formation occurs in the highly functionalized bitter principle limonin (**5.243**) from citrus. A sub-type results from incomplete cyclization from each 'end' of squalene. Thus the conformation (c–c–u–u–c; u = unfolded) leads to ambrein (**5.244**), a prized perfumery compound isolated from ambergris, a secretion of the sperm whale: this cyclization is not oxidatively induced as there is no OH group at C-3. The same conformation yields α-onocerin (**5.245**) which cyclizes to serrantenediol (**5.246**) containing a C_7-ring. The second and third main types involve full cyclization of 2,3-oxosqualene to give ring systems exemplified by lupeol

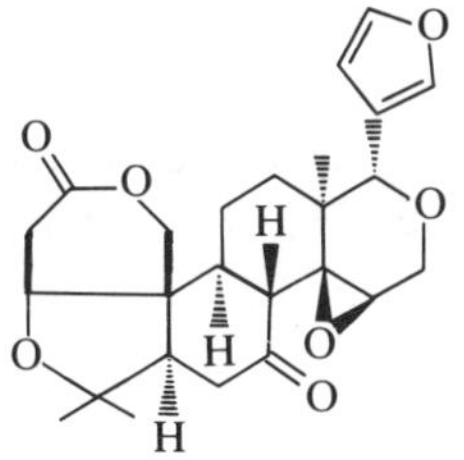

(5.243) 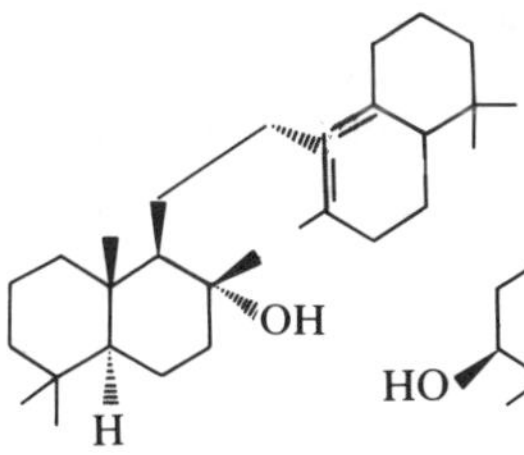   (5.244) 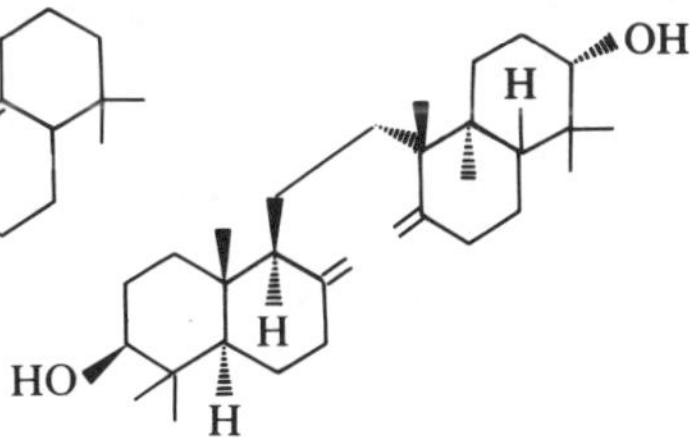   (5.245)

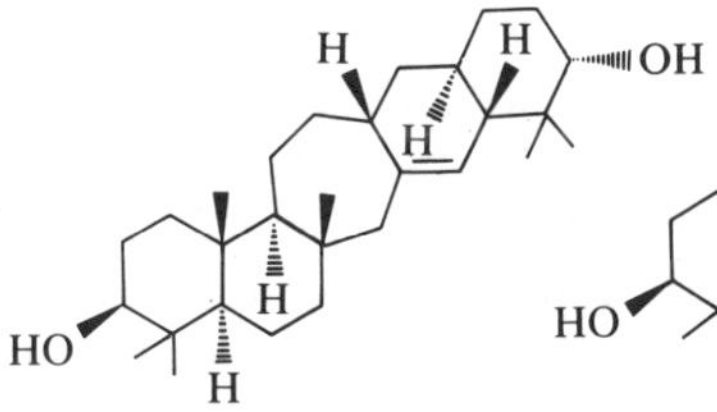

(5.246) 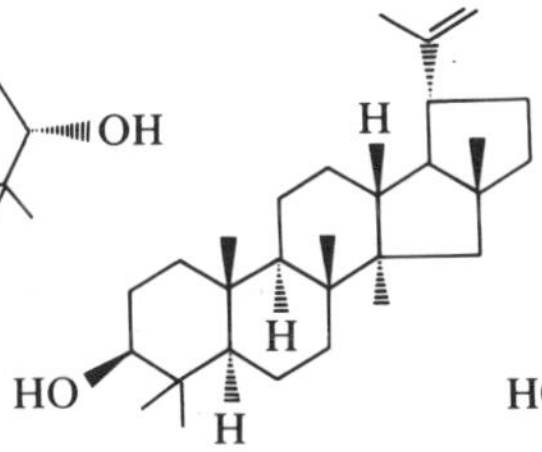   (5.247) 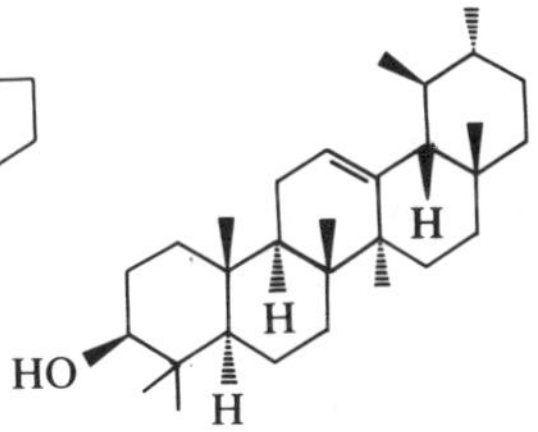   (5.248)

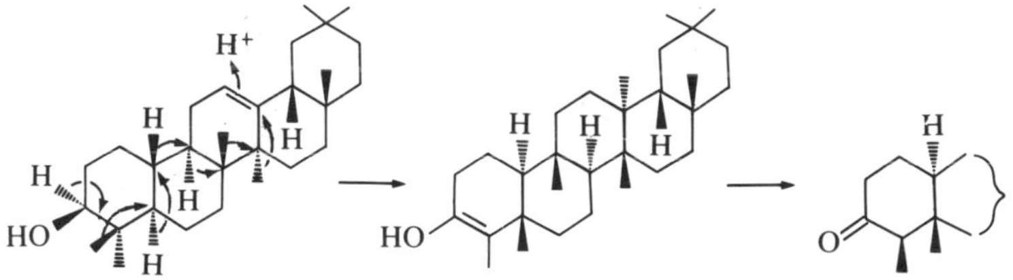

(5.249)   (5.250)

(5.247; from lupin seed; mistletoe) and α-amyrin (5.248; ubiquitous). A 'backbone'-rearrangement of the isomeric β-amyrin (5.249) yields friedelin (5.250), a component of cork which was isolated, although of course not characterized, in 1807. Another important C-30 compound, gossypol, is (structurally) a dimer of 2 cadinane (C-15)-units: it has become important as an anti-tumour, anti-fertility, and anti-HIV compound.

5.11 Carotenoids – C$_{40}$ compounds

The above name is universally adopted for the tetraterpenoids and reflects the discovery of members of the class in carrots in the 19th century, although separation of the individual components was not possible until the advent of column chromatography (by Karrer, Kuhn, Zechmeister *et al.*, 1928 onwards). Carotenoids occur in all photosynthetic tissue (green foliage, photosynthetic bacteria, and algae), in egg yolk, in roots (carrot, swede), and as a pigment in yellow flowers (roses, daffodils, marigolds, tulips, etc.).

Some bacteria produce C_{45} and C_{50}-homologues. The carotenoids are easily oxidized or rearranged during extraction from plant tissue.

The parent of the class is produced, in a manner analogous to that of the triterpenoids, by tail-to-tail coupling of GGPP, cf. (**5.251**). Just as for

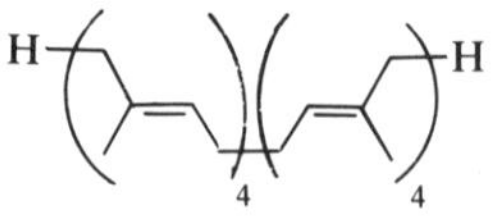

(**5.251**)

squalene a C_3-ring intermediate is formed, but this breaks down in a slightly different way by proton loss rather than by hydride uptake (Fig. 5.67) to yield phytoene (**5.253**; Fig. 5.68) with a central *E* or more usually a *Z*-linkage. The intermediate, prephytoene alcohol (**5.252**; directly analogous to **5.207**), has been isolated from mutants of *Neurospora* fungal species that cannot produce carotenoids. Lycopersene, the direct analogue of squalene, occurs naturally but is not thought to be an obligatory precursor of the class. The formation of the rigid triad of conjugated double bonds at the centre of

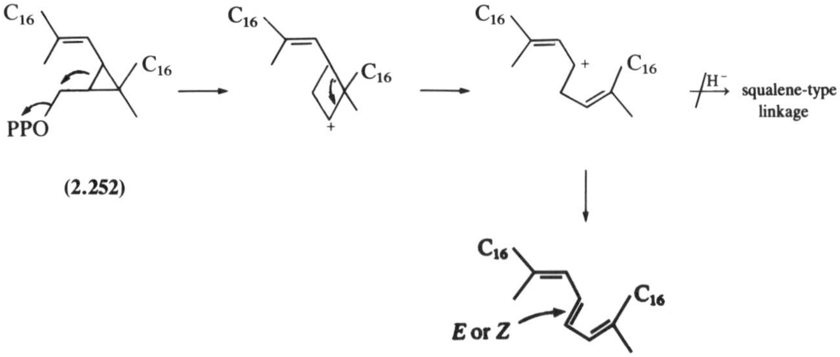

Fig. 5.67.

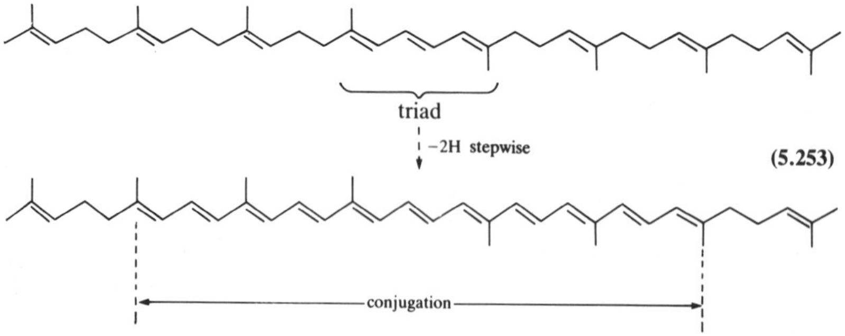

Fig. 5.68.

phytoene has profound stereochemical consequences. Unlike squalene (or notionally lycopersene), this precursor cannot be curled-up on enzyme surfaces to generate conformations leading directly to polycyclic compounds and as a result there is little complexity of structure in the class.

Phytoene (Fig. 5.68; **2.253**; the *E*-isomer is here written for convenience) is oxidized *in vivo* by stepwise removal of two hydrogens from each side of the central triad in turn to yield derivatives with increasing conjugation resulting eventually in lycopene (**5.254**). Phytoene is a colourless oil and occurs in unripe fruits but the increasing conjugation which occurs during ripening results in yellow and red compounds (as the chloroplasts transform into chromoplasts) culminating in (**5.254**) which is the pigment of tomatoes. The unconjugated end groups of (**5.254**) can then cyclize (Fig. 5.69) to form pendant rings (e.g. **5.255**), which especially in certain marine organisms may become aromatic (**5.256**). The main (red) carotenes in higher plants are γ-carotene (**5.257**), β-carotene (**5.258**; especially abundant in carrot root) and α-carotene (**5.261**; Fig. 5.70).

The yellow xanthophylls are ubiquitous ring-oxygenated derivatives of carotenes and common examples are zeaxanthin (**5.259**), violaxanthin

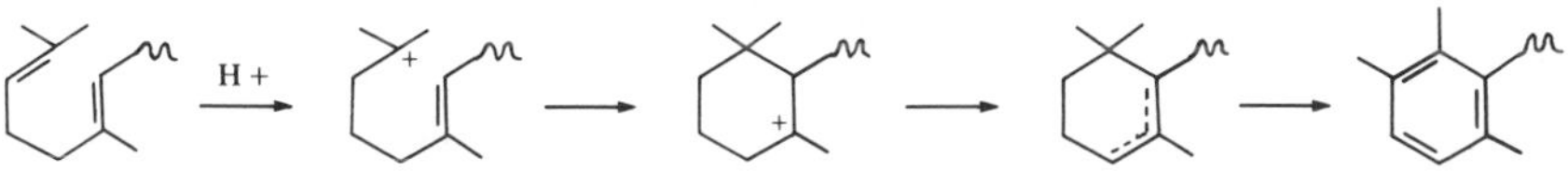

(5.255)   (5.256)

Fig. 5.69.

(5.254) ⟶   (5.257) ⟶

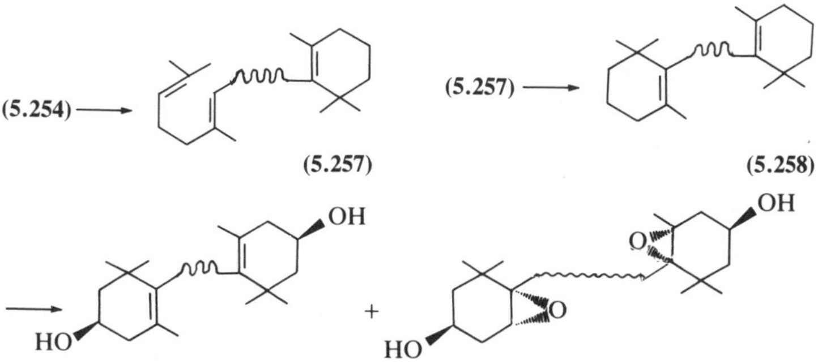

(5.257)   (5.258)

(5.259)   +   (5.260)

(5.257) ⟶   (5.261)   (5.262)

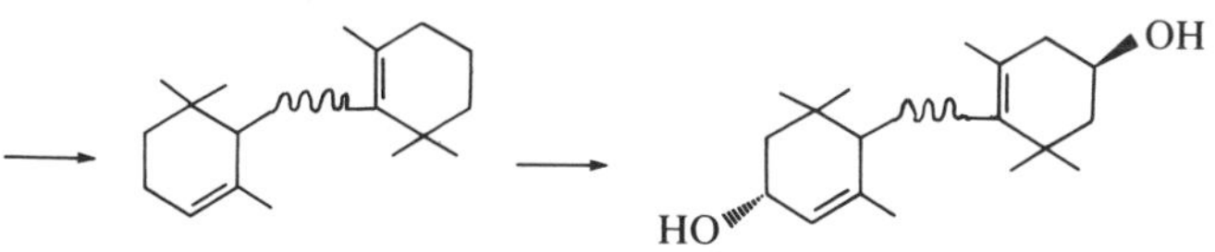

Fig. 5.70.

(**5.260**), and lutein (**5.262**). Rearrangement of the epoxide group leads to capsanthin (**5.263**), the main pigment of red pepper. Xanthophylls predominate (in esterified or protein-bonded form) in autumn leaves where they are degraded less rapidly than are the green chlorophyll and the red carotenes.

(**5.263**)

Carotenoids have profound physiological importance for photosynthetic plants. They occur in intimate association with chlorophyll wherever it is stored and apparently act as scavengers to remove the potentially harmful oxygen-containing radicals and excited states formed concomitantly with the primary processes of photosynthesis, e.g. by quenching singlet (excited state) oxygen, and they also quell the excited state of chlorophyll. Thus the simple carotenes are converted into xanthophylls which are enzymically recycled to carotenes via a 'xanthophyll cycle'. Certain 'bleaching' herbicides block phytoene synthetase and in the absence of continuing synthesis of carotenoids (to replace natural degenerative turnover) the treated plant soon dies, having lost its photoprotection. Similarly, any natural mutation that seriously lowers the level of carotenoids can be lethal. Such mutants can only survive in the dark when fed with nutrients. There is also some evidence (from action spectra) that carotenoids can play a role as photoreceptors, perhaps as an ancillary to the phytochrome system. Neither of these roles accounts for the accumulation of β-carotene in carrot roots, but few such roots except carrot have been found to accumulate noticeable levels of such metabolites.

It is generally accepted that animals lack the ability to biosynthesize carotenoids and that the wide variety of such compounds isolated therefrom are modified from the diet. Nevertheless, animals need the carotene derivative vitamin A (derived from plant sources) for growth and for visual processes, and an early indication of vitamin A deficiency is the impairment of vision in dim light. Vitamin A_1-alcohol (**5.264**, $\equiv$ retinol) is formed by cleavage of β-carotene by bacteria in the intestinal mucosa and is formally a diterpenoid. Vitamin A_2 (3-dehydro-vitamin A_1) occurs in fish liver oils. The aldehyde

(**5.264**)   (**5.265**)

of (**5.264**) (retinal) in the form of the (Z)-C_{11}-isomer reacts with the protein complex rhodopsin (visual purple) in the retina, but after absorption of a photon is liberated as the unbonded all (E)-form. This change in geometrical isomerism is believed to cause a local fluctuation in pH which triggers the nerve impulse. Further degradation of retinol results in the C_{18}-trisporic acids, such as (**5.265**), that are sex markers that determine the mating types of certain fungal classes. These compounds stimulate carotenoid biosynthesis and so their presence is autocatalytic.

5.12 Polyisoprenoids

Repeated addition of IPP moieties to an (all E) GGPP 'starter' leads to rubber (**5.266**) or gutta percha (**5.267**) with repeating Z- or E-linked units respectively. These polymers have significantly different molecular masses (rubber $c.$ 10^3–10^5 residues; gutta $c.$ 10-fold lower) and physical properties: e.g. gutta is supposed to be superior for the manufacture of golf balls. Plants usually produce one or the other – although sometimes (e.g. in chicle, the base of chewing gum) a mixture is obtained, but individually the molecules are (with the exception of the starter unit in rubber) all either E- or Z-linked.

(**5.266**)   (**5.267**)

The commercial source of rubber is the tree *Hevea brasiliensis* and of gutta (far less important) is guayule, *Palaguum gutta* or *Parthenium argentatum*, but over 2000 plant species are known to produce such polyisoprenoids. Dandelion root is a relatively rich source and was grown in Central Asia during World War II when sources from the Far East were cut off. The other species mostly produce small amounts – suitable for sealing of minor wounds – and *Hevea*, with its enormous harvestable output, is a physiological freak that has been selected and bred by growers to eliminate regulatory control and to increase the yield.

Rubber readily oxidizes and for use in tyres and other manufactured goods has to be vulcanized by cross-linkage with sulphur or with boron compounds. On pyrolysis the natural rubbers yield mixtures containing lower terpenoids and such products may become an important energy source as oil stocks decline.

A variety of polyprenols is also known to occur mainly in plants and microorganisms. These are straight chain alcohols (C_{55} to C_{120}) but are not formed by the same enzyme systems as the rubbers since each molecule may now contain Z- or E-linked or even fully reduced C_5-units. These compounds are thought to be important as components of membranes or microorganisms, as such organisms cannot usually synthesize steroids that are the components of membranes of higher species.

5.13   Synthesis

Most important terpenoids have been synthesized in the laboratory by several (sometimes many) routes, and numerous lesser compounds have been similarly prepared either to prove a structure or as an intellectual exercise. The establishment of a synthetic route to a natural product also allows modifications that can result in related non-naturally occurring compounds that may be screened for biological or medical activities. Often, partial syntheses have been devised based on available products of known structure and these can have potential for the (relatively few) higher terpenoids that have to be produced on a large scale and cannot be obtained from natural sources. Partial syntheses on the laboratory scale have been much used for correlating relative and absolute configurations between compounds in the same or different classes.

The two main drawbacks of most syntheses are the length of the routes and the number of stereoisomers that may occur. Even if each conversion were 80 per cent efficient, a 20-step sequence will reduce the overall yield to less than 1 per cent, and a few poor steps could further invalidate the whole scheme. Sometimes this problem can be overcome by the relay procedure of feeding-in quantities of intermediates on the pathway which are obtained by degradation of the target compound, but this obviously is only possible when the latter is readily available. The second restriction necessitates the choice of highly stereoselective processes and optical resolution at some stage if racemic substrates are used. The numerous successful preparations involving many steps are a tribute to the current insight into physical organic principles and mechanisms. In particular, (a) conformation analysis allows an understanding of the relative stabilities of fused (*cis* and *trans*) ring systems of different sizes; (b) a detailed knowledge of the stereoelectronic requirement of reactions allows the course of reactions of unsaturated groups to be predicted with confidence; and (c) the influence of thermodynamic stabilities and of steric effects can be assessed. These principles are briefly exemplified in the following.

The state of the art at *c.* 1905 is illustrated by Perkin's synthesis to prove the structure of α-terpineol (**5.268**; Fig. 5.71). Many monoterpenoids are now compounds of commercial importance in flavouring and perfumery, and geraniol and camphor, which are starting materials for numerous

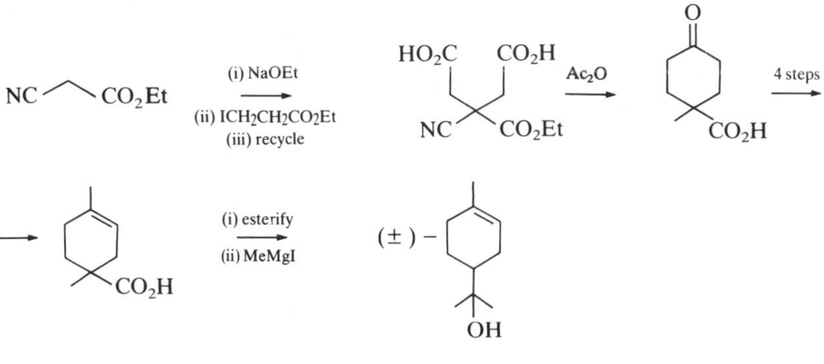

Fig. 5.71.

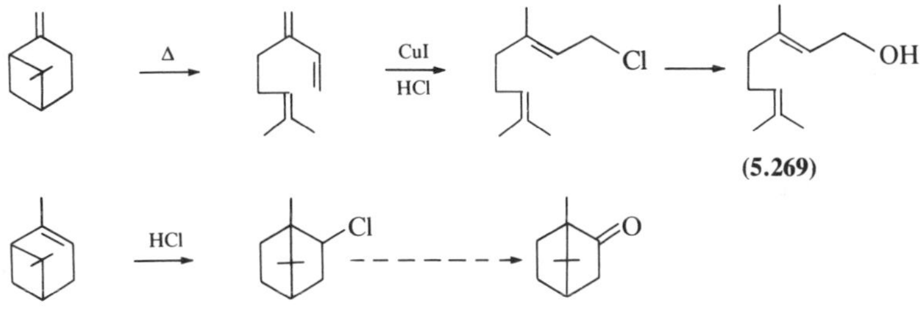

Fig. 5.72.

syntheses, are produced on the 10^3 tonne scale from α- and β-pinenes distilled from turpentine (Fig. 5.72). The regiospecific 1,4-addition of HCl to the diene system of myrcene to yield geranyl chloride is uniquely catalysed by Cu(I) salts. Like the carbohydrates (Chapter 1) monoterpenes offer a cheap source of stereochemically defined starting materials for synthesis, and the construction of (+)-*cis*-planococcyl (**5.273**), the sex pheromone of the citrus mealybug, *Planococcus citri*, is exemplary (Fig. 5.73).

Sesquiterpenoids are rarely synthesized or modified on the industrial scale but have been extensively studied in the laboratory. Internal Michael additions involving incipient carbonionic centres next to keto groups play a crucial role in the key steps of many syntheses; e.g. the routes (Fig. 5.74 and Fig. 5.75) to α-cedrene (**5.271**) and longifolene (**5.272**); and the pathway to the guaiolfuran gnididione (**5.274**; Fig. 5.76) makes repeated use of carbanion chemistry. Of course in these, and other, syntheses considerable ingenuity may be necessary to obtain even the starting materials for the scheme! Another strategy for sesquiterpenoids uses photo-addition as, for example, in the synthesis of caryophyllene (**5.275**; Fig. 5.77). An idea of how routes may be devised is illustrated in Fig. 5.78: analysis of the structure

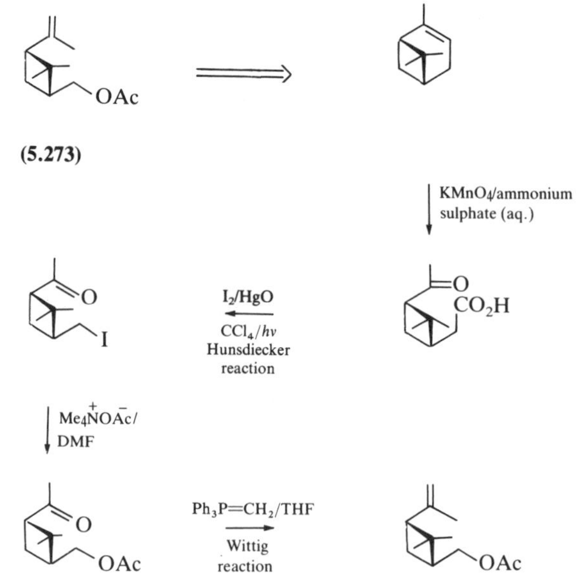

Fig. 5.73.

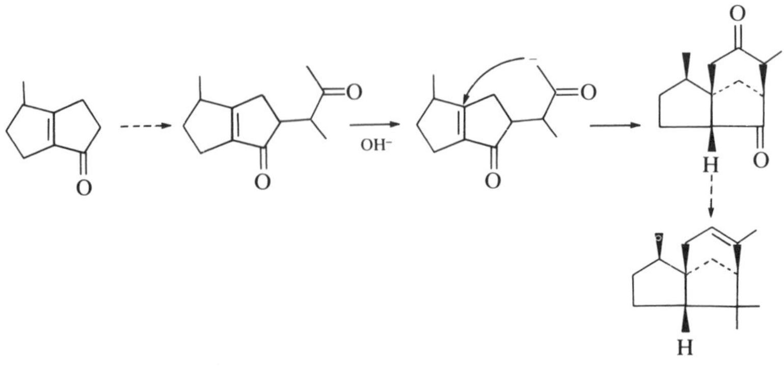

Fig. 5.74.

indicates the cornerstone for the strategy for the synthesis of (+)-quadrone (**5.276**) to be the acid-catalysed rearrangement of (**5.278**) to (**5.277**). The actual synthesis also involved a photochemical step.

The most strenuous efforts have been applied, not unnaturally, to the synthesis of the steroids. Equilenin (**5.215**) and oestrone (**5.213**) were partially synthesized from cholesterol in 1939 and 1948 respectively; and this was followed in 1951 by the completion of two total syntheses of steroids (by

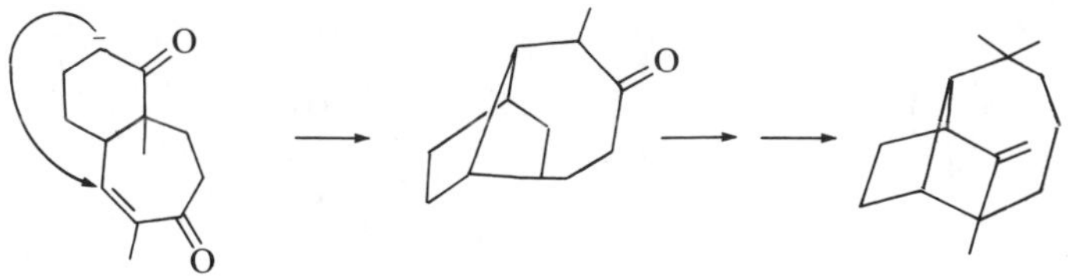

Fig. 5.75.

(5.272)

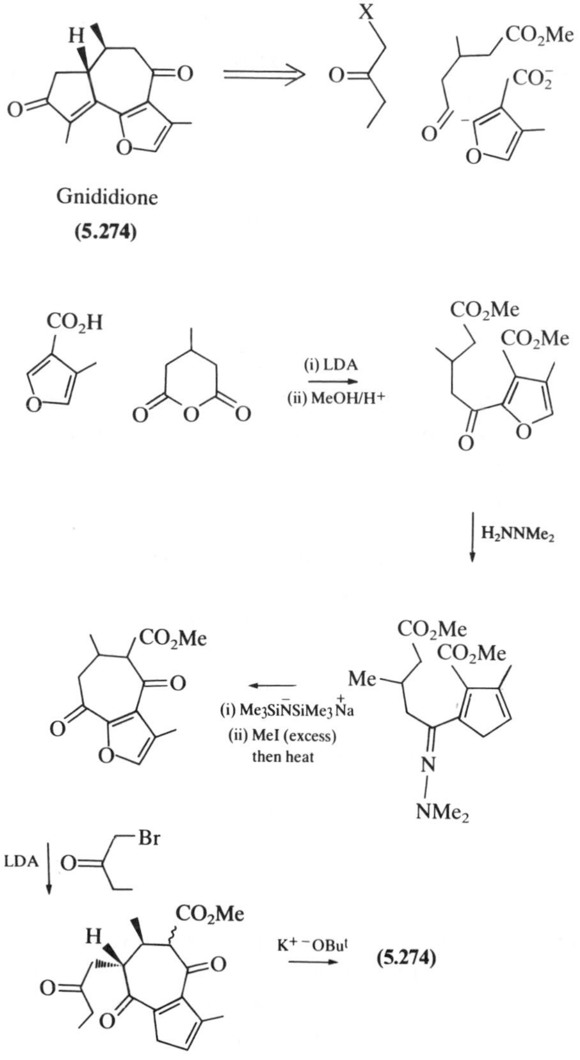

Gnididione

(5.274)

Fig. 5.76.

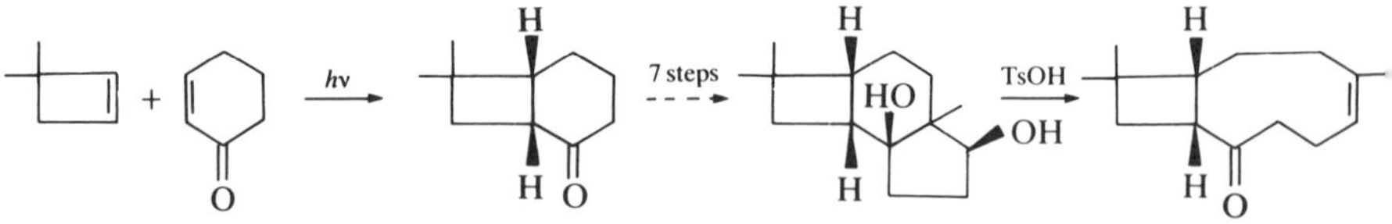

(5.275)

Fig. 5.77.

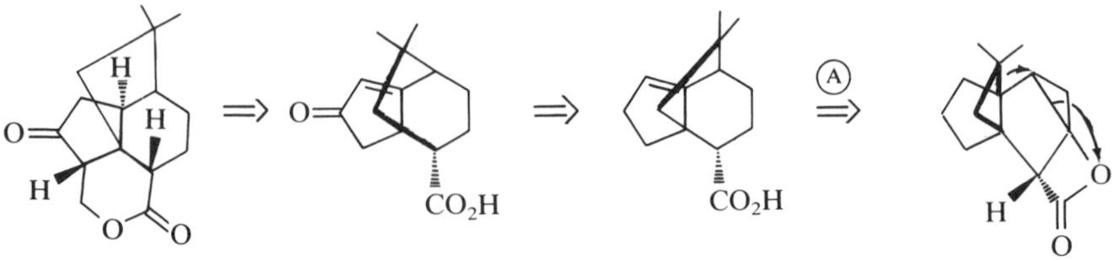

(5.276)   (5.277)   (5.278)

Fig. 5.78.

Woodward and Robinson and their coworkers). The Woodward route, which comprises about 35 steps (several variants are possible), has the advantage that it incorporates unsaturation into ring C, from which a keto group at C-11 can be developed to lead to corticosteroids, and it is also characterized by very strict stereochemical control. It makes use of the ring annelation or its variation that were developed in Robinson's scheme (steps A and B; Fig. 5.79). Numerous other total syntheses of steroids were subsequently published – some based on biomimetic ideas such as the cyclization of polyenes (cf. example in Section 5.7) and some on methods for the generation of ring systems as in Fig. 5.80. However, most of these are not even remotely economic for producing steroids in bulk for medicinal purposes. One total synthesis that has been applied to the production of steroids on the tonne scale (Fig. 5.81; by Velluz in 1960) is noteworthy for utilizing a convergent pathway whereby large fragments are assembled and then put together as late as possible – rather than the linear construction that is usually adopted. Other important features of the process were the use of microbial reduction in a high-yielding stereospecific step and optical resolution (the starting materials were achiral) at an early stage so that the unwanted enantiomer was not carried (with expenditure of reagents and energy) through to the end of the process. Glutaric anhydride (**5.279**) was manipulated to give the fragment (**5.280**) that was coupled to the Grignard reagent (**5.281**) and the product further functionalized to either oestrogen or corticosteroids. Although the route possessed some 20 steps, the processes were commercially viable.

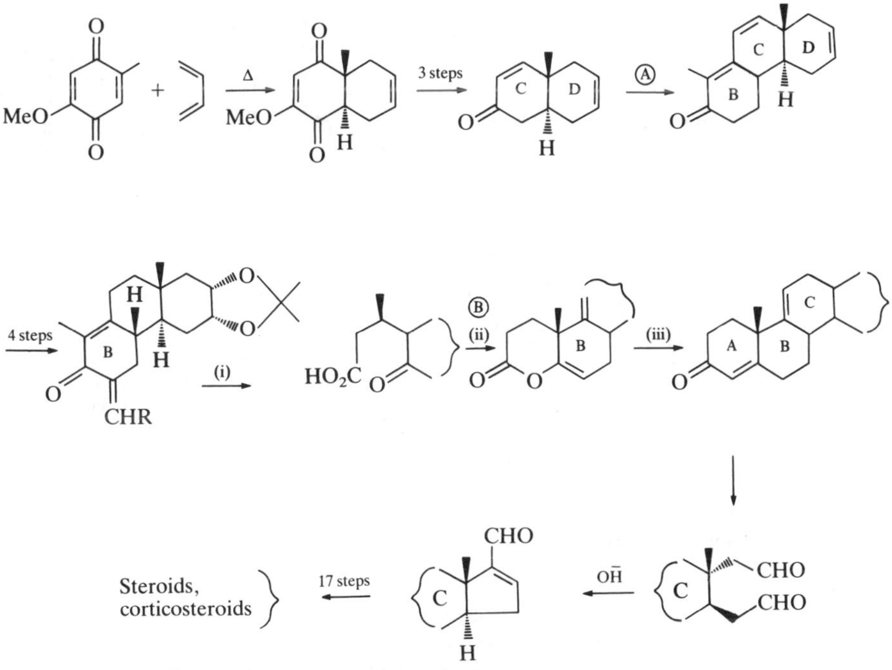

(i) ⌁CN/OH⁻, (ii) Ac₂O, (iii) MeMgI

Fig. 5.79.

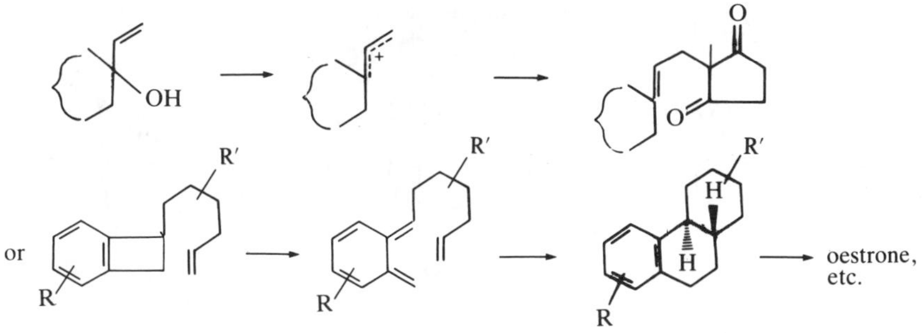

Fig. 5.80.

Despite this success most industrial procedures have been based on partial synthesis from sapogenins usually obtained from plants native to Central America. Diosgenin (**5.227**; from yams) could be converted into pregnenolone (**5.282**) and thence into progesterone (**5.218**) in only eight steps (Fig. 5.82) and another short route to the same hormone was devised from stigmasterol (**5.283**; Fig. 5.83) which (as its esters) is a major component of soyabean oil. Cortisone and its relatives could be obtained from hecogenin ex. *Agave*

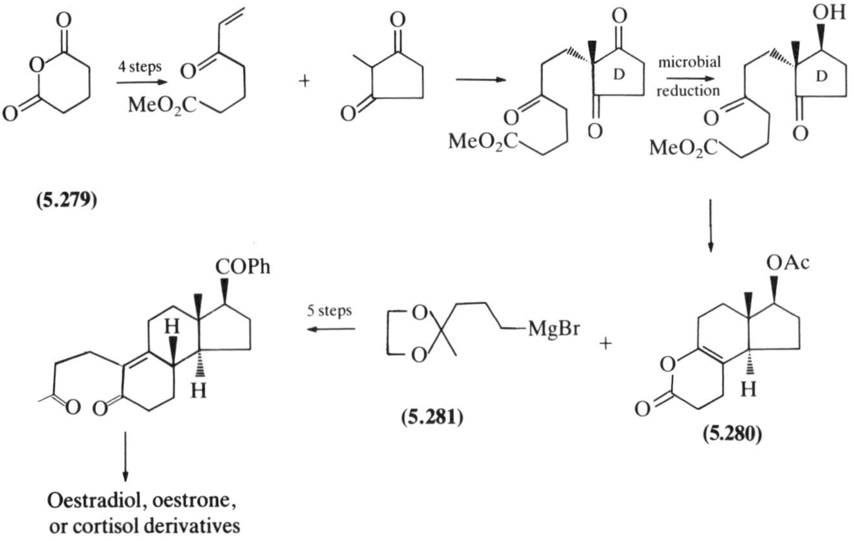
4 steps
MeO2C
microbial reduction
D
OH
OAc
COPh
5 steps
MgBr
(5.279)
(5.281)
(5.280)
Oestradiol, oestrone, or cortisol derivatives

Fig. 5.81.

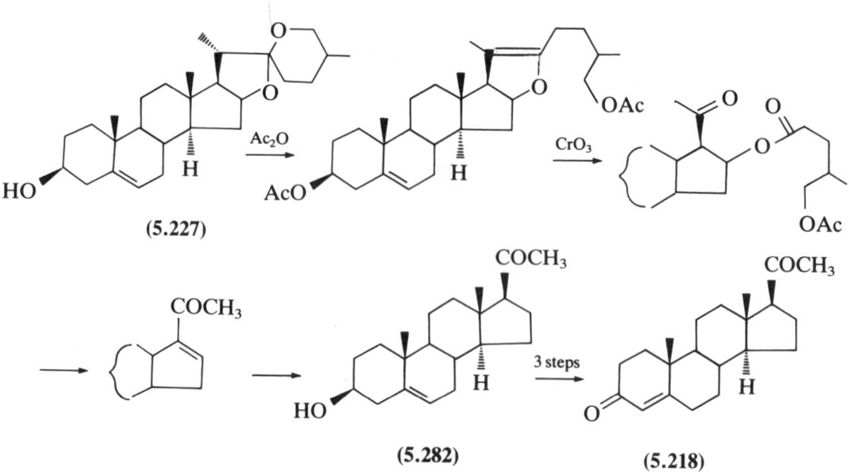
HO
Ac2O
AcO
CrO3
OAc
COCH3
3 steps
(5.227)
(5.282)
(5.218)

Fig. 5.82.

sisalana (**5.284**) by transposition of the oxo group from C-12 to C-11 by α-bromination, hydrolysis, and reduction. Bacterial or fungal species have been isolated that either stereospecifically or highly stereoselectively oxidize every position of the steroid nucleus, and these have been used in the production of corticosteroids. Thus, progesterone can be α-hydroxylated at C-11 (over 90% yield) by a mould of the *Rhizopus* family.

β-Carotene (**5.258**) and certain other carotenoids need also to be synthesized on an industrial scale for production of vitamin A and of food colourings and an efficient route from cheap starting materials was devised

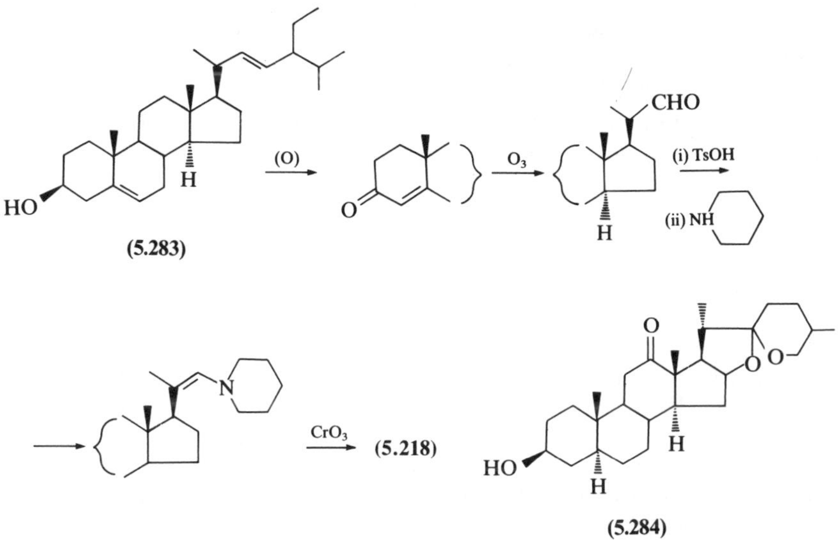

Fig. 5.83.

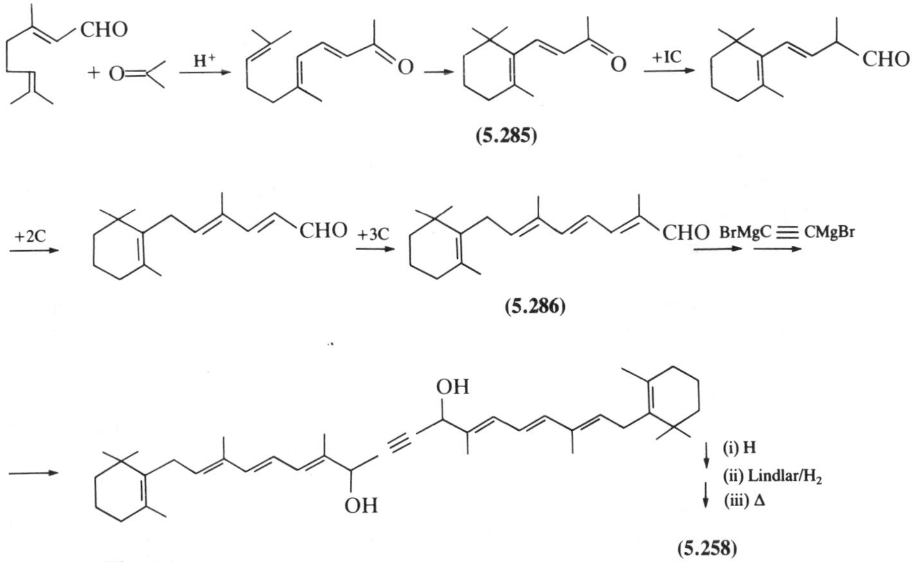

Fig. 5.84.

as long ago as 1956 (by Isler). This uses β-ionone (**5.285**) which is the main product from acid-treatment of the condensation product of acetone and citral (the latter being cheaply obtained by oxidation of geraniol): see Fig. 5.84. A one-carbon unit is first added in a Darzen's reaction followed by two and three carbon units in two aldol-type condensations to yield (**5.286**) which is coupled by a *bis*-Grignard reagent and the product rearranged, partially reduced and equilibrated to the all-*E* form of the target molecule. It is fortunate that the desired product has such a stable arrangement of double bonds that the geometric isomerism and position of bonds falls into line during a procedure which could lead to several possible isomers. The sequence can be adapted to give lycopene or certain other carotenoids.

Further reading

The results of many hundreds of studies have been summarized in the previous sections although only the names of a few eminent leading workers have been quoted. It seems invidious to single out a few dozen research papers for recommended perusal; rather the following is a list (not exhaustive) of books and reviews that will be found useful. Some of the older references are worth while in providing useful discussions and viewpoints not mentioned in later summaries. The best way of locating recent work on a particular compound or class is to follow through the sections and indexes of *Specialist Periodical Reports: Terpenoids and Steroids* (vols 1 to 12; 1971–1983) and its sequel *Natural Products Reports* (vol. 1 on; 1984 on), both of which are published by the Royal Society of Chemistry of London.

1. General references: J.B. Hendrikson, *Molecules of Nature*, Benjamin, New York, 1965; P. Yates, *Structure Determination*, Benjamin, New York, 1967; W. Templeton, *Introduction to the Chemistry of Terpenoids and Steroids*, Butterworths, London, 1969; D.G.H. Crout and T.A. Geissman, *Organic Chemistry of Secondary Plant Metabolism*, Freeman Cooper, San Francisco, 1969; W.I. Taylor and A.R. Battersby, *Cyclopentanoid Terpene Derivatives*, Dekker, New York, 1970; I. Fleming, *Selected Organic Syntheses*, Wiley, London, 1972; A.A. Newman (ed.), *Chemistry of Terpenes and Terpenoids*, Academic Press, London, 1972; K. Nakanishi, T. Goto, S. Iho, S. Natori, and S. Nozoe, *Natural Product Chemistry*, Academic Press, New York, 3 vols, 1974–76; E.A. Bell and B.V. Charlwood (eds), *Secondary Plant Products*, Encyclopaedia of Plant Physiology, new series, vol. 8, Springer, Berlin, 1980; P. Manitto, *Biosynthesis of Natural Products*, Ellis Horwood, Chichester, 1981; J.W. Porter and S.L. Spurgeon (eds), *Biosynthesis of Isoprenoid Compounds*, Academic Press, New York, 2 vols, 1981, 1984; J. ApSimon (ed.), *Total Synthesis of Natural Products*, Wiley-Interscience, New York, vols 1–7, 1973–88; K.B.G. Torssell, *Natural Product Chemistry*, Wiley, Chichester, 1983; S. Dev and R. Misra (eds), *Diterpenoids*, CRC Press, Boca Raton, 4 vols, 1985; T.W. Goodwin and E.I. Mercer, *Introduction to Plant Biochemistry*, 2nd ed., Pergamon, Oxford, 1984; W.D. Nes, G. Fuller, and L.-S. Tsai (eds), *Isopentanoids in Plants*, Dekker, New York, 1984; W.F. German (ed.), *Chemistry of Monoterpenes*, Dekker, New York, 2 vols, 1985; J. Mann, *Secondary Metabolism*, 2nd ed., Clarendon, Oxford, 1987; J.B. Harborne, *Introduction to Ecological Biochemistry*, 3rd ed., Academic Press, London, 1989; B.V. Charlwood and D.V. Banthorpe (eds), *Methods in Plant Biochemistry*, Terpenoids, vol. 7, Academic Press, London, 1991.
2. Isoprene rule: L. Ruzicka, *Proc. Chem. Soc.*, 1959, 541; L. Ruzicka, *Ann. Rev. Biochem.*, 1970, **42**, 7.

3. Monoterpenoids: W.W. Epstein and C.D. Poulter, *Phytochemistry*, 1973, **12**, 737 (irregular); B.V. Charlwood and D.V. Banthorpe, *Prog. Phytochemistry*, 1978, **5**, 65 (regular); R. Croteau, *Chem. Rev.*, 1987, **87**, 929 (regular).

4. Sesquiterpenoids: J.B. Hendrikson, *Tetrahedron*, 1959, **7**, 82; W. Parker, J.S. Roberts, and R. Ramage, *Quart. Rev.* (London), 1967, **21**, 331; G. Rücker, *Angew. Chem. Intern. Edn English*, 1973, **12**, 793; D. Arigoni, *Pure Appl. Chem.*, 1975, **41**, 219; G.A. Cordell, *Chem. Rev.*, 1976, **76**, 425; T.A. Geissman, *Rec. Adv. Phytochemistry*, 1973, **6**, 65 (lactones); F.T. Addicott (ed.), *Abscisic Acid*, Praeger, New York, 1983 (ABA).

5. Diterpenoids: J.R. Hanson, *Fortschr. Chem. Org. Naturst.*, 1971, **39**, 395; J.R. Hanson, *Prog. Phytochemistry*, 1972, **3**, 231; P.A. Heddon, J. MacMillan, and B.O. Phinney, *Ann. Rev. Plant. Physiol.*, 1978, **29**, 149 (gibberellins).

6. Sesterterpenoids: G.A. Cordell, *Phytochemistry*, 1974, **13**, 2343; G.A. Cordell, *Prog. Phytochemistry*, 1977, **4**, 209.

7. Non-steroidal triterpenoids: A. Eschenmoser, L. Ruzicka, O. Jeger, and D. Arigoni, *Helv. Chim. Acta*, 1955, **38**, 1890; J.D. Connolly, K.H. Overton, and J. Polonsky, *Prog. Org. Chem.*, 1970, **2**, 285; T.W. Goodwin, *Rec. Adv. Phytochemistry*, 1973, **6**, 97; C. Grunwald, *Ann. Rev. Plant Physiol.*, 1975, **26**, 209.

8. Steroids: G. Schroepker, *Ann. Rev. Biochem.*, 1981, **50**, 585; 1982, **51**, 555; H. Danielsson and J. Sjovall, *Steroids and Bile Acids*, Elsevier, Amsterdam, 1985; E. Caspi, *Tetrahedron*, 1986, **42**, 3; H.F. Deluca and H.K. Schnoes, *Ann. Rev. Biochem.*, 1983, **52**, 411 (vitamin D).

9. Carotenoids: G. Britton, in *Chemistry and Biochemistry of Plant Pigments* (ed. T.W. Goodwin), vol. 1, Academic Press, London, 1976, 262; T.W. Goodwin (ed.), *Biochemistry of Carotenoids*, 2nd ed., Chapman & Hall, London, 2 vols (1980, 1984).

6 Phenolics

J.B. Harborne

6.1 Introduction

The expression 'phenolic compounds' embraces a vast range of organic substances which are aromatic compounds with hydroxyl substituents. The parent compound is phenol (6.1) but most are polyphenolic; the commonly occurring flavonol quercetin (6.2), for example, has five hydroxyl groups. While a small number of phenolics occur in animals, most are of plant origin. Indeed, the presence of a 'phenolic fraction' is a characteristic feature of all plant tissues.

Among the plant polyphenols, of which over 8000 are known, the flavonoids such as quercetin (6.2) form the largest group. However, phenolic quinones, lignans, xanthones, coumarins, and other classes exist in considerable numbers. In addition to monomeric and dimeric structures, there are three important groups of phenolic polymer – the lignins of the plant cell wall, the black melanin pigments of plants, and the tannins of woody plants. Phenolic units are also encountered among nitrogen compounds; for example the aromatic amino acid tyrosine (6.3) is phenolic. Similarly, there are phenolics with terpenoid substitution. Δ^1-Tetrahydro-cannabinol (6.4), the active hallucinogenic principle of cannabis resin, is a monoterpene-substituted resorcinol derivative.

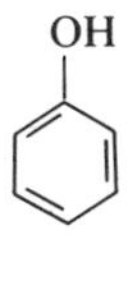

(6.1)

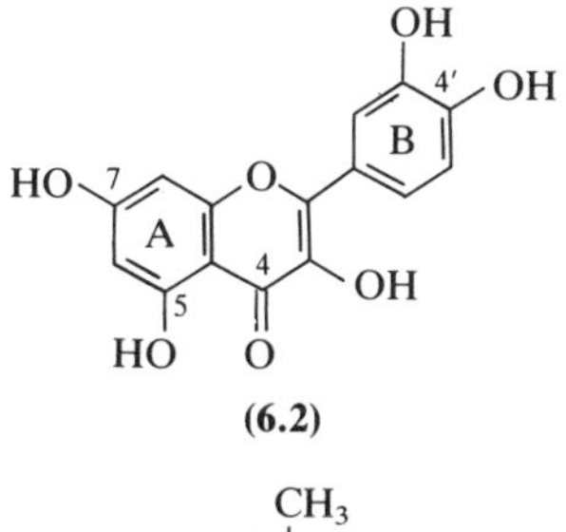

(6.2)

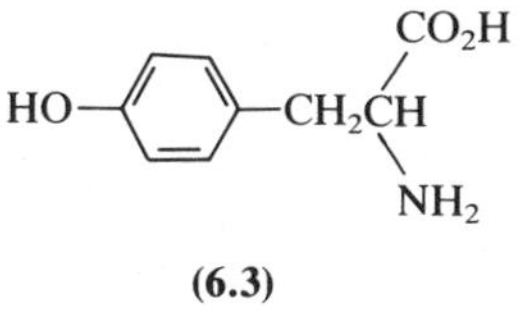

(6.3)

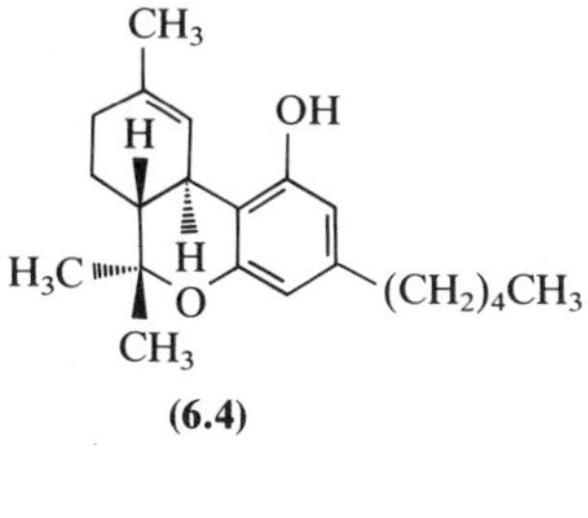

(6.4)

Phenols are chemically reactive and this must be borne in mind when isolating them from plants. They are usually acidic and can often be separated from other plant constituents by their solubility in aqueous sodium carbonate. Unless sterically hindered, all phenols are capable of taking part in hydrogen bonding. This may be intramolecular, as between the 5-hydroxyl and 4-carbonyl groups in most flavonoids (see, for example, quercetin, **6.2**). More importantly, it may be intermolecular and bring about interactions between plant phenols and the peptide links of proteins in the so-called 'tanning' reaction. Another important property of the many phenols which have an *o*-dihydroxy (catechol) grouping is their ability to chelate metals, and such metal chelates may be important in biological systems. Finally, phenols are very susceptible to oxidation. There are special enzymes in plants – the phenolases – which catalyse the oxidation of monophenols to diphenols, of diphenols to quinones, and of these products to polymeric coloured materials, the phlobaphenes.

Plant polyphenols are economically important because they make major contributions to the taste, flavour, and colour of our food and drink. The flavour and taste of tea is related to the fact that the tea leaf contains up to 30 per cent of its dry weight as polyphenol. Likewise, the bitterness of beer is due to the content of the phloroglucinol derivative, humulone (**6.5**), while the colour of red wine is imparted by anthocyanins, such as the pigment malvin (**6.6**). In nature, phenolics have a significant role in protecting plants from being overeaten by herbivores. They also act as chemical signals in the flowering and pollination of plants and in the processes of plant symbioses (e.g. in nitrogen fixation) and plant parasitism.

(**6.5**)

(**6.6**) (Glc = glucosyl)

6.2 Structural types

Phenolics are conveniently classified according to the number of carbon atoms in the basic skeleton (Table 6.1). There are simple phenols, such as phenol (**6.1**), catechol (1,2-dihydroxybenzene), and phloroglucinol (1,3,5-trihydroxybenzene). Then there are derived classes with one, two, or three side-chain carbons represented by salicylic acid (**6.7**), *p*-hydroxyphenylacetic acid (**6.8**), and the hydroxycinnamic acid, caffeic acid (**6.9**), a ubiquitous constituent of higher plant cells.

Table 6.1 The major classes of phenolics in plants.

Number of carbon atoms	Basic skeleton	Class	Example
6	C_6	Simple phenols Benzoquinones	Catechol, hydroquinone 2,6-Dimethoxybenzoquinone
7	C_6-C_1	Phenolic acids	*p*-Hydroxybenzoic, salicylic
8	C_6-C_2	Acetophenones Phenylacetic acids	3-Acetyl-6-methoxybenzaldehyde *p*-Hydroxyphenylacetic
9	C_6-C_3	Hydroxycinnamic acids Phenylpropenes Coumarins Isocoumarins Chromones	Caffeic, ferulic Myristicin, eugenol Umbelliferone, aesculetin Bergenin Eugenin
10	C_6-C_4	Naphthoquinones	Juglone, plumbagin
13	C_6-C_1-C_6	Xanthones	Mangiferin
14	C_6-C_2-C_6	Stilbenes Anthraquinones	Lunularic acid Emodin
15	C_6-C_3-C_6	Flavonoids Isoflavonoids	Quercetin, malvin Genistein
18	$(C_6$-$C_3)_2$	Lignans	Podophyllotoxin
30	$(C_6$-C_3-$C_6)_2$	Biflavonoids	Amentoflavone
n	$(C_6$-$C_3)_n$ $(C_6)_n$ $(C_6$-C_3-$C_6)_n$	Lignins Catechol melanins Flavolans (condensed tannins)	– – –

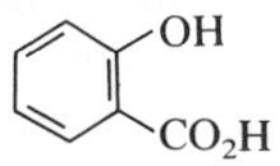

(6.7) 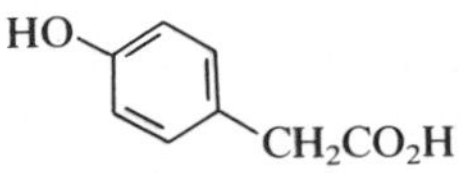   (6.8) 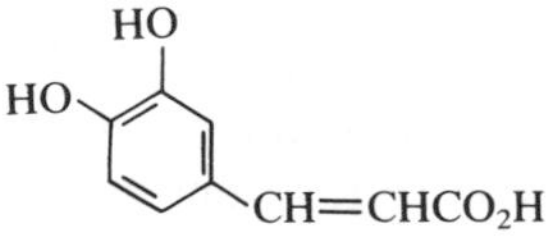   (6.9)

The hydroxycinnamic acids such as caffeic are one of several classes of C_6-C_3 compound known as phenylpropanoids. Lactonization and ring-closure of *o*-hydroxycinnamic acids can give rise to coumarins, a typical member being aesculetin (**6.10**) formed theoretically by ring closure of *o*-hydroxycaffeic. Dimeric phenylpropanoids are known as lignans. One

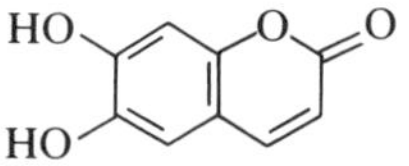

(6.10)

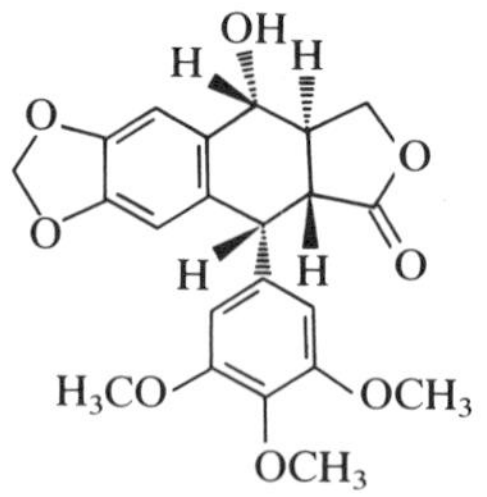

(6.11)

well-known lignan is podophyllotoxin (**6.11**) from the rhizomes of *Podophyllum peltatum*, which is in clinical use in modified form for the treatment of certain cancers.

The flavonoids are formed biosynthetically from a phenylpropanoid precursor (e.g. *p*-hydroxycinnamic acid) linked to three malonyl coenzyme A units. They are structurally derived from the parent compound, flavone (**6.12**) which occurs naturally as the farina on leaves of *Primula* species. Flavonoids are classified according to the oxidation level of the central pyran ring into anthocyanins (e.g. malvin, **6.6**), flavonols (e.g. quercetin, **6.2**), and so on. They vary, within each class, according to the number and position of hydroxyl, methoxyl, and other substituents. A related group are the isoflavonoids, represented by genistein (**6.13**), an oestrogenic substance from clover. Dimeric (biflavonoids) and oligomeric (flavolans) flavonoids are also known.

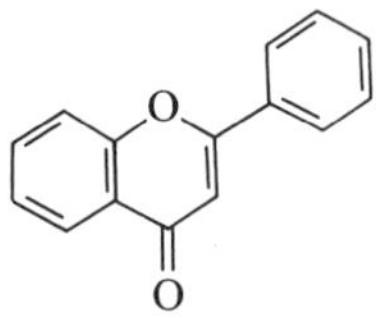

(6.12)

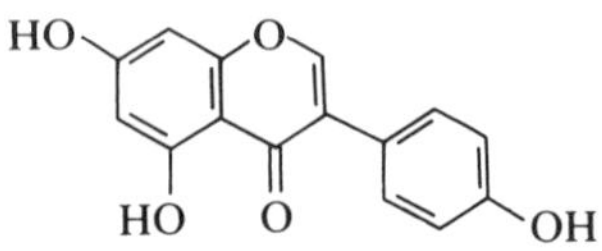

(6.13)

Oxidation of 1,4-dihydroxybenzene (or hydroquinone) (**6.14**) gives rise to 1,4-benzoquinone (**6.15**), the parent compound of the benzo-, naphtho-, and anthraquinones. Representative structures are 2,6-dimethoxybenzo-quinone (**6.16**) from wheat bran, juglone (**6.17**), a naphthoquinone from walnuts, and emodin (**6.18**), a yellow anthraquinone from rhubarb and cascara.

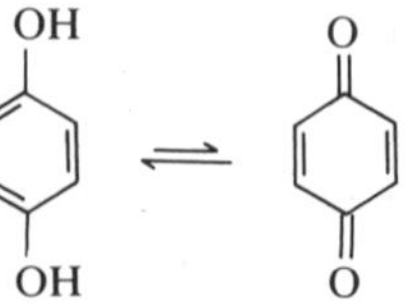

(6.14)   (6.15)

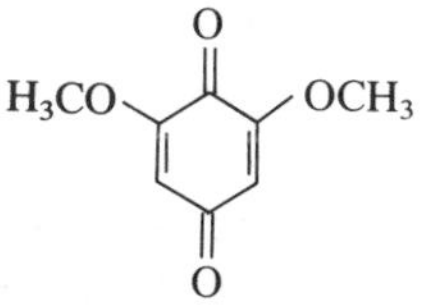

(6.16) 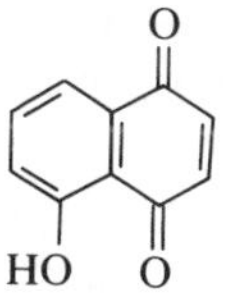   (6.17) 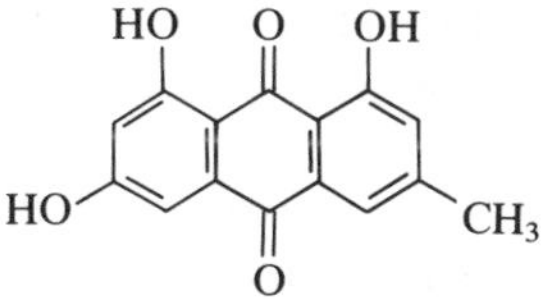   (6.18)

6.3 Natural occurrence

Phenolic compounds are found throughout the plant kingdom, but the type of compound present varies considerably according to which plant group is under consideration (Table 6.2). Phenolics are infrequent in bacteria, algae, and fungi. When present, they may have antibiotic properties. This is true, for example, of griseofulvin (6.19) from *Penicillium griseofulvum*, which is used for the treatment of athlete's foot in humans.

Although fungi and algae on their own do not often synthesize phenols, their symbiotic association, the lichens, characteristically produce special phenolic substances. The most distinctive lichen compounds are depsides and depsidones such as diploicin (6.20), but more widely distributed classes of phenol, notably dibenzofurans, xanthones, and anthraquinone pigments, may be produced. Bryophytes are regular producers of polyphenols (Table 6.2) and a range of simple flavonoids have been recorded in both mosses and liverworts.

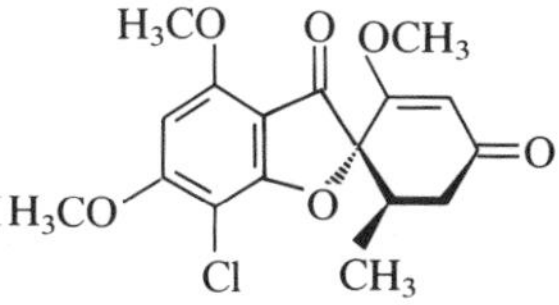

(6.19) 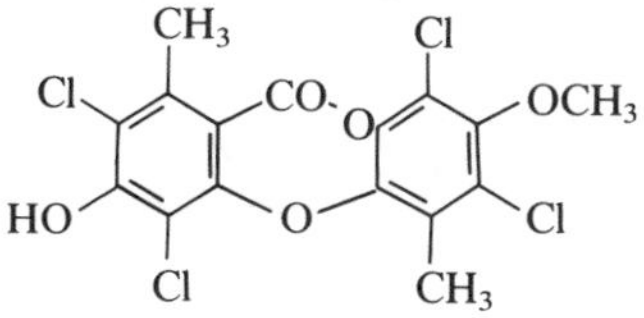   (6.20)

Table 6.2 Distribution of different classes of phenolics in the plant kingdom.

Phylum	Pattern
Bacteria	Polyketide-derived phenols and quinones occasionally present
Fungi	Simple phenols, phenylpropanoids, and quinones regularly reported
Algae	Iodo and bromophenols, phloroglucinol derivatives in cell walls
Lichens	Anthraquinones, depsides, depsidones, and xanthones
Bryophyta	Cell wall phenols, phenylpropanoids, stilbenoids, some flavonoids
Ferns, conifers, and flowering plants	Cell wall lignins, wide range of phenolics of all types

It is in the vascular plants that the full expression of phenolic biosynthesis can be observed. All ferns, gymnosperms, and angiosperms have lignin at the cell wall, the lignin precursors being phenylpropanoid in origin. Hydroxybenzoic acids, hydroxycinnamic acids and flavonoids are universally present. Other classes of phenol have a more discrete distribution. Isoflavonoids are mainly confined to the family Leguminosae, anthraquinones are reported in about six families, while xanthones are recorded principally in the Gentianaceae, Guttiferae, Moraceae, and Polygalaceae.

The chemistry of natural phenols is complicated by the fact that most compounds are present in the plant in conjugated form, principally with a sugar residue linked to one or more phenolic groups. The variety of conjugated forms is considerable and this adds to the task of chemical characterization (see Section 6.4). Monosaccharides associated with phenolics include glucose, galactose, arabinose (in both furano and pyrano forms), rhamnose, xylose, mannose, apiose, allose, glucuronic, and galacturonic acid. There is a further complication that these sugars may be present as di-, tri-, or tetrasaccharide combinations.

Hydroxycinnamic acids occur naturally in a wider group of combined forms than any other group of polyphenol. Caffeic acid (**6.9**) can occur in association with cyclohexane carboxylic acids (e.g. with quinic acid as chlorogenic acid (**6.21**), sugars (e.g. rhamnose), organic acids (e.g. malic), amines (e.g. putrescine), and lipids (e.g. glycerol). It can even occur linked to other phenols (e.g. with salicin) in the compound populoside (**6.22**) from bark of poplar, *Populus grandidentata*.

(6.21)   (6.22)

In the animal kingdom, phenolics are found principally in the defensive secretions of arthropods. Phenol and *o*-cresol (2-methylphenol) are produced, for example, in millipedes and beetles. Hydroquinone (**6.14**) occurs in the bombardier beetle, where it serves as a precursor of the toxic benzoquinone (**6.15**), formed within the beetle by peroxidase oxidation, and expelled at a high temperature at any predator. In the Lepidoptera, some 10 per cent of butterflies contain flavonoids in their wings; these are not synthesized *de novo* but are of dietary origin. In mammals, phenolic amines such as adrenaline have an important function in brain and other tissues. Simple phenols occasionally appear in animals. The temporal gland of the elephant secretes large quantities of *p*-cresol, while the anal sac of the beaver contains a mixture of phenols and phenolic acids.

6.4   Isolation and structure elucidation

6.4.1   Isolation and purification

Those phenolics that are lipophilic and occur in leaf waxes or bud exudates can be extracted by washing the tissues briefly with methanol or chloroform. The majority of phenolics, however, are water-soluble and are extracted by plunging fresh tissues in boiling methanol. Their presence in crude extracts can be confirmed by the intense red, green, or blue colours produced by adding aqueous iron (III) chloride. Alternatively, addition of alkali will turn colourless flavonols yellow and anthocyanins will change from red to greenish blue.

Phenolics are conveniently purified on a small scale by chromatographic separation on filter paper or thin layers of cellulose. Large scale separations are achieved by column chromatography, with adsorbents such as Sephadex LH-20, ion-exchange resin, or polyamide powder. High performance liquid chromatography (HPLC) on reversed phase silica gel columns, using gradient elution with aqueous methanol and ultraviolet detection, is now widely applied to the separation of phenolic constituents.

6.4.2   Chromatography and absorption spectroscopy

Phenolic compounds can be characterized (and identified) by means of this chromatographic behaviour and spectral properties. This is because R_F values are correlated with the number of phenolic (or masked phenolic) groups present in a molecule. Likewise, the absorption bands in the ultraviolet and visible regions shown by phenols are determined by the nature of the resonating system, i.e. the extent of aromatic double bonds linked to the phenolic hydroxyl group(s). Furthermore, all substances with one or more free phenolic groups undergo characteristic bathochromic shifts in the presence of alkali, while those with a chelating catechol group undergo similar shifts in the presence of aluminium ion.

The importance of chromatographic and spectral measurements for characterizing polyphenols is nicely illustrated by reference to the six common anthocyanidins. These pigments occur widely in glycosidic form in flowers, being responsible for nearly all the orange to blue colours of garden plants. They are difficult to characterize by classical procedures, since they do not have sharp melting points and do not form derivatives easily. The six pigments are pelargonidin (**6.23**), cyanidin (**6.24**) and its methyl ether peonidin (**6.25**), delphinidin (**6.26**), and the mono- and di-methyl ethers petunidin (**6.27**) and malvidin (**6.28**). These cationic pigments are normally isolated from the plants as their chlorides.

All six pigments are readily separated by either paper chromatography in an acetic acid–hydrochloric acid solvent or by HPLC on a reversed phase

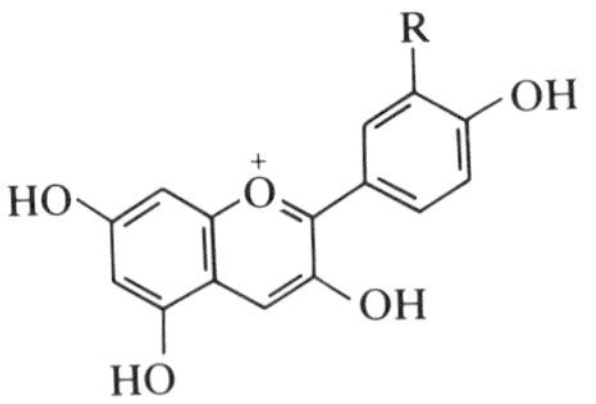

(6.23) R = H
(6.24) R = OH
(6.25) R = OCH₃

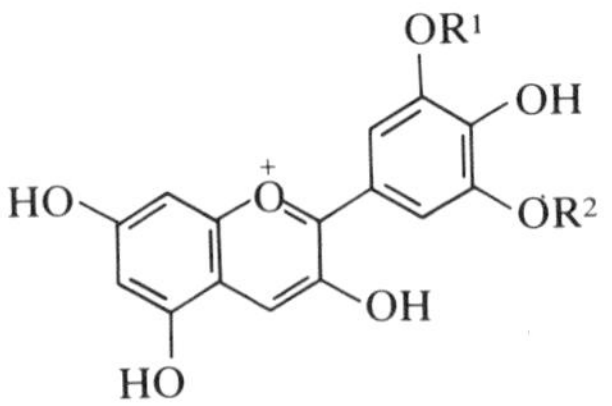

(6.26) R¹ = R² = H
(6.27) R¹ = CH₃, R² = H
(6.28) R¹ = R² = CH₃

silica column (Table 6.3). On paper, increasing hydroxylation lowers the R_F value regularly while increasing O-methylation increases the R_F. On HPLC, the situation is exactly reversed. Here the six pigments, separated as the 3-rutinosides, have a lower retention time as the number of hydroxyl groups increase. Delphinidin with six hydroxyl groups is eluted before cyanidin (five hydroxyls), which is eluted ahead of pelargonidin (four hydroxyls). Masking of the hydroxyl groups by methylation has a reverse effect so that the methyl ethers are retained on the column and the dimethyl ether, malvidin, is the last pigment to be eluted (Table 6.3).

These six pigments are also distinctive in their colours and spectral maxima (Table 6.4). Chelation with aluminium chloride is also useful for detecting anthocyanidins with a catechol grouping. Thus, cyanidin can be distinguished from peonidin by its bathochromic shift with aluminium chloride, and likewise delphinidin and petunidin from malvidin. The colours of these pigments in solution (Table 6.4) are related to the colours they impart in flowers, although there are other factors *in vivo* (pH of the cell sap, copigmentation) which can have modifying effects.

Table 6.3 Chromatographic properties of anthocyanidin pigments on paper and on HPLC.

Pigment (substituents)	HPLC retention time (min)[a]	R_F on paper[b]
Delphinidin (6 OH)	27.0	0.32
Cyanidin (5 OH)	31.1	0.49
Petunidin (5 OH, 1 OMe)	33.6	0.46
Pelargonidin (4 OH)	34.8	0.68
Peonidin (4 OH, 1 OMe)	39.0	0.63
Malvidin (4 OH, 2 OMe)	42.8	0.60

a On a Lichrosorb RP-18 column, eluted with a gradient of water–acetic acid–methyl cyanide, of the anthocyanidins as their 3-rutinosides. *b* Solvent Forestal = conc. HCl–acetic acid–water (3:10:3).

Table 6.4 Colours and spectral maxima of common anthocyanidins.

Pigment	Spectral max. in MeOH/HCl	Aluminium chloride shift	Visible colour
Pelargonidin	520	0	Orange
Cyanidin	535	18 ⎫	Magenta
Peonidin	532	0 ⎭	
Delphinidin	546	23 ⎫	
Petunidin	543	14 ⎬	Mauve
Malvidin	542	0 ⎭	

6.4.3 Structural elucidation

6.4.3.1 The classical era

The first flavonoid to be isolated in a pure state was the flavone glycoside apiin, from parsley seed, in 1845. Structural studies on flavonoids were later carried out by W.H. Perkin in the latter half of the 19th century and by Sir Robert Robinson in the first half of the 20th century. These workers depended on derivatization (acetylation, methylation) to establish the number of phenolic groups present. Alkaline degradation was then carried out to determine the substitution pattern of the flavonoid nucleus. An example of the procedures necessary to establish flavonoid structures at this period is the structural eludication of the yellow biflavone, ginkgetin (**6.29**), carried out by Wilson Baker, and W.D. Ollis in Bristol (Fig. 6.1).

Ginkgetin is one of a number of similar biflavones present in the leaves of the maidenhair tree, *Ginkgo biloba*. This plant is not only a taxonomic curiosity because of its primitive nature, but also an important medicinal plant of current interest for treating thrombosis. Ginkgetin accumulates mainly in the autumnal yellow leaves rather than in the green leaves of midsummer. At the time of this work, dimeric flavonoids were unknown and these studies were important in showing that such dimers could exist in nature.

The structure of ginkgetin (**6.29**) could not be fully established by alkaline hydrolysis. However, one of the products, the acetophenone (**6.30**), confirmed its flavone-like nature and showed that it had a methoxyl in a 7-position. The key experiment turned out to be one in which the hexamethyl ether (**6.31**) was degraded by alkaline hydrogen peroxide. The isolation of the diphenolic acid (**6.32**) established beyond doubt the dimeric nature of the compound, and also the presence of a 3′,8″ or 3′,6″-carbon–carbon linkage.

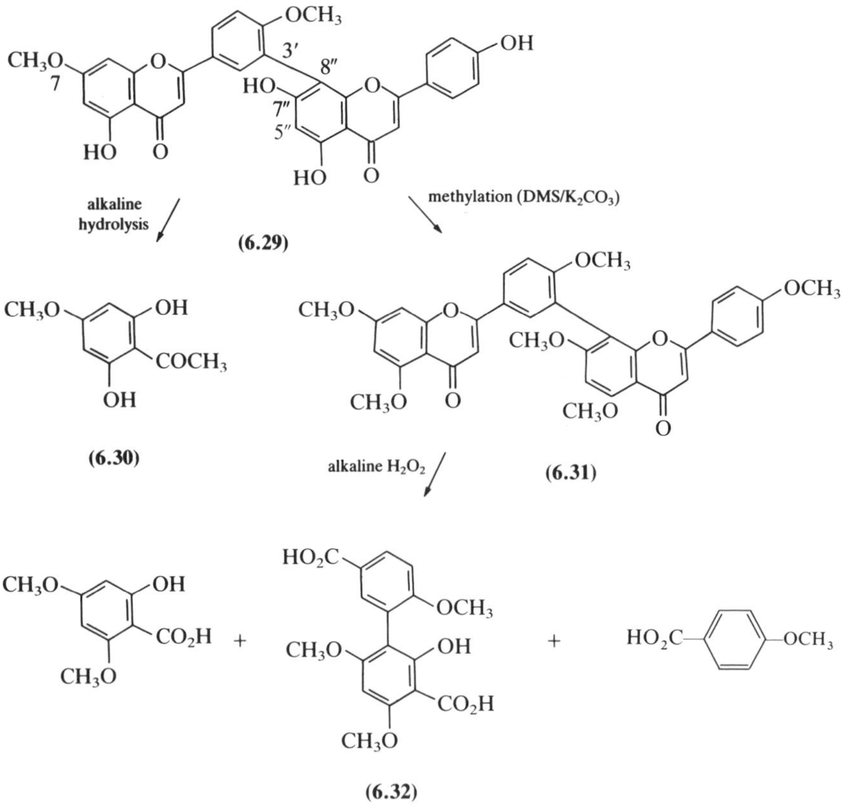

Fig. 6.1 Structural degradation of the biflavone ginkgetin. DMS = dimethyl sulphate.

The 3′,8″-linkage was favoured over a 3′,6″-linkage on steric grounds. Thus, the 5″-hydroxyl of ginkgetin proved to be easily methylated. If it had been next to a bulky 6″-substituent, then steric hindrance would have prevented its easy methylation. The placing of the methoxyl group at the 7- rather than the 7″-position also depended on steric arguments. Measurement of the UV spectral shift of the 7″-hydroxyl in the presence of increasing concentrations of alkali showed it to be a hindered position. If this free hydroxyl had been at 7, there would have been no problem in rapidly ionizing the proton in the presence of weak base.

6.4.3.2   *The modern period*

The elucidation of phenolic structures today relies heavily on the interpretation of ^{1}H and ^{13}C NMR data. Many structures are determined entirely from such information, although confirmation from other sources (e.g. synthesis, X-ray crystallography, chemical conversion) is always desirable. Another useful technique, particularly applicable to flavonoid

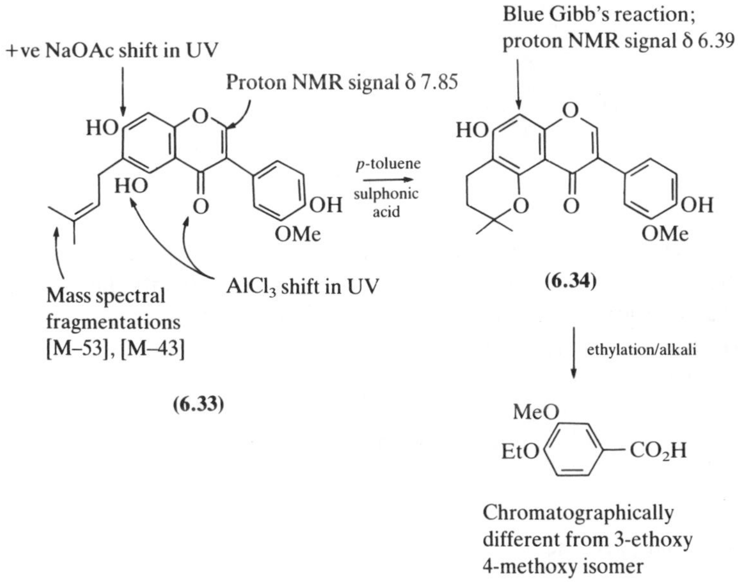

Fig. 6.2 Structural analysis of lupisoflavone.

glycosides, is FAB-MS, not only for determining molecular ions but also for indicating the order of sugars or acyl groups in a complex glycoside.

Two examples of modern analysis have been chosen to indicate the way that different spectral measurements (UV, MS, NMR) can complement each other in establishing new structures. The first is from our own studies at Reading on lupisoflavone (6.33), an anti-fungal compound from leaf washings of *Lupinus albus* (Fig. 6.2). Recognition of its isoflavone nature depended on the shape of the UV spectrum (intense peak at 269 nm with a shoulder at 292 nm) and the NMR signal at $\delta 7.85$ for the 2-proton, which is diagnostic for isoflavones. The isopentenyl substitution was apparent from the molecular weight, determined by MS, from the ^{1}H NMR aliphatic proton signals and from the fragmentation during MS measurement (see Fig. 6.2). Chemical conversion by ring closure of the isopentenyl side-chain onto the 5-hydroxyl to give (6.34) was helpful in establishing the 3,3-dimethylallyl side-chain to be at the 6- and not the 8-position. The presence of hydroxyl groups at the 5- and 7-positions followed from UV spectral shifts, but the B-ring substitution pattern could only be established by ethylation and alkaline degradation to give the isovanillic acid ethyl ether (6.35).

The second example is of the structural elucidation of the differentiation-inducing factor from the slime mould *Dictyostelium discoideum* as the dichlorophloroglucinol derivative (6.36). Only a small sample (50 μg) was available and there were some doubts about its purity. Nevertheless, electron-impact MS proved to be very helpful in establishing the molecular formula as $C_{13}H_{16}O_4Cl_2$ from precise mass measurement of the molecular

OH

Cl——CO(CH₂)₄CH₃

CH₃O——OH

Cl

(6.36)

ion. MS fragmentation also showed that there was a pentyl side-chain adjacent to a carbonyl. ^{1}H NMR spectral measurements then indicated that the aromatic nucleus was fully substituted. At this stage, there were 32 possible isomeric structures. Biosynthetic arguments, however, favoured the 1,3,5-trioxygenation pattern. At this stage, it became necessary to synthesize two isomers, one of which proved to be identical to (**6.36**), and direct (HPLC, GC, MS, NMR) comparison completed this structural analysis.

6.5 Biosynthesis

6.5.1 *General principles*

Phenolic compounds can be formed from primary metabolites of the cell by either of two routes: the shikimate pathway (see Section 6.5.3), which starts from carbohydrate; and the polyketide pathway, which starts from acetyl and malonyl coenzyme A. Many compounds with two or more phenolic nuclei, e.g. flavonoids, are of mixed biosynthetic origin, with one ring being shikimate-derived and the other polyketide-derived. The biosynthetic origin determines the hydroxylation pattern of the subsequent phenol, so that it is usually possible by inspection to determine which pathway is operating.

Shikimate-derived phenols are formed from cinnamic acid (**6.37**), which is always oxidized in the *para*-position to give 4-coumaric acid (**6.38**). When this undergoes further enzymic oxidation in biosynthesis, oxygen is added in adjacent positions to give the 3,4-dihydroxy (catechol) pattern or 3,4,5-trihydroxy (pyrogallol) pattern (Fig. 6.3).

In the initial oxidation of cinnamic acid (**6.37**), a distinctive shift in aromatic protons occurs, the NIH shift named after the National Institute of Health at Bethesda, Washington, DC, where this discovery was made. The proton in the 4-position gives way to an OH group, being moved itself to the 3-position. This NIH shift was established by tritium-labelling experiments and the mechanism is shown in Fig. 6.4. It may be noted that the further oxidation of 4-coumaric acid to caffeic acid (**6.39**) occurs without an NIH shift, since the ketone intermediate is stabilized by expulsion of the 3-proton.

By contrast with the shikimate pathway, the polyketide pathway produces phenols with hydroxyl groups in the 1,3,5-positions, as in phloroglucinol

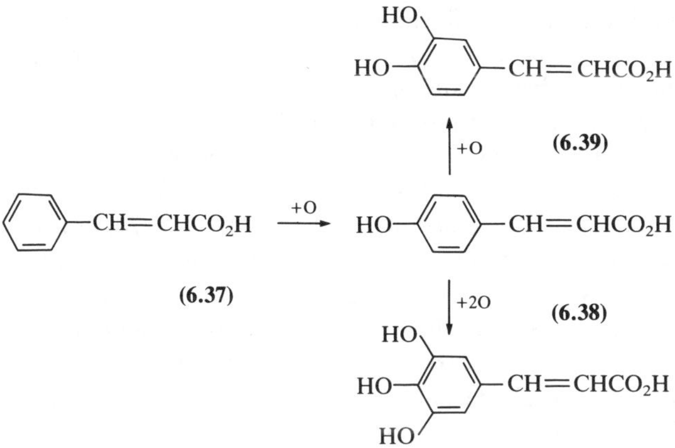

(6.39) (6.37) (6.38)

Fig. 6.3 Oxidation patterns of shikimate-derived phenols.

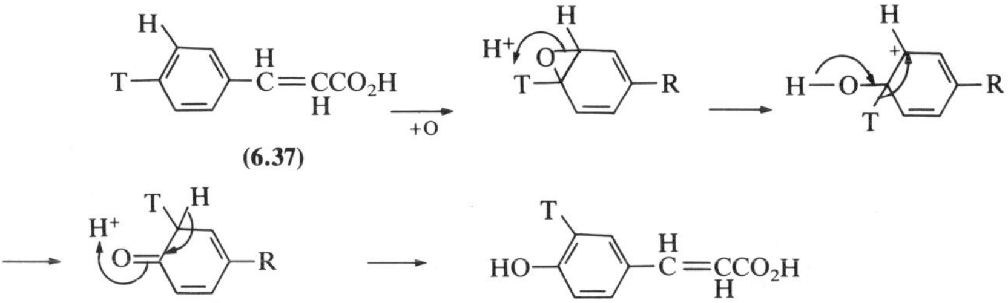

(6.37) (6.38)

Fig. 6.4 Hydroxylation of cinnamic acid to 4-coumaric acid with the NIH shift.

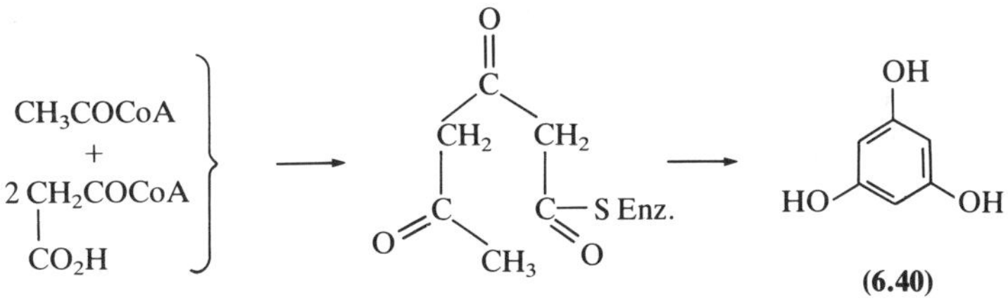

(6.40)

Fig. 6.5 Biosynthesis of phloroglucinol by the polyketide pathway.

(**6.40**; Fig. 6.5). This 1,3,5 pattern is widespread among the many natural polyphenols. It is, however, possible for one or more of these hydroxyl groups to disappear during polyketide biosynthesis. For example, 6-methylsalicylic acid (**6.41**), a fungal metabolite, is clearly derived from acetyl coenzyme A and four malonyl coenzyme A units, but it ends up with only one phenolic group adjacent to the carboxyl substitution (Fig. 6.6).

Biosynthesis of the flavonoids involves both pathways, the A-ring being polyketide-derived and the B-ring shikimate derived. This can be seen clearly in the biosynthesis of the flavanone naringenin (**6.42**), via a chalcone intermediate. The 4-coumaryl CoA is the starter unit, replacing acetyl CoA, and this is linked to three malonyl CoA units (Fig. 6.7).

While all higher plant flavonoids are formed by the above route, a significant variant has been observed in the fungus *Aspergillus candidus*. In fact, flavonoids are very rare in lower plants and this fungus is very unusual in producing the compound chloroflavonin (**6.43**). It is perhaps not surprising that it is formed by a different pathway, involving four, and not three, malonyl CoA units and a C_6-C_1 precursor, benzoic acid (Fig. 6.8).

Another general feature of phenolic biosynthesis is the ability of plants to produce the same compound by more than one route. This is particularly noticeable in the case of the naphthoquinones, which can be produced by any one of four routes (see Section 6.5.6). It is also true of some widely occurring phenolics, such as chlorogenic acid (**6.21**) and scopolin. As

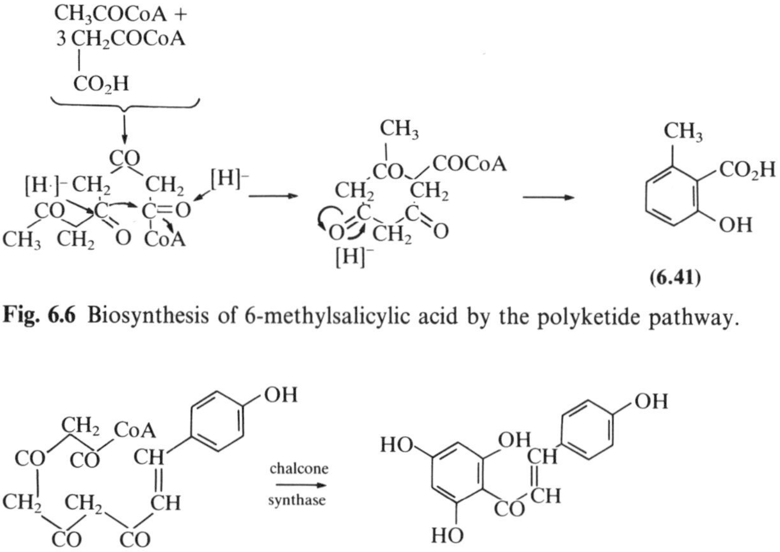

Fig. 6.6 Biosynthesis of 6-methylsalicylic acid by the polyketide pathway.

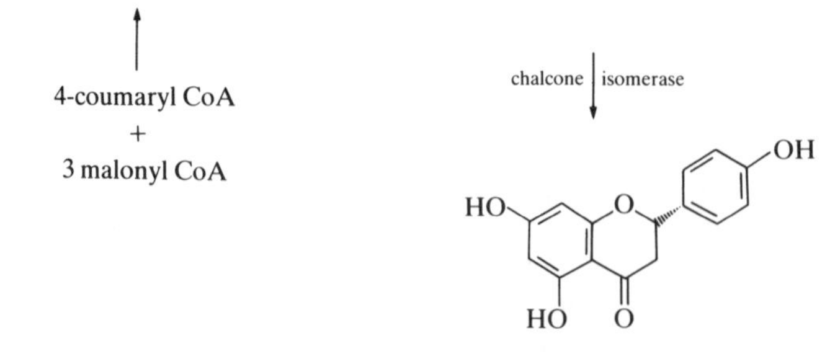

Fig. 6.7 Biosynthesis of the flavanone naringenin.

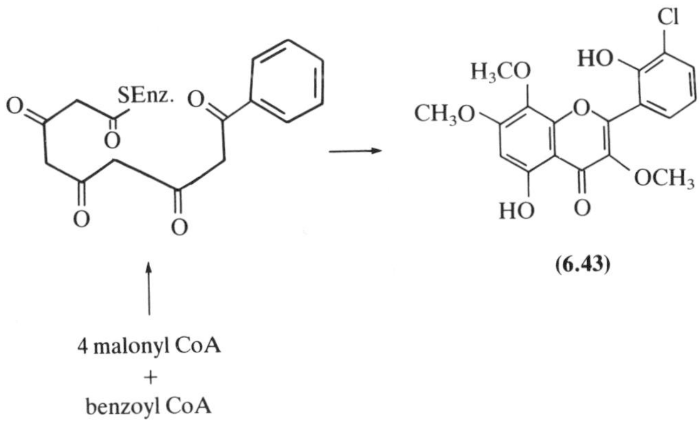

(6.43)

4 malonyl CoA
+
benzoyl CoA

Fig. 6.8 Biosynthesis of the flavonoid chloroflavonin in *Aspergillus candidus.*

illustrated in Fig. 6.9, there are alternative routes to their production. The source of carbon is identical in both routes, but the order in which the final steps in biosynthesis take place may vary.

One final feature of phenolic biosynthesis needs mentioning: the ability of phenolic compounds to link together by oxidative coupling. Peroxidase oxidation of coniferyl alcohol, the immediate precursor of lignin synthesis, will produce various radicals, two of which can readily dimerize to produce pinoresinol (**6.44**), a naturally occurring lignan (Fig. 6.10). This process of free radical coupling of phenylpropanoids is the final step in lignin biosynthesis, where a random polymer is formed on the cellulose matrix of the plant cell wall. Oxidative coupling is an essential feature also of biflavone formation, e.g. ginkgetin (**6.29**), and of condensed tannin synthesis. It is also an important step in the biosynthesis of some aromatic alkaloids, such as morphine (see Chapter 7).

To illustrate phenolic biosynthesis in more detail, the production of different end-products will now be considered. Polyketide biosynthesis is exemplified in the production of usnic acid. A very significant product of the shikimate pathway is the cell wall bound lignin, and the route from carbohydrate to this aromatic polymer will be briefly reviewed. Flavonoid synthesis is illustrated by the formation of a widespread anthocyanin, cyanidin 3-glucoside, and a rare isoflavonoid, the insecticidal principle rotenone. The last illustration will be of the various pathways of naphthoquinone synthesis.

6.5.2 *Biosynthesis of usnic acid*

Usnic acid (**6.45**) is not, in fact, a carboxylic acid but is a yellow anti-bacterial diketone produced by a number of lichens, including *Usnea barbata.*

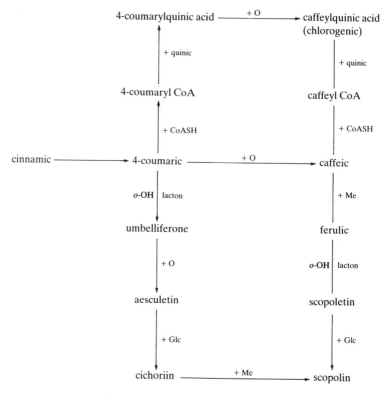

4-coumarylquinic acid ——— + O ———→ caffeylquinic acid
                                               (chlorogenic)

↑ + quinic   │ + quinic

4-coumaryl CoA   caffeyl CoA

↑ + CoASH   │ + CoASH

cinnamic ——→ 4-coumaric ——— + O ——→ caffeic

│ o-OH | lacton   │ + Me

umbelliferone   ferulic

│ + O   │ o-OH | lacton

aesculetin   scopoletin

│ + Glc   │ + Glc

cichoriin ——— + Me ———→ scopolin

Fig. 6.9 Alternative pathways of biosynthesis to chlorogenic acid and scopolin. CoASH = coenzyme A, O = monohydroxylation, Me = O-methylation, Glc = O-glucosylation, o-OH lacton = o-hydroxylation, followed by coumarin formation.

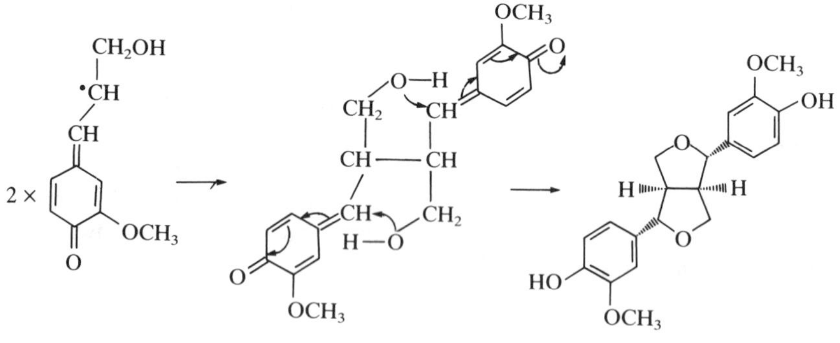

Coniferyl alcohol
radical

(6.44)

Fig. 6.10 Dimerization of coniferyl alcohol to give a lignan.

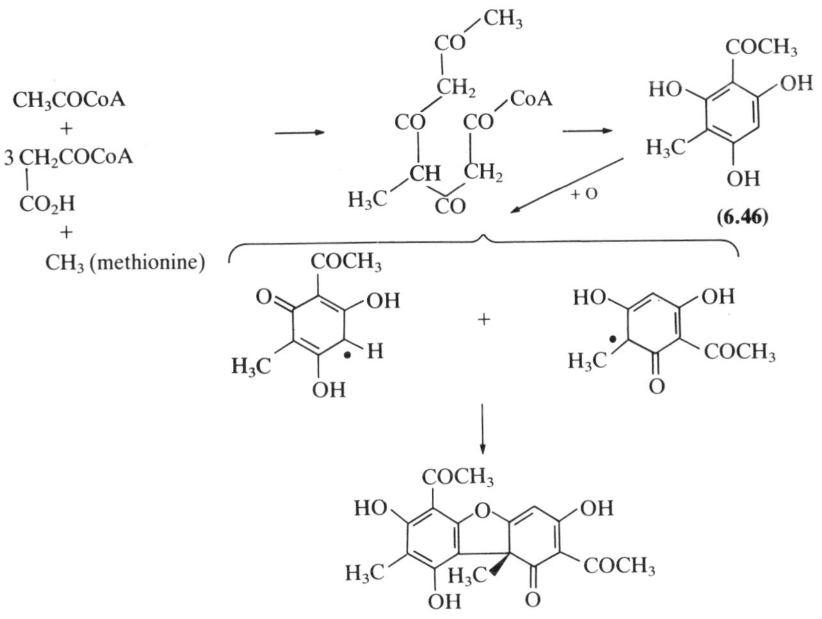

Fig. 6.11 Biosynthesis of usnic acid.

Structurally, it is a dibenzofuran and is polyketide in origin (Fig. 6.11). The first product is 2′,4′,6′-trihydroxy-5′-methylacetophenone (**6.46**), formed from acetyl CoA and three units of malonyl CoA. An extra methyl is inserted in the 5′-position, coming from the amino acid methionine. The next stage in biosynthesis is the oxidative coupling of two radicals of this acetophenone (**6.46**), with the production of an open chain carbon–carbon linked intermediate. This finally undergoes cyclization with the formation of a furan ring between the two aromatic rings and hence the production of usnic acid.

6.5.3 Biosynthesis of lignin

Lignin is the most important phenolic polymer in nature, since it provides strength and structural rigidity to the plant cell wall. Lignification was a key event in plant evolution and it separates the vascular plants from the algae and bryophytes, which are not lignified. It allowed plants for the first time to develop into trees, growing up to 100 m in height. The biosynthesis of lignin from sugar phosphate, formed in the Calvin cycle from carbon dioxide, falls into two linked parts, the more general shikimate pathway followed by the specialized hydroxycinnamic acid route (Fig. 6.12).

The shikimate pathway, named after one of the key intermediates, shikimic

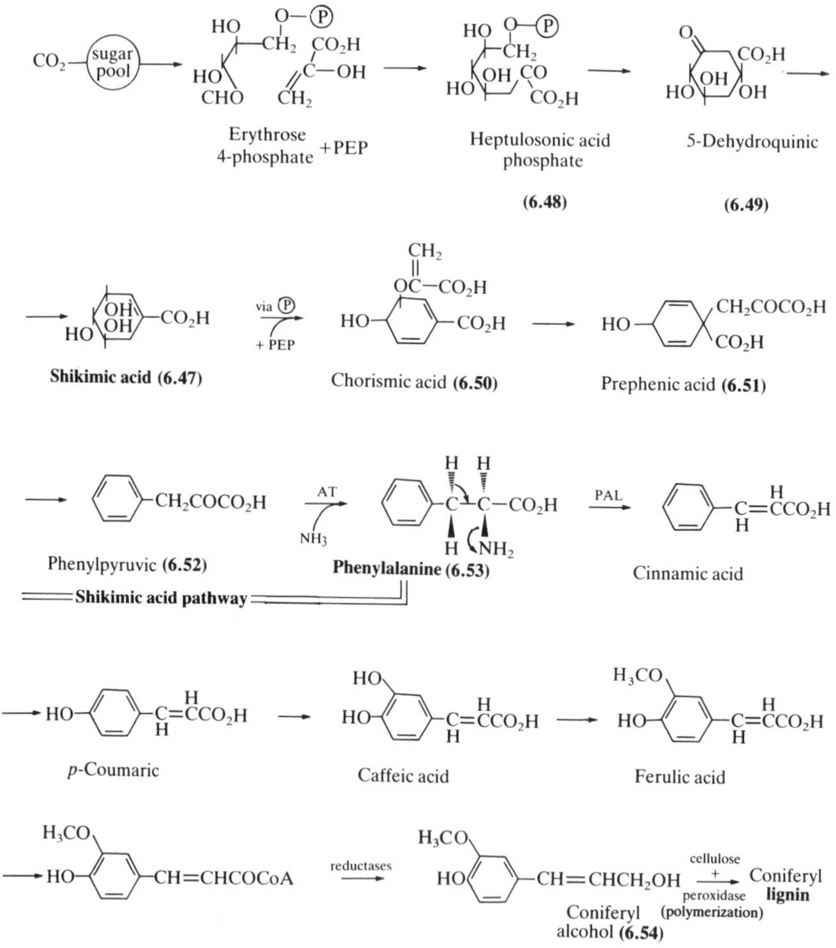

Fig. 6.12 Biosynthesis of lignin.

acid (**6.47**), is a universal process by which aromatic metabolites are produced, but especially the three amino acids, phenylalanine, tyrosine, and tryptophan. Chemically, it is a process by which a cyclohexane ring is aromatized in three easy stages. The first double bond is introduced in the cyclization of heptulosonic acid phosphate (**6.48**) to 5-dehydroquinate (**6.49**). The second is introduced, by splitting out of water, in the conversion of shikimic acid to chorismic (**6.50**), and the third in the conversion of prephenic acid (**6.51**) to phenylpyruvic (**6.52**).

Phenylalanine (**6.53**), the major end-product of the shikimate pathway, is available for protein synthesis and a variety of other uses, one of these being lignin biosynthesis. The conversion of phenylalanine to coniferyl alcohol (**6.54**) is a seven-stage process (Fig. 6.12). The first step, the deamination of phenylalanine to cinnamic acid, is stereospecific and occurs with *trans*-elimination of ammonia from the side-chain carbons 2 and 3 of L-phenylalanine with the formation of the *trans*-cinnamic acid. The next two steps, the oxidation of cinnamate to 4-coumarate and of 4-coumarate to caffeate, have already been mentioned in Section 6.5.1. Caffeic acid then undergoes *o*-methylation to ferulic acid, which is converted to the coenzyme A ester, before the carboxyl group is successively reduced via the aldehyde to the alcohol (**6.54**). The final step in synthesis is the polymerization of this alcohol, in the presence of peroxidase and cellulose, to produce coniferyl lignin. Two other closely related precursors, 4-coumaryl alcohol and sinapyl alcohol, are formed along almost identical pathways, and may be incorporated with coniferyl alcohol in the final polymerization.

6.5.4 Biosynthesis of cyanidin 3-glucoside

The precursors of anthocyanin biosynthesis were established as 4-coumarate and malonyl coenzyme A by a series of radiolabelling experiments carried out by H. Grisebach and others around 1960. The fact that colourless flavonoids are the immediate precursors of the intensely coloured flavylium cations was demonstrated many years ago by chemical reduction of dihydroflavonols using zinc and hydrochloric acid. It was also suggested from early genetic experiments with flower colour mutants in the snapdragon and other garden plants, that a sequence of steps from chalcone to flavone to flavonol to anthocyanin had evolved. Feeding experiments with white flowered mutants, whereby infiltrating dilute solutions of dihydroquercetin (**6.55**; Fig. 6.13) into the stem would produce red colour in the flower, also pointed in the same direction. However, the order of intermediates along a particular pathway to, for example, cyanidin 3-glucoside was only established by characterizing each enzyme along the pathway and determining the substrate specificities.

The initial step in the biosynthesis of all flavonoids is the condensation of 4-coumarate coenzyme A with three malonyl CoA molecules to give 2′,4′,6′,4-tetrahydroxychalcone. This is catalysed by chalcone synthase, an enzyme which has now been isolated from many plants and for which the coding sequence has been determined. The chalcone is then isomerized to the flavanone naringenin, as already illustrated in Fig. 6.7. Naringenin is a key intermediate, which can be converted to several end-products; for example, oxidation of the 2,3-bond, catalysed by flavone synthase, will produce the flavone apigenin. However, in the biosynthesis of cyanidin 3-glucoside, a very widely distributed natural pigment present in copper beech leaves and raspberries, it is oxidized in the B-ring to give eriodictyol (**6.56**). This is then oxidized at the 3-position to yield dihydroquercetin (**6.55**). The 4-carbonyl is then reduced to 4-hydroxyl to produce leucocyanidin (**6.57**), an unstable intermediate which can exist as several stereoisomers. One of these stereoisomers then undergoes oxidation to cyanidin. The enzyme

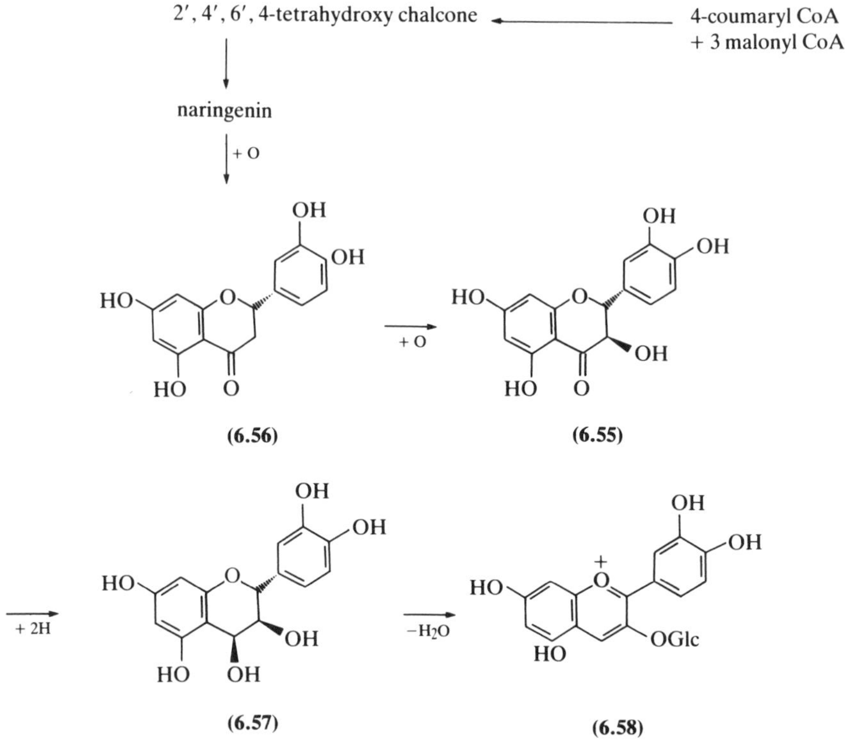

2', 4', 6', 4-tetrahydroxy chalcone ⟵——————— 4-coumaryl CoA
+ 3 malonyl CoA

↓

naringenin

↓ + O

(6.56) → (6.55)

(6.57) → (6.58)

Fig. 6.13 Biosynthesis of the anthocyanin cyanidin 3-glucoside.

or enzymes catalysing this last stage have not yet been fully characterized, but they are almost certainly linked to a glucosyltransferase enzyme, since the final product of anthocyanin synthesis is always a glycosylated pigment, in the simplest case cyanidin 3-glucoside (**6.58**). All the other 270 known anthocyanins are formed by a similar route.

6.5.5 Biosynthesis of rotenone

Rotenone is a well-known insecticide and fish poison, present in derris root and other leguminous plants. Its toxicity is due to its inhibition of the NADH-dependent dehydrogenase of the mitochondrial respiratory chain. It is an isoflavonoid with isoprenyl substitution and its biosynthesis is typical of all isoflavonoids in that an early step involves aryl migration of a flavanone intermediate such as naringenin. The pathway of rotenone biosynthesis has been established by L. Crombie from feeding experiments in young seedlings of *Amorpha fruticosa*. The enzymes catalysing the aryl migration have been characterized from soya bean *Glycine max* by H. Grisebach.

7,4'-Dihydroxyflavanone (**6.59**) is the starting material for rotenone

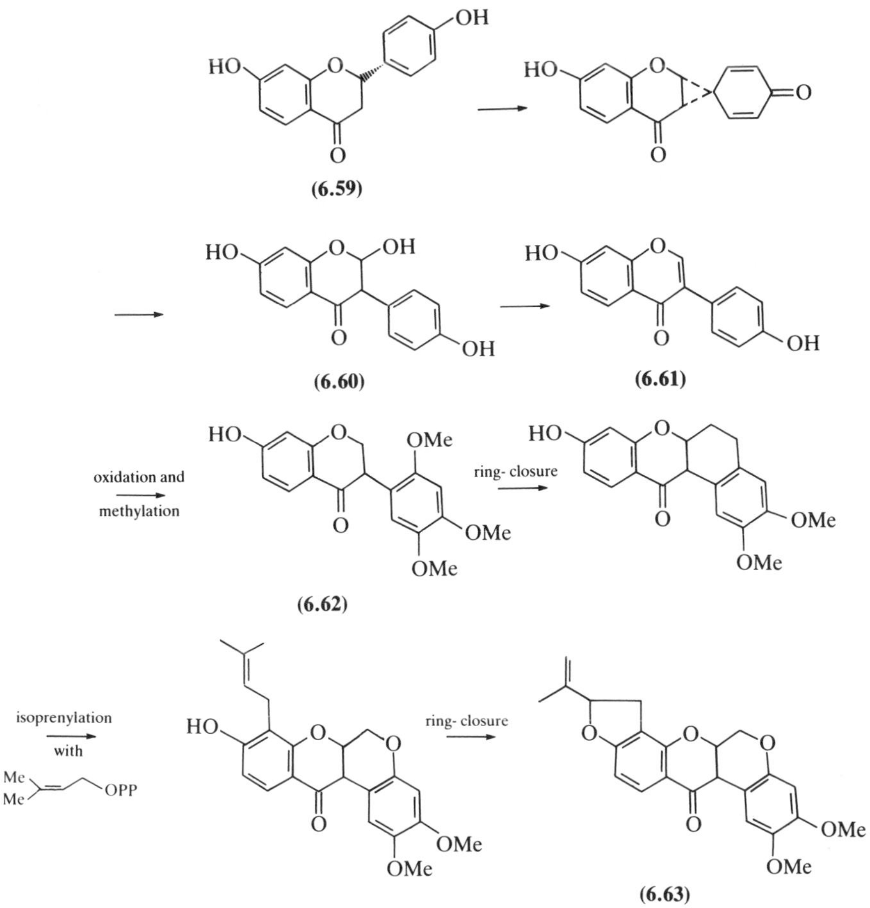

Fig. 6.14 The biosynthesis of the isoflavonoid rotenone.

production. This undergoes a two-stage modification, whereby a transient intermediate with a *p*-quinone system is converted to a 2-hydroxyiso-flavanone (**6.60**) catalysed by isoflavone synthase, and this latter intermediate is then dehydrated to daidzein (**6.61**). Conventional oxidative and methylating enzymes then modify (**6.61**) to 7-hydroxy-2′,4′,5′-trimethoxy-isoflavone (**6.62**). Ring-closure, isoprenylation, and a second ring-closure then produces the final product, rotenone (**6.63**; Fig. 6.14).

6.5.6 *Biosynthesis of plant naphthoquinones*

Biosynthetic diversity of phenolic synthesis is particularly well illustrated by the plant naphthoquinones. These phenolic yellow pigments may be formed by any one of four distinct routes (Fig. 6.15). The first pathway, perhaps the most primitive one, from acetate-malonate, operates in the synthesis of plumbagin (**6.64**), which occurs in the roots of *Plumbago europea*, Plumbaginaceae. The second pathway, known as the *o*-succinylbenzoic acid route, starts from shikimic acid and is followed in the walnut tree, *Juglans regia* (Juglandaceae), which produces juglone (**6.65**). Juglone is exuded from the shell of walnuts, staining the fingers yellowish brown. It is also responsible for the well-known allelopathic (inhibition of seed germination) effects which the walnut tree exerts on neighbouring plant species.

A third pathway to naphthoquinones operates in *Chimaphila* (Pyrolaceae),

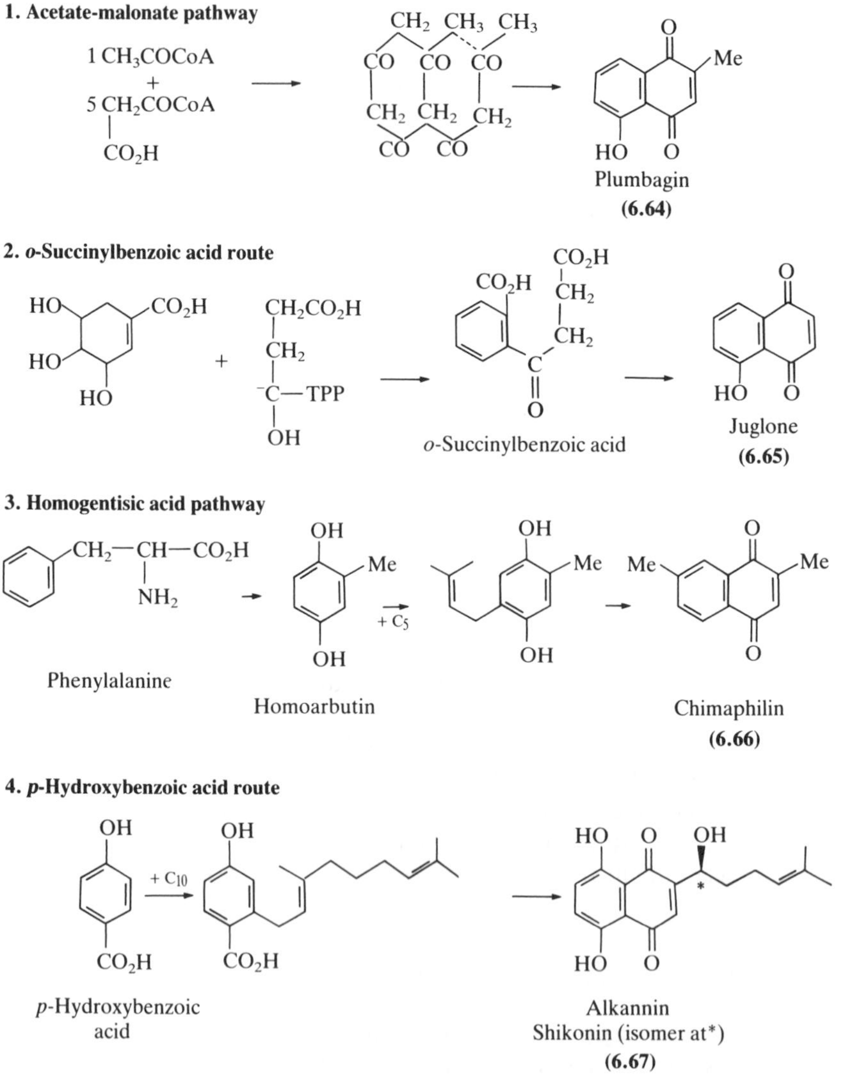

Fig. 6.15 Four biosynthetic routes to naphthoquinones.

a plant genus which produces a dimethylnaphthoquinone, chimaphilin (**6.66**). This is formed from phenylalanine by the homogentisic acid route, with the intermediacy of homoarbutin. The aromatic ring of chimaphilin is isoprenoid in origin.

The fourth pathway to naphthoquinones starts from *p*-hydroxybenzoic acid. In this case, di-isoprenylation is involved, with the addition of geranylpyrophosphate to the phenolic starting material. The product can be either one or two stereoisomers, alkannin or shikonin, in plants of the Boraginaceae. Shikonin (**6.67**) is of particular economic interest, since it is used as a colourant in lipstick and for medicinal purposes. It is now produced commercially in Japan in cell cultures of the borage *Lithospermum erythrorhizon*.

6.6 Laboratory synthesis

Procedures for flavonoid synthesis are still based today on methods originally developed during the classical era by R. Robinson, W. Baker, K. Venkataraman, and T.S. Wheeler, although more effective reagents for some reactions have been introduced in the interim. A common route is the reaction of an *o*-hydroxyacetophenone (C_6-C_2 unit) with an aromatic aldehyde (C_6-C_1 unit) to produce a C_{15} intermediate, which can be subsequently modified to give the desired end-product. Typically, phloracetophenone (**6.68**) can be condensed with *p*-hydroxybenzaldehyde (**6.69**) in the presence of strong alkali to give a chalcone (**6.70**):

(**6.68**)   (**6.69**)   (**6.70**)

The chalcone so formed is readily isomerized to a flavanone, by heating in dilute alcoholic acid, which in turn can be oxidized by selenium dioxide to a flavone. Chalcones are also susceptible to oxidation with alkaline hydrogen peroxide to give flavonols and (in low yield) aurones (Fig. 6.16). Aurones are a class of flavonoid not so far mentioned, which contribute yellow flower colour in plants such as snapdragon and *Coreopsis*. They can be obtained in greater yield by condensing a coumaranone with a substituted benzaldehyde (Fig. 6.16).

An alternative direct synthesis of flavones, illustrated in Fig. 6.17 for the parent compound, is by the Baker–Venkataraman rearrangement of *o*-benzoyloxyacetophenone to *o*-hydroxydibenzoylmethane catalysed by potassium hydroxide in pyridine. In the presence of sulphuric or acetic acid, the dibenzoylmethane undergoes ring-closure and dehydration to give flavone

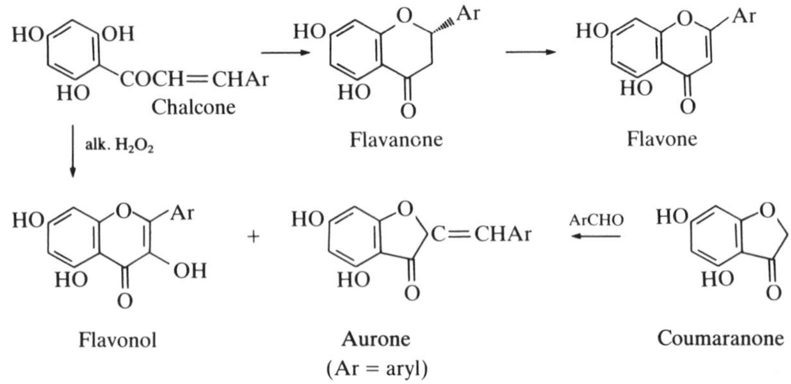

Fig. 6.16 Laboratory interconversions of chalcones to other flavonoids.

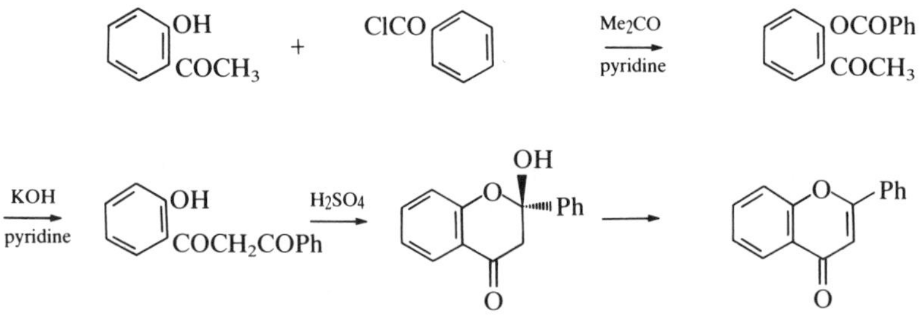

Fig. 6.17 Synthesis of flavone from *o*-hydroxyacetophenone by Baker–Venkataraman rearrangement.

in good yield. With various practical modifications, this method has been widely used. A less popular related procedure is the Allan–Robinson fusion, in which the *o*-hydroxyacetophenone is heated with the anhydride of an aromatic acid and its potassium salt at relatively high temperatures and the flavone can be recovered directly from the reaction mixture.

The total synthesis of most flavonoids requires the addition of glycosyl substituents, using, for example, acetobromoglucose as a glucose donor molecule, since glycosylation is the rule in nature rather than the exception. The synthesis of these glycosides can represent a considerable challenge, particularly to achieve good yields. The synthesis of more than 20 common anthocyanin pigments by R. Robinson at Oxford in the 1930s remains a landmark in the history of modern organic chemistry. A typical synthesis, carried out jointly with A.R. Todd, is that of malvin, a pigment of mallow flowers and of red wine. The two starting materials for the condensation (Fig. 6.18) both carry glucosyl substituents. The condensation is carried out at 0° in ethyl acetate or ether in the presence of hydrogen chloride. The flavylium salt crystallizes out from this reaction mixture and mild treatment with barium hydroxide removes the protecting groups and gives the anthocyanin (**6.71**) in excellent yield.

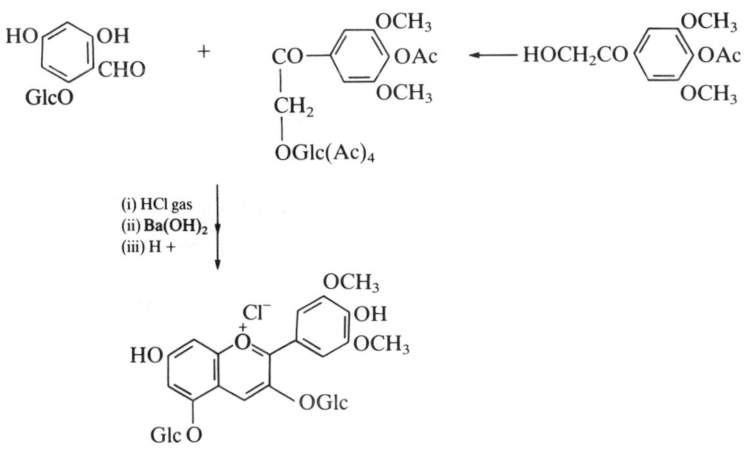

(6.71)

Fig. 6.18 Synthesis of the anthocyanin pigment malvin.

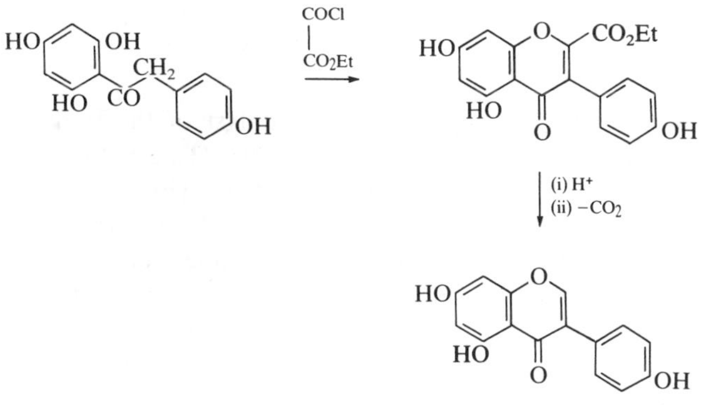

(6.72)

Fig. 6.19 Synthesis of genistein by the ethoxalyl chloride method.

The biosynthesis of isoflavones involves an aryl migration of the sidephenyl substituent of a flavanone (see Section 6.5) but no such isomerization has yet been achieved in the laboratory. Instead, isoflavones are obtained by the reaction of a phenylbenzyl ketone with a reagent capable of providing the 2-carbon atom of the resulting product, such as ethyl formate or ethoxalyl chloride. The latter reaction (Fig. 6.19) works smoothly with ketones containing free hydroxyl groups and the resulting ethyl ester can be hydrolysed and decarboxylated readily to give genistein (**6.72**). In 1970, a much milder reagent, thallium(III) nitrate, was introduced for isoflavone synthesis starting from a chalcone, in which the free hydroxyls are protected by benzylation. The reaction occurs at room temperature in a few minutes and is very convenient if, for example, the related isoflavan is desired. Thus

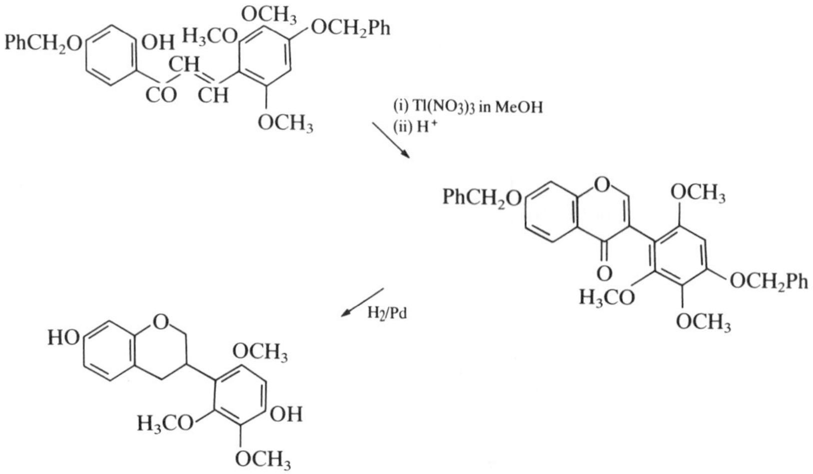

Fig. 6.20 Synthesis of the isoflavan lonchocarpan using the thallium(III) nitrate reagent.

reduction with hydrogen and palladium catalyst of the isoflavone intermediate will yield the isoflavan lonchocarpan (6.73), a naturally occurring substance from the root of *Lonchocarpus laxiflorus* (Fig. 6.20).

The widespread flavonoids are technically 2-phenylchromones and the isoflavonoids 3-phenylchromones, but a number of simple aliphatic substituted chromones are also known in nature. These nearly all have a 2-methyl substituent, in keeping with their biosynthetic polyketide origin. The discovery in the author's laboratory of 5,7-dihydroxy-3-ethylchromone (lathodoratin) as a phytoalexin (plant anti-fungal agent) in the sweet pea *Lathyrus odoratus* was somewhat unexpected. The unique 3-ethyl substitution pattern in (6.74) prompted its synthesis. The 2-ethyl isomer was readily obtained by Allan–Robinson fusion, i.e. by heating phloracetophenone with propionic anhydride and sodium propionate. A study of this product was useful in confirming that it was spectrally different from lathodoratin and that the latter was indeed 3-substituted. Synthesis of the 3-ethyl isomer was achieved by heating 2,4,6-trihydroxybutyrophenone with dimethyl formamide (supplying the 2-carbon) in the presence of boron trifluoride and methyl sulphonyl chloride (Fig. 6.21). Later tracer studies in the sweet pea showed that the aromatic ring of lathodoratin was acetate-derived, while the five aliphatic carbon atoms came from the amino acid isoleucine.

In the synthesis of plant quinones, it is common to produce the corresponding phenolic 1,4-quinols, since oxidation of the phenolic hydroxyl groups with, for example silver oxide, is readily accomplished. Lapachol (6.75), for instance, can be so prepared from 1,4-dihydroxy-2-methoxynaphthalene (Fig. 6.22). Lapachol is a yellow pigment, from the heartwood

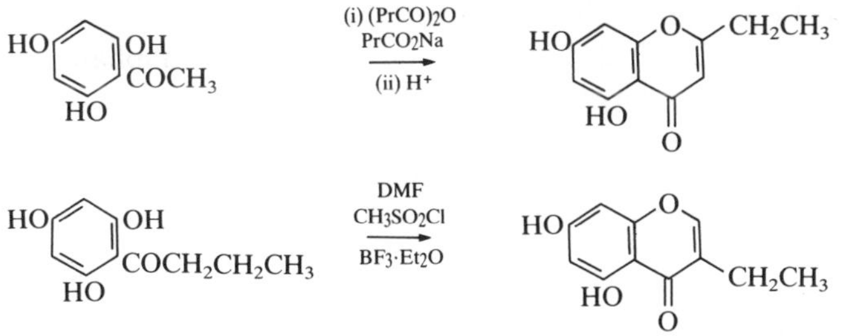

Fig. 6.21 Two syntheses of chromones.

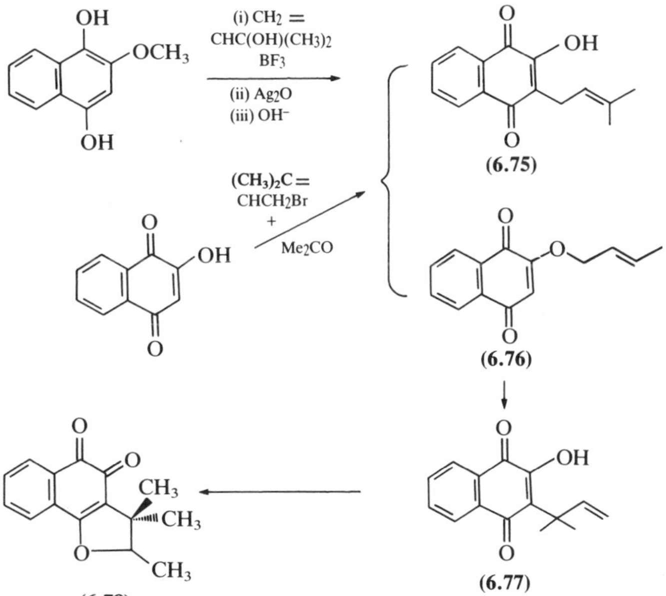

Fig. 6.22 Laboratory synthesis of lapachol and interconversion to dunnione.

of tropical tree *Tecoma* (*Bignoniaceae*), which has economic use in the dyeing of cloth. A more interesting synthesis of lapachol was devised by L. Fieser, who reacted the potassium salt of 2-hydroxynaphthoquinone with 3,3-dimethylallylbromide in hot acetone. He obtained a small yield of lapachol but larger amounts of the isomeric ether (**6.76**). This ether, on heating in alcohol, rearranged (via a Claisen ether rearrangement) to the lapachol isomer (**6.77**), which at first appeared to be a useless by-product. R.G. Cooke then showed that, on treatment with cold sulphuric acid, (**6.77**) underwent ring-closure of the side-chain on the adjacent carbonyl with the production of dunnione (**6.78**). Dunnione happens to be a rare 1,2-quinone, which occurs as an excretory deposit on the under-surface of leaves of *Streptocarpus dunnii*.

Price and Robinson had been studying the structure of dunnione, but they had been unable to decide by degradation whether the aliphatic ring system was dihydrofuran or dihydropyran. This synthesis by Cooke showed beyond doubt that dunnione indeed has the dihydrofuran structure (Fig. 6.22). Thus the laboratory synthesis of a naphthoquinone in this case led to the structural proof of not one, but two natural phenolic pigments.

Further reading

1. F.M. Dean, *Naturally Occurring Oxygen Ring Compounds*, Butterworths, London, 1963.
2. J.B. Harborne, T.J. Mabry and H. Mabry, *The Flavonoids*, Chapman & Hall, London, 1975.
3. J.B. Harborne and T.J. Mabry, *The Flavonoids: Advances in Research*, Chapman & Hall, London, 1982.
4. J.B. Harborne, *The Flavonoids: Advances in Research since 1980*, Chapman & Hall, London, 1988.
5. E. Haslam, *Plant Polyphenols: Vegetable Tannins Revisited*, University Press, Cambridge, 1989.
6. R.H. Thomson, *Naturally Occurring Ouinones*, 2nd ed., Academic Press, London, 1971.
7. R.H. Thomson, *Naturally Occurring Quinones, III. Recent Advances*, Chapman & Hall, London, 1987.

7 Alkaloids

J. Mann

7.1 Introduction

The alkaloids are structurally the most diverse class of secondary metabolites, and over 5000 compounds are known, ranging from relatively simple structures such as coniine (**7.1**) from hemlock to exceedingly complex ones such as that of the neurotoxin batrachotoxin (**7.2**), from the skin of a Colombian frog. They are most commonly encountered in the plant kingdom, but representatives have been isolated from most other orders of organisms ranging from fungi to mammals.

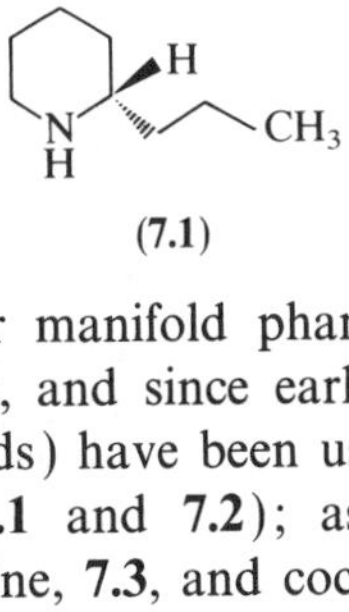

(7.1) 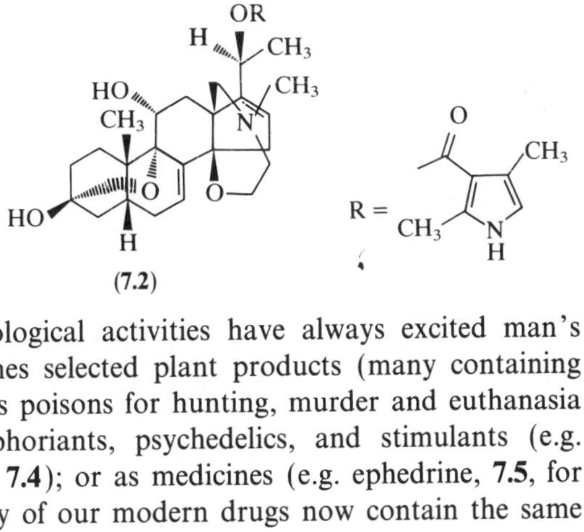   (7.2)

Their manifold pharmacological activities have always excited man's interest, and since early times selected plant products (many containing alkaloids) have been used as poisons for hunting, murder and euthanasia (e.g. **7.1** and **7.2**); as euphoriants, psychedelics, and stimulants (e.g. morphine, **7.3**, and cocaine, **7.4**); or as medicines (e.g. ephedrine, **7.5**, for respiratory problems). Many of our modern drugs now contain the same compounds or synthetic analogues, and the pharmacological and toxicological properties of these compounds are thus of immense interest and importance.

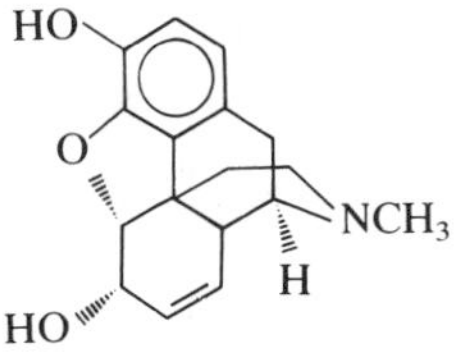

(7.3) 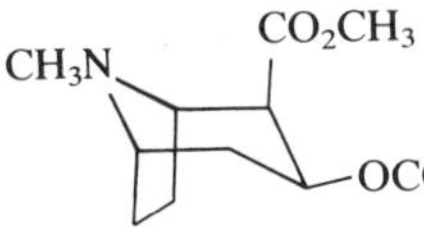   (7.4) 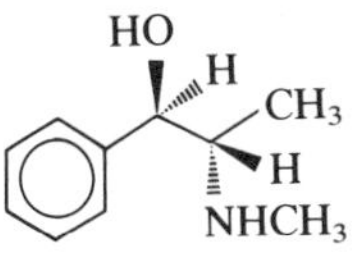   (7.5)

The original definition of an alkaloid, first proposed in 1879 by the pharmacist W. Meissner, encompassed nitrogenous compounds of complex molecular structure and significant pharmacological activity confined to the plant kingdom. This definition is clearly no longer tenable, and a recent definition (1982) due to Pelletier includes cyclic nitrogen-containing molecules which are true secondary metabolites (i.e. of limited occurrence, and produced by living organisms). Simple acyclic derivatives of ammonia and simple amines are thus excluded, and the additional requirement, that the nitrogen atom must have a negative oxidation state, excludes nitro and nitroso compounds. Most compounds covered by this definition derive at least part of the structure from amino acids or their derivatives, and the biosynthetic pathways to the major classes of alkaloids will be covered in later sections.

7.2   Structural types

It is usual to classify alkaloids according to the amino acids (or their derivatives) from which they arise. Thus the most important classes are derived from the amino acids ornithine and lysine (Fig. 7.1); or from the aromatic amino acids phenylalanine and tyrosine (Fig. 7.2); or from tryptophan and a moiety of mevalonoid origin (Fig. 7.3); and a number of compounds are also derived from anthranilic acid (**7.6**) or from nicotinic acid (**7.7**).

This kind of classification is sometimes criticized because it fails to include those alkaloids which are derived from a polyketide or from a terpenoid, with incorporation of a nitrogen atom, ultimately from ammonia. Coniine (**7.1**) and batrachotoxin (**7.2**) are two excellent examples of such alkaloids, and these compounds are often known as 'pseudoalkaloids'.

In addition to these classes of alkaloids there are numerous other compounds which are covered by Pelletier's definition, for example, the antibiotic cycloserine (**7.8**) and mitomycin C (**7.9**); the mushroom toxin muscimol (**7.10**); and the purine alkaloids such as caffeine (**7.11**). These will not be considered further in this chapter.

(7.6)   (7.7)   (7.8)

(7.9)   (7.10)   (7.11)

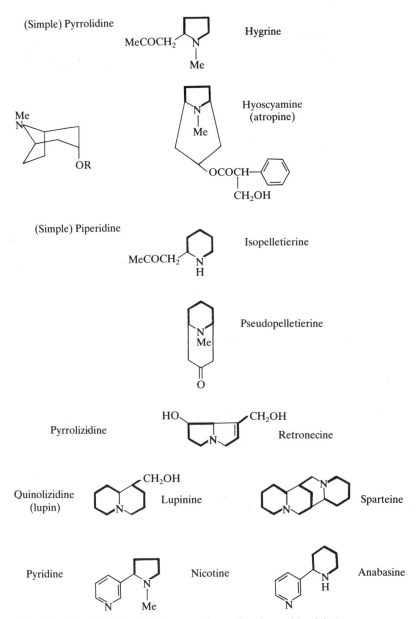

Fig. 7.1. (The bold lines indicate portions of amino acid origin.)

7.3 Occurrence

Of the 5000 or so known alkaloids, the majority occur in flowering plants, though the distribution is far from uniform. Thus, although 40 per cent of all plant families have at least one species containing alkaloids, when the 10 000 plant genera are considered, only about 9 per cent of these have been

(1) SIMPLE MONOCYCLIC COMPOUNDS
ArC_2N

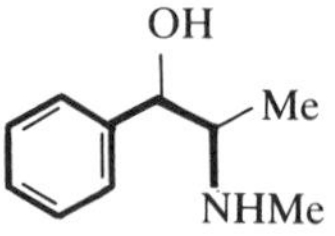

Ephedrine

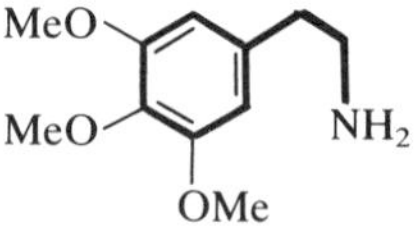

Mescaline

(2) ISOQUINOLINES
$ArC_2N–R_{alkyl}$

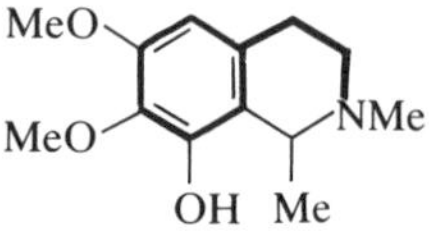

Pellotine

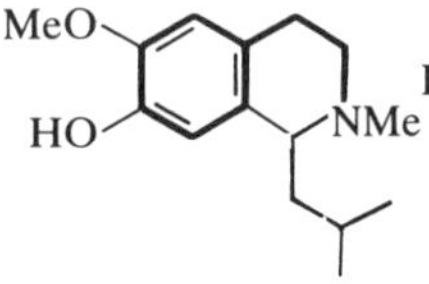

Lophocereine

(3) BENZYLISO QUINOLINES
$ArC_2N–C_2Ar$

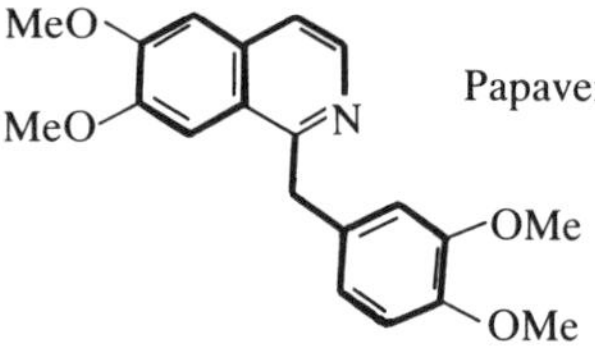

Papaverine

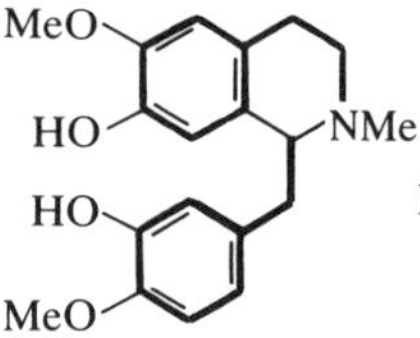

Reticuline

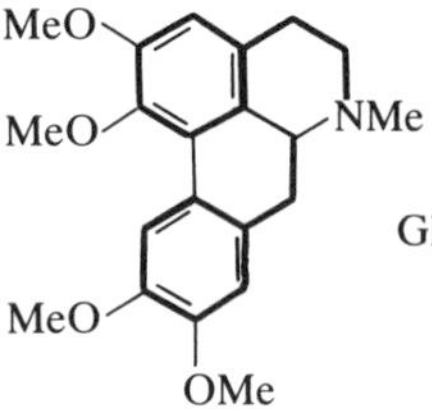

Glaucine

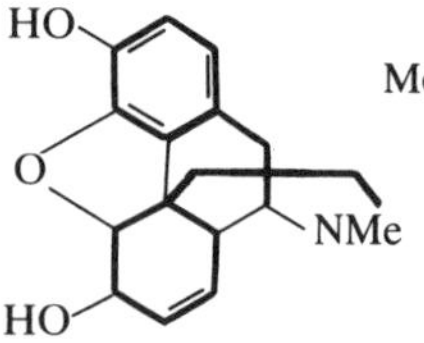

Morphine

(extra carbon atom from methionine)

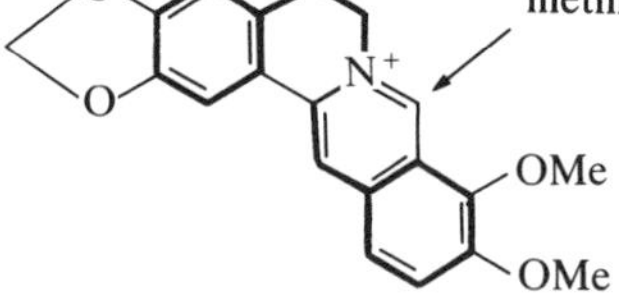

Berberine

(4) AMARYLLIDACEAE ALKALOIDS
$ArC_2N–C_1Ar$

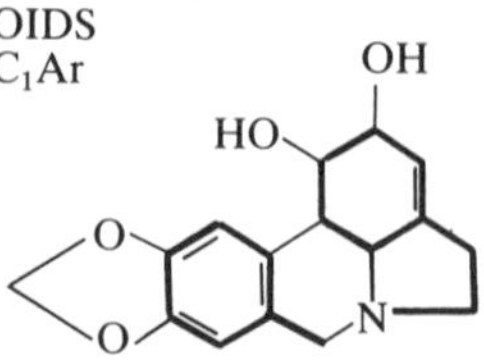

Lycorine

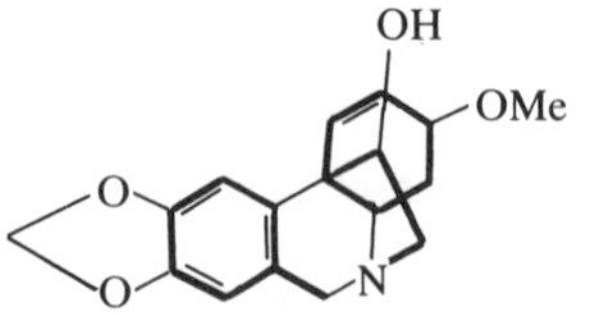

Haemanthamine

Fig. 7.2. (Bold lines indicate portions of amino acid origin.)

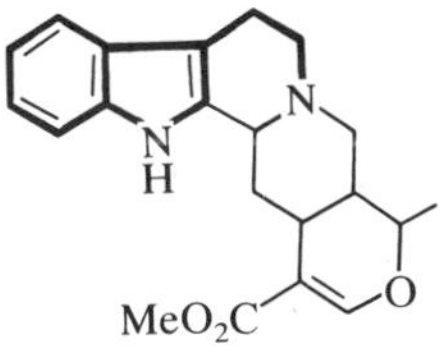

Ajmalicine

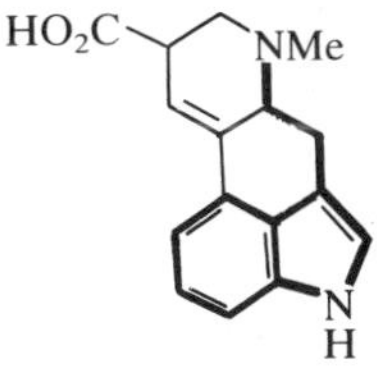

Lysergic acid

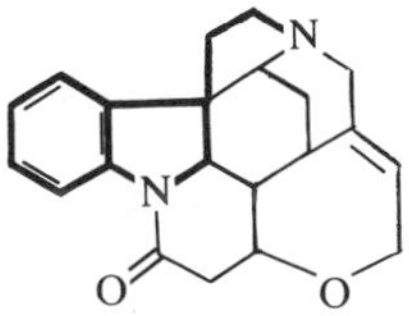

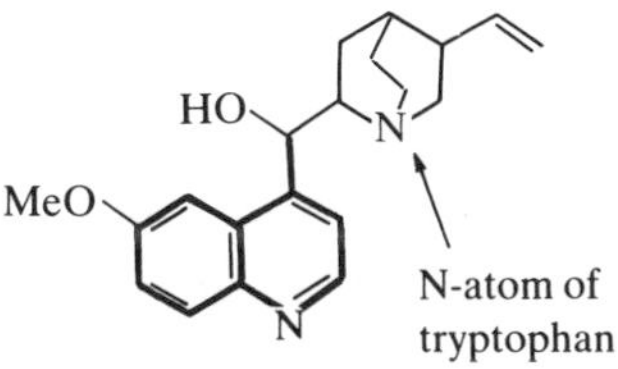

Strychnine

Quinine

N-atom of tryptophan

Fig. 7.3. (Bold lines indicate portion of molecule from tryptophan.)

shown to produce alkaloids. For example, alkaloids are prevalent in members of the Leguminosae family (lupin, laburnum, broom, and gorse), and Solanaceae family (tomato, potato, deadly nightshade, henbane, and tobacco); but they are rarely found in the gymnosperms (e.g. conifers), cryptogams (e.g. ferns, mosses, and liverworts), or in most monocotyledons (e.g. grasses). Very often closely related plant species contain compounds of the same structural type, for example, (−)-sparteine (**7.12**) from broom, and (−)-cytisine (**7.13**) from laburnum. This has obvious utility when a new species must be assigned to a plant family, and is the prime concern of plant taxonomists.

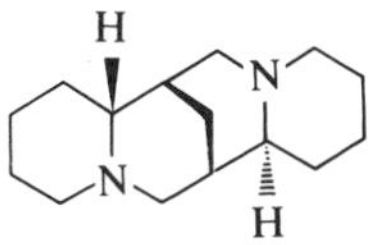

(**7.12**) 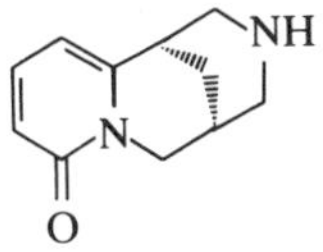   (**7.13**)

In recent years an increasing number of alkaloids have been isolated from animals, insects, and microorganisms. Batrachotoxin (**7.2**) from the skin of a Colombian arrow poison frog, has already been mentioned; and other interesting animal products include samandarine (**7.14**) from the skin of the European salamander, pumiliotoxin C (**7.15**) from a Panamanian frog, and bufotenin (**7.16**) from the common European toad. All of these surely have ecological significance since they act as potent deterrents to would-be predators.

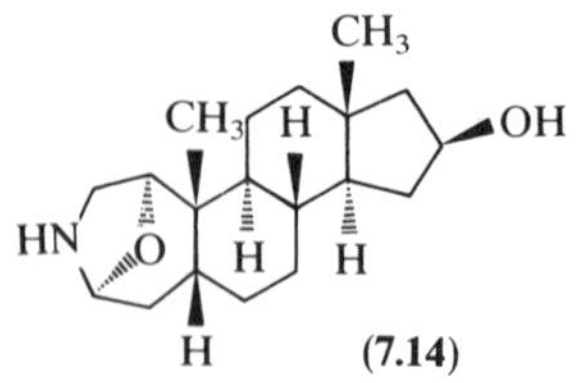

(7.14)

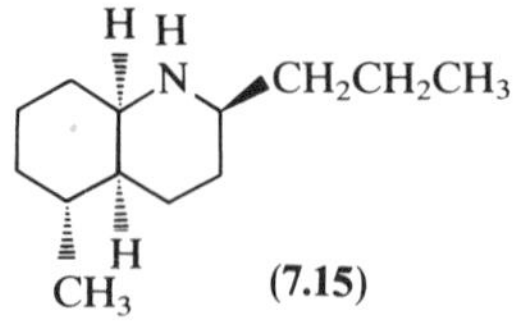

(7.15)

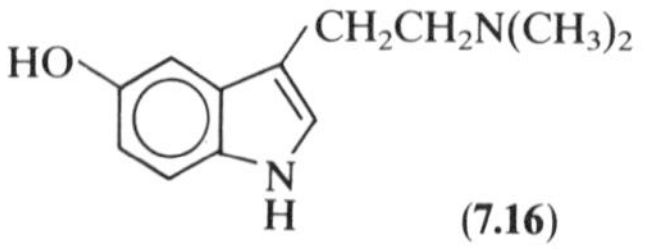

(7.16)

Mammalian alkaloids are rare, but two well-authenticated examples are
(−)-castoramine (**7.17**) from the Canadian beaver, and muscopyridine (**7.18**)
from the musk deer. Both compounds have a role in communication as
territorial marker substances.

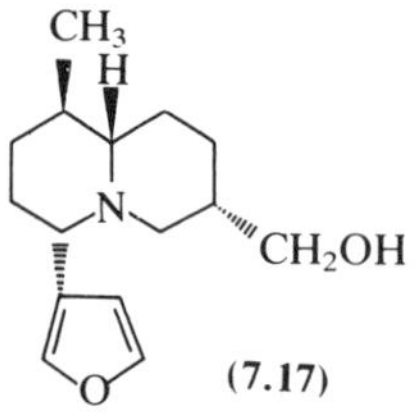

(7.17)

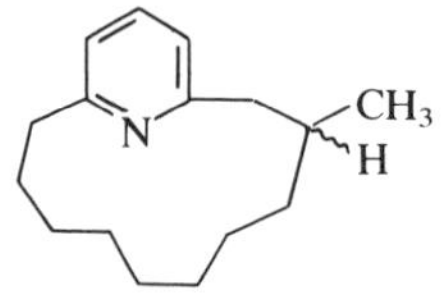

(7.18)

Insects produce a rich variety of structural types which include the
2,6-dialkylpiperidines of the fire ant (*Solenopsis invicata*) (**7.19**), the tricyclic
N-oxides of the ladybird (e.g. **7.20**), and the quinazolinones of the European
millipede (e.g. **7.21**). All of these compounds are used for defence, whilst
the compound (**7.22**) is used as a trail pheromone by the Pharaoh's ant.

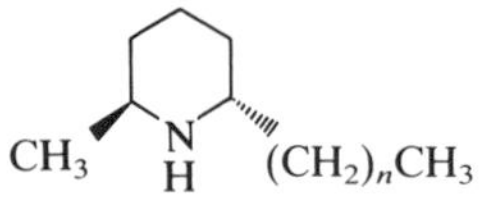

(**7.19**) $n = 10, 12, 14$

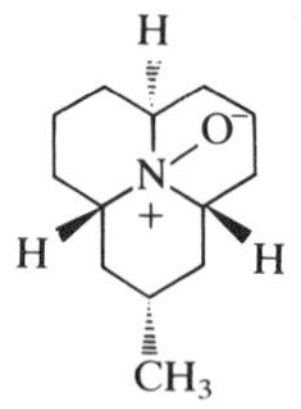

(7.20)

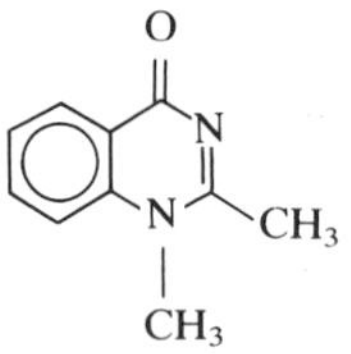

(7.21)

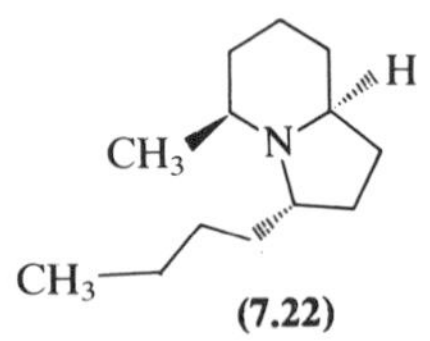

(7.22)

Marine organisms have been much investigated during the last two
decades, and a plethora of secondary metabolites have come to light. Amongst
the alkaloids are the exceedingly complex saxitoxin (**7.23**), produced by a
red-coloured dinoflagellate. The 'red tides' contain mass aggregations of
such organisms, and food poisoning results when the toxic alkaloids are
passed along the food chain to man. The Japanese puffer fish is highly valued
as a culinary delicacy, but is a hazard because its liver and ovaries contain
the highly toxic tetrodotoxin (**7.24**).

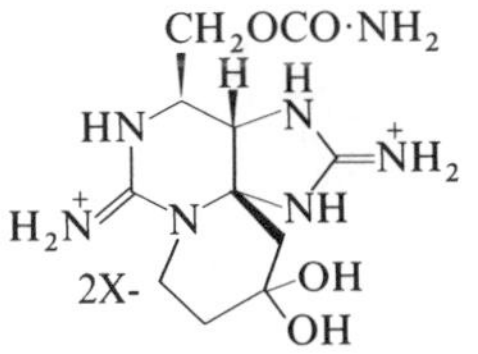

(7.23) 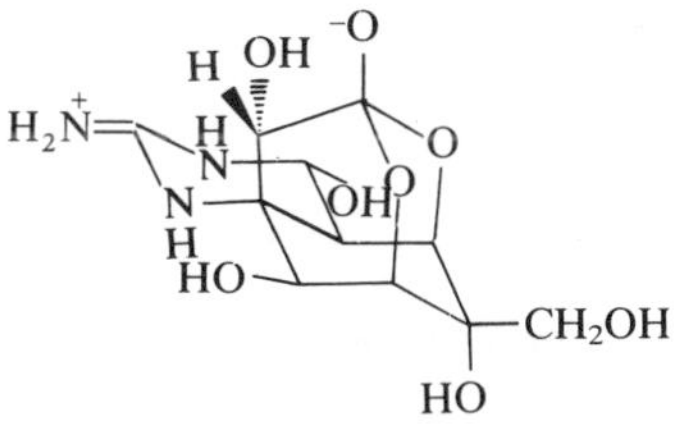   **(7.24)**

Finally, a number of fungi produce alkaloids, and these, too, present potential hazards as food contaminants. The ergot alkaloids, for example chanoclavine I (**7.25**) produced by the fungus *Claviceps purpurea*, were a frequent source of misery and death during the Middle Ages through contamination of rye bread. Certain of these alkaloids were neurotoxic while others caused vasoconstriction, i.e. narrowing of the blood vessels. The resulting 'ergotism' or 'St Anthony's fire' manifested itself as mania and/or gangrene, and there were numerous fatalities (see Section 7.5.3.2). *Penicillium roqueforti*, which is used commercially in the production of blue cheeses, elaborates a structural relative of the ergot alkaloids, isofumigaclavine B (**7.26**), as well as the structurally unrelated roquefortine (**7.27**).

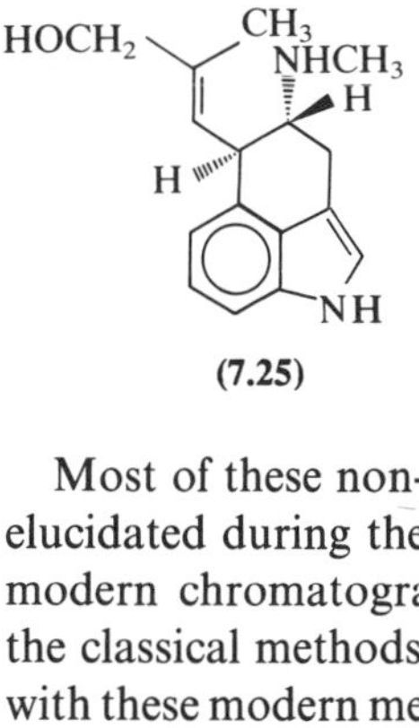

(7.25) 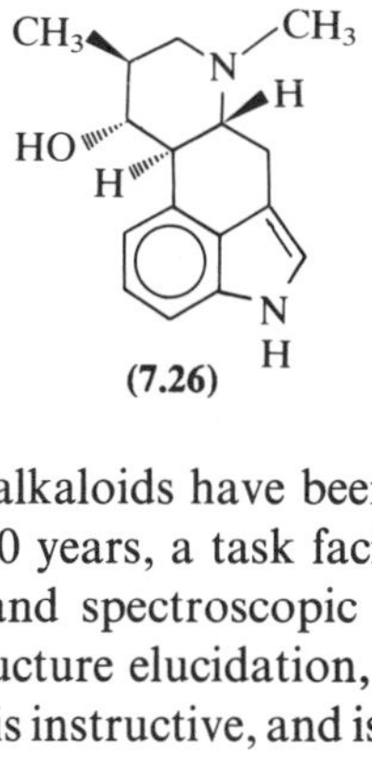   **(7.26)** 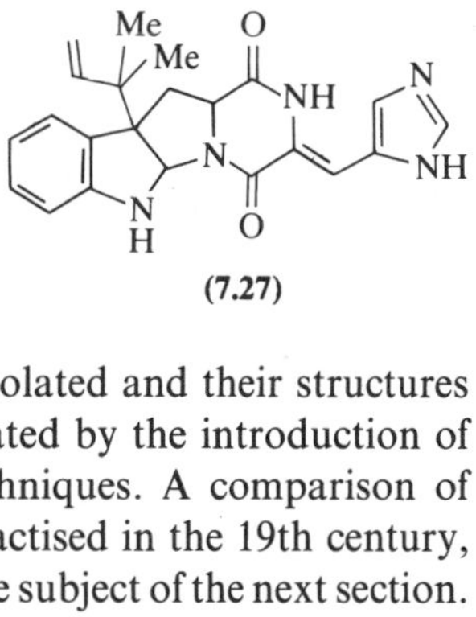   **(7.27)**

Most of these non-plant alkaloids have been isolated and their structures elucidated during the last 20 years, a task facilitated by the introduction of modern chromatographic and spectroscopic techniques. A comparison of the classical methods of structure elucidation, practised in the 19th century, with these modern methods is instructive, and is the subject of the next section.

7.4   Isolation and structure elucidation

7.4.1   *Isolation and purification*

Many alkaloids are basic and occur as salts of 2-hydroxybutane-1,4-dioic acid (malic acid), or of 1,3,4,5-tetrahydroxycyclohexane (quinic acid). They can thus be extracted into acid solution using aqueous HCl, tartaric, or citric acids. Essentially neutral alkaloids like colchicine (**7.28**) or piperine (**7.29**), which are in fact amides, remain in the organic phase, while most other alkaloids are isolated after basification and extraction into ethyl acetate.

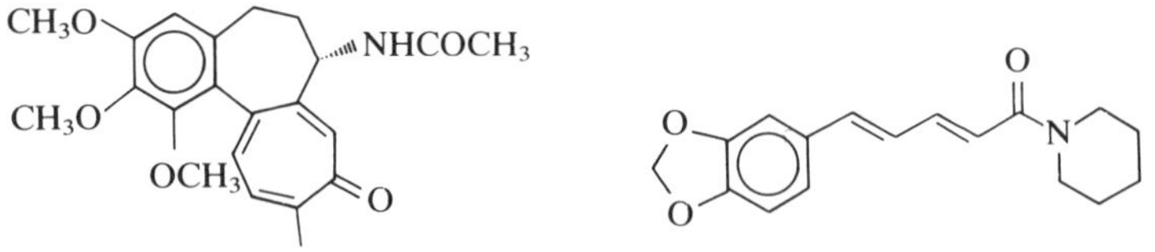

(7.28)   (7.29)

Also, in rare instances steam distillation can be employed, for example with low molecular weight alkaloids such as coniine (**7.1**) and sparteine (**7.12**); but almost invariably subsequent purification of the crude alkaloid mixtures is effected by chromatography using silica or alumina, and then recrystallization of the partially purified compounds from solvent systems like aqueous ethanol, methanol/chloroform, or methanol/acetone.

7.4.2 Structure elucidation

7.4.2.1 Classical era

The classical era for structural studies on alkaloids was the 19th century, though this could be extended to the 1930s (advent of X-ray crystallography), or even to the 1970s (advent of high resolution NMR facilities and modern methods of mass spectroscopy). Two case histories will be discussed, those of morphine and atropine.

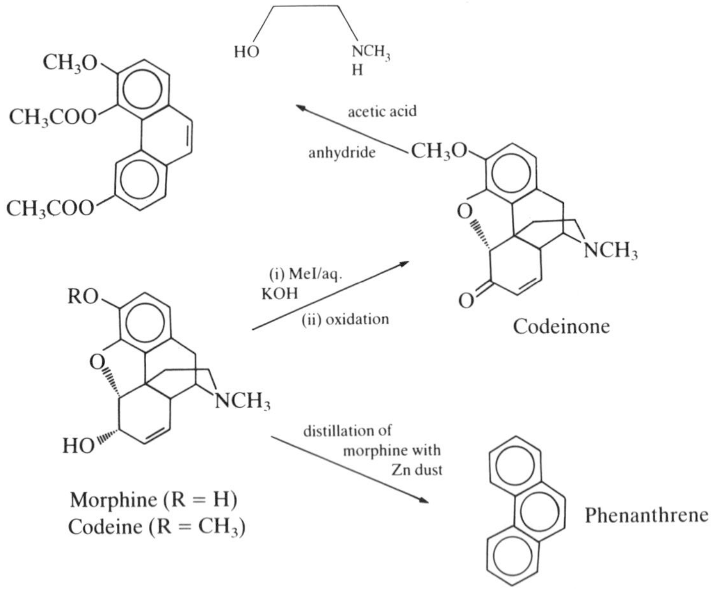

Fig. 7.4 Structure of morphine. Standard tests showed that the nitrogen atom was fully substituted, and that a phenolic hydroxyl was present (positive $FeCl_3$ test). In addition, a diacetate and dibenzoate could be formed, hence two hydroxyls were present. Codeine absorbed one mole of hydrogen, hence both compounds contained one olefinic double bond.

Opium has been used by humans for thousands of years, so it is hardly surprising that the major active ingredient, morphine (**7.3**), was the first alkaloid to be isolated in a pure state (by Sertürner, in 1805). However, 118 years passed before Sir Robert Robinson and coworkers finally established the structure in 1923 and some of the chemical evidence adduced in favour of the structure is set out in Fig. 7.4. This, together with much other chemical evidence, established that a reduced phenanthrene with a two-carbon bridge containing a tertiary nitrogen atom (with methyl as one substituent) was present, and the structures of morphine and codeine were first proposed by Robinson and Gulland in 1923 and 1925 respectively. The first synthesis of morphine was completed by Gates and coworkers in 1956.

Atropine is not generally a natural product, but arises through racemization of (−)-hyoscyamine (**7.30**) during isolation and purification,

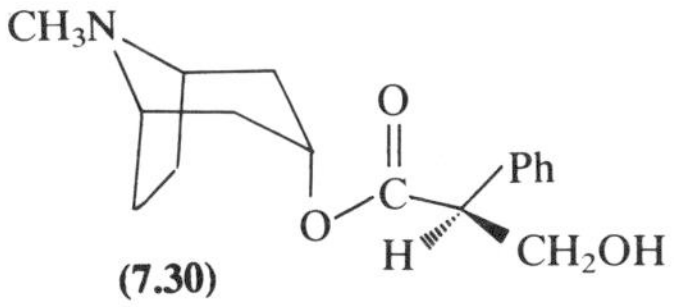

(**7.30**)

and is thus (±)-hyoscyamine. (−)-Hyoscyamine is the most common tropane alkaloid and is present in, amongst other species, *Atropa belladonna* (deadly nightshade), *Hyoscyamus niger* (henbane), and *Mandragora officinarum* (mandrake). Crude extracts of these plants have a long association with witchcraft and folklore, but it was not until 1833 that atropine was isolated in a pure state from *Atropa belladonna*. Hydrolysis with warm barium hydroxide solution produced racemic tropic acid and tropine (see Fig. 7.5(a)).

The structure of tropic acid was established by degradative studies, and then through synthesis (Fig. 7.5(b)). Exhaustive degradation of tropine, carried out by Willstätter between 1895 and 1901, provided evidence for the bicyclic structure of tropine, and some of this chemistry is also shown in the figure. Willstätter then went on to prove the structure by means of a rather tortuous total synthesis (1900 to 1903) – the elegant 'one pot' synthesis by Robinson (1917) will be discussed later.

A more subtle point of structural interest concerns the stereochemistry of (−)-tropic acid from (−)-hyoscyamine, and this was established through correlation with the known structure for (+)-alanine (Fig. 7.6). Racemic 3-chloro-2-phenylpropanoic acid was resolved using codeine, and the resultant pure (−)-enantiomer was hydrolysed to yield (−)-tropic acid, or reduced using catalytic hydrogenation to produce (−)-2-phenyl propanoic acid. In a second sequence of steps the (+) form of 2-phenyl propanoic acid was degraded via the Curtius reaction (known to proceed with retention of configuration) to yield (−)-phenylethylamine, and thence (+)-alanine after oxidation. Since the latter was known to have the (*S*)-configuration, it follows that (−)-2-phenyl propanoic acid must be (*R*), and thus (−)-tropic acid has the (*S*)-configuration.

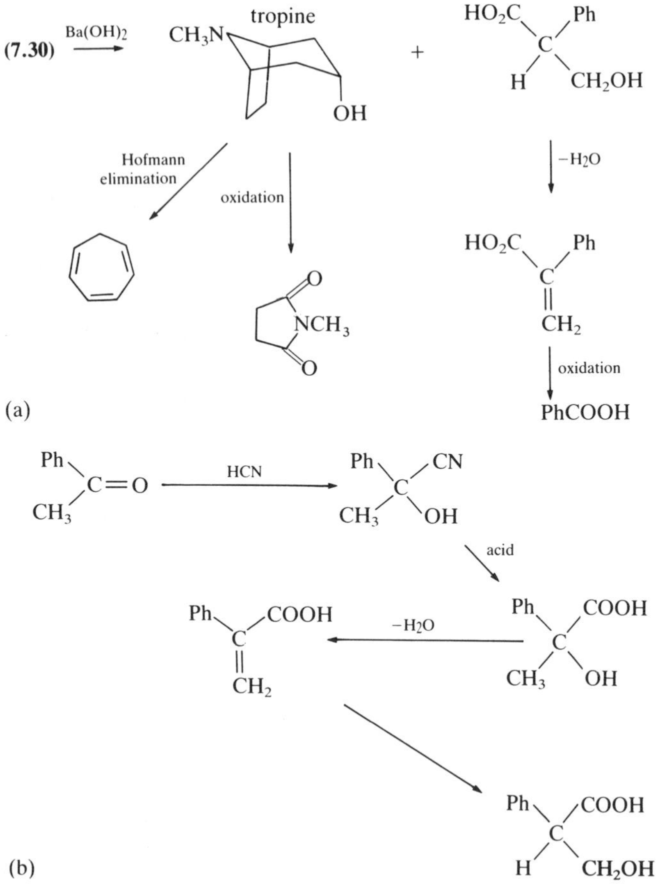

(a)

(b)

Fig. 7.5.

The configuration at carbon-3 was determined by consideration of the respective chemistries of nor-tropine and its geometrical isomer nor-pseudotropine (Fig. 7.7). Upon heating the *N*-acetyl derivatives at 160°C, rearrangement occurred with pseudotropine but not with tropine, and this is only possible if an axial hydroxyl is available for oxine formation. The boat form of nor-pseudotropine satisfies this requirement, and in the chair form it must then have been an equatorial hydroxyl. Nor-tropine then possesses an axial hydroxyl in its chair form.

7.4.2.2 Modern era

During the last 25 years structure elucidation has been facilitated by the use of mass spectrometry, and ^{1}H and ^{13}C NMR techniques. Such is now the level of sophistication of the associated instruments that it is possible to determine a structure in a matter of days given a few milligrams (or less) of

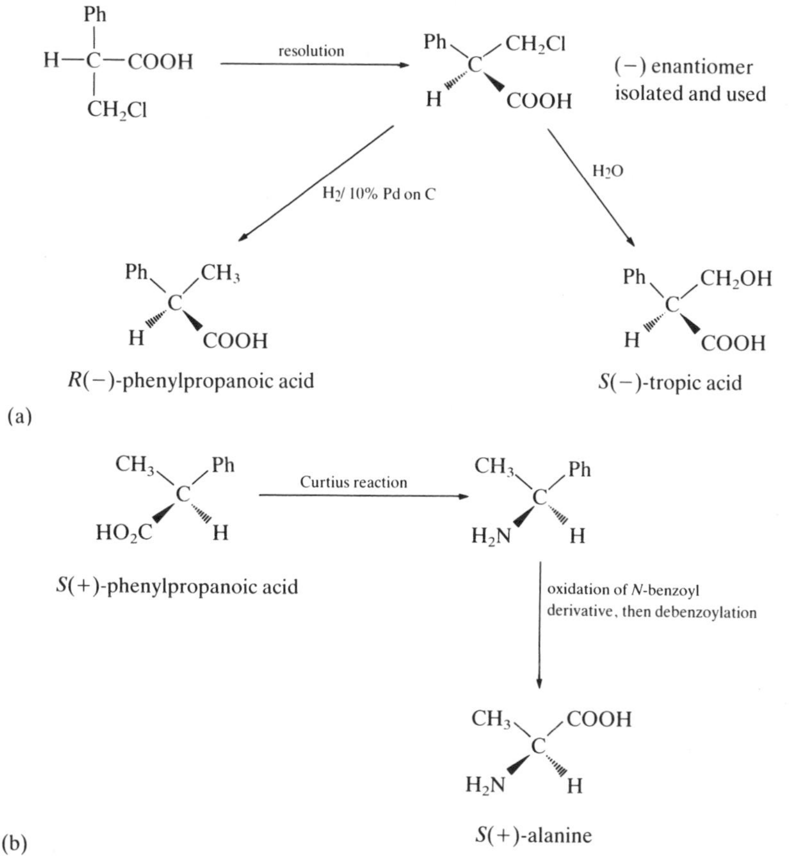

Ph
|
H—C—COOH   →(resolution)→

Ph\ /CH₂Cl
   C
H/ \COOH   (−) enantiomer
CH₂Cl   isolated and used

H₂/ 10% Pd on C

H₂O

Ph\ /CH₃
   C
H/ \COOH

R(−)-phenylpropanoic acid

Ph\ /CH₂OH
   C
H/ \COOH

S(−)-tropic acid

(a)

CH₃\ /Ph
    C
HO₂C/ \H   →(Curtius reaction)→

CH₃\ /Ph
    C
H₂N/ \H

S(+)-phenylpropanoic acid

oxidation of N-benzoyl
derivative, then debenzoylation

CH₃\ /COOH
    C
H₂N/ \H

S(+)-alanine

(b)

Fig. 7.6.

a pure compound. This should be compared with the lengthy period
(118 years) required for elucidation of the structure of morphine, and the yet
more extreme example of strychnine (**7.31**) which was isolated in 1819 (by
Pelletier and Caventou), and whose structure was proved in 1946 by
Robinson and coworkers.

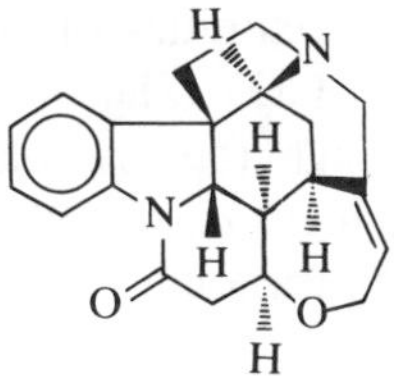

(**7.31**)

It is interesting to consider how modern spectroscopic methods could have
helped with structure elucidation of compounds such as atropine (**7.30**) and
morphine (**7.3**). The ¹H and ¹³C NMR spectra for tropinone are shown in

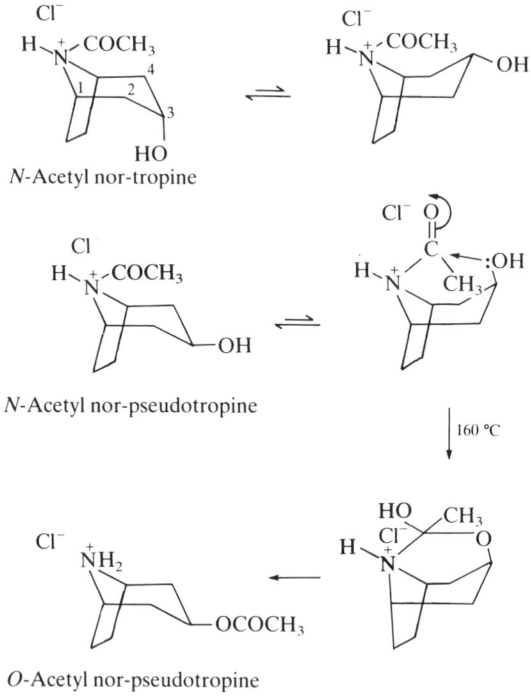

N-Acetyl nor-tropine

N-Acetyl nor-pseudotropine

160 °C

O-Acetyl nor-pseudotropine

Fig. 7.7.

Figs 7.8 and 7.9 respectively, and are both highly informative. The mass spectrum would also have been of great benefit, and key features are shown in Fig. 7.10. The corresponding data for morphine are shown in Fig. 7.11.

A more recent example of structure elucidation is provided by the compound eudistomin K from the Caribbean and New Zealand ascidians (sea squirt family) *Eudistoma olivaceum* and *Ritterella sigillinoides*. This and related compounds are of particular interest because they exhibit potent anti-viral activities against Herpes simplex type 1 and Polio type 1 viruses. The structure is shown in Fig. 7.12 and associated ^{1}H and ^{13}C NMR data, together with the main nuclear Overhauser effects (n.O.e), are shown in Tables 7.1 and 7.2.

Most significant were the proven close spatial proximities of H-1 with H-3$_\alpha$ and H-11$_\alpha$; H-11$_\alpha$ and H-13$_\alpha$; and H-3$_\beta$ and H-13$_\beta$. The molecule is thus shown with a 2-α oxygen configuration, though the low barrier to inversion of stereochemistry at trivalent nitrogen will ensure that there is an equilibrium that also involves the 2-β oxygen configuration.

Once structure elucidation has been accomplished, partial or total syntheses can be attempted, and several examples of classical and modern syntheses will be considered in the last section of this chapter. However, since many of the more modern approaches have involved biomimetic

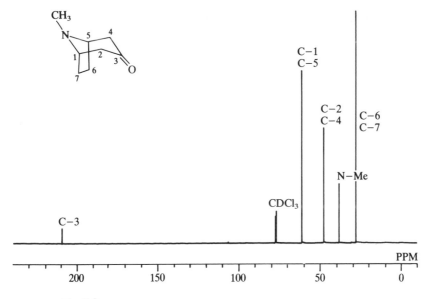

Fig. 7.8.

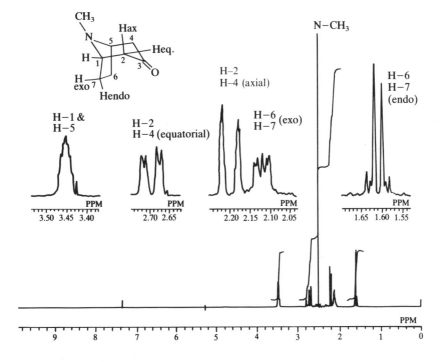

Fig. 7.9.

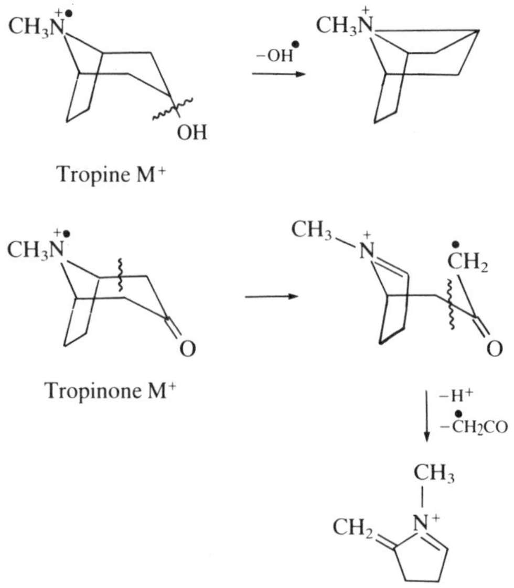

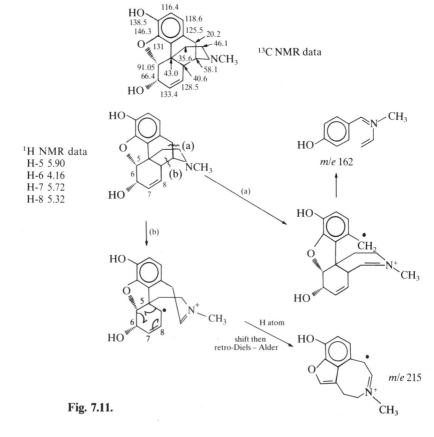

Fig. 7.10.

Fig. 7.11.

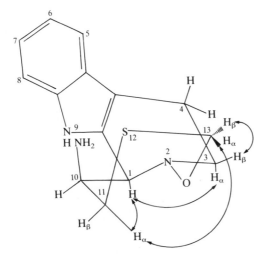

Fig. 7.12 Eudistomin K.

Table 7.1 ¹H NMR data.

H-1	4.10 (br.s)
H-3	3.11 (ddd, *J* 11.5, 5.9, 4.0)
H-3	3.62 (ddd, *J* 9.8, 5.0, 2.2)
H-4	2.83 (m, 15.8, 4.0, 2.2, 0.9)
H-4	2.86 (m, *J* 15.8, 11.5, 5.0, 2.4)
H-5	7.39 (d, *J* 8.5)
H-6	7.1 (d, *J* 1.8, 8.5)
H-8	7.57 (d, 1.8)
H-11	3.31 (d, 14.4)
H-11	2.82 (dd, *J* 14.4, 5.6)
H-13	4.96 (d, *J* 9.2)
H-13	4.82 (d, *J* 9.2)

Table 7.2 ¹³C NMR data.

C-1	70.3
C-3	54.1
C-4a	110.0
C-4b	129.6
C-4	20.6
C-5	119.7
C-6	122.3
C-7	114.55
C-8	114.2
C-9a	131.6
C-9b	138.3
C-10	50.6
C-11	35.0
C-13	71.4

strategies, that is the synthesis follows a similar course to the secondary metabolic pathway by which the alkaloid is produced, it is worth spending some time considering the biosynthetic pathways. These are the subject of the next section.

7.5 Biosynthesis

It is sometimes possible to identify, at a glance, the amino acid from which an alkaloid is derived. Thus for chanoclavine I (**7.25**) an indole-C_2N moiety

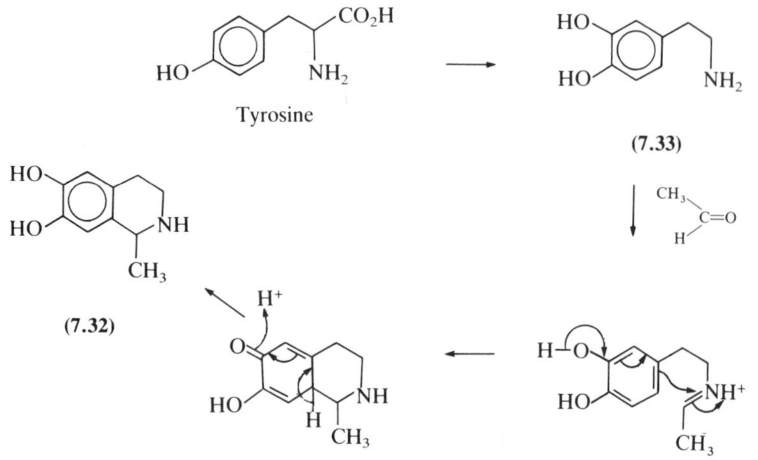

Fig. 7.13.

(derived from tryptophan) is obviously involved, together with a five-carbon component which must be of mevalonoid origin (see Chapter 5). Prior to the ready availability of the radio-isotopes ^{14}C and tritium, and more recently the stable isotopes ^{13}C and ^{15}N, it was not possible to do other than speculate about likely biosynthetic pathways. However, the intuitive approach was sometimes surprisingly successful, as for example with the suggested pathway to the isoquinoline alkaloids (Fig. 7.13). Radio-tracer studies have established the accuracy of this prediction, and the alkaloid pellotine (**7.32**), from the peyote cactus, is derived from dopamine (**7.33**) essentially as shown in Fig. 7.13, except that pyruvic acid (2-ketopropanoic acid) replaces acetaldehyde as the source of a two-carbon unit. The *in vivo* pathways thus involve a decarboxylation at some stage, and this is a common feature of alkaloid biosynthesis.

It is convenient to subdivide alkaloid biosynthesis into two categories according to whether the products are obtained from the amino acids ornithine (**7.34**) and lysine (**7.35**), or from the aromatic amino acids phenylalanine (**7.36**), tyrosine (**7.37**), and tryptophan (**7.38**).

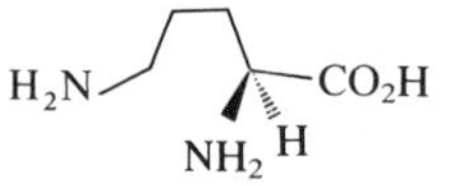

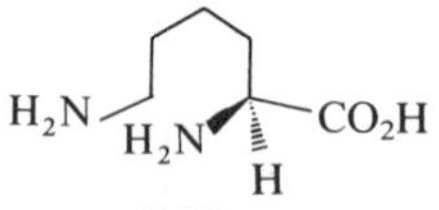

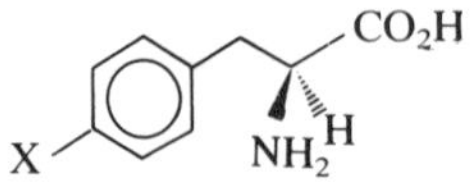

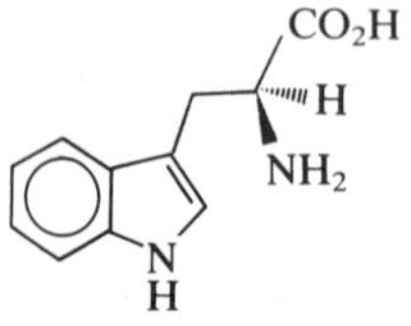

7.5.1 *Alkaloids derived from ornithine and lysine*

The main structural types were shown in Fig. 7.1, and include the pyrrolidine, piperidine, pyrrolizidine, quinolizidine, and pyridine alkaloids. Much of the chemistry involved in the biosynthetic pathways is relatively straightforward, and involves Schiff base formation and subsequent reactions of the Mannich type (for example equation 7.1):

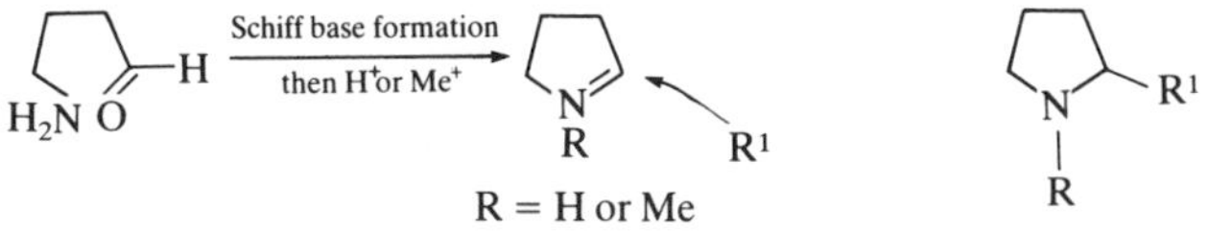

$$R = H \text{ or } Me \qquad [\text{eqn } 7.1]$$

7.5.1.1 *Pyrrolidine alkaloids*

The alkaloid hygrine (**7.39**) can be isolated from the Peruvian shrub *Erythryloxum truxillense* (which also contains cocaine), and is probably formed from ornithine essentially as shown in equation 7.1, where the side-chain (R^1) is derived from acetyl-coenzyme A (Fig. 7.14). The alkaloids tropinone and cocaine are formed by an extension of this pathway (Fig. 7.14).

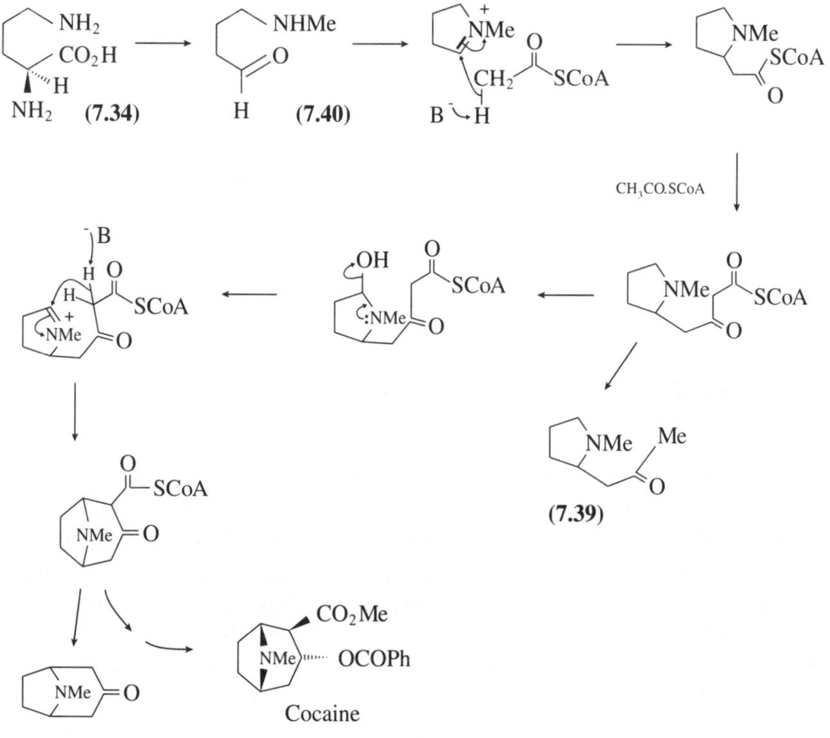

Fig. 7.14.

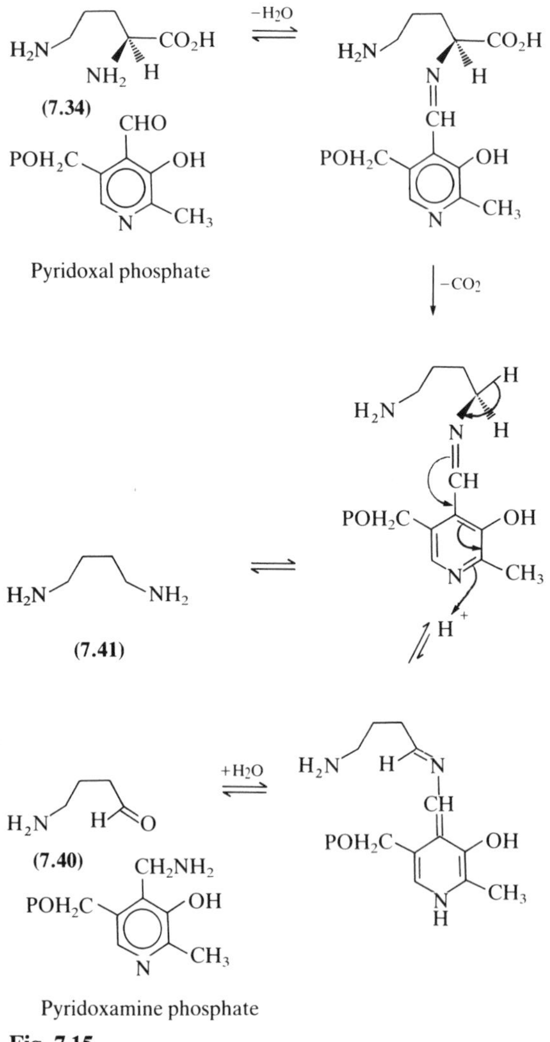

Fig. 7.15.

The requisite amino aldehyde (**7.40**) is derived from ornithine (**7.34**) via the symmetrical diamine putrescine (**7.41**) (Fig. 7.15). The pathway shown in the figure was first suggested by Spenser to account for the fact that the two nitrogen atoms of ornithine (and of lysine) were apparently non-equivalent, since most isotopic labelling studies with 2-^{14}C-lysine or 2-^{14}C-ornithine resulted in incorporation of only one ^{14}C-label. Involvement of free symmetrical intermediate (**7.41**) would have resulted in scrambling of the label between two carbon centres. It is thus likely that the amino function at C-2 is attached to pyridoxal phosphate throughout the reaction sequence, hence allowing the possibility of differentiating between the two nitrogen atoms.

The typical tropane alkaloids, hyoscyamine (**7.30**) and cocaine (**7.42**), are formed by an extension of this pathway. Oxidation and cyclization produce tropinone or its α-carbonyl derivative, and then reduction and ester formation with the appropriate alcohols yield hyoscyamine and cocaine (equation 7.2):

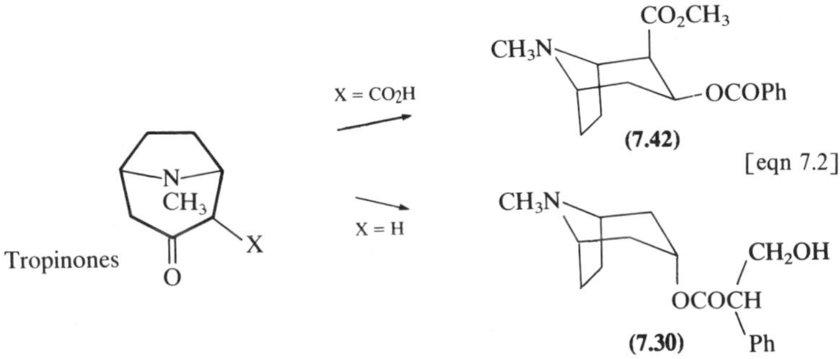

The anaesthetic properties of cocaine were first recognized by Köller in 1882, but it has now been largely replaced in the clinic by synthetic analogues, primarily due to its widespread abuse as a narcotic. It is, however, still much used as a stimulant by Andean Indians. After chewing the leaves they are less easily fatigued and can go for long periods without food.

Atropine and scopolamine (**7.43**) have a long association with witchcraft and soothsaying, and Homer mentions the use of henbane in magical preparations which caused a state of apparent insanity in the consumer, but nonetheless allowed them to become prophetic! The usual witch's brew

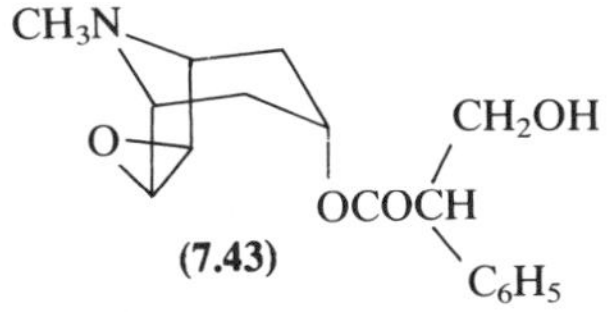

contained extracts of henbane, mandrake (*Mandragora* species), and deadly nightshade in combinations with fats (from 'a stillborn child'); and with this the witches anointed a staff (i.e. broomstick) and applied the extract under their arms or (as the contemporary treatises rather delicately put it) to 'other hairy places', and then 'flew off' to the sabbat. The pharmacology is of interest, since oral administration of tropane alkaloids is less efficient than sub-cutaneous administration, because the latter allows direct access to the bloodstream without the necessity of absorption from the gastro-intestinal tract. The most potent alkaloid in these brews was undoubtedly scopolamine (hyoscine) and this causes intoxication, then a form of narcosis during which hallucinations are experienced, and finally sleep.

7.5.1.2   *Piperidine alkaloids*

There are piperidine alkaloids with structures and biosynthetic derivations analogous to those of the corresponding pyrrolidine alkaloids, for example, pelletierine (**7.44**) from lysine (equation 7.3):

(**7.44**)   [eqn 7.3]

Once again, specific incorporation of radiolabel was observed, suggesting that the symmetrical intermediate cadaverine (**7.45**) is not released from pyridoxal phosphate during biosynthesis.

Other alkaloids of this class, such as the pepper alkaloid piperine (**7.29**), and (−)-lobeline (**7.46**), clearly have a more complex biogenesis. The latter is of interest because of its marked expectorant properties (induces coughing), and has been used for this purpose since the 18th century.

Certain piperidine derivatives are not derived from lysine at all, and the pathway to coniine (**7.1**) via 5-keto-octanal (**7.47**) is typical (Fig. 7.16). This alkaloid, from hemlock, is a poison of some renown and Plato gives a dramatic account of its effects on Socrates in his book *Phaedo*. The fire ant venoms (e.g. **7.19**) probably have a similar biogenesis.

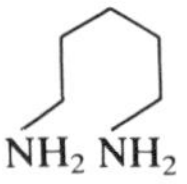

(**7.45**) 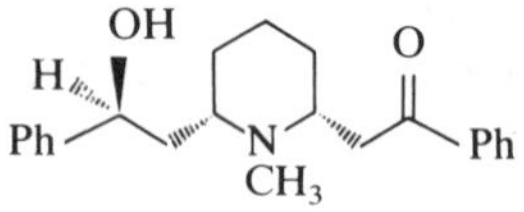   (**7.46**)

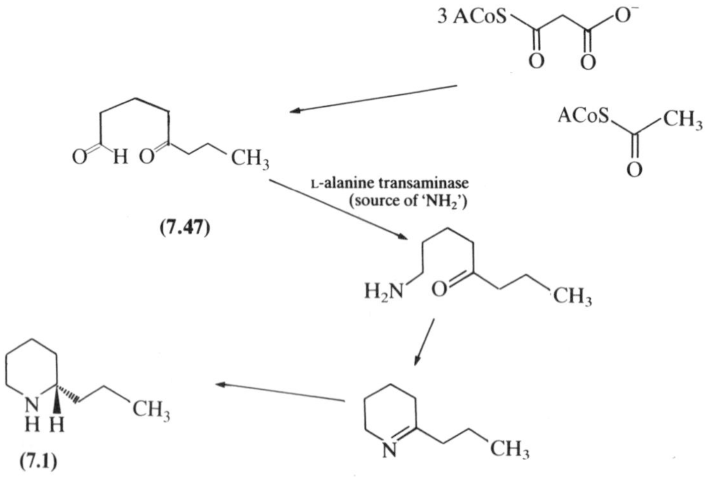

Fig. 7.16.

7.5.1.3 Pyrrolizidine alkaloids

The pyrrolizidine alkaloids are most commonly encountered in the plant species *Senecio* and *Heliotropium*, and retronecine (**7.48**) and senecionine (**7.49**) are typical examples. Many occur as esters of so-called necic acids, e.g. senecionine, and this portion of the molecules appears to be derived from the amino acids isoleucine (**7.50**) and valine (**7.51**).

The biosynthetic pathway to the pyrrolizidine ring system proceeds by way of putrescine (**7.41**) and homospermidine (**7.52**), though other intermediates have yet to be identified (Fig. 7.17). Metabolism of these alkaloids in the mammalian liver produces pyrroles such as (**7.53**), and these can induce a form of cirrhosis if sheep or cattle consume plants containing such alkaloids. It is even said that bees pollinating these plants incorporate the alkaloids into their honey, so even humans may be at risk. Such toxicity does have some potential, however, and indicine *N*-oxide (**7.54**) is being evaluated as an anti-cancer agent.

Finally, some of the alkaloids have an ecological role over and above any use by the plant as feeding deterrents. Several species of butterflies feed upon plants containing pyrrolizidine alkaloids, incorporate them into their tissues, and thus become unpalatable. Male butterflies of several danaid species can, in addition, convert these dietary alkaloids into the pyrrole (**7.55**), which apparently has aphrodisiac properties when applied to the female prior to mating.

7.5.1.4 Quinolizidine alkaloids

Compounds of this kind were mentioned earlier in the chapter, for example, sparteine (**7.12**) and cytisine (**7.13**), and other commonly occurring members of this family are (−)-lupinine (**7.56**) and matrine (**7.57**). A biosynthetic pathway to lupinine can be proposed based upon the suggested route to the pyrrolizidine alkaloids, and these proposals have been largely substantiated through the use of isotopically labelled precursors. The presently accepted routes are shown in Fig. 7.18, and it is quite clear from the isotopic studies that a symmetrical intermediate, cadaverine (**7.45**), is involved.

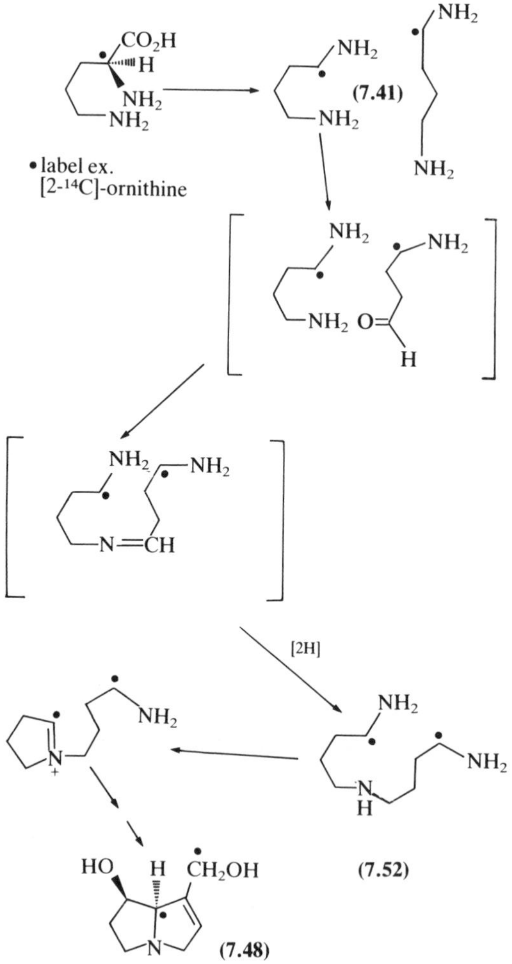

Fig. 7.17.

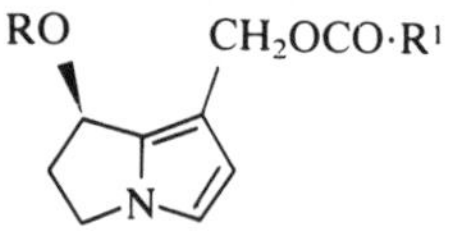

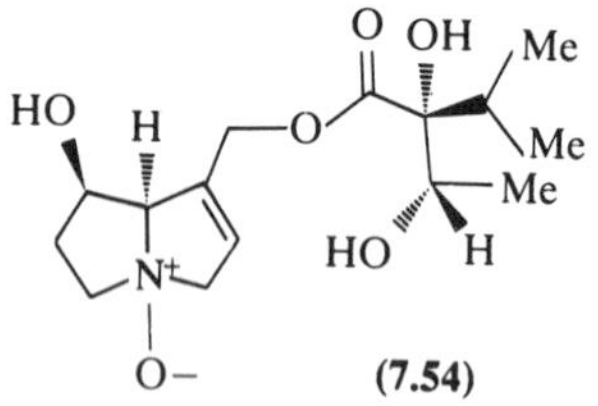

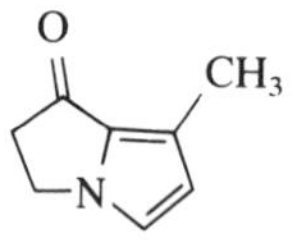

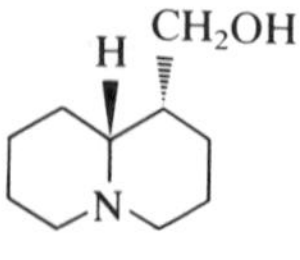

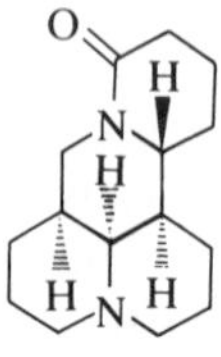

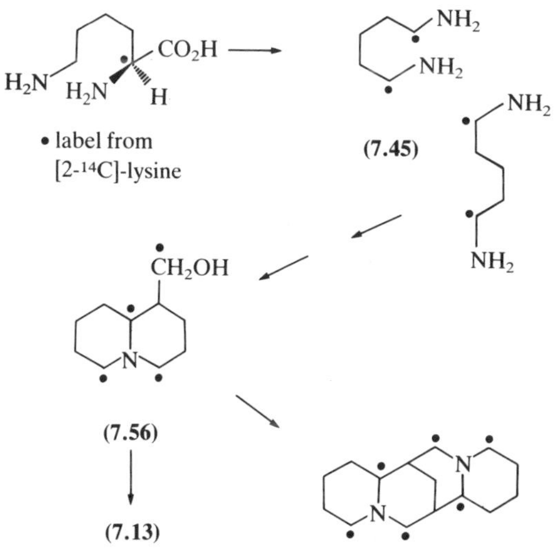

Fig. 7.18.

Of special value in this research has been the use of doubly isotopically labelled precursors such as $[1,2-^{13}C_2]$-cadaverine (**7.58**). Incorporation of this into lupanine and sparteine resulted in pairs of coupled ^{13}C NMR signals, and the proven connectivity shown in equation 7.4:

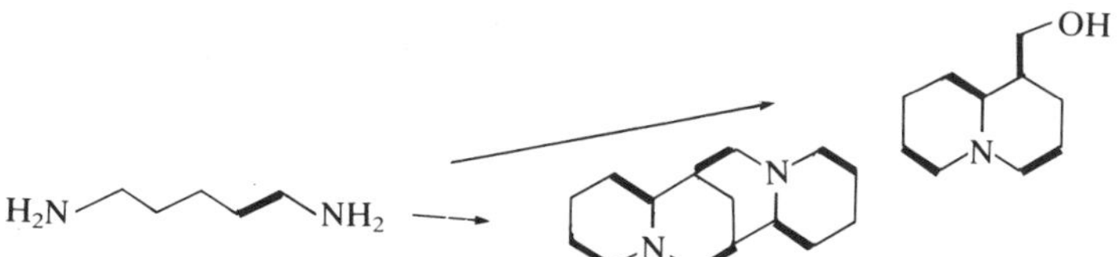

[eqn 7.4]

(**7.58**)

All of these alkaloids are toxic, but also bitter, so may act as feeding deterrents for both mammals and insects.

7.5.1.5   Pyridine alkaloids

The three most important alkaloids in this class are nicotine (**7.59**), anabasine (**7.60**), and anatabine (**7.61**), and these are produced by a number of *Nicotiana* species. As anticipated $[2-^{13}C]$-ornithine is incorporated into the pyrrolidine ring of nicotine, while $[2-^{14}C]$-lysine is incorporated into the piperidine ring of anabasine. However, a symmetrical labelling pattern is observed for the former, and a non-symmetrical pattern for the latter (Fig. 7.19), reflecting subtle differences in the pyridoxal phosphate-mediated steps.

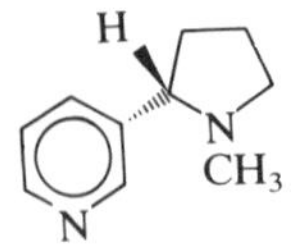

(7.59)

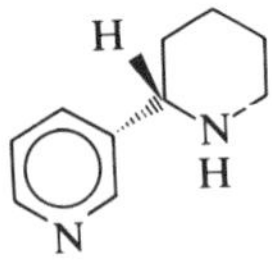

(7.60)

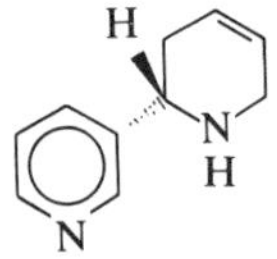

(7.61)

The pyridine ring is derived from nicotinic acid (**7.62**), which is itself derived from aspartic acid and glyceraldehyde-3-phosphate (equation 7.5):

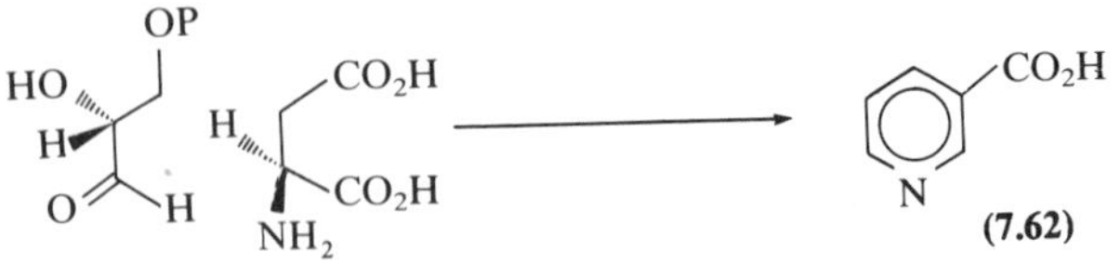

(7.62)

[eqn 7.5]

The remaining steps *en route* to nicotine are shown in Fig. 7.19 Anabasine probably has a similar biogenesis but, interestingly, anatabine is derived solely from nicotinic acid.

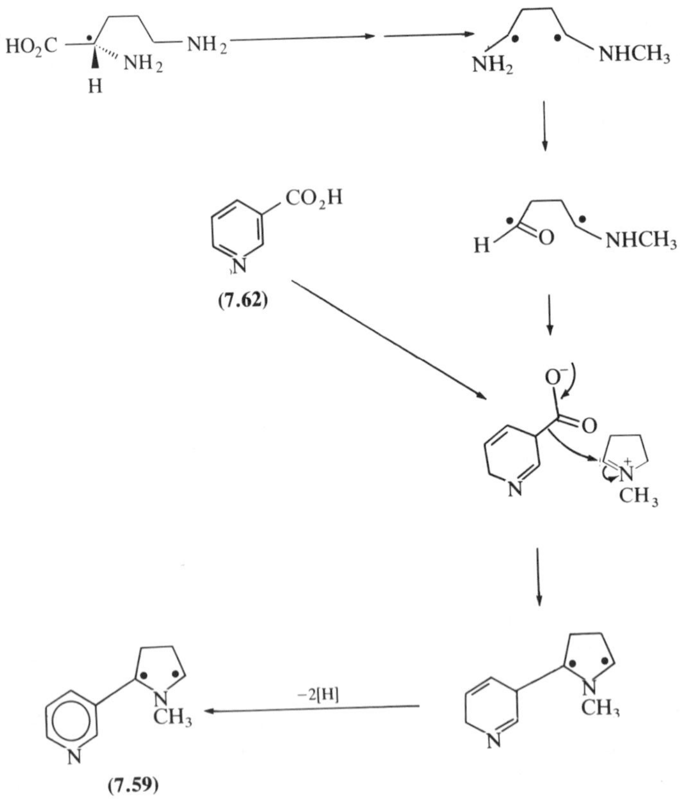

(7.62)

(7.59)

Fig. 7.19.

As well as potent central nervous system activity, nicotine also possesses useful insecticidal activity and is used commercially for this purpose. Its role in the plant may well be as an insect deterrent.

7.5.2 *Alkaloids derived from phenylalanine and tyrosine*

The main classes are shown in Fig. 7.2, and the parts of the structures which are derived from the two amino acids have been emphasized. Usually the actual precursors are the decarboxylation products of the amino acids, e.g. dopamine (3,4-dihydroxyphenylethylamine, **7.33**) or arylpyruvic acids, e.g. phenylpyruvic acid (**7.63**) from phenylalanine via transamination. The initial stages of most biosynthetic pathways closely follow the chemistry displayed in Fig. 7.13, but there is considerable diversity in the later stages, thus providing a large number of complex structural types.

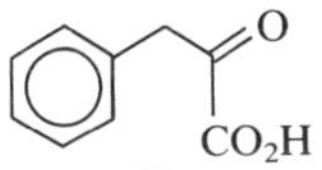

(**7.63**)

7.5.2.1 *Monocyclic compounds*

Several biologically interesting alkaloids merit inclusion in this section. Hordenine (**7.64**) is produced by barley (*Hordeum vulgare*) and appears to act as a germination inhibitor, thus preventing the growth of other plants in stands of barley. Secondary metabolites of this kind are usually known as 'allelopathic agents'. The biosynthesis of hordenine proceeds via hydroxylation of phenylalanine to produce tyrosine, followed by decarboxylation yielding tyramine (**7.65**), and thence, after methylation, to hordenine (equation 7.6):

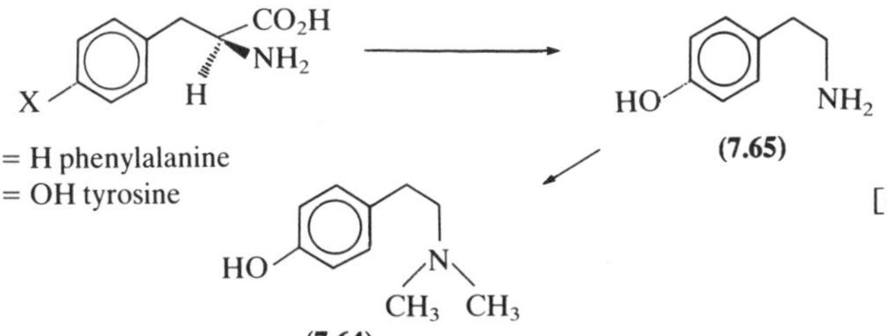

X = H phenylalanine
X = OH tyrosine

[eqn 7.6]

Noradrenaline (norepinephrine) (**7.66**), a key mammalian neurotransmitter, is derived from tyrosine via dopamine (**7.33**), and can be further converted into adrenaline (epinephrine) (**7.67**), the so-called 'fight or flight' hormone, which initiates a rise in blood pressure and heart rate (equation 7.7):

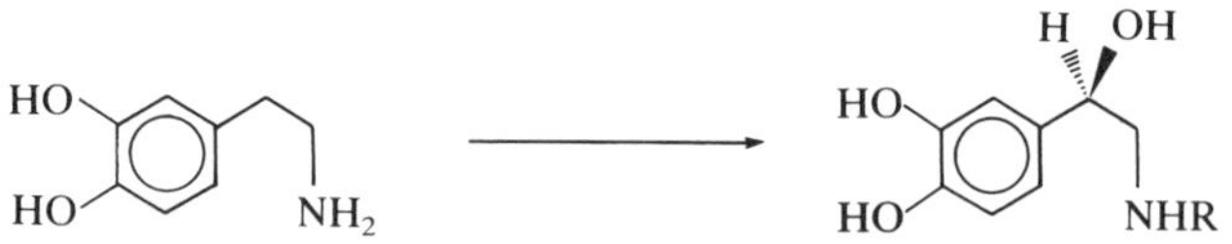

(7.66) R = H
(7.33)   (7.67) R = CH$_3$

Dopamine is also on the pathway to mescaline (**7.68**), a psychedelic from the peyote cactus, *Lophophora williamsii*. This plant is a native of the deserts of Mexico and the southern states of the USA, and has a long association with folklore. Its hallucinogenic properties were first described officially in the 16th century, but its use certainly predates the arrival of the conquistadores. In common with other hallucinogens it probably exerts its effects by interfering with the metabolism of essential brain amines like serotonin (**7.69**), a product of a tryptophan pathway.

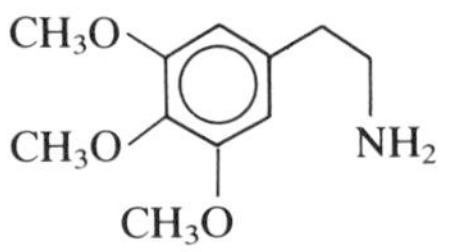

(7.68) 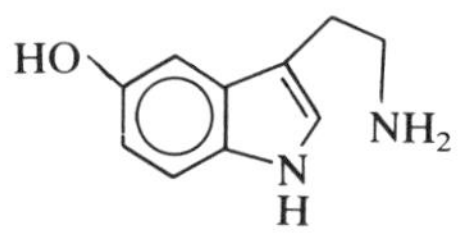   (7.69)

7.5.2.2 Tetrahydroisoquinoline alkaloids

The peyote cactus contains over 50 alkaloids, many of which have a tetrahydroisoquinoline structure, e.g. pellotine (**7.32**), and their biosynthesis is closely analogous to the *in vitro* synthesis shown in Fig. 7.13. This type of chemistry also provides access to the precursors of the benzylisoquinoline alkaloids, and the biosynthetic pathway to morphine (Fig. 7.20) is exemplary. The assembly of the basic building blocks (a 2-arylethylamine and a 2-arylethanal) into a benzylisoquinoline is followed by a complex series of reactions to produce the final alkaloid. The pathway to morphine has been fully substantiated through the use of radiolabelled precursors, and by the isolation of most of the intermediates. One of the particularly interesting steps is the oxidative phenolic coupling that produces salutaridine. This kind of chemistry is prevalent on the routes from benzylisoquinolines to various classes of alkaloids.

The opium alkaloids, primarily morphine, codeine, thebaine, noscapine (**7.70**) (formerly narcotine), and papaverine (**7.71**), are present in con-

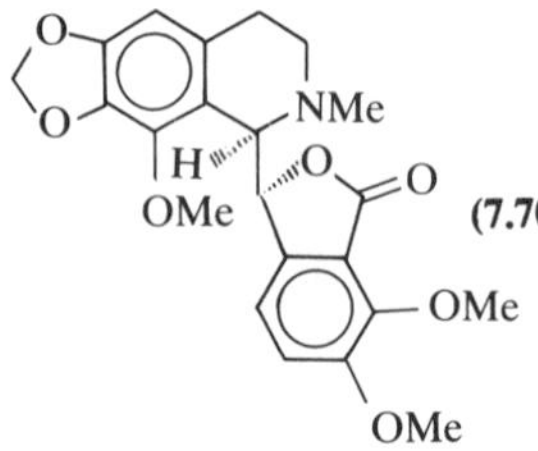

(7.70) 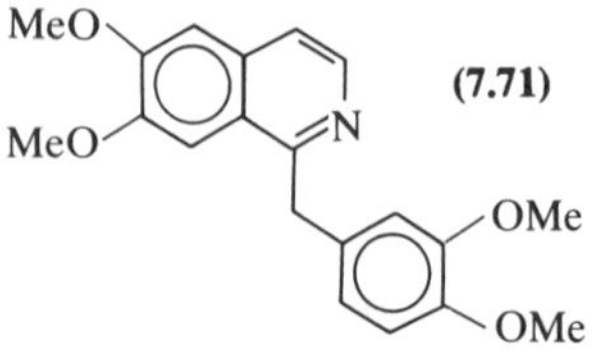   (7.71)

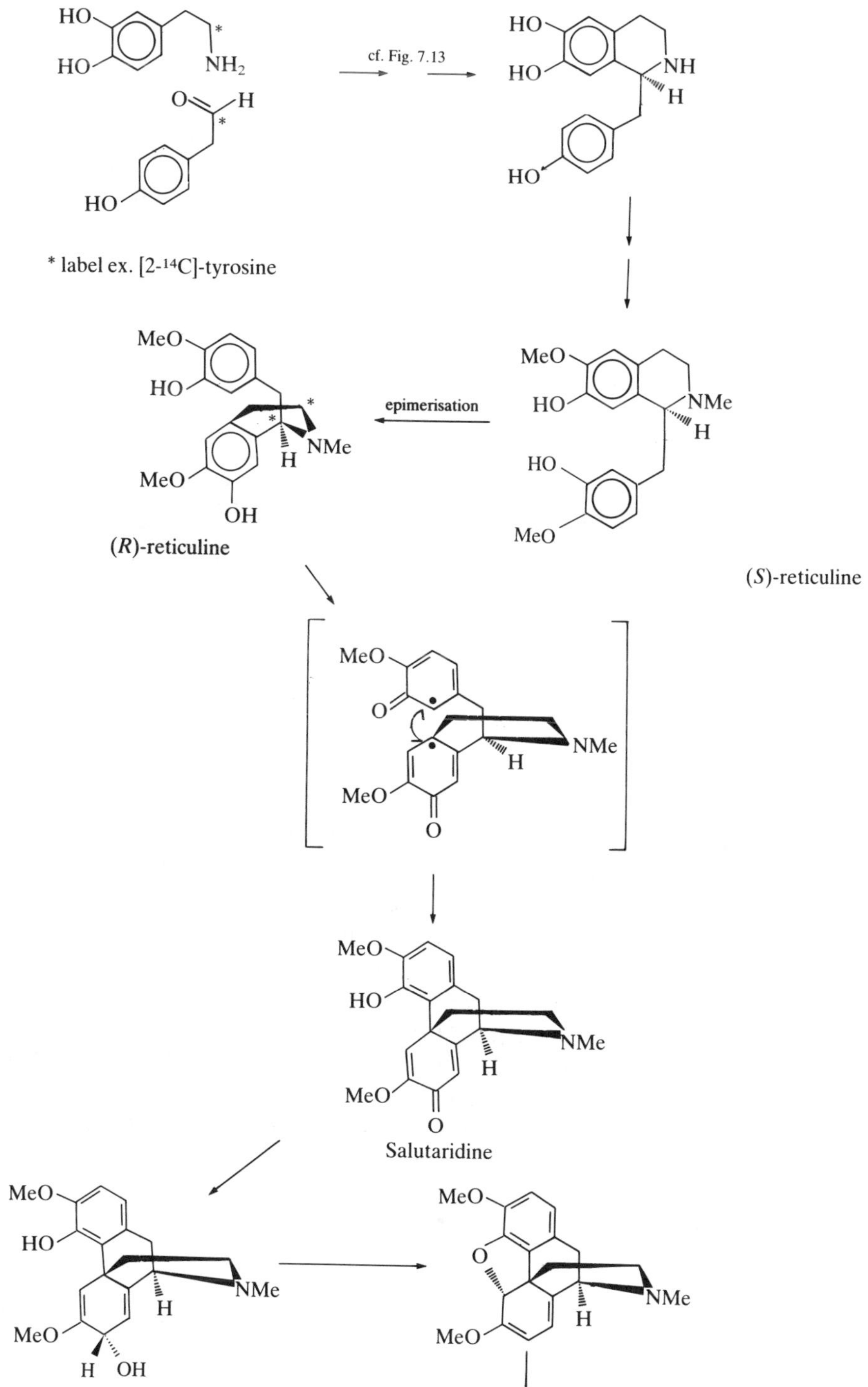

* label ex. [2-¹⁴C]-tyrosine

(R)-reticuline

(S)-reticuline

Salutaridine

Fig. 7.20.

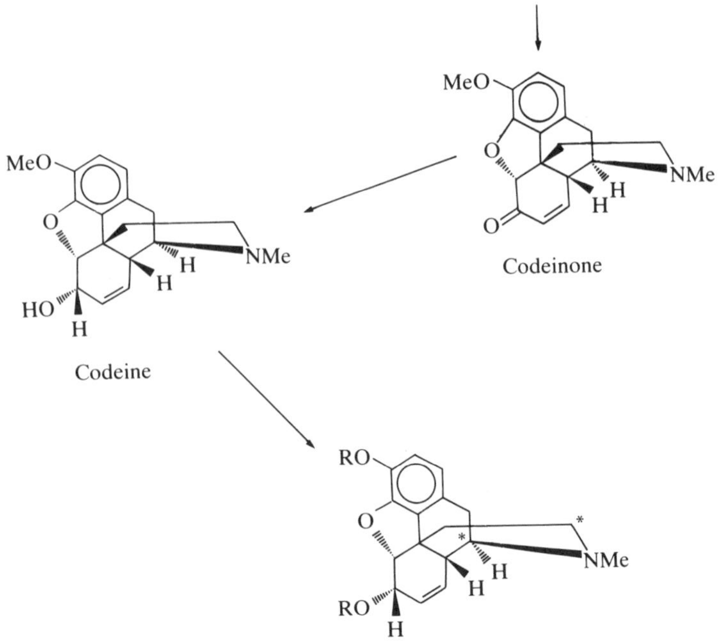

Codeinone

Codeine

R = H morphine
R = CO.Me heroin

(7.72)

Fig. 7.20 *continued*

siderable amounts in crude opium. This is defined as the air-dried, milky exudate from unripe fruit capsules of *Papaver somniferum*, and has been a much prized commodity for thousands of years. More recently, heroin (7.72), a synthetic derivative of morphine (the diacetate), has achieved even greater notoriety due to its extra potency as a euphoriant and increased tendency to produce addiction. Other, totally synthetic, analogues have been produced with the aim of maintaining the powerful analgetic (pain-suppressing) effects of morphine and heroin, while reducing the addictive and respiratory depressant effects. The compound naloxone (7.73) is of particular interest because it is a narcotic antagonist, and although taken up by the opiate receptors in the brain, thus denying access to morphine or heroin, it does not elicit an opiate-like response.

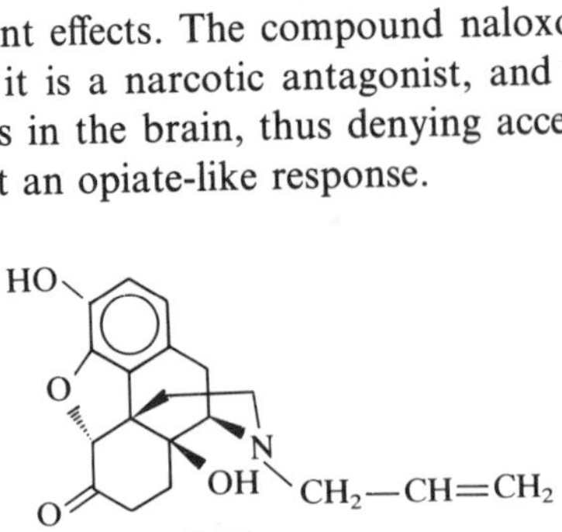

(7.73)

H_2N-Tyr-Gly-Gly-Phe-Met-CO_2H

(7.74)

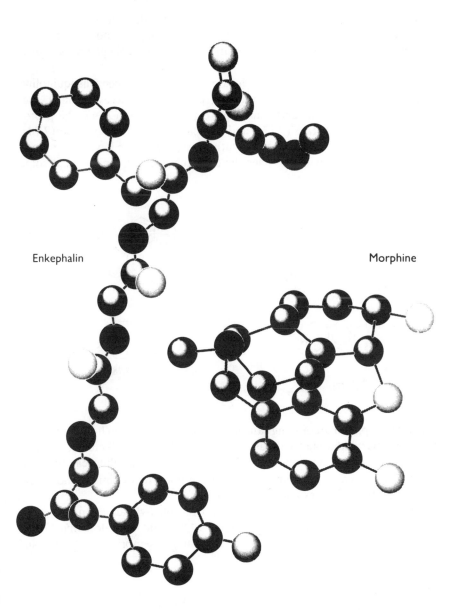

Enkephalin

Morphine

Fig. 7.21. A comparison of the structures of morphine and met-enkephalin. Reproduced by permission of Oxford University Press, from J. Mann, *Murder, Magic and Medicine*, Oxford University Press, Oxford, 1992.

Finally, since the opiate receptor was surely 'designed' for a natural substrate, it was not too surprising when the enkephalins (e.g. the pentapeptide met-enkephalin, **7.74**) were identified as endogenous brain peptides with opiate-like activity. They bind to the same receptors as morphine or heroin, and have a role in the control of pain and the emotions. A comparison of the structures of morphine and met-enkephalin is shown in Fig. 7.21.

Several other types of alkaloids are derived from benzylisoquinoline precursors, and these include the aporphines (the largest group), e.g. (S)-glaucine (**7.75**); the erythrina type, e.g. erythraline (**7.76**); the proto-berberines, e.g. berberine (**7.77**), and the pavines, e.g. argemonine (**7.78**). The complexities of the various biosynthetic pathways are beyond the scope of this chapter, but oxidative phenolic coupling is usually involved at some stage.

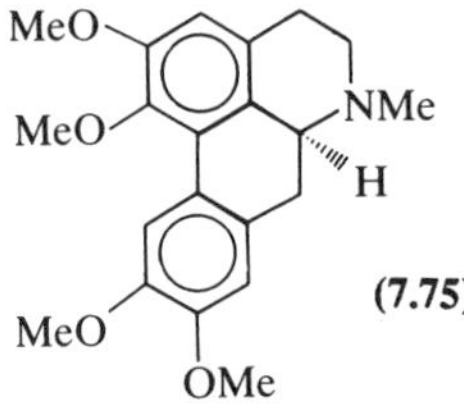

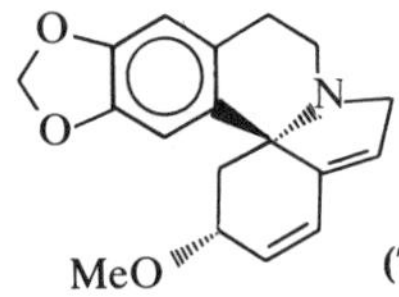

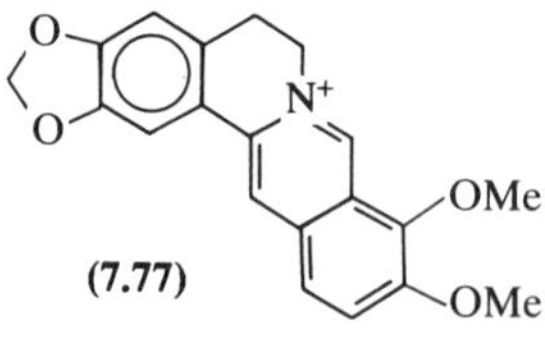

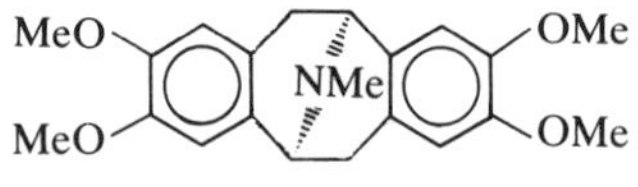

Oxidative phenolic coupling can also be invoked to explain how the various types of Amaryllidaceae alkaloids arise (Fig. 7.22). These alkaloids, of which there are more than one hundred, only occur in the Amaryllidaceae (daffodil family), and such findings are of great use to plant taxonomists (those who assign plants to families and genera, etc.).

Numerous other alkaloids are derived from phenylalanine and tyrosine, and representative examples are nitidine (**7.79**) and cephalotaxine (**7.80**), both of which are potent anti-leukaemic agents, mesembrine (**7.81**), emetine

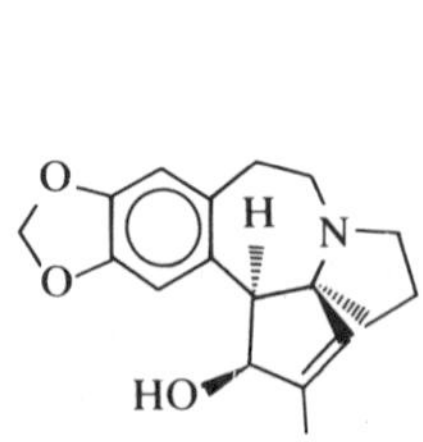

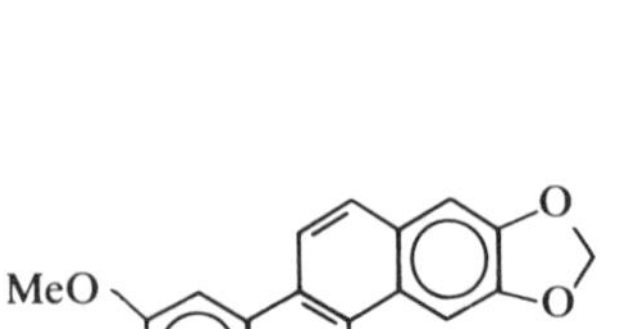

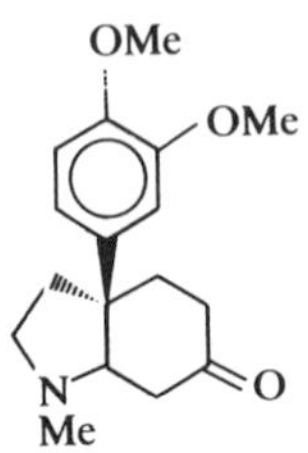

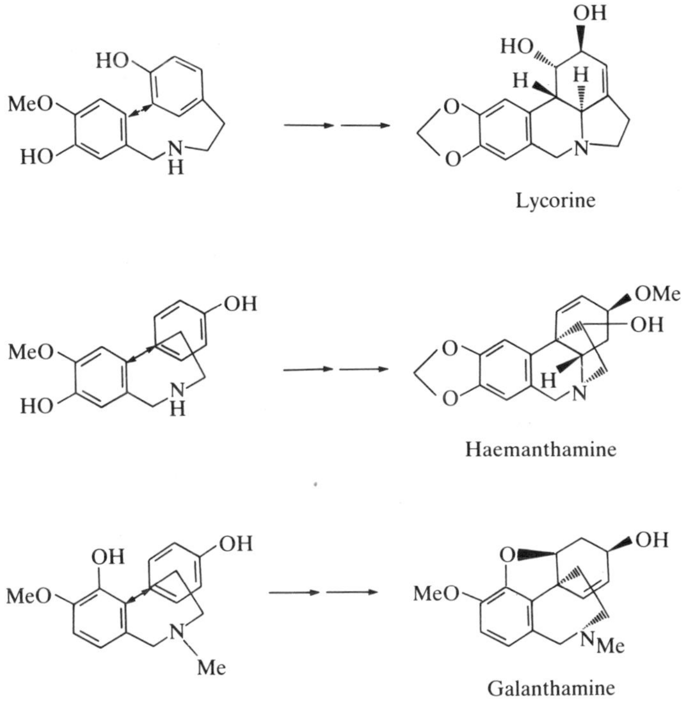

Lycorine

Haemanthamine

Galanthamine

Fig. 7.22.

(**7.82**), one of the ipecac alkaloids, which is used in the treatment of amoebic dysentry, and colchicine (**7.28**). This last alkaloid, from the autumn crocus, *Colchicum autumnale*, also has a long association with folk medicine, and was used by the ancient Romans as an effective treatment for gout – it is palliative rather than curative. More recently it has been evaluated as an anti-cancer agent, but is probably too toxic for clinical use. Elucidation of the biosynthetic pathway is still incomplete, but such data as are available are given in Fig. 7.23, together with speculations (primarily by Battersby and coworkers) concerning the final stages.

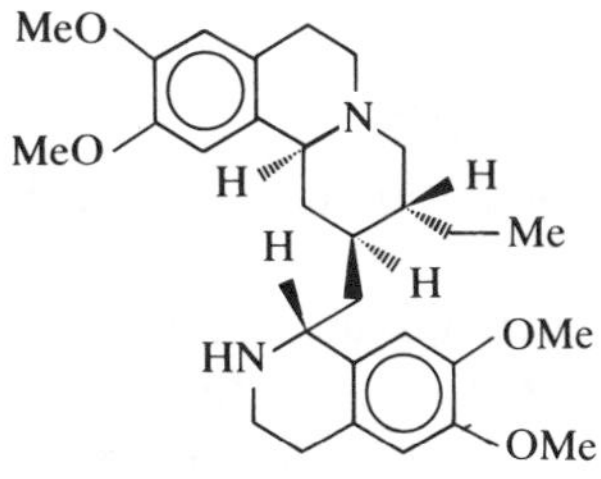

(**7.82**)

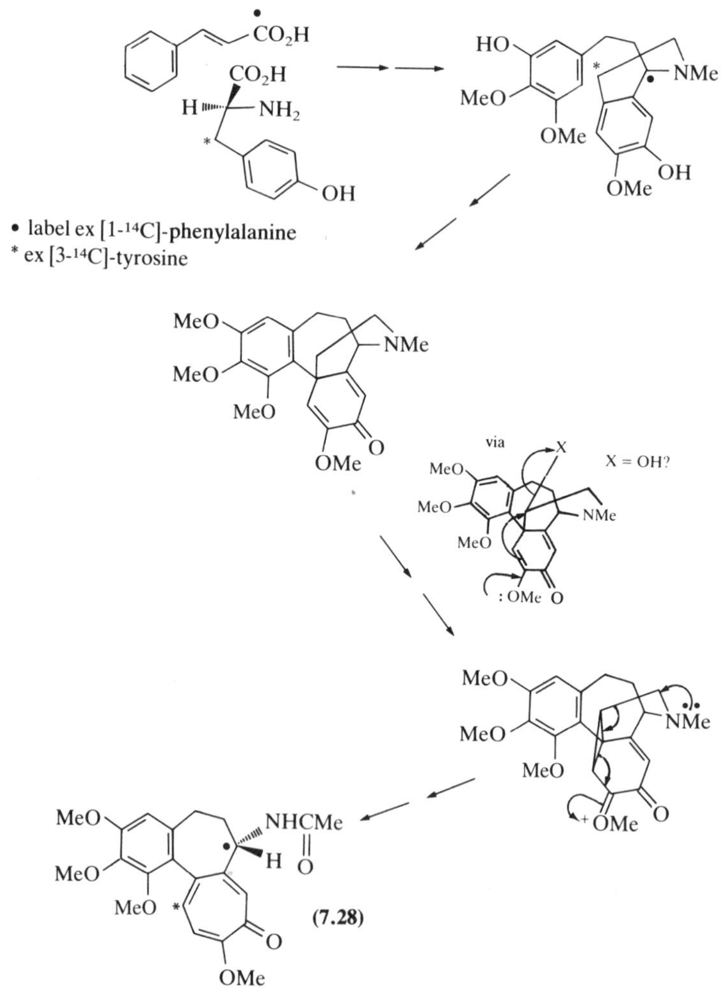

Fig. 7.23.

7.5.3 *Alkaloids derived from tryptophan*

Once again the alkaloids fall into two main classes:

(i)   simple products of the metabolism of the amino acid which contain solely an indole C_2N moiety; and

(ii)   more complex compounds which contain an indole C_2N moiety together with other carbon atoms derived from either the acetate or the mevalonate pathway. Representative examples of the two groups are bufotenine (**7.16**) and chanoclavine I (**7.25**).

7.5.3.1 Simple indole derivatives

Many of these have marked activity on the central nervous system due to the similarity to the natural neurotransmitter 5-hydroxytryptamine (serotonin, **7.69**). Psilocybin (**7.83**) from mushrooms of the *Psilocybe* genus, and dimethyltryptamine (**7.84**; R = CH$_3$) from a variety of South American plants, have been used as hallucinogens, albeit as components of crude extracts, since pre-Columbian times. They are derived from tryptamine (**7.84**; R = H), the decarboxylation product of tryptophan (**7.38**).

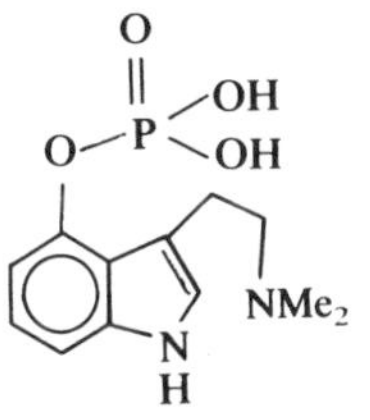

(7.83) 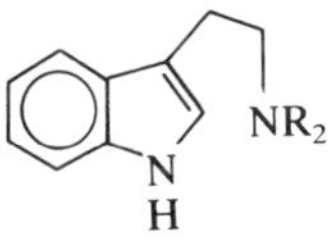   (7.84)

The more complex alkaloid physostigmine (**7.85**) is of unknown biogenesis, but may arise via the route shown in equation 7.8. The methyl group transfer agent *S*-adenosylmethionine probably supplies the three methyls on the rings. The pharmacology of this alkaloid is of interest since it is a rare example of a naturally occurring inhibitor of the enzyme acetylcholine esterase. This enzyme deactivates (through ester hydrolysis) the neuro-transmitter acetylcholine (**7.86**) after transmission of a signal, and several insecticides and pharmaceuticals have structures and activity profiles modelled on physostigmine. It was originally used in trials by ordeal in West Africa, and since it is a powerful emetic agent, innocent victims may have been saved if they gulped the poison then vomited, while guilty parties probably sipped the brew and thus allowed uptake by the gastrointestinal tract with consequent fatal results.

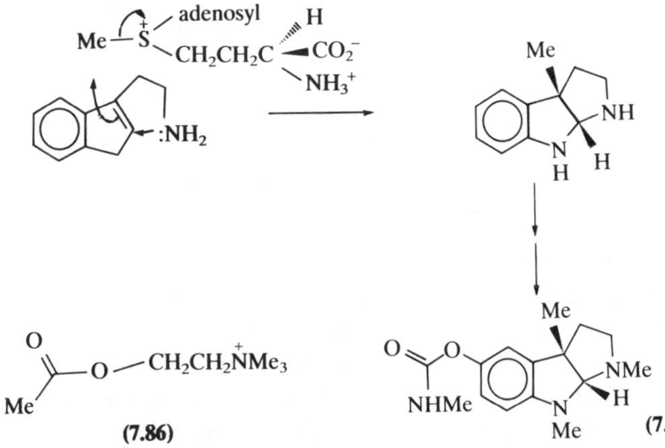

[eqn 7.8]

(7.86)   (7.85)

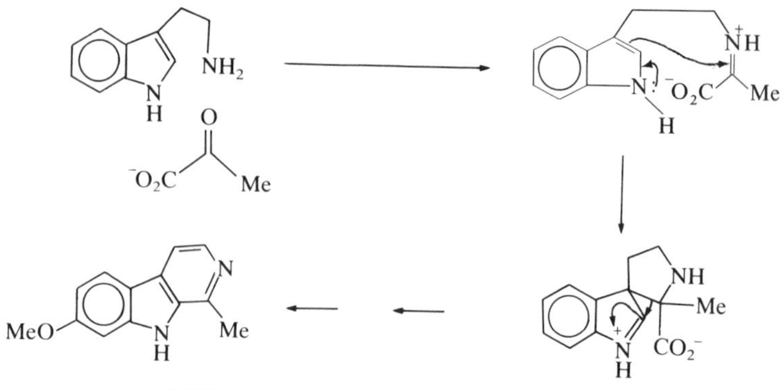

(7.87)

Fig. 7.24.

7.5.3.2   *Complex indole alkaloids*

Most of these alkaloids incorporate a moiety derived from mevalonate, and usually contain an extra five or ten carbon atoms. A few have an extra two-carbon unit derived from pyruvate, and the postulated pathway to harmaline (**7.87**) is shown in Fig. 7.24. The route has obvious similarities to the one employed in the formation of the tetrahydroisoquinolines. The harmala alkaloids are the basis of numerous South American hallucinogenic snuffs and magical potions.

The diketopiperazine, echinulin (**7.88**), from the fungus *Aspergillus echinulatus*, has three five-carbon units probably derived ultimately from dimethylallylpyrophosphate, and in addition a molecule of alanine. Tracer studies have identified cyclo-L-alanyl-L-tryptophan (**7.89**) as an intermediate (equation 7.9):

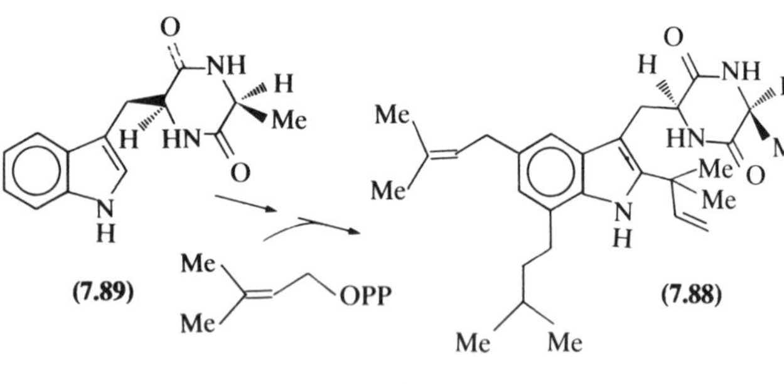

(7.89)   **(7.88)**   [eqn 7.9]

A five-carbon unit is also easily discerned within the structures of the ergot alkaloids. Chanoclavine I (**7.25**) has already been mentioned, but it is the amides of lysergic acid (**7.90**) which have had the greatest impact on mankind. The fungus *Claviceps purpurea*, a common contaminant of rye, produces various ergot alkaloids, and consumption of contaminated rye bread produces the state known as ergotism. During the Middle Ages in Europe,

R = HO— **(7.90)**

R = (structure: HO—CH(Me)—NH—, with H) **(7.91)**

the suffering and deaths caused by ergotism were only surpassed by the ravages of bubonic plague. In some parts of Europe, hallucinations and severe mental aberrations were the prime symptoms, but a more serious manifestation was gangrene of the extremities caused by the vasoconstrictive actions of many of the alkaloids, and hence the term 'St Anthony's fire' which described the blackening of limbs that was observed. This property is of clinical utility, and crude extracts of the fungus were used as long ago as 1100 BC to control postpartum bleeding. One of the most vasoconstrictive alkaloids, ergonovine (ergometrine, **7.91**), is still used for this purpose today.

The biosynthesis of these alkaloids is complex and is still not completely understood. However, on the basis of extensive tracer studies using cell-free systems from *Claviceps purpurea* it is now certain that two *cis–trans* isomerizations take place *en route* from 4-dimethylallyltryptophan to lysergic acid (Fig. 7.25).

The most infamous synthetic derivative of lysergic acid is of course LSD (the diethylamide), and the stories surrounding its discovery by Albert Hoffmann are now part of folklore. Other hallucinogenic ergot alkaloids have been in use for centuries, and the magical preparation of the Aztecs, ololuiqui, from seeds of the plant *Ipomoea violaceae* and *Rivea corymbosa*, contained the simple amide of lysergic acid and chanoclavine I.

Other indole alkaloids incorporate an entire monoterpenoid moiety (derived from secologanin, **7.92** – see Chapter 5), and this major group comprises more than 1000 of the most complex and structurally diverse structural types. A few representative examples are shown in Fig. 7.26, and those atoms of terpenoid origin are emphasized.

Although much progress has been made in the last few years towards elucidation of the various biosynthetic pathways, few are known with absolute certainty. The route to ajmalicine is shown in Fig. 7.27, and key intermediates which figure in other pathways as well include strictosidine and dehydrogeissoschizine.

The indole alkaloids shown in Fig. 7.26 all have interesting pharmacological properties. Ajmalicine is a major constituent of *Rauwolfia serpentina* and *Catharanthus roseus*, and has marked hypotensive activity (lowers blood pressure). Camptothecine, from the Chinese ornamental tree *Camptotheca acumunata*, is used as an anti-cancer agent in China, but has not found favour in the West owing to its serious side-effects. Catharanthine is said to have hypoglycaemic activity (lowers blood sugar levels), while the

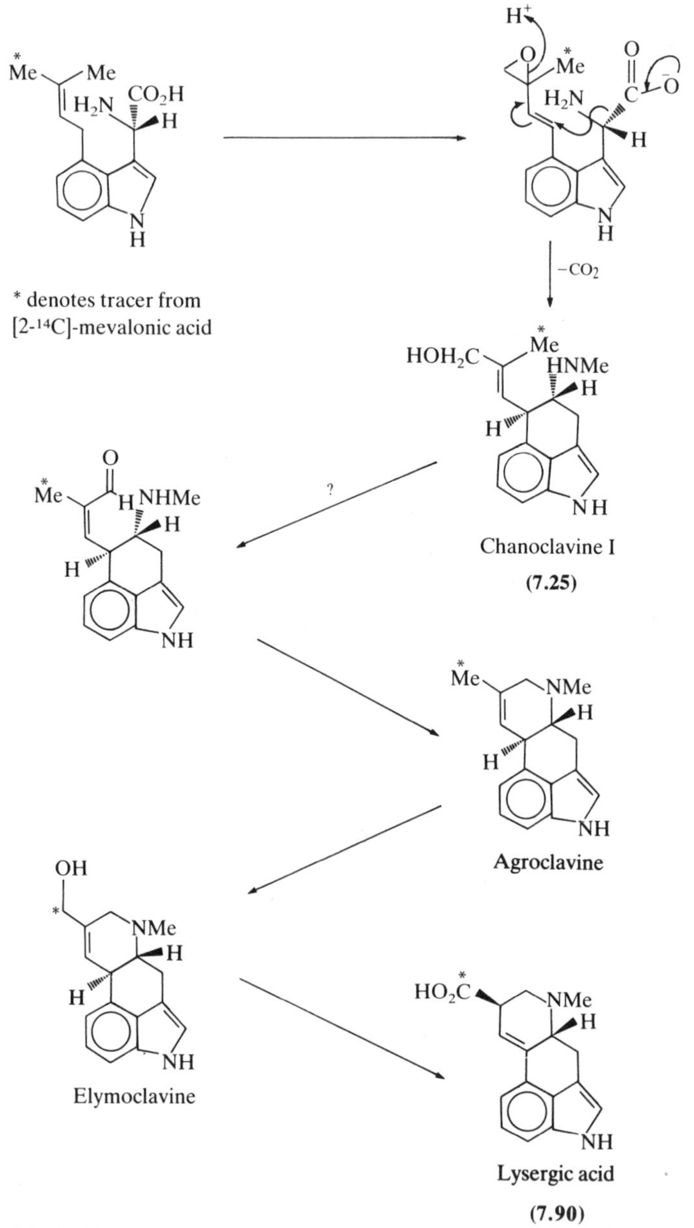

* denotes tracer from
[2-¹⁴C]-mevalonic acid

Chanoclavine I

(7.25)

Agroclavine

Elymoclavine

Lysergic acid

(7.90)

Fig. 7.25.

structurally related compound (−)-ibogaine (10-methoxyibogamine, **7.93**), from the West African plant *Tabernanthe iboga* is claimed to be a hallucinogen. Certainly the natives use extracts of the roots as a stimulant in much the same way as the Peruvians employ cocaine. Strychnine is a potent toxin, and death due to asphyxia follows a general excitation of the

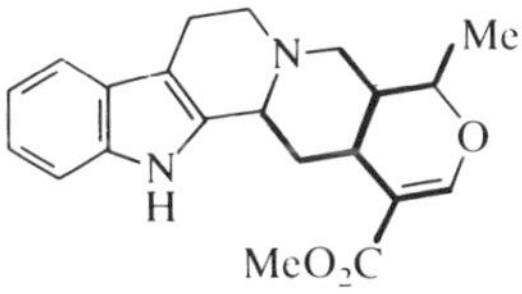

Ajmalicine

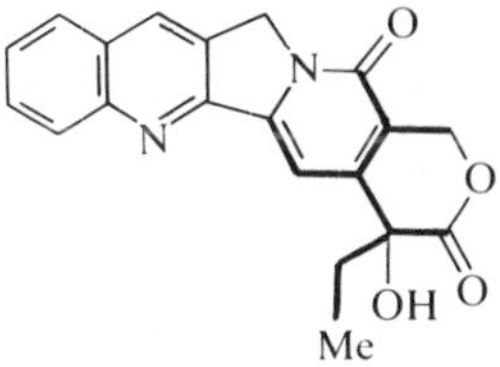

Camptothecine

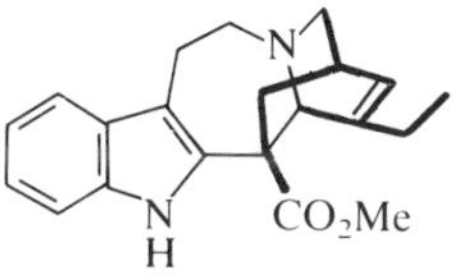

Catharanthine

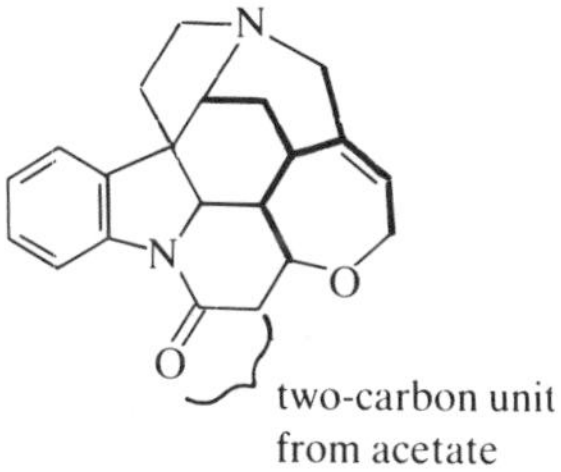

Strychnine

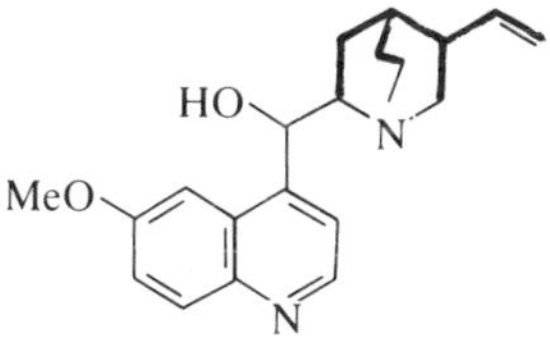

two-carbon unit
from acetate

Quinine

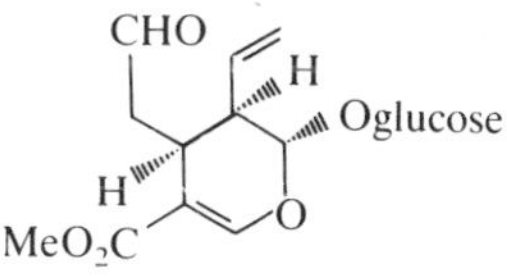

(7.92)

Fig. 7.26.

central nervous system. But of all these compounds, quinine has pride of place as a medicinal agent. Extracts of the bark of the Cinchona tree were recommended as a treatment for malaria as long ago as 1633, and the English physician, Thomas Sydenham, prescribed 'an ounce of bark with two pints of claret, in doses of eight or nine spoonsful, every four hours'. In fact, quinine was the only anti-malarial substance known until totally synthetic drugs became available during the mid 1940s.

Finally, the bisindole alkaloids vincaleukoblastine (vinblastine, **7.94**, R = CH₃) and leucocristine (vincristine, **7.94**, R = CHO) deserve mention, not least for their structural complexity, but also for their very marked anti-cancer activity. They too are derived from *Catharanthus roseus*, albeit in small quantities. They exert their anti-cancer effect by inhibiting cell division (mitosis), and most of the successful cancer chemotherapy regimes incorporate one or other of these alkaloids.

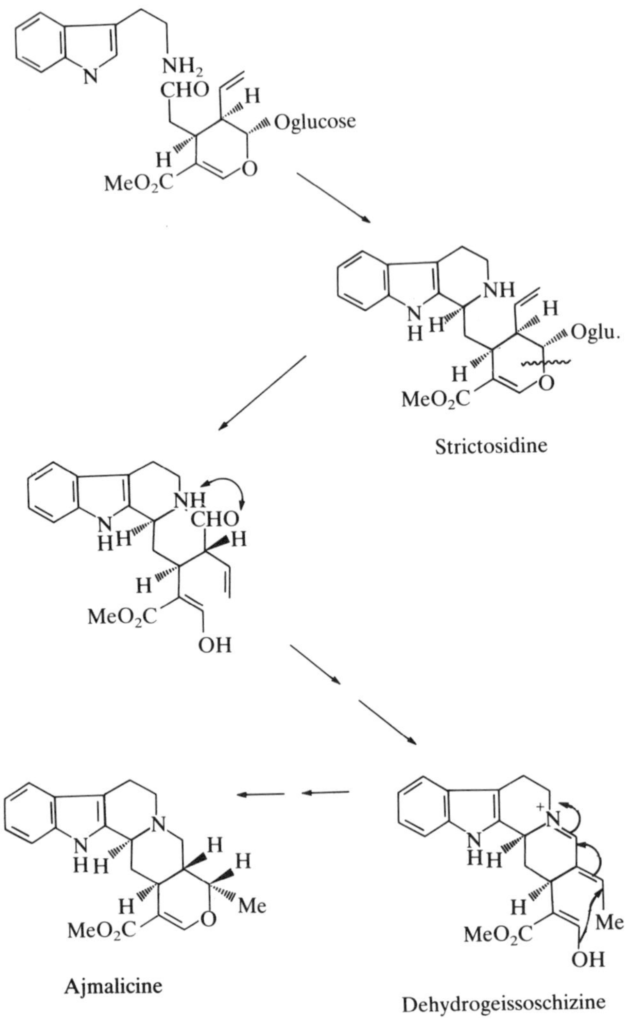

Strictosidine

Ajmalicine

Dehydrogeissoschizine

Fig. 7.27.

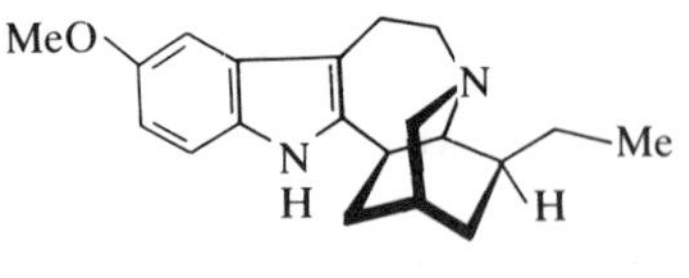

(7.93)

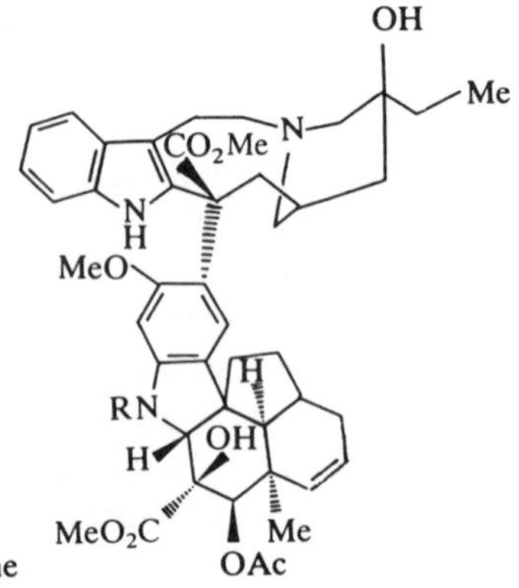

R = −Me:vinblastine
R = −CHO:vincristine

7.6 Synthesis

If the 19th century was the heyday for structural studies on the alkaloids, then the 20th century has been notable for the large number of elegant total syntheses that have been accomplished. Virtually all of the major alkaloids have now been synthesized, and some of the routes will be described in the following sections. Apart from two examples that are taken from the 'classical era' of alkaloid synthesis, most of the other examples have been chosen because they represent modern methodology for alkaloid synthesis.

7.6.1 Classical era

A number of syntheses carried out in the early part of this century stand out as supreme achievements, not least because they were executed without the benefit of modern chromatographic and spectroscopic techniques.

Of these, the routes to tropinone and cocaine devised by Robinson and Willstätter are probably the most famous. In addition, a biomimetic synthesis of *N*-norlaudanosine will be described, because this parallels the biosynthetic pathway, but was executed before that pathway was delineated.

7.6.1.1 Tropinone and cocaine

It was previously mentioned that Willstätter worked on a rather tortuous synthesis (*c.* 20 steps) of tropinone between 1900 and 1903, but in 1917 Robinson reported his 'one-pot' synthesis and also provided what was probably the first example of a formal retrosynthetic analysis. He stated: 'By imaginary hydrolysis at the points indicated by the dotted lines, the substance may be resolved into succinaldehyde, methylamine, and acetone.' This is shown in equation 7.10. The yield of this particular reaction is very

poor, but by using the calcium salt of acetone decarboxylic acid he was able to improve upon this. Subsequently, Schöpf and Lehmann reported optimized conditions (a buffered solution at pH 5 and 25°C) which allowed obtention of tropinone in 83 per cent yield. A variety of mechanisms are possible, and the one shown in equation 7.11 envisages reaction between the enol form of acetone dicarboxylate and the condensation product from succinaldehyde and methylamine:

$$[\text{eqn } 7.11]$$

The subsequent ring-closure involves a repetition of this chemistry albeit facilitated by the intramolecular nature of the process.

Willstätter made his own contributions based on this type of chemistry, and completed a simple synthesis of tropinone in 1921 (Fig. 7.28), and a synthesis of cocaine in 1923 (Fig. 7.29). In the former route, the diketodiester was produced via a Kolbe electrolysis, and this was then condensed with methylamine. Catalytic hydrogenation provided the pyrrolidine diester with side-chains *syn* with respect to the bridging nitrogen atom, since even if the double bonds are reduced one at a time, a *cis* delivery of hydrogens from the least hindered face of the molecule will ensure that this stereochemistry results. Finally, Dieckmann cyclization and decarboethoxylation produced tropinone.

In the cocaine synthesis, a Robinson-type reaction yielded the expected azabicyclo structure but with a fortuitous axial stereochemistry for the carbomethoxyl group. Subsequent reduction of the ketone and benzoylation of the mixture of alcohols gave racemic cocaine after separation of the diastereoisomeric products.

Robinsons's route is of importance from another aspect, since it represented the first biomimetic synthesis of an alkaloid. He thus provided inspiration for others to consider possible (or proven) biosynthetic pathways, before planning their synthetic routes to alkaloids.

7.6.1.2   Biomimetic synthesis of N-norlaudanosine

The biosynthesis of tetrahydrobenzylisoquinolines was discussed in Section 7.5.2.2, and several syntheses were in fact completed before this pathway was delineated. For example, Späth and Berger prepared *N*-norlaudanosine from 3,4-dimethoxyphenylethylamine and 3,4-dimethoxy-phenyl ethanal as shown in Fig. 7.30. This condensation and subsequent ring-closure sequence is usually known as the Picter–Spengler reaction, and has been widely used for producing this kind of alkaloid.

7.6.2   Modern era

7.6.2.1   Corey synthesis of porantheine

Although Robinson was clearly thinking about retro-synthetic analysis in 1923 when he discussed his plan for a tropinone synthesis, it was E.J. Corey

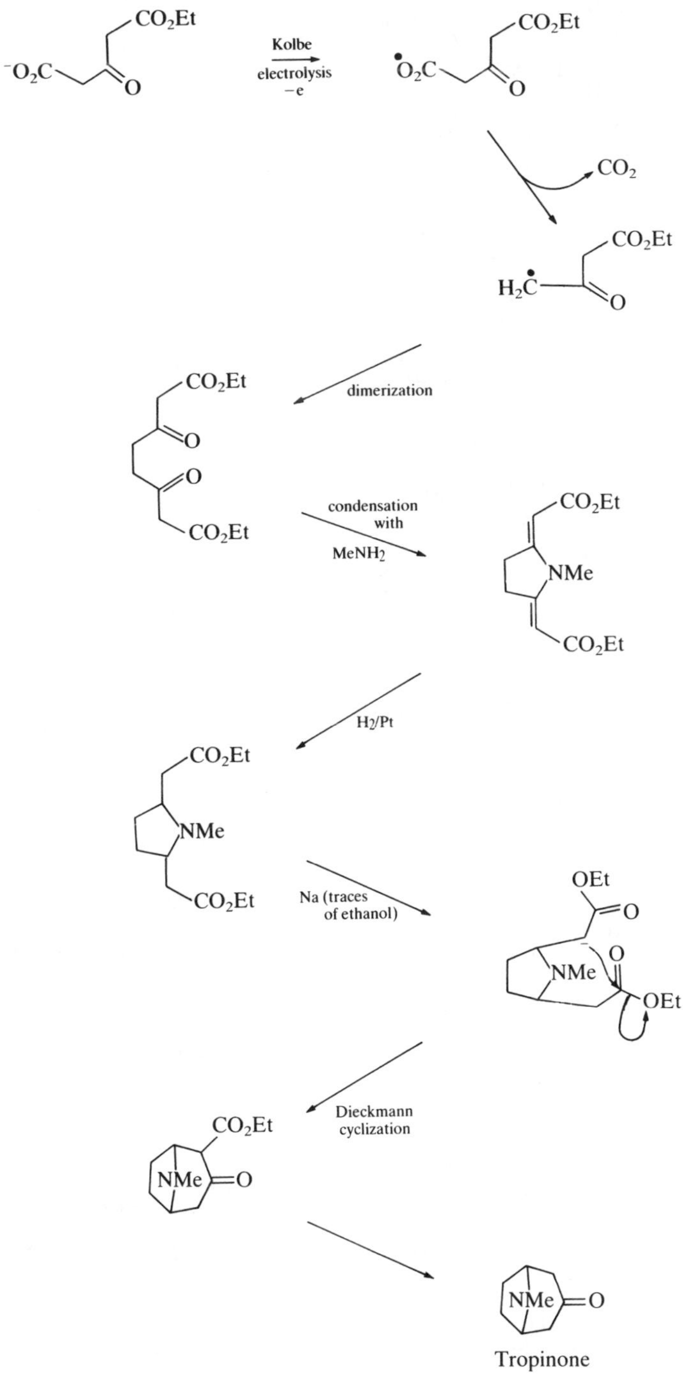
CO_2Et
^-O_2C
O
Kolbe
electrolysis
−e
CO_2
$H_2\dot{C}$
dimerization
condensation
with
$MeNH_2$
NMe
H_2/Pt
Na (traces
of ethanol)
OEt
Dieckmann
cyclization
Tropinone

Fig. 7.28.

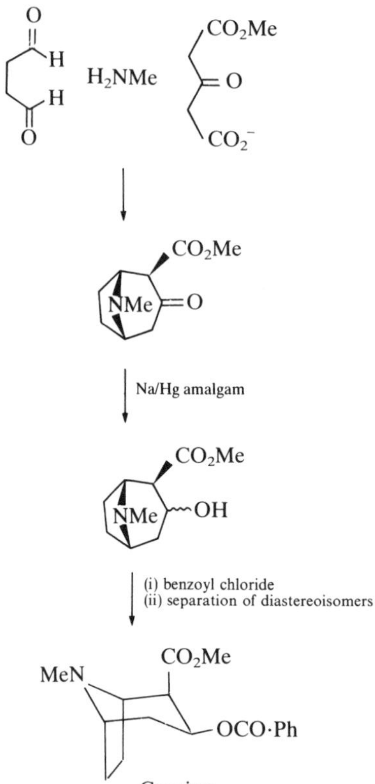

Cocaine

Fig. 7.29.

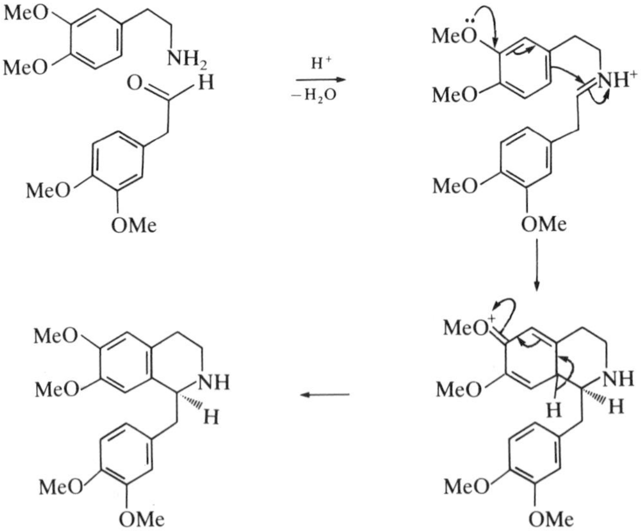

N-norlaudanosine

Fig. 7.30.

who wrote the first definitive account of the logic of retro-synthetic analysis in 1967. He has since produced computer programs that identify all possible disconnections, and provide suggestions for syntheses; and one such analysis, for the alkaloid porantherine (**7.95**), is shown in Fig. 7.31. The intermediate (**7.96**) has symmetry, which is always desirable since it obviates the necessity for site-selective chemistry. The synthesis (Fig. 7.32) as far as the monocyclic compound (**7.97**) requires no comment, and the second ring is then formed via a Mannich-type process involving the enol acetate and iminium ion of (**7.98**). The *N*-methyl group is oxidized prior to oxidative cleavage of the double bond, then protection of both carbonyls, deformylation, and deprotection ensue. A second Mannich-type reaction provides the desired cyclic system (**7.99**); and reduction and dehydration complete the synthesis.

The alkaloid is a major constituent of the woody shrub *Poranthera corymbosa*, a native of New South Wales, and is structurally related to coccinelline (**7.20**), the ladybird deterrent substance.

7.6.2.2 *Formation of alkaloids via intramolecular Diels–Alder reactions*

The Diels–Alder reaction has been used in numerous alkaloid syntheses, but its most elegant applications have involved the intramolecular version of the

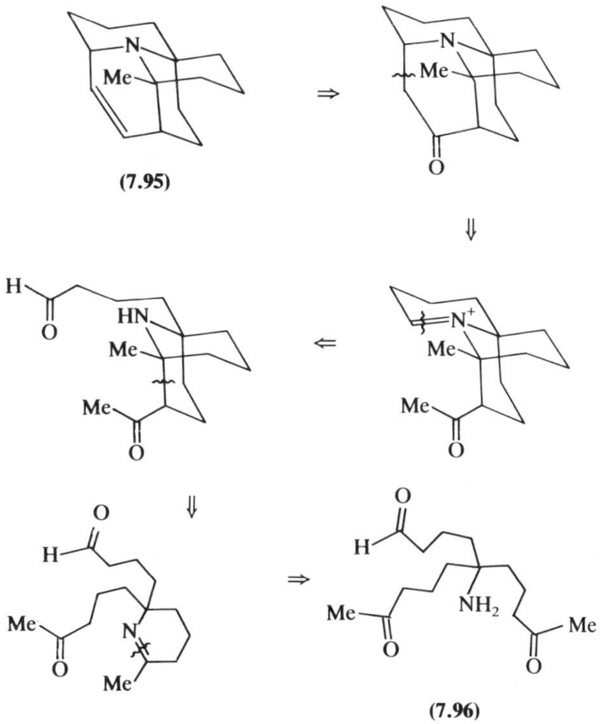

Fig. 7.31.

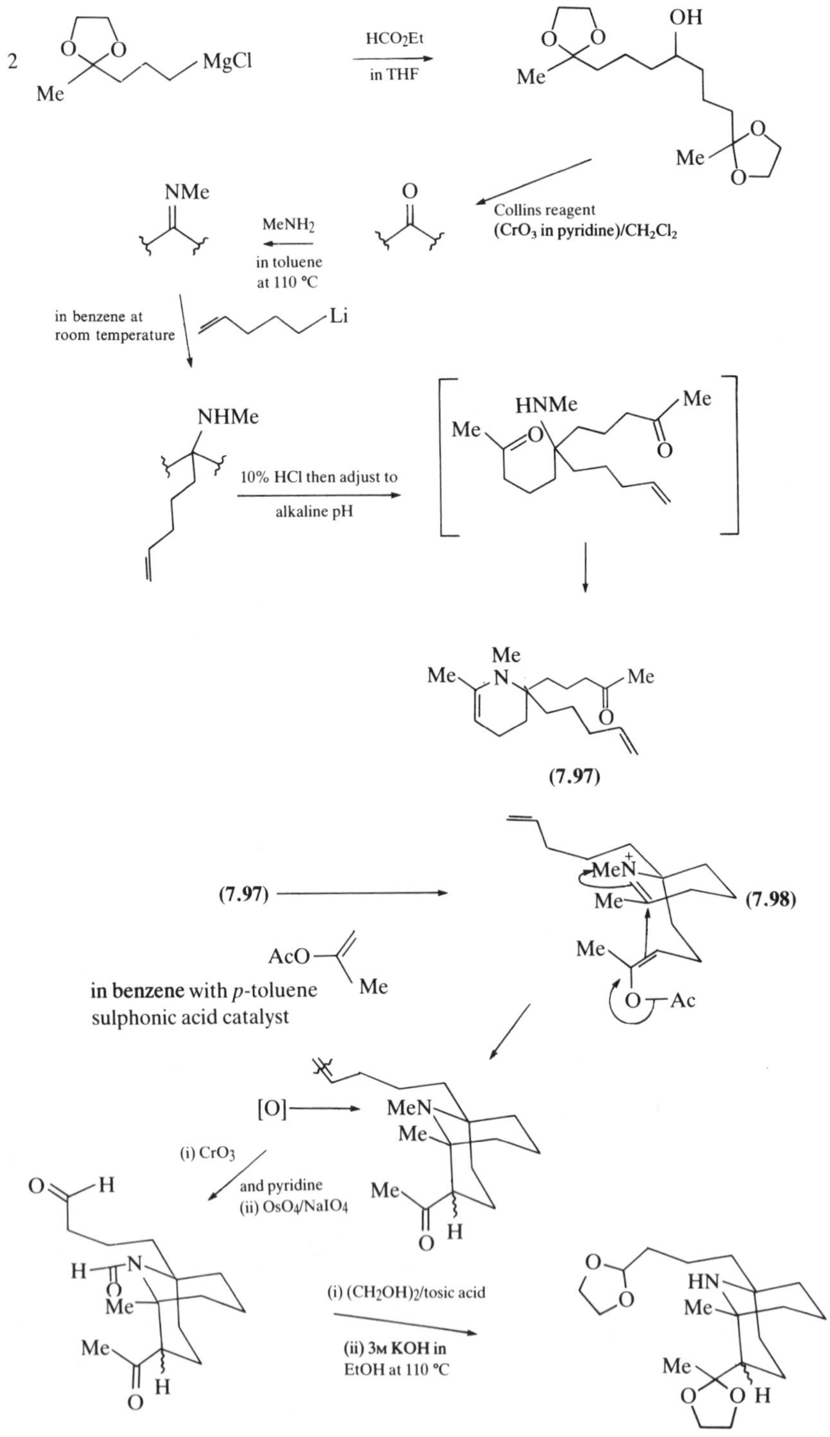

Fig. 7.32.

(*continued*)

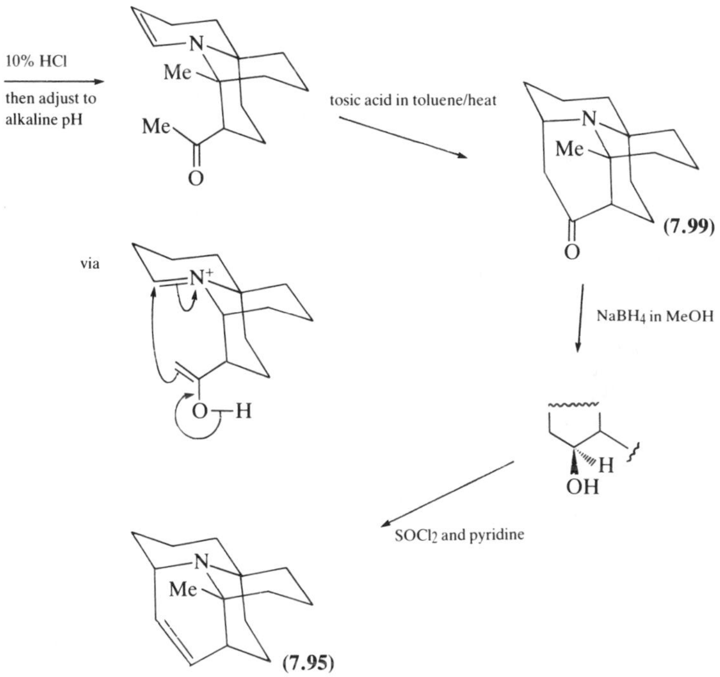

10% HCl

then adjust to
alkaline pH

Me

N

Me

O

tosic acid in toluene/heat

N

Me

O

(7.99)

via

N⁺

O—H

NaBH₄ in MeOH

H

OH

SOCl₂ and pyridine

N

Me

(7.95)

Fig. 7.32 *continued*

process. For example, Oppolzer synthesized lysergic acid (as its racemate) (**7.90**) via the route shown in Fig. 7.33. Standard methods were used to convert 4-hydroxymethyl-1-tosylindole into the desired phosphonium salt (bromide formation using CBr_4/Ph_3P, and then reaction with Bu_3P in refluxing benzene), and after conversion into the corresponding ylid, this underwent a Wittig reaction with the bicyclic aldehyde (itself available by formylation of the Diels–Alder product from cyclopentadiene and methacrylate). After removal of the tosyl group, the nitroethyl side-chain was introduced, via a Michael reaction between the indole nucleus and nitroethene. Formation of the key oxime methyl ether (**7.100**) was achieved in one operation using $TiCl_3/NH_4OAc$ and O-methylhydroxylamine and a possible mechanism is shown in equation 7.12:

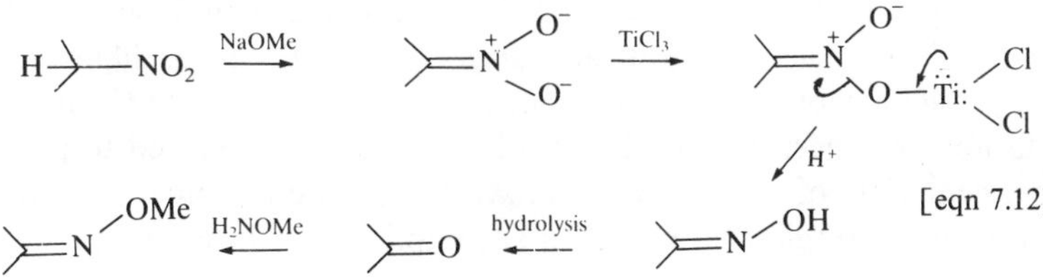

H—NO₂ → (NaOMe) → =N⁺ with O⁻, O⁻ → (TiCl₃) → =N⁺ O⁻, O—Ti: Cl Cl

↓ H⁺

=N—OMe ← (H₂NOMe) ← =O ← (hydrolysis) ← =N—OH   [eqn 7.12]

The crucial intramolecular Diels–Alder cycloaddition took place after a prior retro-Diels–Alder reaction which produced the diene ester (**7.101**).

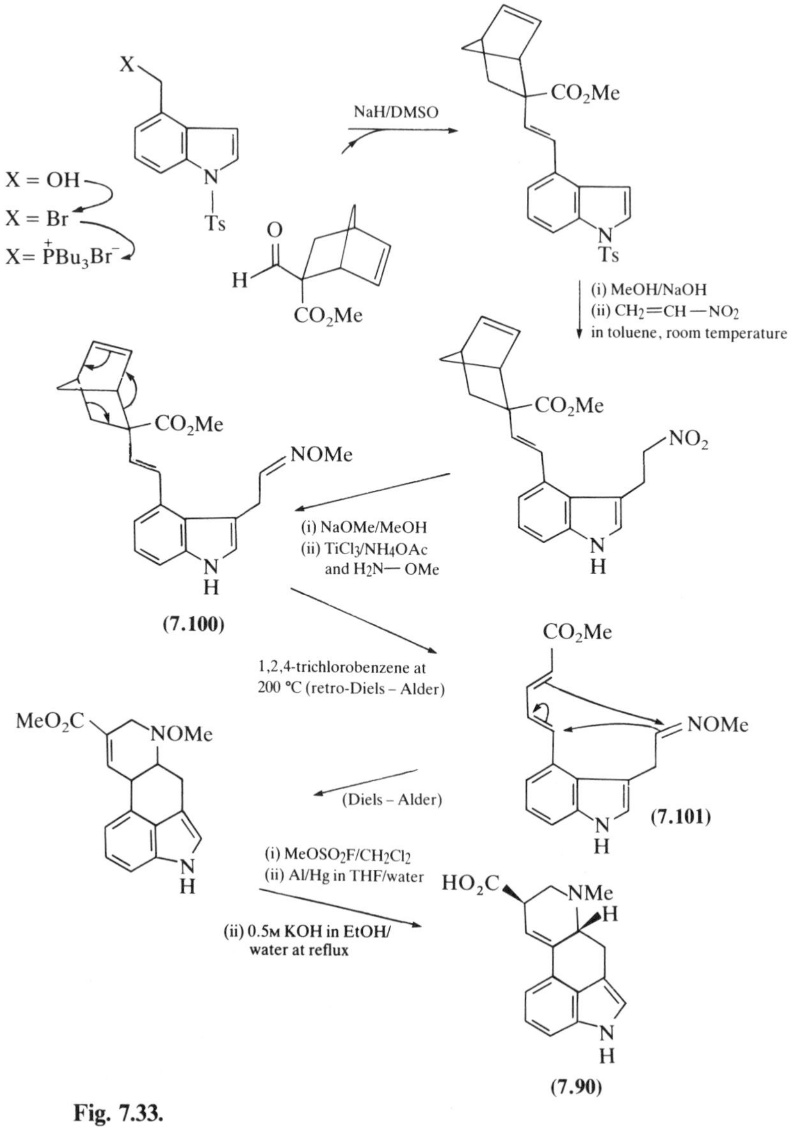

Fig. 7.33.

Although neither the diene nor the dienophile would be expected to be particularly reactive, slow addition of (**7.100**), as a dilute solution in 1,2,4-trichlorobenzene, to the same solvent held at 200°C ensured a low stationary concentration of (**7.101**), and reaction occurred to produce a 67 per cent yield of cycloadducts (mixture of diastereoisomers). Conversion of these into lysergic acid was then accomplished in three steps:

(a)   *N*-methylation with methanesulphonyl fluoride;
(b)   hydrogenolysis of the resultant salt; and finally
(c)   isomerization of the double bond, and concomitant ester hydrolysis.

The stereochemically more demanding alkaloid dendrobine (**7.102**), isolated from a Chinese ornamental orchid, and believed to be of medicinal value, has been synthesized by Roush (and others). His retrosynthetic analysis is shown in Fig. 7.34, and this involves a double disconnection of the lactone and pyrrolidine rings, followed by removal of the amine side-chain. Roush reasoned that the angular methyl group (R in the figure) could be introduced in a base-mediated alkylation, and that base would favour the formation of a *cis*-fused ring system. Hence prior to this alkylation, a *cis*- or *trans*-fused system could be present, and since the latter is more favourable for the envisaged intramolecular Diels–Alder reaction, intermediate (**7.103**) was proposed. Practical considerations then ruled out pathway (a) since this would involve an *exo*-transition state, though this would provide the desired β-CO_2 in (**7.103**); while pathway (b) should proceed via the preferred *endo*-transition, but would provide an α-CO_2Me

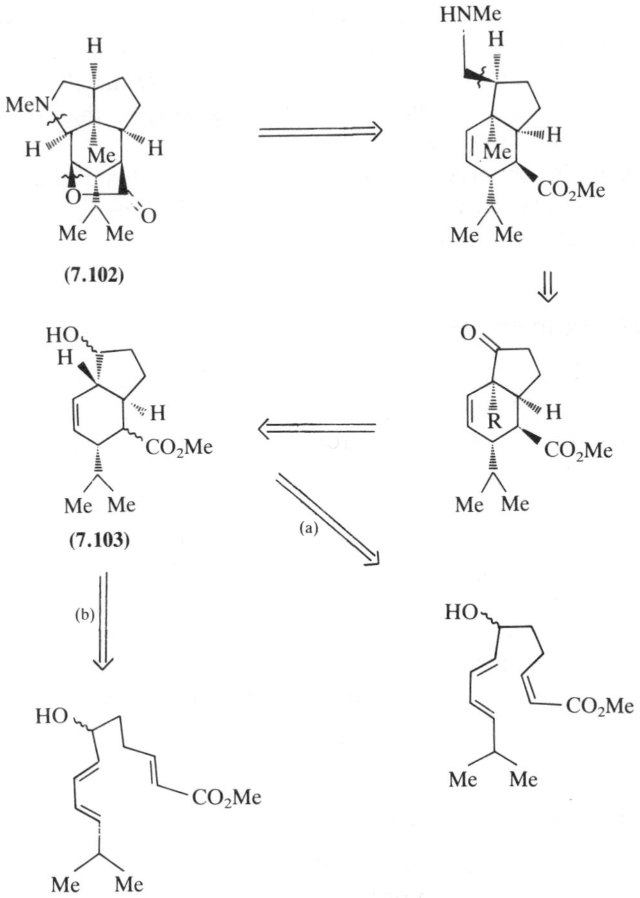

Fig. 7.34.

in (**7.103**). In the event, the synthesis was executed essentially according to this analysis, and the reader is referred to the paper by Roush for details.

7.6.2.3   *The use of quinodimethanes*

A related approach that has been much employed in recent years utilizes species such as (**7.104**), so-called ortho-quinodimethanes. These may be obtained in a variety of ways, but one favoured by many workers is shown in equation 7.13, and involves thermolysis of benzocyclobutenes:

$$\text{RO-} \quad \xrightarrow{\Delta} \quad \text{RO-} \qquad \text{[eqn 7.13]}$$

(**7.104**)

The synthesis of racemic chelidonine (**7.105**), shown in Fig. 7.35, is illustrative of this approach. The route is relatively short, and the actual cycloaddition of quinodimethane and nitroolefine proceeds with high regio- and stereoselectivity. This feature has been explained in terms of a favoured transition state (**7.106**) (with *exo*-NO$_2$) rather than involvement of transition state (**7.107**) (with *endo*-NO$_2$).

The alkaloid is produced by *Chelidonium majus*, and is of interest due to its marked cytotoxic activity. It has been used in Europe for the treatment of warts for many centuries.

7.6.2.4   *The use of nitrones*

Although the Diels–Alder reaction has been the pre-eminent method of cycloaddition in alkaloid syntheses, a number of dipolar cycloadditions using species such as nitrile oxides and nitrones have become increasingly popular of late. For example, the quinolizidine alkaloid lupinine (**7.56**) has been prepared by the short sequence shown in Fig. 7.36 which includes a cycloaddition step involving a nitrone. The cycloadduct was not isolated but underwent a second ring-closure with displacement of mesylate. Reductive cleavage of the N—O bond and dehydration yielded an unsaturated ester, which was hydrogenated and reduced to provide lupinine.

The more complex alkaloid chanoclavine I (**7.25**) was synthesized via the route shown in Fig. 7.37, and this also incorporates a nitrone cycloaddition as the key step.

7.6.2.5   *Oxidative phenolic coupling*

Numerous biomimetic syntheses of alkaloids have been carried out using oxidative phenolic coupling in order to provide the polycyclic framework of the compounds. The synthesis of the tricyclic system of maritidine (**7.108**)

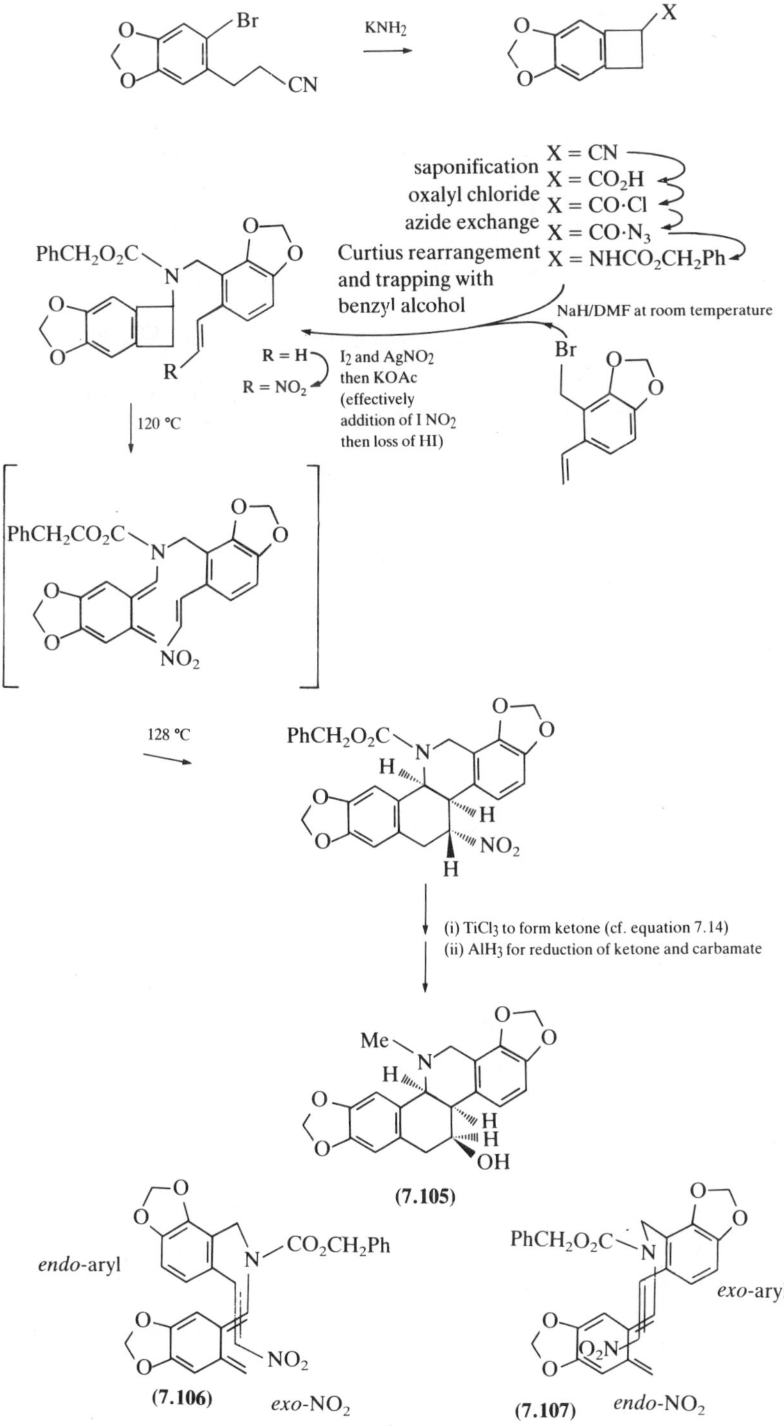
Br
CN
KNH_2
X
X = CN
saponification
X = CO_2H
oxalyl chloride
X = CO·Cl
azide exchange
X = $CO·N_3$
Curtius rearrangement
and trapping with
benzyl alcohol
X = $NHCO_2CH_2Ph$
NaH/DMF at room temperature
Br
$PhCH_2O_2C$
N
R
R = H
R = NO_2
I_2 and $AgNO_2$
then KOAc
(effectively
addition of I NO_2
then loss of HI)
120 °C
$PhCH_2CO_2C$
N
NO_2
128 °C
$PhCH_2O_2C$
N
H
H
NO_2
H
(i) $TiCl_3$ to form ketone (cf. equation 7.14)
(ii) AlH_3 for reduction of ketone and carbamate
Me
N
H
H
H
OH
(7.105)
endo-aryl
CO_2CH_2Ph
N
NO_2
(7.106)
exo-NO_2
$PhCH_2O_2C$
N
exo-aryl
O_2N
(7.107)
endo-NO_2

Fig. 7.35.

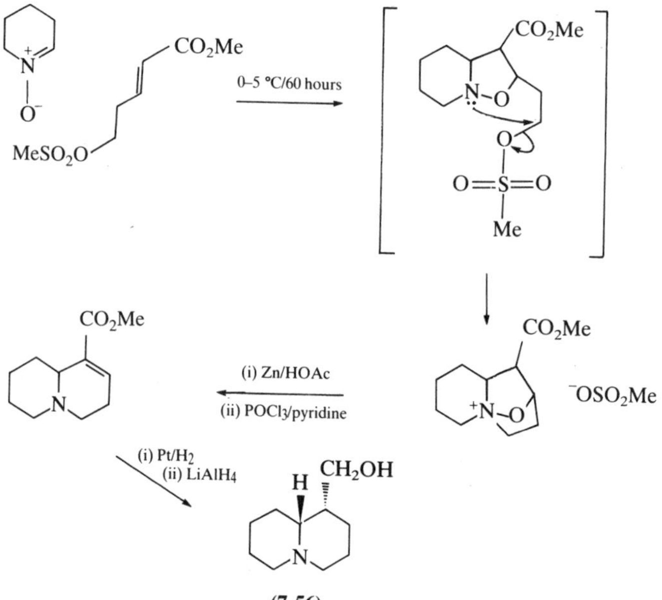

(7.56)

Fig. 7.36.

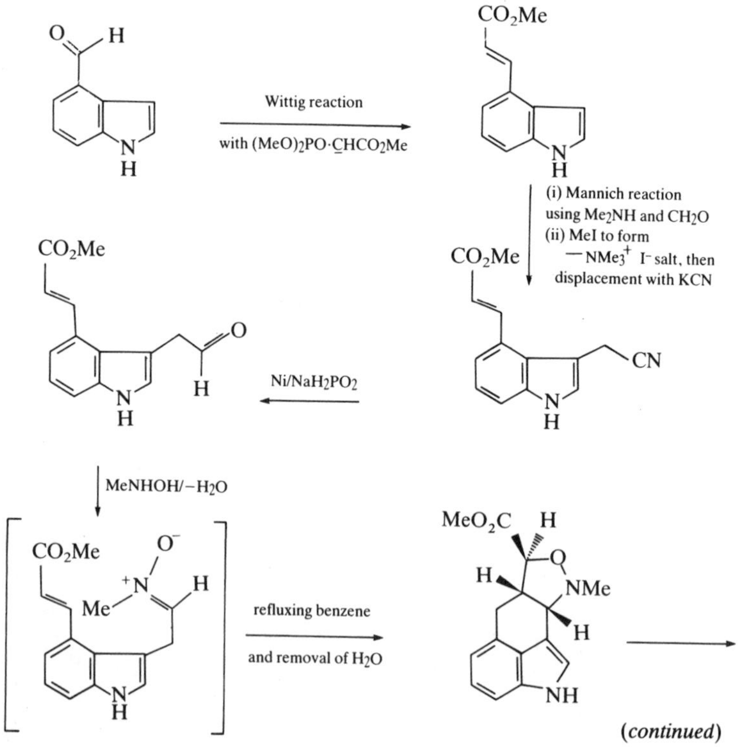

(*continued*)

Fig. 7.37.

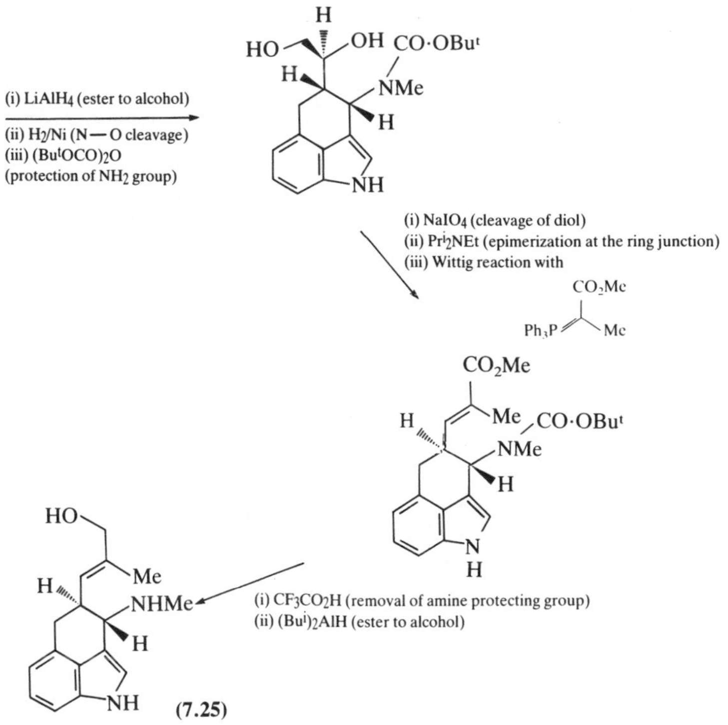

(i) LiAlH4 (ester to alcohol)

(ii) H2/Ni (N—O cleavage)
(iii) (Bu^tOCO)2O
(protection of NH2 group)

(i) NaIO4 (cleavage of diol)
(ii) Pr^i_2NEt (epimerization at the ring junction)
(iii) Wittig reaction with

(i) CF3CO2H (removal of amine protecting group)
(ii) (Bu^i)2AlH (ester to alcohol)

(7.25)

Fig. 7.37 *continued*

supplies a good example of this type of strategy (Fig. 7.38). The final ring bond is formed via a Michael reaction of the tertiary amine with the unsaturated carbonyl moiety in (**7.109**).

7.6.2.6 *The use of cyclopropyl enamines*

In 1977, R.V. Stevens suggested that alkaloids containing pyrrolidine rings might be prepared via annulations of vinyl ketones to endocyclic enamines (equation 7.14). The endocyclic enamines were in turn prepared by acid-catalysed rearrangement of cyclopropyl enamines (equation 7.15):

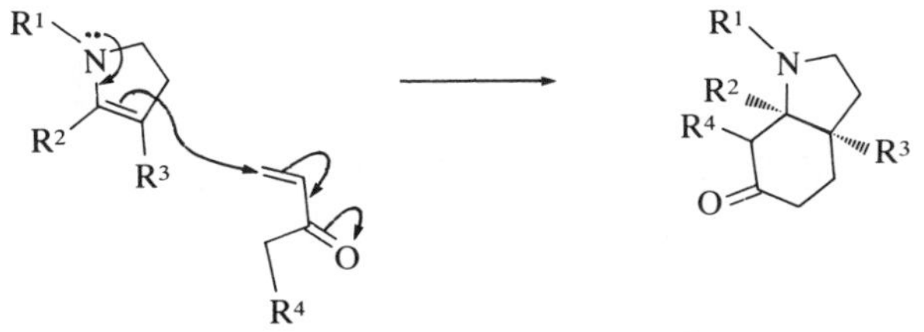

[eqn 7.14]

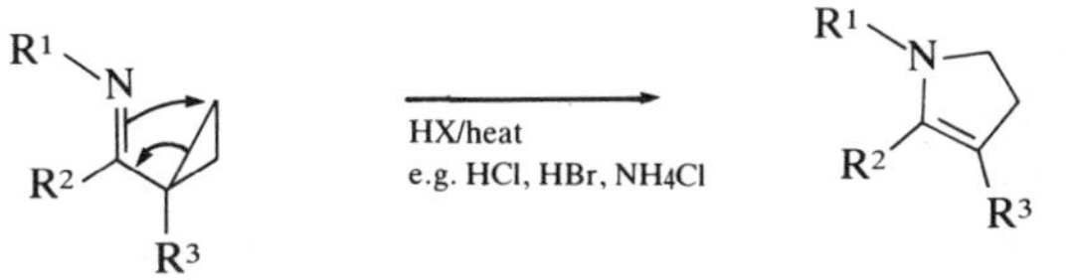

HX/heat
e.g. HCl, HBr, NH4Cl

[eqn 7.15]

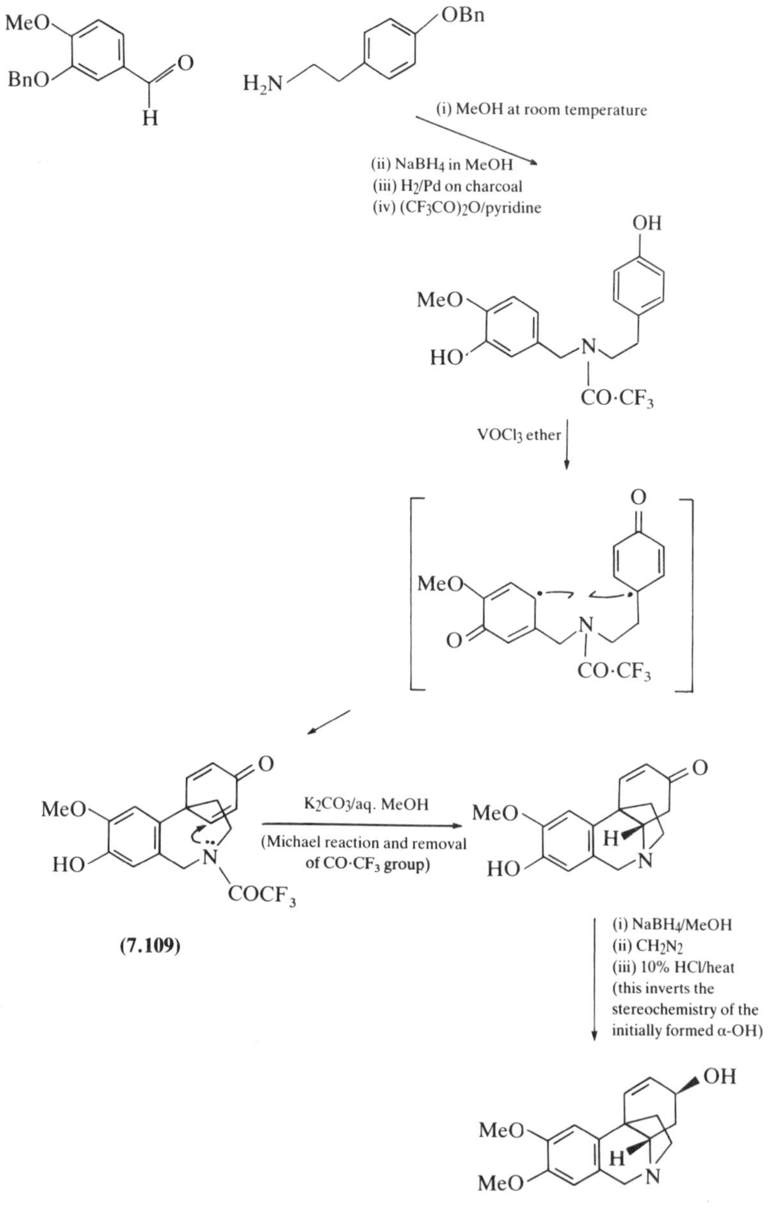

Fig. 7.38.

Numerous alkaloids have since been prepared using this general methodology, and the synthesis of mesembrine (**7.81**) provides a particularly good example of the efficiency of this approach (Fig. 7.39) proceeding via cyclopropyl enamine (**7.110**) and dihydropyrrole (**7.111**).

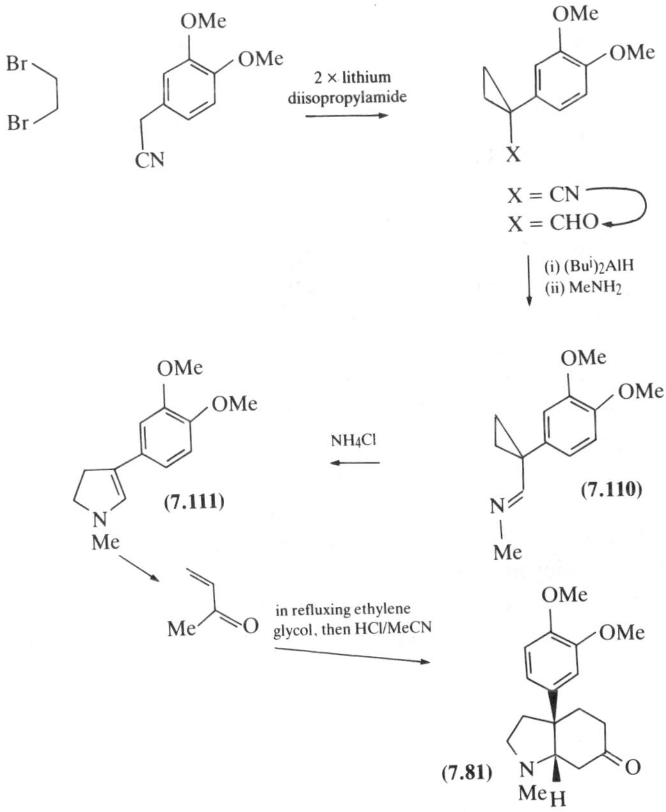

Fig. 7.39.

7.7 Asymmetric synthesis

All of these syntheses described thus far have provided racemic alkaloids, and a resolution step has been required in order to obtain the natural products. Contemporary alkaloid synthesis usually involves chiral intermediates obtained from stereochemically pure starting materials (carbohydrates, amino acids, or terpenoids), or via asymmetric induction, and the alkaloids are then produced in their natural stereochemical configurations.

7.7.1 Synthesis of deoxynojirimycin

Deoxynojirimycin (DNJ) (7.112) is one of a family of simple alkaloids that are nitrogen analogues of carbohydrates – DNJ resembles glucose. They are of considerable contemporary interest because they are potent inhibitors of glycosidases. DNJ and its N-butyl analogue are effective inhibitors of the

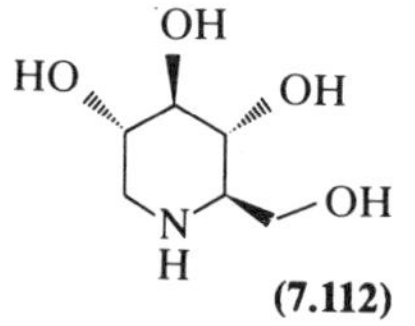

(7.112)

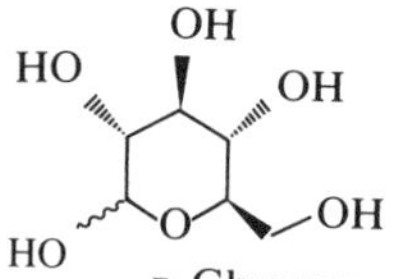

D-Glucopyranose

glycoprotein processing required by the human immunodeficiency virus (HIV) (cf. Section 1.3) and the latter compound is under clinical evaluation as an agent for the chemotherapy of AIDS.

The synthesis by Fleet and coworkers illustrated in Fig. 7.40 commences with the cheap and readily available diacetonide of glucose (**7.113**), and proceeds via a series of hydroxyl protections and deprotections to yield the carbonate, acetonide (**7.114**). Cleavage of the acetonide provided a mixture of methyl glucofuranosides with the alpha-anomer predominating, and this was converted into its trifluoromethanesulphonate prior to nucleophilic displacement with azide. Cleavage of the carbonate, selective formation of the primary tosylate, reduction of the azide, and subsequent cyclization yielded the key bicyclic compound (**7.115**; R = H). This compound was converted into a number of natural and novel carbohydrate mimics, but the synthesis of DNJ required an inversion of configuration of the secondary

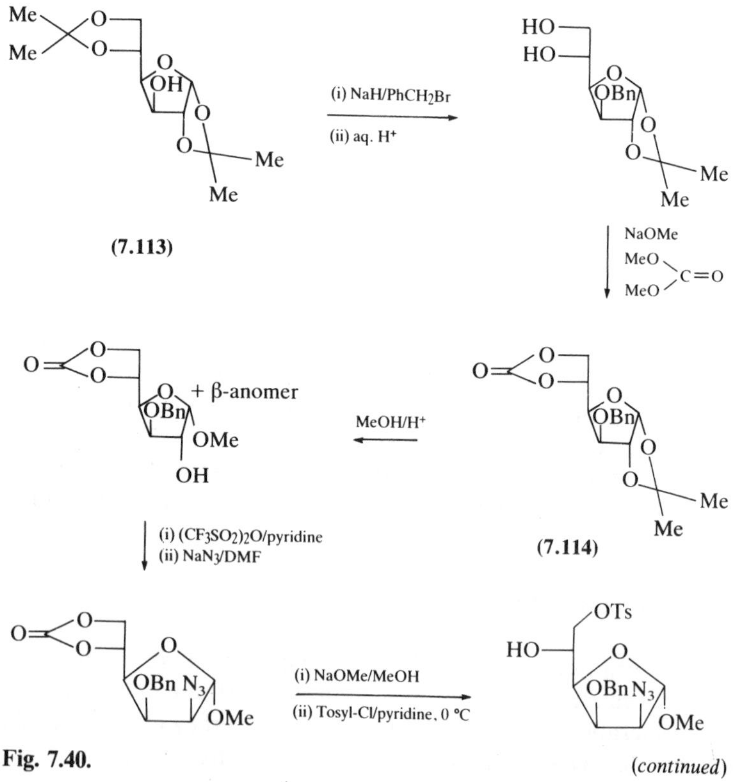

Fig. 7.40.

(*continued*)

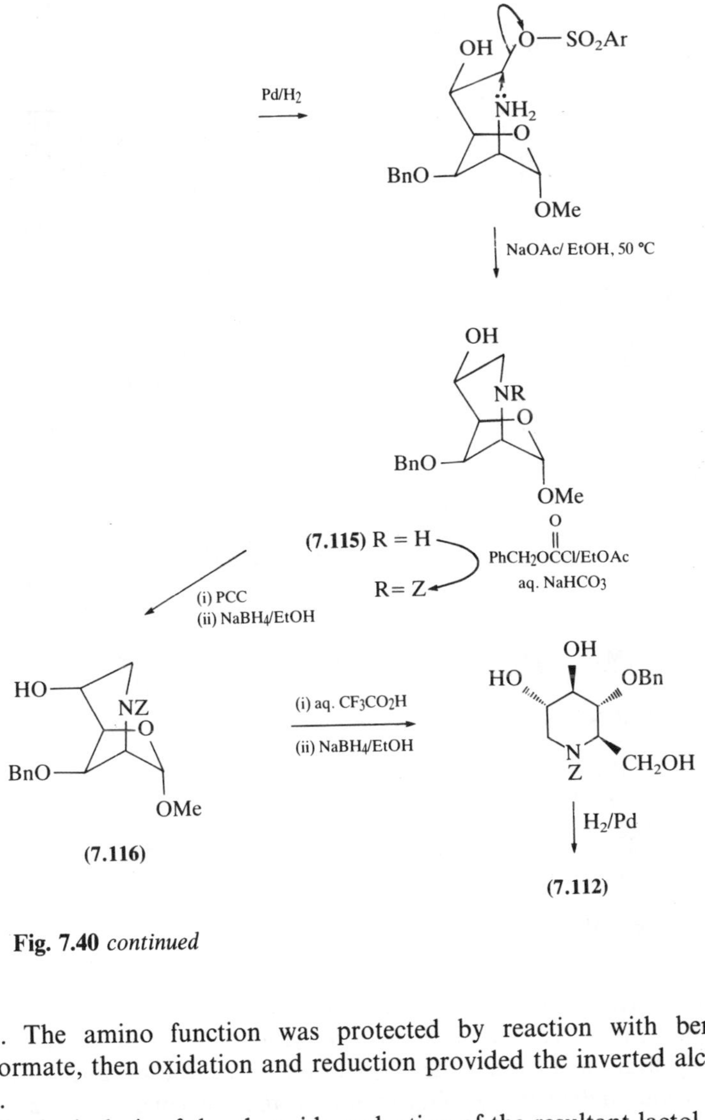

Fig. 7.40 *continued*

alcohol. The amino function was protected by reaction with benzyl-chloroformate, then oxidation and reduction provided the inverted alcohol (**7.116**).

Finally, hydrolysis of the glycoside, reduction of the resultant lactol, and removal of both the benzyl and benzyloxycarbonyl protecting groups using catalytic hydrogenation yielded deoxynojirimycin (**7.112**).

7.7.2 *Synthesis of indolactam V*

Indolactam V (**7.117**) and a number of structurally related alkaloids from actinomycetes are also of considerable contemporary interest due to their potent tumour-promoting activities. The synthesis of indolactam V by Moody and Mascal that is illustrated in Fig. 7.41 is both elegant and brief.

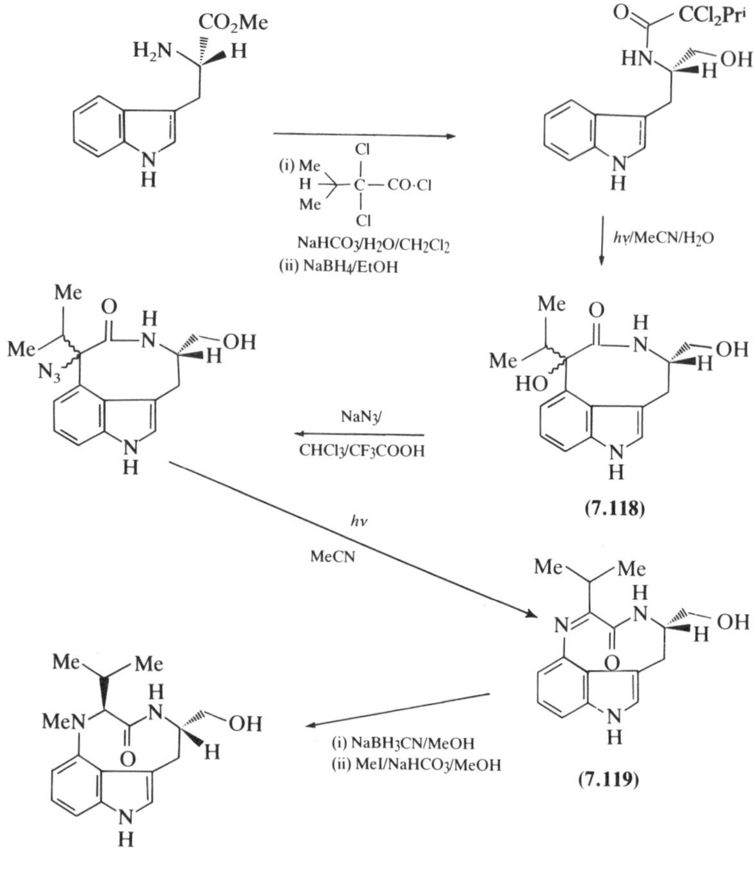

(7.118)

(7.119)

(7.117)

Fig. 7.41.

Tryptophan methyl ester was reacted with 2,2-dichloro-3-methylbutanoyl chloride prior to reduction of the acid with sodium borohydride. A neat photocyclization provided the diols (**7.118**), and the product of reaction with azide underwent a photo-induced rearrangement (via a nitrene, equation 7.16) to yield the key intermediate (**7.119**). This was reduced with a high degree of stereoselectivity by the use of sodium cyanoborohydride, and *N*-methylation completed the synthesis of indolactam V (**7.117**).

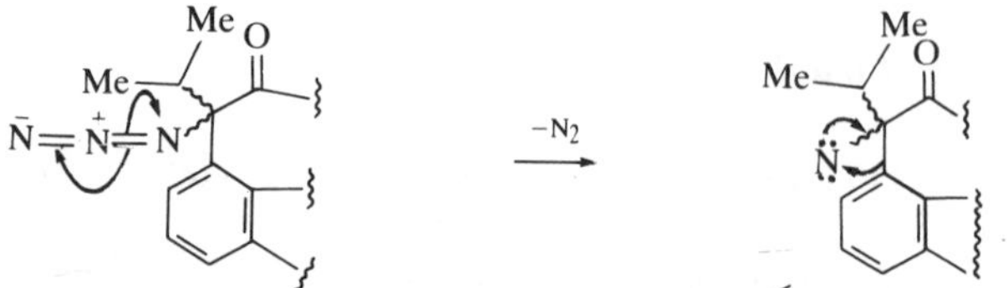

(7.119)

[eqn 7.16]

Further reading

General references

1. I.W. Southon and J. Buckingham, *Dictionary of Alkaloids*, Chapman & Hall, London, 1989; R.H.F. Manske (founding editor), A. Brossi (editor from vol. 21), *The Alkaloids*, vols 1–36, Academic Press, New York.
2. General chemistry and biosynthesis of alkaloids: vols 1–9, *Natural Product Reports*, RSC, London (abbreviated hereafter to *NPR*).
3. Biosynthesis of alkaloids: R.B. Herbert, *NPR*, vol. 9, 1992, 507; J. Mann, *Secondary Metabolism*, Oxford University Press, Oxford, 1987.
4. Ecological significance of alkaloids: J.B. Harborne, *NPR*, vol. 10, 1993, 327.
5. Use of NMR in structure elucidation of natural products: A.E. Derome, *NPR*, vol. 6, 1989, 111.

Research references

6. Structure elucidation and general history of opium alkaloids: H.L. Holmes in *The Alkaloids*, vol. 2 (ed. H.L. Holmes and R.H.F. Manske), Academic Press, New York, 1952.
7. Synthesis of morphine: M. Gates and G. Tscudi, *J. Amer. Chem. Soc.*, 1956, **78**, 1380.
8. Structure elucidation of tropane alkaloids: H.L. Holmes in *The Alkaloids*, vol. 1 (ed. H.L. Holmes and R.H.F. Manske), Academic Press, New York, 1950.
9. Stereochemistry of tropic acid: G. Fodor and G. Csepreghy, *J. Chem. Soc.*, 1961, 3222; M.B. Watson and G.W. Youngson, *JCS Perkin Trans I*, 1972, 1597.
10. Structure elucidation of eudistomin K: J.W. Blunt, R.J. Lake, and M.H. Munron, *Tetrahedron Lett.*, 1987, 1825.
11. Pyrrolidine alkaloid biosynthesis: E. Leete, J.A. Bjorklund, M.M. Couladis and S.H. Kim, *J. Amer. Chem. Soc.*, 1991, **113**, 9286.
12. Piperidine alkaloid biosynthesis: T. Hemscheidt and I.D. Spenser, *J. Amer. Chem. Soc.*, 1990, **112**, 6360.
13. Pyrrolizidine alkaloid biosynthesis: D.J. Robins, *Chem. Soc. Rev.*, 1989, **18**, 375; D.J. Robins, *NPR*, vol. 9, 1992, 313.
14. General chemistry of pyrrolidine, piperidine, and pyridine alkaloids: A.R. Pinder, *NPR*, vol. 9, 1992, 491.
15. Isoquinoline alkaloids: K.W. Bentley, *NPR*, vol. 7, 1990, 245.
16. Morphine biosynthesis: S. Loefler, R. Stadler, N. Nagakura, and M.H. Zenk, *JCS Chem. Commun.*, 1987, 1160; W. De Eknamkul and M.H. Zenk, *Tetrahedron Lett.*, 1990, 4855.
17. Enkephalins and other neuropeptides: F.E. Bloom, *Sci. Amer.*, 1981, **245**, October, 114.
18. Colchicine biosynthesis: R.B. Herbert and E. Knagg, *Tetrahedron Lett.*, 1986, 1099.
19. Indole alkaloids: J.E. Saxton, *NPR*, vol. 10, 1993, 349.
20. Ergot alkaloid biosynthesis: A.P. Kozikowski, J.-P. Wu, M. Shibuya, and H.G. Floss, *J. Amer. Chem. Soc.*, 1988, **110**, 1970; M. Shibuya, H.M. Chu, M. Fountoulakis, S.-U. Kim, H. Kobayashim, H. Otsuka, E. Rogalska, J.M. Cassadyu, and H.G. Floss, *J. Amer. Chem. Soc.*, 1990, **112**, 297.
21. Tropinone synthesis: R. Robinson, *J. Chem. Soc.*, 1917, 762, 876; C. Schöpf, G. Lehmann, and W. Arnold, *Angew. Chem.*, 1937, **50**, 783.
22. Cocaine synthesis: W. Willstätter, O. Wolfes, and H. Mader, *Annalen*, 1923, **434**, 111.

23. Reserpine synthesis: R.B. Woodward, F.E. Bader, H. Bickel, A.J. Frey, and R.W. Kierstead, *Tetrahedron*, 1958, **2**, 1.
24. Porantheine synthesis: E.J. Corey and R.J. Ballanson, *J. Amer. Chem. Soc.*, 1974, **96**, 6516.
25. Lysergic acid synthesis: W. Oppolzer and E. Francotte, *Helv. Chim. Acta*, 1981, **64**, 478.
26. Dendrobine synthesis: W.R. Roush, *J. Amer. Chem. Soc.*, 1980, **102**, 1390.
27. Use of quinodimethanes for alkaloid synthesis: T. Kametani, M. Takemura, M. Ogasawara and K. Fukumoto, *J. Hetero. Chem.*, 1974, **11**, 179.
28. Chanoclavin I synthesis: W. Oppolzer and J.I. Grayson, *Helv. Chim. Acta*, 1980, **63**, 1706.
29. Maritidine synthesis: M.A. Schwartz and R.A. Holton, *J. Amer. Chem. Soc.*, 1970, **92**, 1090.
30. Use of cyclopropylenamines in alkaloid synthesis: R.V. Stevens, *Acc. Chem. Res.*, 1977, **10**, 193.
31. Deoxynojirimycin synthesis: G.W.J. Fleet, L.E. Fellows, and P.W. Smith, *Tetrahedron*, 1987, **43**, 979.
32. Indolactam V synthesis: M. Mascal and C.J. Moody, *JCS Chem. Commun.*, 1988, 587, 589.

Index

Abietic acid, 328
ABO blood-group substances, 23
Abscisic acid, 318, 330
Absolute configuration, 132
ACE (angiotensin-converting enzyme), 141
Acetoacetyl-S-ACP, 244
Acetylcholine, 421, 56
Acetylcholine esterase, 421
Acetyl CoA, 244, 246
Actinomycin D, 189
Actinomycins, 190
Acyl carrier protein (ACP), 244
Acylglycerols, 239, 240
Adenine, 70
Adenosine, 70
Adrenaline, 413
Adrenocortical hormones, 338
Adriamycin, 19, 37, 57
Aesculetin, 363
AIDS, 24, 30, 81, 107, 442
Ajmalicine, 393, 423, 426
Alanine, 397, 422
Aldaric acids, 34
Aldonic acids, 34
Aldoses, 9, 12
Aldosterone, 338
Alginic acids, 25
Alkannin, 382
Alkaptonuria, 149
Amaryllidaceae alkaloids, 418
Amatoxins, 188
Ambergris, 344
Ambrein, 344
Ames test, 152
Aminoacyl adenylate, 191
Aminoacyl-tRNA synthetase, 191
Aminopeptidases, 206
Ampicillin, 125
AMP (adenosine monophosphate), 93

c-AMP (cyclic AMP), 95
α-Amyrin, 345
β-Amyrin, 345
Anabasine, 391, 411
Anatabine, 411
Androgens, 336
Androsterone, 336
Angiotensin, 185
Anhydrosugars, 27, 46, 56
Anomeric centre, 10, 28, 35, 40
Anomeric effect, 29
Antamanide, 187, 189
Anthranilic acid, 390
Anti-fungal compound, 328
Anti-inflammatory drugs
    non steroidal, 264, 277
    steroidal, 277, 339
Antimalarial drug, 325
Antisteromycin, 86
Arachidonic acid, 43, 241, 248, 274
Arrow poisons, 340, 389
Artemisinin, 325
Ascaridole, 312
Aspartame, 149, 182
Aspartic acid, 412
Aspirin, 264, 277
ATP (adenosine triphosphate), 5, 14, 29, 91, 147
Atropine, 397, 407
Automatic sequencer, 209
Auto-oxidation, 252
AZT (azidothymidine), 81, 88, 107

Baeyer–Villiger oxidation, 86, 266, 269
B.R. Baker's rule, 78
Batrachotoxin, 389, 390, 393
1,4-Benzoquinone, 364, 366
Berberine, 392, 418
α-Bergamotene, 319
Betnovate, 339

Bile acids, 336
Bile salts, 336
Bisabolenes, 319
Borneol, 307, 314
Bornyl pyrophosphate, 310
α,β-Bourbonenes, 327
BPOC (2-(4-biphenyl)-
    isopropyloxycarbonyl group), 158
Brevicomin, 53
Bufadienolides, 340
Bufotalin, 340
Bufotenine, 420
Bufotoxins, 340
Bulnesol, 324
t-Butoxycarbonyl group (BOC), 157
Butyric acid, 240
Butyryl-S-ACP, 245

Cadaverine, 408, 409
Cadinane, 345
Caffeic acid, 362, 366
Caffeine, 390
Calciferol, 341
Camphene, 314, 315
Camphenilol, 319
Camphor, 289, 290, 307, 311, 350
Camptothecine, 423, 425
Capping step (in oligonucleotide
    synthesis), 115, 116
Capsanthin, 318
Capsidil, 326
Caratol, 321
Carboxypeptidases, 207
Cardenolides, 340
Cardiac glycosides, 340
Car-3-ene, 296, 307
Carissone, 300
α-Carotene, 347
β-Carotene, 347, 348, 356
γ-Carotene, 347
Carvone, 307, 312
Caryolan-1-ol, 327
Caryophyllene, 322, 327, 351
epi-Caryophyllene, 323
Castoramine, 394
Catechol, 362
Catharanthine, 423, 425
CD4, 30
α-Cedrene, 299, 300, 319, 351
Cedrol, 299, 319
Cell membranes, 239, 242
Cellulose, 22, 30
Cembrene, 328
Cephalotoxin, 418
Chalcones, 384

Chanoclavine I, 395,422, 424, 436
Chaulmoogric acid, 243
Chelidonine, 436
Chicle, 349
Chimaphilin, 382
(*P*-) Chiral nucleotides, 97
Chiron, 44
Chitin, 22, 138
Chloroflavonin, 375
Chlorogenic acid, 366
Cholanic acid, 336
5α, 5β-Cholesterols, 332
Cholesterol, 301, 332, 352
Cholic acid, 333, 336
Chromones, 387
Chrysanthemyl alcohol, 313
Chymotrypsin, 210
1,8-Cineol, 307
Cinnamic acid, 373
Citral, 290, 307, 316, 357
Citronellal, 300, 302
Cloning of genes, 125
Clovene, 327
C-nucleosides, 17, 71, 74, 81,88
Cocaine, 389, 405, 427
Coccinelline, 394, 431
Codeine, 396, 414, 416
Codeinone, 396, 416
Coenzyme A, 91
Colchicine, 395, 419
Conformation of nucleosides, 72
Coniferyl alcohol, 378
Coniine, 389, 390, 395, 408
Copaane, 322
Coprostanol, 332
Corey's lactone, 266, 268
Coronamic acid, 138
Corticosteroids, 338, 354
Cortisol, 338, 339
Cortisone, 338, 339, 355
Corynomycolic acid, 243, 246
4-Coumaric acid, 373
Crepenynic acid, 243, 249
o-Cresol, 366
Cryptone, 294
CTP (cytidine triphosphate), 92
Cubebane, 322
Cuparene, 319
Cyanidin, 368
Cyanidin-3-glucoside, 379, 380
Cyanogen bromide, 199
Cyclic nucleotides, 91
Cycloartenol, 342
Cyclonerodiol, 319

Cyclonucleosides, 80
Cyclo-oxygenase, 262, 264, 274, 277
Cycloserine, 125, 390
Cyclosporins, 188
Cymene, 308
Cyperone, 300
Cytisine, 409
Cytochrome P-450, 338
Cytokinins, 71
Cytosine, 70
Cytidine, 70

Dansyl derivatives, 206
Daunosamine, 20, 57
DCC (dicyclohexylcarbodiimide), 167
Deadly nightshade, 397, 407
Defensive secretion, 309
Dehydroarachidonic acids, 286
Dehydrocholesterol, 341
Dehydrogeissoschizine, 423, 426
Delphinidin, 368
Dendrobine, 435
2-Deoxy-D-ribose, 17
2-Deoxyglucose, 53
Deoxynojirimycin, 441
Deoxyribonucleosides, 70
Deoxyadenine, 70
Deoxyguanosine, 70
Deoxythymidine, 70
Deoxycytidine, 70
DEPT spectrum, 225
Depurination, 108, 109
Desaturase, 246, 247, 248
Dichlorophloroglucinol, 371
Diels–Alder reaction, 86, 286, 431, 433, 455, 436, 437
Diels hydrocarbon, 302
Dienone–phenol rearrangement, 333
Diethylstilboestrol, 337
Digitalis, 340
Digitonin, 340
Digitoxigenin, 340
Dihydroquercetin, 379
Dihydrosterculic acid, 243
Diisopropylidene glucose, 31, 54, 442
2,5-Diketopiperazines, 186
2,6-Dimethoxybenzoquinone, 364
Dimethylallyl alcohol, 306
Dimethylallyl pyrophospate, 292, 305, 422
Diosgenin, 339, 355
Diphenylmethyl esters (ODpm), 162
Diploicin, 365
Disulphide bridges, 209
DNA, 69, 70, 102, 103

DNA ligase, 105, 106, 119, 125
DNA polymerase, 104, 105, 126
DNA replication,104
DNA sequencing, 88, 119
L-DOPA, 145
Dopamine, 404, 413
Double helix, 69, 104
Drimenol, 318
Dunnione, 387

α-Ecdysone, 343
Ecdysterols, 343
Echinalin, 422
Echinomycin, 190
Ecological chemistry, 53, 307, 317, 325, 328, 338, 340, 343, 344, 394, 413
Edman degradation, 203
Eicosapentaenoic acid, 275
Elastase, 210
Electron ionization (EI) mass spectrometry, 212
Emetine, 418
Emodin, 364
Endopeptidases, 183
Enkephalins, 185, 417
Enzyme cofactors, 2, 3, 5, 14, 90, 91, 408, 421
Ephedrine, 389, 392
Equilenin, 336, 352
Eremophilane, 323
Ergometrine, 423
Ergosterol, 302, 341, 342
Ergot alkaloids, 395, 422
Ergotism, 395, 422
Eriodictyol, 379
Erucic acid, 241
Erythraline, 418
Erythromycin, 17
Esters of carbohydrates, 28
Ethers of carbohydrates, 26
Eucarvone, 315
Eudismane, 323
Eudistonin K, 400, 403
Euphol, 344
Exons, 107

FAB (fast atom bombardment mass spectrometry-FABS), 221
FAD (flavine adenine dinucleotide), 90
Farnesol, 317
Fats, 239, 242
Fatty acids
    branched chain, 242
    mono-unsaturated, 246
    polyunsaturated, 241, 247

Fatty acids *continued*
  saturated, 240
  unsaturated, 241, 242, 250
Fatty acids of unusual structure, 248
α-Fenchyl alcohol, 314
Flavone, 384
Fluorescent labels, 199
FMN (flavine mononucleotide), 90
FMOC (fluorenylmethoxycarbonyl),
    109, 119, 157, 201
Formycin, 40, 72, 85
FPP (farnesyl pyrophosphate), 292, 316
Friedelin, 345
D-Fructose, 8
L-Fucose, 87
Fucosterol, 342
Furanose conformation, 11

Gabriel synthesis, 132
D-Galactose, 20
Galanthamine, 419
Gel sequencing technique, 70
Genetic code, 69
Genistein, 364, 385
Geraniol, 307, 309, 350
Germacrene D, 327
GFPP (geranyl farnesyl
    pyrophosphate), 293, 331
GGPP (geranyl geranyl
    pyrophosphate), 292, 327, 349
Gibberellic acid, 330
Gibberellins, 318, 330
Ginkgetin, 369, 375
Glaucine, 392, 418
Gliotoxin, 187
Glucogenic, 145
Glutathione (γ-glutamylcysteinyl-
    glycine), 279
Glyceraldehyde 3-phosphate, 412
Glycogen, 25
Glycoside digitonin, 340
Glycosides,19, 36, 74, 295, 309, 318
D-Glucose, 7, 11, 12, 13, 26, 30, 32, 57
Glycopeptides, 24, 62
Glycoproteins, 24, 30
Glycosylation (in nucleoside synthesis),
    74
GMP(GTP, guanosine phosphates), 93
Gnididione, 351
Gossypol, 345
GP 120, 30
GPP (geranyl pyrophosphate), 292, 307
Gramidicin S, 187
Griseofulvin, 365
Guaiol, 298

Guanine, 70
Guanosine, 70

$^1H/^{13}C$ COSY, 226
$^1H/^1H$ COSY, 223
Haemanthamine, 392, 419
Harmine, 422
Hecogenin, 355
Helminthosporal, 322
Hemlock, 389, 408
Henbane, 397, 407
Heparin, 30
Hernandulcin, 327
Heroin, 416
Hesperidin, 7, 37
5-HETE (hydroxytetraenoic acid), 277,
    285, 286
Heterodetic peptides, 187, 189
HIV, 24, 30, 88, 107, 442
HMG-CoA (3-hydroxy-3-methyl-
    glutaryl-Coenzyme A), 304, 335
HMG-CoA reductase, 305
Homodetic peptides, 187
Homologation of fatty acids, 254, 259
Homomavalonate, 326
Homospermidine, 409
Hordenine, 413
5-HPETE (hydroperoxyeicosatetraenoic
    acid), 285, 286
11-HPETE, 263
HPLC, 112
Humulene, 324
Humulone, 306, 362
Hyaluronic acid, 25
Hydroquinone, 364, 366
1-Hydroxybenzotriazole, 167
3(R)-Hydroxybutyryl-S-ACP, 245
p-Hydroxyphenylacetic acid, 362
N-Hydroxysuccinimde, 167
Hygrine, 391, 405
Hyoscyamine, 391, 397, 407
Hyoscine (scopolamine), 407
Hypothalamic thyrotropic hormone
    releasing factor, 185

Ibotenic acid, 55
Illudin S, 324
Indanomycin. 48
Indicine N-oxide, 409
Indolactam v, 443
Indole alkaloids, 420
Indomethacin, 264, 277
Inosine, 93
Inosinic acid, 93
Insulin, 123

Inulin, 25
Interferon, 123
Introns, 107, 123
α-Ionone, 318
β-Ionone, 294, 318, 356
Ipomeaerone, 326
IPP (isopentenyl pyrophosphate), 292, 305
Iridodial, 309
Isoflavonoids, 366
Isofumigaclavine B, 395
Isoleucine, 409
Isopelletierine, , 391
Isopentenol, 306
Isoprene, 306
Isoprene rule, 290, 292
Isovaleryl CoA, 246

Juglone, 364, 382
Juvabione, 326
Juvenile hormones, 326

(ent)-Kaurene, 330
β-Ketoacyl-ACP synthase, 244
Ketoses, 10
Koenigs–Knorr method of glycoside formation, 37, 38, 64

β-Lactam antibiotics, 23
Lactobacillic acid, 243
Lactones, 324, 325
Lactose, 21, 22
Lanosterol, 334, 342, 344
Lapachol, 387
Lathodoratin, 386
Lavandulol, 313
Leucotrienes, 239, 241, 248, 258, 264, 277ff, 285
  biological effects, 280
  synthesis, 281
Lewis acid, 75
Lignin, 377
Limonene, 307, 312
Limonin, 344
Linalool, 309
Lincosamine, 60
Linoleic acid, 241, 248, 252
α-Linolenic, 241
γ-Linolenic, 241, 248
Liposomes, 239
Lipoxygenases, 252
Loganin, 309
Lonchocarpan, 386
Longifolene, 304, 318, 323, 351
Lophocereine, 392

LPP (linolyyl pyrophosphate), 310
LSD, 423
LT-A_4, 279, 282, 283
LT-B_4, 278, 279, 284
LT-C_4, 279, 280
LT-D_4, 279
LT-E_4, 279, 280
Lumisterol, 341
Lupeol, 344
Lupinine, 391, 409, 411, 436
Lupisoflavone, 371
Lutein, 348
Lycopersene, 346
Lycorine, 392, 419
Lysergic acid, 393, 422, 424, 433,
Lysine, 390, 404, 408, 411

Major groove, 103
Malformin A, 187
Malonyl CoA, 244, 246
Maltose, 21, 22
Malvidin, 368
Malvin, 362, 385
Mandrake, 397, 407
D-Mannitol, 56
D-Mannose, 49, 57
Maple Syrup urine disease, 147
Maritidine, 436
Matricin, 298, 352
Matrine, 409
McClafferty rearrangement, 214
p-Menthadiene, 307
Menthofuran, 312
Menthols, 312, 313
Menthones, 312, 313
Meroterpenoids, 294, 306
Merrifield solid phase synthesis, 172
Mescaline, 392, 414
Mesembine, 418, 440
Metabolic grid, 334
α-Methylbutyryl CoA
Methyl α-D-glucopyranoside, 30, 31, 45
Methyl α-D-mannopyranoside, 58
Methyl L-ribofuranoside, 59, 60
6-Methylsalicyclic acid, 374
Mevalonic acid, 422
Micelles, 239
Minor groove, 103
Miroestrol, 337
Mitomycin C, 390
Mixed function oxygenase, 252
Modified bases, 71
Molluscicides, 339
Morphine, 389, 392, 396, 402, 414, 416
Moulting hormones (insect),343

Multidimensional NMR, 222
Multinuclear Fourier transform
    spectrometers, 222
Muscarine, 55
Muscimol, 55
Muscopyridine, 394
MVA (mevalonic acid), 292, 304
MVAP (mevalonic acid 5-phosphate),
    305
MVAPP (mevalonic acid
    5-pyrophosphate), 305
Myrcene, 53, 307, 315, 316, 351

NADH (nicotinamide dinucleotide), 147
NADPH (nicotinamide dinucleotide
    phosphate), 5, 14, 90, 93
Naloxone, 416
Naphthoquinones, 381
Naringenin, 374
Neoclovene, 327
Nepetalactone, 309
Nerol, 309
Nerolidyl pyrophosphate, 317
Nicotine, 391, 411
Nicotinic acid, 390, 412
NIH-type shift, 148, 149
Nikkomycin B, 138
Ninhydrin, 201
Nitidine, 418
nOe (nuclear Overhauser effect), 42, 400
Non-classical carbocations, 314
Noradrenaline (Norepinephrine), 413
Norethindrone, 337
N-Norlaudanosine, 427, 428
Nortriterpenoids, 344
N-terminal exopeptidases, 183
Nucleic acid, 69, 90
Nucleoside diphosphates, 89, 95
Nucleoside monophosphates, 89, 93
Nucleoside transformation, 80
Nucleoside triphosphates, 89, 95

Oenanthic acid, 240
Oestradiol, 336
Oestriol, 336
Oestrogens, 336, 354
Oestrone, 303, 336, 352
Oils, 239
Okazaki fragments, 104, 105
Oleic acid, 241, 248
Oligonucleotides, 101ff, 108ff
Oligoribonucleotides, 117 ff
Oligosaccharides, 23, 64
Olivose, 42
α-Onocerin, 344

Ophiobolin A, 331
Opium, 397, 414
Ornithine, 390, 404, 405, 411
Oroselone, 306
Orotic acid, 91
Orotidyllic acid, 91
Ovalicin, 319
α-Oxidation, 250
β-Oxidation, 250
ω-Oxidation, 252
Oxidative cleavage of diols, 34, 35
Oxidative phenolic coupling, 415, 418,
    436, 440
Oxytocin, 184

Palmitic acid, 240, 245
Palmitoleic acid, 247
Papaverine, 392, 414
Parthenolide, 325
β-Patchoulene, 324
Patchouli alcohol, 324
Pavoninin, 344
Pectic acids, 25
Pelargonic acids, 240
Pelargonidin, 368
Pelletievine, 408
Pellotine, 392, 404, 414
Pentachlorophenyl ester, 166
Pentafluorophenyl ester, 166
Peonidin, 368
Peptide synthesis, 171
Petunidin, 368
PG A_1, 261
PG A_2, 261, 277
PG E_1, 260, 273
PG E_2, 261
PG E_3, 260
PG F_{1a}, 260
PG F_{2a}, 260, 261
PG F_{3a}, 260
PG G_2, 262, 274
PG I_2, 260, 262, 264, 274
Phallotoxins, 188
β-Phellandrene, 294
Phenol, 361
Phenolases, 362
Phenylalanine, 49, 378, 390, 404, 413,
    418
Phenylketonuria, 149
Pheremone, 53, 257, 306, 307, 327, 351,
    394
Phloroglucinol, 362, 373
Phosphatidylglycerols, 239, 240
Phosphite procedure (of nucleotide
    synthesis), 114

Phosphoamidates, 95
Phosphodiester method (of nucleotide synthesis), 110
Phosphotriester method, 112
Phosphoglycerides, 242
4'-Phosphopantetheine, 244
H-Phosphonate method of oligonucleotide synthesis, 116
Phosphorothionates, 97
Photocitrals A and B, 316
Photosynthesis, 7, 14ff
Phyllocladine, 328
Physostigmine, 421
Phytoalexins, 325
Phytoecdysones, 343
Phytoene,293, 346, 347
Phytol, 327
Phytosterols, 342
Pig liver esterase, 59, 86
α-Pinene, 290, 291, 297, 307, 311, 312, 314, 351
β-Pinene, 53, 307, 310, 315, 351
Pinoresinol, 375
Piperidine alkaloids, 408
Piperine, 395, 408
Plants' defence system, 343
Plasmids, 123, 125
Pleuromutilins, 328
Plumbagin, 382
^{31}P-NMR, 98, 100
Podophyllotoxin, 364
Polyacrylamide resins, 121, 179
Polygodial, 318
Polyisoprenoids, 293, 349
Polymerase chain reaction, 126
Polyprenols, 350
Populoside, 366
Porantheine, 431
Pregnenolone, 355
Prephytoene alcohol, 346
Presqualene alcohol, 333
Presqualene alcohol pyrophosphate, 333
Progesterone, 337, 356
Progestogens, 336
Promoter site, 105
Propionyl CoA, 246
Prostacyclin, 260, 262, 264
Prostaglandins, 43, 239, 241, 248, 259ff
  biosynthesis of, 262
  catabolism, 264
  synthesis, 264
Prostaglandin synthetase, 262
Prosthetic group, 197
Protecting groups, 33, 108, 155, 175

Protein engineering, 184
Protosterol, 342
Primary metabolism, 1
PRPP (phosphoribofuranosyl pyrophosphate), 91, 92
Pscilocybin, 421
Pseudopelletierine, 391
Pseudouridine, 71, 85
Pulegones, 312
Pumilotoxin C, 393
Puromycin, 72
Putrescine, 405, 409
Pyranose conformation, 11, 12, 13
Pyrazofurin, 85
Pyrethrin I and II, 313
Pyrethrosin, 298
Pyridine alkaloids, 411
Pyridoxal phosphate, 146, 408
Pyrrolidine alkaloids, 405
Pyrrolizidine alkaloids, 409

Qinghaousu, 325
Quadrone, 352
Quaternary structure, 198
Quercetin, 361
Quinine, 393, 425
Quinolizidine alkaloids, 409

Radiolabels, 199
Rancidity, 241
Recombinant DNA, 105, 123, 183
REMA (respective excess mixed anhydride), 171
Restriction endonucleases, 70, 123
Reticuline, 392
Retinal, 349
Retinol, 348
Retronecine, 391, 409
Retrosynthetic analysis, 44, 49, 53, 427, 428, 431, 435
Retroviruses, 257
Reverse Hoogsteen (base pairs), 102
Reverse transcriptases (RNA directed DNA polymerases), 107, 124
L-Rhamnose, 57
Rhodopsin, 349
Ribonucleosides, 70
D-Ribose, 7, 17, 60
Ribulose 1,5-diphosphate, 15, 16
Rickets, 341
Rishitin, 326
RNA (ribonucleic acid), 69, 70, 106
m-RNA, 69, 106
r-RNA, 196
t-RNA, 69, 106

RNA splicing, 107
Roquefortine, 395
Rotenone, 380
Rubber (gutta percha), 293, 349
Ruzicka isoprene rule, 313

S-Adenosyl methionine, 249, 342, 421
Salicin, 17, 37
Salicylic acid, 362
Salutaridine, 415
α-Santalene, 319
α-Santonin, 290, 325
Sapogenin, 339, 355
Saponins, 339
Saxitoxin, 394
Sclareol, 328
Secologanin, 309, 423
Secondary structure, 198
Secondary metabolism, 1, 389
Senecionine, 409
Sequencing (of nucleic acids), 119ff
Serotonin, 414
  5HT (5-hydroxytryptamine), 421
Serrantenediol, 344
Sesquibornane, 319
Sesquiisocamphane, 319
Sesquimenthane, 319
Sesquipinane, 319
Sesquiterpenoids, 324
Sex hormones, 336
Sharpless epoxidaton, 62, 282, 285, 286
Shikimate pathway, 377
Shikimic acid, 49
Shikonin, 382
Showdomycin, 51, 60, 82, 86
Silyl ether (prot) groups, 110, 117
Silyl Hibbert–Johnson method (of
    glycosylation), 76, 77, 84
β-Sinesal, 317
Sinigrin, 39
Sitosterol, 342
Soaps, 239
Solid phase synthesis of
    oligonucleotides, 114
Somatostatin, 123
Sparteine, 391, 396, 409, 411
Spectrofluorimetry, 201
Sphingolipids, 239, 240, 242
Squalene, 293, 333
Squalene oxide, 334, 344
SRS-A (slow reacting substance of
    anaphylaxis), 277
Starch, 25, 47
Stearic acid, 240, 245
Sterculic acid, 243

Steroids, 239
Sterol carrier protein, 336
Stigmasterol, 342
Strecker synthesis. 132
Stritosidine, 423, 426
Structure elucidation, 119ff, 194, 296,
    396, 400
Strychnine, 393, 399, 424
Sucrose, 21, 22
Sylvestrene, 296
Synthesising peptides, 170
Synthetic genes, 129

Taxinine, 328
Terpene alkaloids, 344
α-Terpene, 312
α-Terpinol, 297, 307, 314, 350
Tertiary structure, 198
Testosterone, 337
Tetracyclines, 126
Tetrahydrocannabinol, 295, etals, 39
Thioglycosides, 39
Thromboxanes, 239, 241, 248, 264, 274ff
    synthesis, 275
α-Thujene, 311
Thujenes, 315
Thujone, 291, 307
Thujopsene, 319
Thymine, 70
Thymidine, 70
Thymol, 308
α-Tocopherol, 295
Transamination, 136, 151
Trans-β-farnesene, 317
Transketolization, 15, 18
Trehalose, 21, 22
Triacylglycerols, 239
Trichothecin, 319
2,4,6-Trimethylbenzyl ester (OTmb),
    162
Triostin, 190
Trisporic acids, 294, 349
Tritylation, 108
Trityl group, 159
Tropic acid, 397
Tropine, 397
Tropinone, 399, 402, 427
Trypsin, 210
Tryptophan, 390, 404, 420, 421
Trytamine, 421
Tubercidin, 72
Tuberculostearic acid, 243, 248
TXA$_2$, 45, 274, 275
TXA$_3$, 274
TXB$_2$, 43, 45, 274, 277

TXB₃, 274, 277
Tyramine, 413
Tyrosine, 49, 361, 390, 404, 413, 418

Ubiquinones, 295
Uracil, 70
Uridine, 70
Uridyclic acid (UMP,UDP, UTP
    uridine phosphates), 91, 92
Usnic acid, 377

Valeric acid, 240
Valine, 409
Valinomycin, 189
Vasopressin, 184
β-Vetivone, 323
Vinblastine,1, 425
Vincristine 1, 425
L-Vinylglycine, 140
Violaxanthin, 347
Vitamin A, 348, 356
Vitamin A₃, 348
Vitamin B6, 135
Vitamin K₁, 295

Vitamin D₁, 341
Vitamin D₂, 341
Vitamin D₃, 296, 342
Vitamin D₄, 341
Vitamin E, 252

Wagner–Meerwein rearrangements, 311,
    314, 319
Watson and Crick, 69
Watson and Crick base pairs, 102
Waxes, 239, 240
Widdrol, 319
Wilforibose (structure elucidation of),
    41
Wittig reaction, 82,255, 257, 258, 266,
    269, 281, 282, 284, 285, 286, 433,
    438
Woodward–Hoffmann Rules, 342

Xanthones, 366
Xanthophylls, 347

Zeaxanthin, 347